ORIGINS OF BIODIVERSITY

ORIGINS OF BIODIVERSITY

*An introduction to **macroevolution** and **macroecology***

LINDELL BROMHAM | MARCEL CARDILLO

Research School of Biology, Australian National University

OXFORD

UNIVERSITY PRESS

Great Clarendon Street, Oxford, OX2 6DP,
United Kingdom

Oxford University Press is a department of the University of Oxford.
It furthers the University's objective of excellence in research, scholarship,
and education by publishing worldwide. Oxford is a registered trade mark of
Oxford University Press in the UK and in certain other countries

Published in the United States of America by Oxford University Press
198 Madison Avenue, New York, NY 10016, United States of America

British Library Cataloguing in Publication Data
Data available

Library of Congress Control Number: 2019934135

ISBN 978–0–19–960871–3

Printed in Great Britain by
Bell & Bain Ltd., Glasgow

For Beverley & Barry Bromham

and

Annemarie & Tom Cardillo

for their steadfast support and endless encouragement.

Preface

New ways of answering old questions

Macroevolution and macroecology aim to describe large-scale evolutionary and ecological patterns, and to understand the processes that produced them. For as long as people have been able to travel far afield, they have been interested in biological patterns at large scales. Indeed, these are some of the most fascinating puzzles of the natural world. But for many people, it is not obvious how we can do science at the macro-scale. How can we devise experiments or test hypotheses for events that happened millions of years ago, for patterns that emerge at global spatial scales, or processes that affect thousands of different species?

This is a challenging field because it often concerns processes and patterns that cannot be directly witnessed. We can't go back in time to witness the diversification of the animal kingdom, or make direct observations on the causes of dinosaur extinctions, nor can we do direct experiments on the latitudinal diversity gradient of species richness, or manipulate species traits to test their effect on speciation or extinction rates. But there is a wealth of scientific resources we can draw on to generate hypotheses, design elegant tests, and provide satisfying answers to many intriguing questions. The tools used to test hypotheses for patterns of biodiversity include experiments, models, and statistical analyses, but often the most important tool we have is the ability to frame a clear question, from which we can make simple predictions that can be put to robust tests.

Although macroevolution and macroecology are often taught separately, they share an intellectual framework, rest on many common concepts, and share many analytical tools. The development of macroevolution and macroecology as fields of scientific enquiry have been driven by new ways of compiling and analysing data that have allowed big-picture questions about biodiversity to be tackled. For example, the massive growth of DNA sequence databases has led to a proliferation of methods for using molecular phylogenies to understand the origins of lineages and their patterns of diversification. Molecular data have been combined with global databases of species distributions and environmental features and with large palaeontological databases to analyse patterns across large spatial and temporal scales. Most of this 'big data' is now freely available for anyone to explore.

Although many of the new methods are technically complex and analytically sophisticated, the basic principles of sound scientific reasoning and inductive inference are as important in macroevolution and macroecology as in any other area of science. Our aim in this book is to introduce you to some of the useful approaches to asking questions and solving problems in macroevolution and macroecology. The underlying logic of these techniques does not require you to be competent in mathematics, statistics, or computing (although these intellectual tools may help you frame and answer questions).

What is in this book?

This book is not meant to be an exhaustive treatment of all problems, questions, and methods associated with macroevolution and macroecology. Rather, the idea is to give you an introduction to some of the key issues, a taste of the research that is being done in these fields, and a glimpse of some of the approaches that researchers are currently using to make progress in these areas. There is no clear distinction between macro- and micro-levels of evolution and ecology, and there is no definitive list of patterns or processes that should be included in an introductory book on macroevolution and macroecology.

This book takes a case study led approach. Each chapter focuses on one particular question as a way of introducing some core concepts in macroevolution and macroecology, and as a way of exploring some useful sources of data and modes of analysis. This does not mean that we think these are the only important questions to be asked in macroevolution and macroecology. We have chosen these particular issues because each serves as a useful focus for thinking about important concepts that run through macroevolution and macroecology (such as levels of selection or diversification rates), or important analytical techniques (such as null models or phylogenetic comparative methods). The book emphasizes the critical appraisal of evidence, techniques, and assumptions in testing macroevolutionary and macroecological hypotheses, and encourages you to form your own opinions on important debates.

Each chapter has a number of features designed to help you get the most from your reading.

Roadmap

Why study macroevolution and macro

Most textbooks on evolution and ecology large operating within populations. But many of the qu about patterns of biodiversity are not easily answ on the population level, such as: How does nove mass extinctions? Does selection act at the level of

- **Roadmap** sets out the path we will navigate in each chapter, giving a brief introduction to the examples we will use to illustrate the topic, the key concepts we will cover, some of the analytical techniques we will consider, and the case study included at the end of the chapter.

Key points

- Macroevolution and macroecology focus on patterns of biodiversity that are shaped by processes operating over large time-scales, across wide spatial scales, and affecting a wide-variety of lineages.
- Most researchers assume that processes

- **Key points** boxes act as waymarkers, highlighting the main take-home messages of the section.

What do you think?

The disagreement between selectionists and mutationists partly reflected different attitudes to science: mutations could be generated and observed in the laboratory, whereas the role of natural selection in the formation of new species is inferred rather than directly

- **What do you think?** boxes encourage you to stop and think about the broader implications of what you are reading. They prompt you to take a moment to think critically about the core concepts you have just learned, and invite you to extend your intellectual reach beyond the material and form your own ideas.

Points for discussion

1. Some people have proposed that life may have evolved and then colonized earth, an idea sometimes referred Does this provide a plausible answer to questions sur of life? How would you test the panspermia hypothes

2. Should increases in the coding capacity of replicato crease in genome size) slow the rate of evolution in ord

- **Points for discussion**, at the end of each chapter, provide a focus for more in-depth discussion. These can be used for self-study, to prompt discussion in the classroom, or as a focus for a group investigation about the key issues presented in the chapter.

References

1. Darwin C (1859) *On the origin of species by means or the preservation of favoured races in the struggle fo* Murray, London.

2. Nilsson D-E, Pelger S (1994) A pessimistic estimate o for an eye to evolve. *Proceedings of the Royal Society cal Sciences* 256(1345): 53–8.

- **References** include the details of studies cited in the chapter. This isn't a comprehensive guide to literature on the topic, or a list of must-read classics. But if you want to look further into the details of studies we mention, this is the place to find their publication details.

Case Study 1
Testing hypotheses about past ev uncovering the evolutionary orig a new virus

While scientific training might involve the acqu cal skills, such as those associated with laboratory analysis, or field studies, the most important sets tist needs are those associated with asking good q

- Each chapter also includes a **Case Study** in which we explore in a bit more depth one or more examples of research in macroevolution and macroecology. Each case study includes the following features.

Questions to ponder

1. We have seen how a range of corroborating evid to support or refute hypotheses. What if a ne and we only have DNA sequences, no other hist records. Can we trust phylogenetic estimates o own?

2. How can we test whether viral genomes evolve v rate of change?

3. Are there limits to reconstructing past events?

- **Questions to ponder** encourage you to think deeply and critically about some of the issues raised by the research.

Further investigation

Ancient DNA derived from human remains is increa to trace the origins of major epidemics, such as plag sis. What are the advantages of using these 'molecu might be the dangers? Can these old sequences b same way as the modern sequences? How can res their samples are not contaminated with modern vi zles might we be able to solve if we could find app Which diseases will be most likely to benefit from t

- **Further investigation** invites you to extend your investigation beyond the case study to a related topic. This section could serve as a starting point for further independent study or group discussion on the topic.

References

1. Bromham L (2016) Testing hypotheses in macro in *History and Philosophy of Science Part A* 55: 47–59

2. Sharp PM, Hahn BH (2010) The evolution of HIV-1 AIDS. *Philosophical Transactions of the Royal Society*

3. Hahn BH, Shaw GM, De Cock KM, Sharp PM (2000 osis: scientific and public health implications. 607–14.

- **References** provide publication details of studies mentioned in the Case Study.

online resources
www.oup.com/uk/bromham-biodiversity/

Online resources provide lecturers with outlines for discussion-based tutorials and computer workshops to run in conjunction with a lecture course as well as downloadable figures from the textbook to use in lecture presentations and teaching materials. Find these at www.oup.com/uk/bromham-biodiversity/

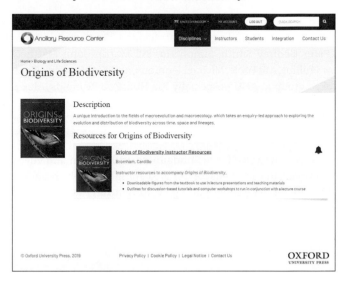

Who is this book for?

This book is aimed primarily at budding biologists who want to apply what they have learned about evolution and ecology to investigating and understanding the world around them. The reader we have in mind has already taken some introductory courses in evolution and ecology, so will be familiar with core processes of evolution, such as mutation, selection, drift, and speciation, and will understand basic concepts in ecology, such as competition, population growth, and the way that environmental conditions shape species distributions. This book builds on these basic concepts to explore and explain patterns in biodiversity over space and time. We will revise some of these core concepts along the way—for example, considering the operation of natural selection in Chapter 2—but if you want to brush up on these background concepts, it might be a good idea to have a look at a general evolution or ecology text first.

Many case studies in this book utilize analyses of molecular phylogenies. To be an intelligent user of phylogenetic analyses, you need to understand the nature of molecular data, how phylogenies are produced, and the way that phylogenetic inference is affected by data selection, methods used, and assumptions made. Although we touch on these areas in many chapters, we cannot cover phylogenetics in detail in this book. However, there is a companion text that provides a 'from the ground up' description of molecular phylogenetics, from data generation to phylogenetic inference and analysis. You may find *An Introduction to Molecular Evolution and Phylogenetics* (Lindell Bromham, Oxford University Press, 2016) a useful base from which to gain an understanding of the role of molecular phylogenies in evolution and ecology.

Acknowledgements

Our greatest thanks are to Gulliver, Alexey, Arkady, and Asha for their patience, curiosity and fortitude, and for making life interesting and exciting while this book was being written, with never a dull moment along the way.

We are grateful to our colleagues for generously giving their time and expertise to read and comment on chapters, including Tim Barraclough, Graham Budd, Brett Calcott, Adrian Currie, Peter Godfrey Smith, Simon Ho, Bill Martin, John Matthewson, Arne Mooers, Matthew Phillips, Ant Poole, Michael Jennions, and Kim Sterelny. Thanks are due to the MacroEvoEco group at ANU—especially Xia Hua, Zoe Reynolds, Alex Skeels, and Russell Dinnage—for their encouragement and support. Thanks also to Rampal Etienne for supplying data, and Rod Peakall, Robin Eckerman, and Russell Dinnage for providing beautiful photos and figures.

We owe much to Jonathan Crowe for his undaunted enthusiasm and cheerful refusal to give up on this project, to Lucy Wells for deftly keeping the book on track, against the odds, and to Sal Moore for her patience and good humour in the production stage. We could not be more surprised to see this book finally become a reality, thanks to their persistence.

Contents

What is macroevolution? What is macroecology?

1

Roadmap

Why study macroevolution and macroecology?

Most textbooks on evolution and ecology largely focus on processes operating within populations. But many of the questions we want to ask about patterns of biodiversity are not easily answered by only focusing on the population level, such as: How does novelty arise? What causes mass extinctions? Does selection act at the level of genes, individuals, species, or lineages? Why is biodiversity distributed so unevenly in space? Questions like these require us to build upon our knowledge of micro-level processes, but consider them over large spatial scales, long periods of evolutionary history, and many different evolutionary lineages. Studying the intellectual development of the fields of macroevolution and macroecology is important for several reasons. Scientific ideas evolve over time, building on what has come before, so we can't fully understand what people think today without appreciating how that point of view has been constructed by research and contemplation over long periods. Many of the same areas of debate come up again and again, so a familiarity with history helps us to recognize and evaluate different hypotheses. This chapter will also show how the individual case studies covered in the rest of the chapters connect to the bigger picture of evolutionary and ecological ideas.

What are the main points?

- Macroevolution focuses on changing patterns of biological diversity across time, space, and lineages.
- Macroecology concerns broad-scale patterns in the abundance and distribution of species.

Photograph: Nothofagus forest. © Lindell Bromham.

- Key debates resurface again and again, such as whether change happens gradually by accumulation of small changes or quickly by larger 'leaps', and the relative roles of chance, selection, and history in shaping biodiversity.

What techniques are covered?

- **Development of ideas:** what we think now builds upon what scientists have thought in the past, so we need to understand past debates as well as current questions.
- **Hypothesis testing:** framing testable questions, experimental design, and statistical analysis are critical in macroevolution and macroecology.

What case studies will be included?

- Testing hypotheses in time and space: reconstructing the origins of HIV.

"General views lead us habitually to regard each organic form as a definite part of the entire creation, and to recognise, in the particular plant or animal, not an isolated species, but a form linked in the chain of being to other forms living or extinct. They assist us in comprehending the relations which exist between the most recent discoveries, and those which have prepared the way for them. They enlarge the bounds of our intellectual existence, and . . . they place us in communication with the whole globe."

Alexander von Humboldt (1847). *Cosmos: Sketch of a Physical Description of the Universe*, Vol. I. Trans: E. Sabine. John Murray, London.

Why study macroevolution and macroecology?

Evolution and ecology are central to the study of biology, and the key to understanding the world around you. If you look out of the window, most of what you see is the product of evolution. Not only the species themselves—such as animals, plants, and fungi—but also the soil and atmosphere, even the buildings, roads, and cars, are ultimately products of the biological world. So the fundamental principle we need to bear in mind if we want to understand the world around us is that all biological phenomena are the product of a long evolutionary history, shaped by interactions between organisms and their environment, and subject to both chance events and selection. Not surprisingly, evolution and ecology are taking an increasingly prominent role in biological education.

Introductory courses in evolution and ecology typically focus on population-level processes, such as selection, drift, migration, and competition. Most biologists assume that these population-level processes underlie all evolutionary phenomena, and that any patterns in biodiversity in space and time can ultimately be traced back to changes in the genetic composition of lineages over time. But it may not be obvious to everyone how we get from changes in gene frequencies in populations to the grand patterns of biodiversity across space and time. There are many interesting questions that we cannot adequately answer using only our knowledge of population-level processes. Consider the following questions:

- Why did the dinosaurs go extinct?
- Why are there so many different kinds of beetles?
- Why are the tropics more biodiverse than the temperate zones?

These questions involve observation of patterns in the numbers and kinds of organisms (which we will refer to as species richness), whether over time, or space, or among lineages. When we look for answers to questions like these, we will find our focus is not always on what is happening to individuals, or changes in gene frequencies within populations. Instead, we might find ourselves considering lineage-level processes like diversification rates and geographical distribution. In doing so, we are confronted by a dilemma. Does a focus on 'higher-level' observations, made by comparing different lineages over long time periods, imply that there are special 'higher-level' processes that cannot be explained simply in terms of the 'lower-level' processes occurring in populations over generations? Or does it simply reflect that sometimes we need to take the long view in order to appreciate the way that simple underlying processes can play out over long timescales and large areas? Many of the discussions in this book will explore this interaction between levels of observation and the scaling up of evolutionary and ecological processes to large-scale and long-term patterns.

Progress in macroevolution and macroecology is becoming increasingly relevant to conservation efforts, owing to recognition of the importance of large-scale processes. For example, current patterns of species diversity and extinction risk are the product not only of population-scale processes, such as genetic isolation, but also the product of time (evolutionary history), space (regional-scale ecological factors), and lineage-level processes (different rates of change, speciation, and extinction). Biodiversity is increasingly being viewed as a dynamic phenomenon rather than a static pattern. We need macroevolutionary and macroecological ideas and tools to develop and enrich this new perspective.

So what is meant by the word 'macroevolution'? We can compare 'microevolution', which typically is used to describe heritable changes within a population, with 'macroevolution', which is broadly concerned with variation in patterns of species over space, time, and lineages. These simple definitions are descriptive; they refer to patterns we can observe. But what causes macroevolutionary patterns? Do they emerge only from the microevolutionary processes that occur in every population? Or are there mechanisms that operate in addition to population processes? Are there any special macroevolutionary processes that operate only occasionally in particular lineages? We won't attempt to provide a definitive answer to these questions, because opinions differ quite widely, as we will discover throughout this book. While most scientists consider that macroevolution and macroecology describe broad-scale patterns generated by the population-level processes of microevolution and ecology, some scientists think that there are special macroevolutionary processes that shape biodiversity. Fundamental disagreements such as these make the field of macroevolution exciting and intellectually stimulating.

The field of macroecology is closely associated with macroevolution, to the extent that there is substantial overlap between the two. That is why the two fields are presented together in this book. We have just defined macroevolution (patterns in the distribution of species across space, time, and lineages) by contrasting it with microevolution (change in gene frequencies in populations over generations). Can we do the same for macroecology? You probably won't hear people using the term 'microecology' (unless they are referring to the ecology of microbial organisms). However, we might contrast local-scale ecology, which is concerned with interactions among species and their environments within small areas, with broad-scale ecology which considers how species diversity is distributed at regional, continental, or global scales.

While macroecology as a scientific discipline is relatively young, many of the core questions about the processes governing the distribution and abundance of species have been asked for centuries. Most macroecological studies focus on explaining current (or recent) patterns in abundance, distribution, and diversity of species. However, we can't understand current species distributions without an appreciation of both evolutionary and environmental history, and we will expect current patterns of biodiversity to have been shaped by processes that operate over long timescales. So macroecology can be thought of as the study of present-day

patterns in the distribution of biodiversity that are (at least partly) the product of macroevolutionary processes.

Macroevolution and macroecology are dynamic fields in a constant state of flux. Ideas change, new evidence emerges, novel analyses are devised. This makes them exciting fields to study, but it also means that we cannot always offer undisputed facts or universally agreed explanations. If you want to study macroevolution and macroecology, you will need to become accustomed to seeing both sides of a debate, weighing up competing explanations, and seeking data to test different ideas. That is why the chapters of this book have been presented as a set of questions rather than answers. Learning to ask good questions is the most important skill a scientist can have. The second most important skill is designing creative and effective tests to compare the plausibility of different explanations (see Case Study 1). That is why this book emphasizes research tools rather than a catalogue of 'facts'. Our focus is on the process of scientific enquiry, not just its outcomes. We hope that this book will help give you the confidence you need to tackle big ideas, the background you need to ask interesting questions, and the tools you need to seek answers.

Key points

- Macroevolution and macroecology focus on patterns of biodiversity that are shaped by processes operating over large timescales, across wide spatial scales, and affecting a wide variety of lineages.

- Most researchers assume that processes operating at the population level (microevolution, population ecology) can be scaled up to explain long-term and large-scale macroevolutionary and macroecological patterns. But some scientists invoke additional mechanisms that operate at the lineage or ecosystem level to explain these phenomena.

Research tools for big questions

66 *One of the strengths of scientific enquiry is that it can progress with any mixture of empiricism, intuition and formal theory that suits the convenience of the investigators.* 99

George C. Williams (1966). *Adaptation and Natural Selection: A Critique of Some Current Evolutionary Thought.* Princeton University Press, Princeton, NJ.

Framing and testing hypotheses in large-scale patterns of biodiversity over space, time, and lineages presents some challenges. To demonstrate this, first consider some examples of research tools used to study population-level processes. Research into evolutionary processes can involve manipulative experiments, where replicate lines are subjected to different treatments and the effects monitored. For example, experimental populations of bacteria exposed to different concentrations of antibiotics can be used to test the genetic basis of the evolution of resistance. Not all experiments take place in a laboratory. For example, plots sown with different combinations of grass species can be used to measure the relationship between diversity and biomass production over many years. Many studies in evolution and ecology don't involve any manipulation, but instead rely on close observation of nature. For example, taking blood samples from wild individuals and sequencing their DNA can reveal high levels of extra-pair mating in socially monogamous birds, and can be used to estimate fitness by tracking descendants over several generations.

Can we use the same kinds of research tools to ask questions about the patterns of species diversity among lineages, over time, or across the earth? The answer to this question is 'yes and no'. We can use the results of studies like these to infer mechanisms underlying macroevolutionary change or macroecological patterns, but we could not use them to study those processes directly. For example, we can breed populations of fruit flies under different temperature regimes, and then examine the changes in genes underlying thermal tolerance, and compare the survival and reproduction

of the experimental and control lines at different temperatures. We might use those observations to infer that this process of genetic change in the experimental population would, if left for a very, very long time, result in the formation of distinct species with specific adaptations to particular thermal niches. But, in general, we cannot directly observe the evolution of lineages characterized by major new adaptations, or families with different numbers of genera, or communities with different numbers of co-adapted species, because these phenomena occur over vast timescales. For the same reason, we can't directly observe the geological process of mountain building or the astronomical phenomena underlying the formation of stars, which occur on timescales much longer than a human lifetime. But we can make inferences about how mountains are built or stars born, for example by studying uplift along geological fault lines or comparing existing stars at different stages of formation. Evolutionary biology, like geology or astronomy, is sometimes referred to as a 'historical science', because we often wish to explain past events that we cannot directly witness.

Because the subjects of analysis in macroevolution and macroecology cannot be experimentally manipulated, researchers must infer past events and processes through observations of the outcomes of those events and processes. These observations might involve comparing the number or type of organisms present in different time periods (see Chapter 3), comparing variation in current species richness between lineages (see Chapter 9), or asking why species richness varies so dramatically between different geographical regions (see Chapter 10). However, it is important to note that principles of sound experimental design, and a clear understanding of statistical analysis, are as fundamental to research in macroevolution and macroecology as in experimental and field studies of population-level processes. This is a point we will be emphasizing throughout this book: to weigh hypotheses about the origins and patterns of biodiversity, we need very carefully designed scientific tests and clearly formulated statistical analyses.

A long history of ideas

“ *It is better that a truth, once perceived, fight a long time without obtaining the attention it deserves, than that everything produced by men's keen imaginations be easily accepted.* ”
Jean-Baptiste Lamarck (1809). *Philosophie Zoologique* (Translation by Ian Johnston, 1999)

Large-scale evolutionary change and broad-scale patterns of species distribution and diversity have long been on the intellectual agenda of people interested in the natural world. It is important that we consider the history of these ideas for two reasons. Firstly, scientific fields of enquiry, like biological lineages, build upon what has come before. Theodosius Dobzhansky famously said that nothing in biology makes sense except in the light of evolution, because to understand the current features of organisms, or how species are distributed in time and space, we have to understand their evolutionary history. To make sense of ideas in evolutionary biology we need to understand where those ideas have come from and what has shaped them. Secondly, many of the central ideas surface again and again throughout the history of the field, albeit framed by different debates. So one of the best ways to get a grip on the key ideas in macroevolution is to see how the ideas have been reshaped throughout history, as new data are gathered and hypotheses are reconsidered.

What follows is not a comprehensive timeline for the development of evolutionary biology. Instead, it is a brief, selective account that highlights some key ideas relevant to contemporary research in macroevolution and macroecology. For example, throughout the history of evolutionary biology, there has been a tension between two alternative ideas—one, that evolutionary change happens gradually by small increments; the other, large changes can happen suddenly by big leaps—but the terms under which this debate has been conducted have changed with the times and with the available data.

The evolved world

The possibility that species change over time had been discussed for centuries before Darwin. Many philosophers recognized evidence for biological change, such as the modification of domestic varieties and the occasional occurrence of 'monstrosities' or 'sports' (individuals born with distinct differences such as extra digits). For example, Pierre Louis Maupertuis (1698–1759) developed ideas about heredity and variation, and gave a rough sketch of the mechanism of natural selection. Erasmus Darwin (Charles's grandfather, 1731–1802) also considered the descent of species from a common ancestor, including the influence of mate choice and competition.

But how did new species arise, and how did they become adapted to diverse ways of life? The great French naturalist Jean-Baptiste Lamarck (1744–1829) (**Figure 1.1**) attempted to answer these questions by developing a theory of species transformation. His experience in classifying plants and animals led him to declare that the more you examined the natural world in detail, the harder it was to draw clear boundaries between varieties and species. He explained this continuum of biodiversity as the result of ongoing transformation of species: organisms had an innate drive towards increasing perfection and complexity, developing new habits and novel traits in response to their environment. Lamarck's evolutionary theory was one of continuous, incremental change.

Lamarck's theory of continuous improvement of species was rejected by many naturalists as speculative and unsupported by evidence. The rejection of this particular evolutionary theory cast a pall over other attempts to develop a theory of species transmutation. Two key reasons for the rejection of Lamarck's hypothesis of evolutionary change were the perceived lack of evidence of intermediate states between species, and the absence of a plausible mechanism for consistent change over generations. The answers to both these criticisms would come largely from the study of geology. Lamarck had studied geology and described fossil diversity, yet he made very little use of fossil evidence in expounding his evolutionary ideas. But, in the century following the publication of Lamarck's ideas, the development of geology as a scientific discipline not only provided evidence of change in biodiversity over time, but also laid the groundwork for a mechanistic explanation of that biotic change.

Fossils reveal the history of the living world

Fossils had been unearthed throughout history, in many different parts of the world, but the interpretation of these objects varied widely. Although it seems obvious to us now that fossils are the remains of long disappeared species from past eras, the antiquity of fossils might not be apparent to someone who doesn't already know that the natural world has a long history of change. Instead, fossils were sometimes interpreted as 'sports' of nature, formed as the result of some unknown geological process, such as a petrifying fluid that turned living matter into stone.

However, fossils were sometimes interpreted as revealing a change in the biological composition of an area over time. For example, Shen Kuo (沈括, 1031–95) used his study of fossilized forms to suggest that species distributions had changed in response to a gradual change in environment. He realized that the presence of fossilized bamboo in a region far removed from its contemporary distribution might indicate a change in climate over time, and that marine shells found in deposits high in the mountains were evidence that the sea had changed position over time. Fossilized creatures distinctly different from known species were sometimes taken as evidence of the existence of wonderful beasts not yet observed alive; for example, the exposed remains of ceratopsid dinosaurs in the Gobi desert may have led to tales of gryphons (eagle-headed winged lions) guarding the goldfields (**Figure 1.2**).[1]

DE LAMARCK.

Figure 1.1 Jean-Baptiste Pierre Antoine de Monet, Chevalier de la Marck (better known as Lamarck). Lamarck was a talented naturalist who published on botany and invertebrate classification, and was praised by Thomas Henry Huxley as having 'possessed a greater acquaintance with the lower forms of life than any man of his day, Cuvier not excepted, and was a good botanist to boot'.[12] Lamarck's theory of evolution, published in his *Philosophie Zoologique* in 1809, was not translated into English until 1914, so many English-speaking biologists only knew his work through Sir Charles Lyell's less than complimentary description of the theory. Lyell was deeply troubled by Lamarck's inclusion of humans in his theory of species transformation, and he ridiculed Lamarck's concept of humans evolving gradually from orang-utans. After the publication of *The Origin of Species*, Lyell wrote to Darwin that 'When I came to the conclusion that after all Lamarck was going to be shown to be right, that we must "go the whole orang", I re-read his book, and remembering when it was written, I felt I had done him an injustice'.[13] Today many people associate Lamarck's name with the idea of inheritance of acquired characteristics, but this idea was common among naturalists (including Charles Darwin) up until the early 1900s.

Photo from Wellcome Library, London via Wikimedia Commons (CC4.0 license). Wellcome Images http://wellcomeimages.org Jean Baptiste Pierre Antoine de Monet Lamarck. Stipple engraving by A. Tardieu, 1821, after J. Boilly.

Figure 1.2 Dinosaurs in disguise? In her book *The First Fossil Hunters*, Adrienne Mayor (2000) suggests that fossilized remains of extinct animals may have been uncovered by ancient Greeks and interpreted in light of unknown beasts or mythical beings. For example, she suggests that remains of ceratopsid dinosaurs (a) could have given rise to the legendary gryphon (b) as large quadrupeds with beaks. She also suggested that revered relics of heroes, such as Pelops' shoulder, may have been remains of ice age beasts. Similarly, in other parts of the world, dinosaur bones have sometimes been interpreted as the remains of dragons.

In the closing decades of the eighteenth century, the fossil record became increasingly well studied. The development of geology as a systematic discipline was at least partly driven by economic development. A systematic approach to the study of geological strata was needed to make the exploitation of mineral resources less haphazard. The digging of roads and canals revealed consistent strata that could be correlated across different areas. Not only could the various rock layers be recognized and correlated over the landscape, but each layer was characterized by particular fossil forms, and the older the stratum, the more the fossils within it differed from today's species. Rapid expansion of fossil collections in the 1800s, and the growing number of professional geologists and fossil enthusiasts, gave rise to the new field of palaeontology.

One of the most influential naturalists in this field was the French zoologist Georges Cuvier (1769–1832). Cuvier studied the growing collections of specimens of animals at the Paris Natural History Museum, garnered from around the globe by exploration, trade, and conquest (**Figure 1.3**). He developed the principles of comparative anatomy, correlating body parts between different species, to show how related species were essentially modifications of the same basic forms. His skills were such that he could describe a species given a single bone, even predicting its mode of life. This allowed him to reconstruct previously unknown species, such as giant sloths and mastodons, from fossil bones.

Cuvier argued that these animals represented extinct species, since such conspicuously large animals had never been seen alive. Thus Cuvier was one of the first scientists to argue strongly for the process of extinction. Furthermore, Cuvier showed that the fauna he studied from the rocks around Paris was entirely different from the living species of the region, demonstrating that whole faunas had become extinct. This convincing evidence for extinction showed that the biological world was not static, but had changed dramatically over time, such that the species of the past were of a different kind than those found alive today.

Figure 1.3 Georges Cuvier served as director of the Muséum national d'Histoire naturelle in Paris in the early 1800s. The collections of specimens served as a powerful bank of biodiversity information for researchers, allowing them to compare anatomy across many species, extant and extinct. This is an early example of the power of cross-species data collection in driving advances in biodiversity research.

© Lindell Bromham.

> **Chapter 11** will look at extinction, and compare rates of species loss now with those in the past.

Cuvier explained the dramatic changes in the world's biota as the result of occasional revolutions. He proposed that catastrophic changes in the natural environment, such as floods or sudden changes in climate, wiped out whole faunas. Local species were extirpated and then replaced by migration of species from elsewhere, producing a sudden change in the regional fauna. Cuvier considered that these past catastrophes were of a magnitude or type not witnessed today; therefore the nature of biological change in the past was entirely different from the present. Although Cuvier saw no evidence that the species themselves changed over time, he used comparative anatomy to demonstrate a pattern of progression in the fossil record, with fossils in younger strata being more similar to species alive today. Both of these ideas—catastrophic extinctions and biological progression—were initially rejected by one of the most influential geologists of all time, Sir Charles Lyell (1797–1875).

> We will consider the role of fossil evidence, and catastrophes in evolution, when we look at dinosaur extinctions in **Chapter 5**.

Using the present to explain the past

66 *No unnecessary intervention of unknown or hypothetical agent* **99**
Charles Lyell in a letter to Charles Darwin, 29 June 1856.

Lyell aimed to put the study of geology on a firm scientific footing. Following in the footsteps of the pioneering geologist James Hutton (1726–1797), Lyell argued that geologists should strive to explain phenomena using only those processes they could witness in operation. He applied three basic rules of reasoning to explain geological phenomena: one, assume basic laws of nature have not changed over time; two, explanations should invoke only mechanisms we can witness today; and three, assume that these forces were of the same strength in the past as they are today. Lyell's doctrine is best summarized by the aphorism 'the present is the key to the past'.

Lyell rejected explanations of the changing earth based on past catastrophes of unknown cause. Instead, he insisted that the processes we can witness today were sufficient to explain the formation of all geographic features. The rain that erodes sediment into rivers will gradually wear down mountains. The water flowing down the river will eventually carve out a valley. The uplift that raises the earth by a metre could ultimately raise a mountain by degrees. To contrast it with 'catastrophism', where past changes were attributed to rare extraordinary events of devastating effect, this doctrine of continuous change was referred to as 'uniformitarianism', because it assumed uniformity of processes operating over time. However, uniformitarianism does not require rejection of changes of large magnitude or rapidity. For example, the gradual process of erosion of a natural barrier over time could lead to a catastrophic flood occurring in one single terrible moment when the barrier is finally breached.[2]

Uniformitarianism can be considered as both a statement of natural process and a guideline for scientific endeavour. We can't go back in time, so we cannot make direct observations on past events and processes. Scientists who wish to explain the earth's history can only use observations made in the present day, such as documenting the current distribution of species, discovering the remains of long-dead creatures in exposed rocks, or sequencing the genomes of living organisms. From observations made today, we construct explanations of past events. Given that we cannot directly observe processes operating in the past, we need a way of identifying plausible explanations for present diversity. How should we direct our imagination? And how do we distinguish a scientific explanation from a fantasy? The uniformitarian solution to this dilemma is to restrict our explanations to those that assume that the same basic laws of operation applied in the past as they do today. If we can observe mechanisms today, then we can regard them as plausible contributors to past events.

Lyell was initially reluctant to accept that species were also subject to change over geological time (see Figure 1.1). But through his insistence that the present is the key to the past, Charles Lyell aided the development of modern evolutionary biology by his profound influence on Charles Darwin (1809–1882). Darwin read Lyell's three-volume masterpiece *Principles of Geology* on his five-year voyage around the world on the *Beagle*. Just as Lyell had wished to put geology on a scientific footing, so Darwin created the science of evolutionary biology by applying Lyell's approach to the living world. Darwinian gradualism is uniformitarianism applied to the biological world. By drawing on the variation observable in living populations, and the evident possibilities for differential survival and reproduction, Darwin invoked everyday processes to gradually build large-scale evolutionary change. He suggested that long periods of accumulation of small heritable changes, each of which may be nearly insubstantial, could generate evolutionary novelty and diversity. In doing so, he did what no-one had done before—he gave a plausible mechanism for evolutionary change.

Evolution by natural selection

By the middle of the nineteenth century the idea that the biological world was not fixed but had changed over time was a matter of much discussion, both amongst scientists and by interested members of the public. Indeed, a best-selling popular science book, *Vestiges of the Natural History of Creation*, written and published anonymously in 1844 by the journalist Robert Chambers (1802–1871), described the history of the biosphere as one shaped by transformation and extinction, and claimed that all species were the product of evolution. The *Vestiges* was scorned by many respectable scientists of the time as sensationalist and uncritical of questionable claims (such as spontaneous generation of animals from inanimate matter), though it inspired others to seriously consider species transformation (notably Alfred Russel Wallace). Although he amassed much evidence for evolution, the author of the *Vestiges* offered no tangible mechanism for the transformation of species over time. Instead, he invoked the constant action of unspecified universal laws that promoted the transformation of species.

The idea of transformation of species over geological time had been discussed for decades, but the first observable mechanism for evolutionary change was provided by two gifted naturalists who independently described how the differential reproductive success of individuals with heritable variations could lead to the generation of new species. Like Darwin, Alfred Russel Wallace (1823–1913) drew on comparisons with artificial selection, in addition to observations of the natural world conducted over many years of fieldwork in South America and South East Asia. Wallace, too, was inspired by Lyell's uniformitarian approach, and he saw the potential for 'progression, by minute steps, in various directions, but always checked and balanced by the necessary conditions', and that the gradual accumulation of those variations that increased chances of survival in the struggle for existence could ultimately explain 'all the phenomena presented by organized beings, their extinction and succession in past ages, and all the extraordinary modifications of form, instinct and habits which they exhibit'.[3]

Wallace wrote a manuscript outlining his theory of evolution while collecting natural history specimens in the Malay Archipelago, and then sent it by mail steamer to the respected naturalist Charles Darwin. Unbeknownst to Wallace, Darwin had arrived at a similar conclusion from his own long-standing studies of the natural world and domesticated varieties. Their two papers were presented together to the Linnean Society of London in July 1858. Both Darwin and Wallace emphasized that many more individuals were born than could survive and reproduce, so any individual having a slight advantage would have a greater chance of contributing to the next generation. Both also considered that this change in the characteristics of a population over generations would, over time, lead to divergence between populations, such that naturally occurring species were descended, like varieties, from other species. Like Lamarck, Darwin and Wallace both suggested that there was often no clear dividing line between species, but a continuous gradation of differences between varieties, races, sub-species, and species. But, as Wallace pointed out in his 1858 paper, the principle of natural selection made Lamarck's attempts to explain the pattern in terms of some vital force unnecessary: change would happen simply as a result of variation between individuals influencing their chances of success in the 'struggle for existence'.

The Origin of Species

The following year, Charles Darwin published an outline of the theory he had been working on for over two decades, under the title *On the origin of species by means of natural selection, or the preservation of favoured races in the struggle for life*. Darwin used patterns from the distribution of species in time (palaeontology) and space (biogeography) to suggest that species were modified gradually and continuously over long time periods. Similar species tended to be clustered in space; neighbouring areas often contained related species; and species suited to particular habitats are not always found in the same habitats on other continents. Similar species also tend to be clustered in time; fossil species tend to be more similar to those found in adjoining strata than they are to those in more widely separated strata. Darwin claimed that the fossil record bears witness to the continuous appearance and disappearance of species, such that biodiversity turnover was not confined to occasional catastrophes in which whole faunas were removed and replaced.

Unlike earlier works on evolution, such as Lamarck's *Philosophical Zoology* and the *Vestiges*, Darwin's *Origin* did not rely on unseen forces to drive the transformation of species over time. He argued that we could see the potential for change in species over time all around us. For example, a naturalist seeing two breeds of fancy pigeons for the first time would probably call them different species, yet they were clearly descended from a single ancestral type (**Figure 1.4**). Darwin gave examples of naturally occurring varieties of plants and animals, and showed that in many cases there was no clear distinction between varieties, races, and species. This pattern could be explained if natural varieties were, like domestic varieties, the products of descent. If such large differences in appearance and behaviour could be produced in a few centuries, might not a much longer period of time produce distinctly different species? But, in the natural world, what would take the place of the breeder selecting the breeding stock?

Darwin considered the consequences of naturally occurring variation between individuals in wild

Figure 1.4 'Believing that it is always best to study some special group, I have, after deliberation taken up domestic pigeons'.[5] Darwin used pigeon breeding to illustrate the important principles of his theory of descent with modification. He presented evidence that the hugely varied modern breeds were all descended from the wild rock pigeon (centre). From conversations with pigeon breeders, he concluded that the breeders selected, knowingly or unconsciously, breeding stock which carried slight variations that distinguished them from others. Over generations of selective breeding, the differences between breeds became more and more pronounced until forms with distinct forms and behaviours were produced, such as (from top, clockwise) the fantail, shortfaced tumbler, bar, frillback, trumpeter, English carrier, Jacobin, scandaroon, pouter, turbit, and nun. These are all breeds that Darwin himself kept at Down House.

populations. Through exhaustive study, he was able to provide evidence that most traits varied to some extent within natural populations, and no individuals were exactly alike. He proposed that, as a result of these naturally occurring differences, some would be more likely to survive and reproduce than others. The offspring of the successful individuals might inherit their parents' favourable traits, and thus also have an advantage over other members of the population. By this means, heritable traits that enhanced survival and reproduction would increase in frequency in the population. In this way, Darwin illustrated how a population could potentially undergo change driven by no other force than a naturally occurring variation between individuals which affects their chances of reproduction.

> We will look closely at natural selection in **Chapter 2** when we consider the origin of life.

Key points

- Evolutionary change in the biosphere over time has been discussed for centuries.
- Darwin and Wallace provided the first plausible mechanism for evolutionary change, using observations of both wild and domesticated varieties, combined with information on biogeography and fossil evidence, to argue for continuous gradual change through increased representation of favourable heritable variations.

The case for Darwinian gradualism

Darwin turned the tide of scientific opinion in favour of evolution. By the time the sixth edition of the *Origin* was published in 1872, Darwin could state that 'almost every naturalist admits the great principle of evolution'. But his argument for the application of Lyell's uniformitarian principles to explaining evolutionary change was not as widely accepted. Just like Lyell, Darwin proposed that processes we can witness today (the continuous production of heritable variation in populations, and the 'struggle for existence' in which some variants are more successful than others) were sufficient to explain the great changes of the past (the divergence of distinct species from ancestral stock). His revolutionary idea was that the modest changes we can observe at the population level could gradually accumulate over vast time spans to produce substantial differences between species, genera, classes, and even kingdoms. It is important to recognize that 'gradualism' refers to cumulative change by many steps, and does not necessarily imply a constant rate of change.

Darwin's uniformitarian explanation—that observable population-level processes (microevolution) were all that were needed to explain the types and distributions of species over space and time (macroevolution)—failed to convince even some of his staunchest supporters. Some objections arose from contemporary limitations on knowledge of key concepts such as the principles of heredity. For example, some critics questioned the efficacy of natural selection to produce permanent changes in populations by suggesting that newly arisen variants would be diluted away through interbreeding. But other criticisms were more substantive, particularly those that highlighted the difficulties in proving that the processes we can witness today are responsible for all changes in the past, without needing to invoke any additional mechanisms. For example, Thomas Henry Huxley (1825–1895), known as 'Darwin's bulldog' for his pugnacious defence of Darwin's theory of evolution, remained agnostic on the power of natural selection on the grounds that no-one had yet proved that Darwin's mechanism of gradual accumulation of variation within populations could indeed produce all of the features of the natural world:

There is no fault to be found with Mr. Darwin's method, then; but it is another question whether he has fulfilled all the conditions imposed by that method. Is it satisfactorily proved, in fact, that species may be originated by selection? that there is such a thing as natural selection? that none of the phenomena exhibited by species is inconsistent with the origin of species in this way? If these questions can be answered in the affirmative, Mr. Darwin's view steps out of the rank of hypotheses into those of proved theories; but, so long as the evidence at present adduced falls short of enforcing that affirmation, so long, to our minds, must the new doctrine be content to remain among the former—an extremely valuable, and in the

highest degree probable, doctrine, indeed the only extant hypothesis which is worth anything in a scientific point of view; but still a hypothesis, and not yet the theory of species.[4]

Can nature make jumps?

Much of the controversy over Darwin's hypothesis, then and now, concerned his insistence on the cumulative effect of small changes. Darwin proposed that natural selection was the main engine of evolutionary change, but he did not claim that it was the only mechanism, and he acknowledged the potential for population-level changes due to chance (now referred to as genetic drift). But Darwin argued firmly that all of the characteristic inherited features of organisms, without exception, were the product of the gradual accumulation of small changes over many, many generations. Darwin's insistence that *natura non facit saltum* (nature does not make jumps) arose from his commitment to the Lyellian approach to scientific explanation.

Just as it was for Lyell, it was important to Darwin that he did not need to invoke any unknown mechanisms to explain natural features: 'If it could be demonstrated that any complex organ existed, which could not possibly have been formed by numerous, successive, slight modifications, my theory would absolutely break down. But I can find out no such case.'[5] Darwin demonstrated that individuals within populations varied in many ways, and provided evidence that there was a 'struggle for existence' because more individuals were born than survived to adulthood or reproduced. By basing his theory of evolutionary change on the differential propagation of these ever-present variations, Darwin provided a hypothesis for descent with modification that relied only on observable phenomena.

Since Darwin's hypothesis of gradual evolutionary change relied on the constant presence of heritable variation in populations, it rested critically on the frequency and type of mutations that arose in natural populations. Therefore the development of genetics in the early 1900s was of critical importance to the maturation of evolutionary biology. However, while the early geneticists were largely convinced of the reality of evolution, many did not accept Darwin's proposed mechanism. Natural selection could clearly cause deleterious mutations to disappear from populations, as their carriers failed to survive or reproduce. But could new species arise simply through the accumulation of small advantageous variations? There was little evidence available to suggest that they could, and many arguments were offered against the power of natural selection to generate new species. If selection only operated on the variation present in populations, then how did novel characters arise that allowed the evolution of a new species adapted to a different way of life? Since species are typically separated by distinct differences, and cannot interbreed to form viable hybrids, many biologists predicted that species must form by discontinuous 'jumps' rather than by constant accumulation of small variations.

One of the most outspoken proponents of an alternative view of evolution was William Bateson (1861–1926), one of the pioneering geneticists who put Mendelian inheritance centre stage in the early decades of the twentieth century. Whereas Lamarck and Darwin emphasized cases in which there was continuous variation between varieties and species, Bateson asserted that these cases were in the minority and that most species were separated by discontinuous variation, with no known transitional forms. Since most of the differences between species recognized by naturalists were not directly associated with adaptation to their particular way of life, Bateson doubted whether natural selection could explain the formation of new species. If evolution of new species was built upon differences between individuals, then Bateson expected two kinds of variation: minor differences between individuals within populations, and discrete mutations that could form the basis of new species.

Because of the emphasis on mutation as a primary force driving evolution, this view of evolution is commonly referred to as mutationism, to contrast it with gradualism which explained the formation of species through the continuous accumulation of small variations, ever present in populations. The two schools of thought differed in their view of the heritable variation that drove evolutionary change. Supporters of the Darwinian hypothesis of gradual evolution tended to emphasize that mutation was random, ubiquitous, and continuous. Mutationists, on the other hand, emphasized the discontinuous recurrence of particular mutations which might give rise to new species without the need for natural selection. Mutationists did not deny selection occurred, but doubted that it could provide the main creative role in evolution. In other words, mutationists felt that there were two separate evolutionary processes: the microevolutionary processes that resulted in fluctuations in the frequency of minor variants within populations and the removal of harmful mutations, and macroevolutionary processes that resulted in the formation of new species.

> We will consider evolutionary explanations based on large discontinuous changes in **Chapter 4** when we examine the 'Cambrian explosion' of animal body plans.

For several decades, evolutionary biology consisted of parallel research programmes. One, dominated by naturalists, concentrated on variation within natural populations. Another, championed largely by geneticists, focused on the generation of substantial new variants that could potentially form the basis of distinct lineages. In addition, many palaeontologists focused on directional trends in the fossil record. These separate foci on different aspects of evolution led to something of a split in the way that evolutionary biology was studied, with some biologists focusing on the changes that could occur within populations, and others on the origin of new lineages. Following terminology introduced by Yuri Filipchenko (Юрий Александрович Филипченко, 1882—1930), these two levels became known as microevolution and macroevolution. Like Bateson, Filipchenko felt that the differences that separated higher systematic categories of living things were of a different kind than the variations found within populations, and therefore explaining the origins of higher taxa was not simply a matter of extrapolating population-level processes over long timescales.

What do you think?

The disagreement between selectionists and mutationists partly reflected different attitudes to science: mutations could be generated and observed in the laboratory, whereas the role of natural selection in the formation of new species is inferred rather than directly witnessed. If we can't witness natural selection creating species, can we still treat it as a scientific hypothesis that can be tested? Can we ever prove the Darwinian hypothesis that new species arise from selection of small variants accumulated over long timescales?

The return of Darwinism

In the early 1900s, evolution was widely accepted in the biological community, but the majority of biologists did not believe that natural selection was the prime mechanism for evolutionary change. However, in the 1920s and 1930s, several lines of research began to turn the tide of opinion by demonstrating that natural selection did have the power to create substantial changes over time simply by favouring slight variants present in the populations.

Experimentalists, working on organisms such as the fruit fly *Drosophila,* challenged the earlier view that mutation created occasional large changes which could result in the formation of a new species. Instead, it was suggested that mutations occur frequently, in a stochastic fashion; each gene has a chance of undergoing mutation each generation, so given enough individuals and sufficient time, the production of variation on which selection can act is assured. Furthermore, these mutations are random with respect to fitness, unaffected by particular changes in the environment or the needs of the organism. Because mutations are essentially errors, they are usually deleterious. But, by chance, some mutations will be advantageous, depending on the environment they occur in.

This view of the continuous supply of many new mutations fed into a new view of selection. A number of scientists developed the mathematical framework for understanding the fate of mutations in populations. This population genetic framework demonstrated that even tiny selective differences can, given a large enough population and a sufficient number of generations, lead to new traits becoming fixed in a population, or to directional change in the average value of a trait over time. One of the key figures in the development of the population genetic framework for evolutionary biology was R.A. Fisher (1890–1962). Fisher used mathematical modelling to reconcile Mendelian genetics, which had focused on the inheritance of discrete traits, with population-level studies, which described continuous variation between individuals. Fisher showed that even slight heritable modifications that conveyed very small selective advantages could be steadily accumulated in populations under selection.

However, these mathematical models would have been of limited value if it had not been shown that the processes they described could occur in natural populations. One of the arguments levelled against evolutionary theory was that the mutations studied by geneticists were artificial productions of the laboratory, which produced strange and unhappy individuals with a reduced chance of surviving to reproduce. Sergei Chetverikov (Сергей Сергеевич Четвериков, 1880–1959) demonstrated the ubiquity of natural variation in the wild by sampling individuals from natural populations of fruit flies (*Drosophila*). When crossed in the laboratory, these flies revealed surprisingly high numbers of variable heritable traits. Chetverikov concluded that mutations constantly accumulate in populations, so that virtually all individuals carry heritable variants.[6] Random mating redistributes this variation between individuals every generation. Selection, on the other hand, reduces variation by removing deleterious mutations, and promoting advantageous variations until they replace alternative traits in the population. Dividing a population into isolated subpopulations drives diversification, because different mutations will become fixed in each population.

By careful observations of wild populations, field biologists were able to demonstrate that heritable variation between individuals in natural populations

Figure 1.5 Populations of day-flying scarlet tiger moths (*Callimorpha dominula*) living in Cothill, Oxfordshire, have played an important role in the debates around the relative influence of selection and chance on allele frequencies. To test theoretical predicitions of population genetic models in the field, researchers needed to study the fate of alleles in populations of known size, but reliable estimates of population size are rare. Scarlet tiger moths provided an ideal test case because they have one generation per year, the conspicuous adults are active during the day for a short period in summer, and they are polymorphic, differing in colour and the number, size, and position of colour patches on the wings. These populations have been monitored for over 60 years, making this one of the longest running natural history studies.[14]

Denis F. Owen and Cyril A. Clarke (1993) 'The medionigra polymorphism in the moth, *Panaxia dominula* (Lepidoptera: Arctiidae): a critical re-assessment', *Oikos*. 67: 393–402. Republished with permission of John Wiley and Sons, Inc. Illustration by Derek Whiteley.

phenomenon. For example, R.A. Fisher worked with one of the founders of ecological genetics, E.B. Ford (1901–1988), to painstakingly record the number of different morphs within a single population of moths in Oxfordshire with the aim of demonstrating that allele frequencies do fluctuate in the wild, and that these fluctuations are far more than would be expected under random sampling alone (**Figure 1.5**).

During the first half of the twentieth century, evidence was growing that variation was continuously generated in populations, and that, given a sufficiently large number of individuals and enough time, these differences could accumulate to make a measurable difference to the characteristics of a population. These studies established the plausibility of Darwin's hypothesis that the accumulation of small variations in populations could lead to large-scale evolutionary change. This population genetic view of evolution laid the foundations for a return to Darwinism. Not only did biologists embrace Darwin's dictum that evolution worked through accumulation of small, ever-present variation, they increasingly viewed natural selection as the primary force driving evolutionary change.

Key points

- Although Darwin convinced scientists that the living world had been produced by evolution, his proposed mechanism of change was not widely accepted until the twentieth century.

- Laboratory experiments showed that new heritable variations arose spontaneously and could be selected.

- Mathematical models showed that mutations with a small selective advantage could rise in frequency and become fixed in a population.

- Field observations showed that variations between individuals in natural populations did make a difference to their chance of survival and reproduction in the wild.

could influence survival. For example, it was shown that variation in colour and banding patterns in snails was correlated with the environment in which the snails were found, and that these patterns affected the chance of a snail being eaten by birds. The results of field experiments like these were combined with the predictions of population genetics to show that natural selection was a common natural

The modern synthesis

66 *It was in this period, immediately prior to the [first world] war, that the legend of the death of Darwinism acquired currency. The facts of Mendelism appeared to contradict the facts of palaeontology, the theories of the mutations would not square with the Weismannian views of adaptation, the discoveries of experimental embryology seemed to contradict the classical recapitulationary theories of development. Zoologists who clung to Darwinian views were looked down on by the devotees of the newer disciplines... The opposing factions became reconciled as the younger branches of biology achieved a synthesis with each other and with the classical disciplines: and the reconciliation converged upon a Darwinian theme.* 99

Julian Huxley (1942). *Evolution: The Modern Synthesis.* George Allen & Unwin.

Charles Darwin was a geologist, systematist, marine biologist, botanist, coleopterist, pigeon fancier, and more. Darwin's breadth of knowledge was remarkable, and his works are strengthened by drawing on many different forms of evidence that today would be considered to belong to separate professional domains. But in the early decades of the twentieth century, biologists became increasingly divided into separate disciplines such as genetics, embryology, systematics, palaeontology, and ecology. Each of these disciplines approached evolutionary biology from a different perspective. Early Mendelian geneticists focused on the production of dramatic mutants, rejecting the Darwinian hypothesis of change by accumulation of tiny heritable variations. Biometricians, on the other hand, focused on change in the value of continuous characters over time, and rejected the Mendelian view of particulate inheritance. Some palaeontologists had used observation of directional trends or shifts in character in response to environmental change to reject Darwinian mechanisms, on the grounds that random mutation could not produce a sustained directional response. Many systematists also downplayed the significance of natural selection, because many of the traits that distinguished species and higher taxa had no obvious role in adapting populations to different niches.

The 'modern synthesis' of evolution countermanded this trend, drawing the disciplines together and uniting them under a common set of assumptions about the evolutionary process: that variation was continuously generated by mutation; that evolution proceeded by the gradual change in the frequencies of these genetic variants in populations over time; that the main driver of this change was natural selection; and that this process could explain all evolutionary change without the need to invoke directional trends, 'jumps' due to large mutations, or non-Mendelian inheritance. The unifying premise was that observable microevolutionary processes were sufficient to explain all macroevolutionary patterns.

The modern synthesis embraced Darwinism, codified in a population genetic framework. Theories based on inheritance of acquired characteristics or vitalistic forces were comprehensively rejected. And as the modern synthesis gathered momentum, it was marked by an increasing acceptance that observable population processes—mutation, migration, selection, drift—could explain all evolutionary change. These processes were invoked as evolutionary explanations even if they could not be observed directly, once again aligning evolutionary biology with uniformitarian principles that past changes should be explained using processes occurring in the present. More specifically, the synthetic view was marked by a confidence that natural selection was, as Darwin supposed, the primary driver of evolutionary change.[8]

For example, palaeontologist George Gaylord Simpson (1902–1984) re-examined a classic example of a linear fossil trend—the evolution of horses—and re-interpreted it as diversification due to the selective response of a suite of characters to changing environmental conditions, and competition to exploit new and existing resources. Considering the palaeontological evidence not only as a temporal series, but also in a biogeographical and environmental context, the change in characters over time could be viewed as the result of populations evolving across the landscape, in response to and altering the environment, and not just as single lineages undergoing directional change over time. Simpson emphasized that different lineages would respond at different rates and in different modes, depending on circumstance. Patterns in the fossil record, such as gaps in the continuity of change, were interpreted using population genetic principles, such as rapid evolution in small populations due to genetic drift. Simpson's work served to emphasize that a commitment to Darwinian gradualism (continuous change by many small steps) did not imply uniformity of rate of evolutionary change across space, time, and lineages.

> The evolution of horses will provide a convenient case study for us to consider evolutionary trends over time in **Chapter 3**.

The modern synthesis also served to highlight the areas of gravest ignorance. It was acknowledged that, just as a lack of understanding of genetics had impeded the progress of evolutionary theory in Darwin's day, imperfect knowledge of the processes of development (expression of genetic information and the way it is influenced by environmental conditions) was a serious impediment to understanding evolution. And while the modern synthesis united biologists around a shared commitment to Darwinian principles, the possibility that mechanisms other than natural selection might shape evolution could not be discounted.[7]

The gene's eye view of evolution

The modern synthesis aligned many fields that had previously been working at tangents. It also marked the move towards a more quantitative approach to evolutionary biology. Simpson's synthetic palaeontology provides a good example of this, moving from a qualitative description of trends over time to quantitative analysis of character shifts, and comparison of numbers of species across space and time. It also heralded close attention to the underlying genetic basis of change, whether studied directly or simply assumed. This focus on the gene was rejected by some as unhelpful reductionism: if all evolution was microevolutionary change

in gene frequencies, where did that leave the study of macroevolution? An even greater challenge arose from placing genes at the centre of evolutionary theory: where did that leave the individual?

The gene-centred view considered evolution in terms of the competition between alternative alleles (variants of the same gene) in a population, mediated through their effect on phenotype and behaviour, and even the effect of those alleles on other individuals and the wider environment. The gene's eye view was a natural extension of the population genetic approach to evolution, considering the fate of alternative alleles in a population and asking which will rise in frequency over generations due to selection. If a particular allele has properties that make it more likely to be copied to the next generation, then it will tend to increase in representation in the population until it replaces all other variants. Any 'selfish gene' which replicates at the expense of others will come to predominate.[8]

By focusing on the fate of alternative alleles, rather than individuals, this view shed light on some issues that had been mysterious. For example, W.D. (Bill) Hamilton (1936–2000) demonstrated how the apparent altruism of cooperative breeders like honey bees, where workers sacrifice their own reproduction to raise the queens' offspring, could be understood in terms of the relative reproduction of alleles. The worker bees share their genes with their sister queens, so an allele that causes them to help raise thousands of her offspring can rise in frequency through natural selection (**Figure 1.6**).

The gene's eye view also gave rise to new ways of considering the action of selection in populations. John Maynard Smith (1920–2004) introduced the concept of game theory to evolutionary biology, considering how the success of different alleles in a population could depend upon the alternative 'strategies' in the population. A classic example is the limitation of within-species combat to a non-lethal level. In a battle between males for territory, why does the conflict resemble a ritual, such as the clashing of antlers, rather than a fight to the death, inflicting fatal wounds? Instead of ending with one individual backing down while the other claims

Figure 1.6 Bees, and other eusocial insects, presented a challenge to Darwinism, so much so that Darwin referred to the problem as 'one special difficulty, which at first appeared to me insuperable, and actually fatal to my whole theory. I allude to the neuters or sterile females in insect-communities: for these neuters often differ widely in instinct and in structure from both the males and fertile females, and yet, from being sterile, they cannot propagate their kind'.[5] If the vast majority of bees are workers that do not reproduce, then how can natural selection act on worker bee morphology and behaviour? A solution to this problem was offered over a hundred years later by Bill Hamilton's work on 'inclusive fitness'[15] (considering the fitness of an allele in terms of related individuals who share the same allele). Hamilton considered the fate of alleles that caused altruistic behaviour (helping others despite costs to yourself). He argued that helping relatives, who are also likely to have inherited the same allele, can be favoured by natural selection as long as the relative benefit to the recipient (*b*) sufficiently outweighs the cost to the helper (*c*), adjusted to their degree of relatedness (*r*), which reflects the chance of carrying the same allele. This can be written as *rb>c*, now known as Hamilton's law.

different strategies could persist stably without the 'doves' being extirpated by the 'hawks'. In these and other ways, the gene's eye view enriched the role of selection in shaping evolutionary change, leading to a complicated array of outcomes.

This view of evolution reinforced the Darwinian view that competition between heritable variants in populations (microevolution) can lead to complex outcomes when run over long timescales (macroevolution). The 'gene's eye view' also led to clarification of ideas around levels of selection by throwing into focus the basic requirements for a system to evolve. Any system that consists of entities that have heritability (offspring resemble parents) and variation in copy success (some heritable traits confer greater likelihood of replication) will lead to an accumulation of heritable variants that increase replication success. This 'universal' formulation of Darwinian evolution allows us to consider the way selection can operate at different levels of biological organization, including not only the gene, but also the genome, individual, population, and lineage. It also allows us to consider the application of evolutionary models developed in biology to other evolving systems, such as languages, computer programs, and cultures.

> Different perspectives on the levels of selection—from the gene to the lineage—will be investigated in **Chapter 7** where we meet the challenges of explaining the evolution of sexual reproduction.

Neutral theory

The population genetic framework also led to the emergence of an alternative view of evolutionary change that some initially regarded as 'non-Darwinian'. Heritable variation in populations had been studied primarily by its effect on phenotype (observable characteristics of individuals). But the development of new molecular techniques allowed biologists to consider variation at the level of genes and proteins. And what they saw was initially shocking. Almost all proteins varied between individuals in a population. If all of this variation was under selection, then there should typically be only

victory, why would the stronger not simply kill the weaker, thereby permanently eliminating the competition? This limitation of hostility had often been explained as being of benefit to the species as a whole, resolving competition between males without unnecessary deaths in combat. But Maynard Smith and others showed that a 'limited war' strategy could, in some cases, outcompete a strategy of naked aggression. Under certain conditions, populations containing a mix of individuals with one variant that was the fittest version, and so all the others should be at a reproductive disadvantage and therefore declining in frequency. Given the observed levels of variation, few individuals would have the fittest variant of all genes, suggesting that populations would be burdened with an intolerable genetic load, dragging down the average fitness.

But the excess genetic variation could be explained if most of the observed variants were selectively neutral, so that it didn't make any different to fitness which variant of the gene an individual happened to carry. In this case, as Darwin had recognized, the neutral variants would fluctuate in frequency randomly over generations, and occasionally might be lost or, by chance, increase until they replaced other variants in the population. The neutral theory was dismissed by many biologists as providing little explanatory traction for traits beyond nucleotide variations in the genome. But, as we shall see in later chapters, not only is genetic drift now considered an important force shaping the genome, neutral processes are increasingly being credited with the power to generate major patterns in macroevolution and macroecology.

> We will consider neutral molecular evolution in **Chapter 6**, when we look at the role of molecular data in testing macroevolutionary and macroecological hypotheses.

Key points

- The modern synthesis of evolutionary biology brought different disciplines in evolutionary biology together through an agreement that evolutionary patterns were best explained by the accumulation of small heritable variants, primarily through natural selection.

- Building on the population genetic framework of the modern synthesis, the gene's eye view reinforced the view that all evolutionary change could be understood in terms of the accumulation of variations which are present in every population.

- Mathematical population genetic models, combined with empirical observations of genetic variation, led to increasing recognition of genetic drift as a force in evolution.

Challenges to Darwinian gradualism

Essentially all biologists today agree that all species are descended from other species, and that the microevolutionary processes of selection and drift of allele frequencies within populations lead to evolutionary change over time. And most biologists agree with Darwin that macroevolutionary patterns in biodiversity are generated by microevolutionary change in allele frequencies within populations. Yet Darwin's hypothesis—that microevolutionary processes are all that we need to explain the evolution of biodiversity—has been questioned from the time it was proposed to the present day. Challenges to this doctrine have come from two fields in particular: palaeontology and developmental biology.

Punctuated equilibrium, proposed by two palaeontologists in the early 1970s, was one of the most prominent challenges to Darwinian gradualism. Stephen Jay Gould (1941–2002) and Niles Eldredge (b. 1943) suggested that the fossil record did not show continuous and gradual change transitions, as predicted by Darwinian gradualism. Instead, new species often appeared suddenly and then remained unchanged for long periods. While Darwin had attributed the lack of transitional fossils to the imperfection of the fossil record, Gould and Eldredge attributed these abrupt transitions to an uneven process of evolutionary change. They proposed that most species formed in small isolated populations which underwent rapid evolutionary change. Because species began in small populations and the rate of change was fast, there would be little chance of a transitional form being represented in the fossil record. Once a species was established, and the population grew, it would undergo very little evolutionary change. So all we would see in the fossil record would be species remaining unchanged over long periods, followed by the sudden appearance of a new related species without any evidence of transitional forms being preserved.

One proposed mechanism for rapid morphological change at speciation is through changes in development. Development is the process whereby a single cell divides again and again to give rise to a multicellular organism with many different cell types, modulated through differences in gene expression between cell lines. Every cell has the same genetic information, so differentiation relies on regulation of gene action, with different genes turned on in different tissues at different times. Mutations in gene regulatory sequences can result in changed gene expression. A single mutational change to a regulatory sequence can lead to a large change in morphology, as an entire genetic program could be switched on or off. For example, regulatory genes specify the development of different body segments in fruit flies, and changes to these regulatory genes can result in the growth of an extra pair of wings or a change in the positioning of appendages. Because dramatic changes in morphology can be caused by a single genetic change, regulatory changes have been suggested to play a role in rapid evolution. Alternatively, it has been suggested that developmental plasticity will slow the rate of evolutionary divergence in morphology by preventing small genetic changes from altering the phenotype. So developmental evolution has been put forward as a possible mechanism for a large amount of change at speciation events, followed by periods of stasis.

> **Chapter 4** examines the role of developmental genes in large macroevolutionary changes.

The emergence of macroecology

In addition to changes in heritable traits (as studied by field biologists and agriculturalists), and modification of form over time (as captured in fossil series), the development of macroevolution relied heavily on the consideration of spatial distributions of species. In parallel with the development of scientific methods for investigating the natural world, exploration for trade, science, or conquest fuelled an awareness of how biodiversity varies from region to region. Travellers from Europe to colonial outposts and beyond collected specimens of exotic plants and animals, and wrote detailed natural history accounts of their travels. These global travellers contributed to the growing field of systematics, the naming and classification of species into a nested hierarchy—a body of knowledge that would later provide ample evidence for descent with modification (see Figure 1.5).

Just as the study of macroevolution preceded the invention of the term (by Yuri Filipchenko in the 1920s), so the study of macroecology long predates its current label. Although macroecology was first presented as a distinct approach in 1989,[9] the term 'macroecology' was frequently used by other authors as early as the 1960s. However, we might say that macroecology as a field began with scientists who moved beyond cataloguing species and began to examine how diversity was distributed, why certain species were found in some places and not others, and how large-scale processes could generate distinct assemblages of species in different parts of the world. Alexander von Humboldt (1769–1859) was one such explorer and thinker. With an unstoppable enthusiasm for scientific investigation and a passion for exploration, Humboldt travelled widely in Europe and the Americas (**Figure 1.7**). While his natural history collections were extensive, he moved beyond collection and classification and began to think about why species of different kinds were distributed as they were. Humboldt considered the way that geography and climate shaped a region's flora and fauna, and how human impacts on ecosystems were rapidly changing aspects of the climate and environment.

With his evocative descriptions of far-off lands, emphasizing the little-known riches of the natural world, and his holistic approach to understanding nature, Humboldt's books were an inspiration to many other naturalists. The young Charles Darwin was captivated by Humboldt's accounts of his travels, which inspired him to undertake his own

Figure 1.7 Alexander von Humboldt and his travelling companion, botanist Aimé Bonpland, in the Amazon jungle. Humboldt and Bonpland travelled extensively in South America, including a bold crossing of the Andes and an attempt to climb Chimborazo (then thought to be the highest mountain on earth). They travelled overland by foot and horseback, and along mighty rivers by canoe (at one point, when their supplies ran low, they had to subsist on wild cacao beans). Humboldt's scientific interests were broad, and he conducted investigations into geology, astronomy, geography, botany, and zoology. He brought all of these threads together in his major multi-volume work, *Kosmos*, which viewed all these aspects of the earth as parts of an interconnected whole.

Humboldt and Bonpland by the Orinoco River, 1870, by Eduard Ender. Akademie der Wissenschaften, Berlin. © akg-images.

journey. He took Humboldt's published narratives of his travels with him on the *Beagle*, as guidance for his own exploration and as a reference brimming with useful facts on tropical nature. Darwin rated Humboldt the greatest travel writer in history. On publication of Darwin's own travel narrative from his years on the *Beagle*, Humboldt repaid the compliment: 'Considering the importance of your work, Sir, this may be the greatest success that my humble work could bring. Works are of value only if they give rise to better ones.'[10]

Darwin's travels were critical to the formulation of his theory of evolution, not just because he sampled a wider variety of species than he could have done if he had stayed at home. It was his travels that prompted Darwin to puzzle over species distributions. Why would each island of the Galapagos have its own species of finch? Why were the peculiar mammals of South America like the fossil animals found on the same continent? Writers such as Humboldt had discussed how climate and landscape combined to generate habitats suited to particular species, but why should regions with similar climates be populated by very different groups of species? Why were oceanic islands missing animals that would clearly be suited to their habitats, such as frogs and terrestrial mammals?

Alfred Russel Wallace likewise encountered many puzzles in species distributions as he travelled through South America and South East Asia. He noted that species boundaries were often defined by natural barriers, such as rivers or mountains, even if there was suitable habitat and climate for them across a wider area. Both Darwin and Wallace came to realize that the sometimes perplexing distribution of species could be understood if species arose by descent from other species. Since each species has a single point of origin, their distributions may be limited not just by habitat suitability but also by opportunities to disperse away from their ancestral ranges. In other words, history can be as important as environment and adaptation in explaining species distributions.

Explanations for the distributions of species in the context of evolutionary theory were explored in detail by Wallace, and developed into the new discipline of biogeography. Species distributions could be considered in both temporal and spatial dimensions. For example, in the 1920s John Willis (1868–1958) developed a theory to account for the structures and sizes of species geographic distributions.[11] According to Willis, the ranges of taxa follow predictable trajectories, expanding from a small localized area of origin to become more widespread as the taxon increases in age. Under this 'age and area' hypothesis, narrow-range endemic species are indicative of recent evolutionary origins. Such biogeographic theories were not necessarily tied to particular models of evolution—while Wallace was a committed Darwinian gradualist, Willis rejected natural selection and favoured a 'saltationist' view of speciation by occasional macromutation.

> **Chapter 9** combines macroevolutionary and macroecological mechanisms to investigate why some lineages are so much more diverse than others.

The development of biogeography, with its broad-scale approach to understanding patterns of distributions and diversity, was based heavily on ecology and natural history. However, during the twentieth century ecology itself became increasingly focused on small-scale, population-level patterns and processes. As ecology developed from its origins in descriptive natural history, a body of mathematical theory grew and methods of analysis became increasingly quantitative, model-based, and experimental. This approach brought predictive power to ecology. For example, theories of relative

abundance gave insight into the prevalence of common and rare species in communities, and even allowed ecologists to predict the true number of species in an under-sampled community.

By the 1960s, much of ecological theory was focused on explaining population dynamics, diversity, or community structure within relatively small areas consisting of sets of potentially interacting individuals of different species (the local community). Ecological processes operating at these local scales, such as interspecific competition, were widely thought to be the primary drivers of community structure and dynamics, with membership of the local community determined by the filtering (or 'sorting') of species from a broader regional species pool. But observations began to accrue that could not easily be interpreted in terms of local-scale control of diversity patterns. For example, it became apparent that communities with climatically and structurally similar habitats on different continents could contain very different numbers of species. Temperate forest communities of the west coast of North America typically contain 20–40 tree species, but similar sized communities at the same latitudes in east Asia contain hundreds of tree species. These communities are in regions with different levels of diversity—the Pacific Slope region of North America has around 70 tree species, while the coastal region of east Asia has over 700. These differences in regional diversity seem unlikely to be wholly due to differences in climate or landscape, but are more likely to reflect different histories. Much of North America was under an ice sheet at the Last Glacial Maximum (around 20,000–25,000 years ago), but most of east Asia remained ice free. The implication is that the North American forests have not yet been fully recolonized from lower latitudes following the retreat of the ice sheets.

> We will investigate regional differences in species richness in **Chapter 10** when we ask why tropical regions support a greater number of different species than temperate regions.

During the 1980s, the growing appreciation of the important role of historical events and regional diversity in shaping ecological patterns shifted the focus of ecology back towards evolutionary and bio-geographic processes operating over large scales of time and space, such as speciation, extinction, and colonization and range expansion. The importance of history also helped to shape an important debate over whether equilibrium or non-equilibrium dynamics prevail in ecological systems. Much of community-level ecological theory from the 1950s to the 1980s assumed that diversity was at equilibrium—that communities are 'saturated,' with all available niches occupied, so that the number of species is close to the maximum that can be supported by each particular kind of environment. During the 1980s, an alternative non-equilibrium ecology became more widely accepted, suggesting that diversity in many systems has not yet reached saturation. The equilibrium versus non-equilibrium debate continues to the present day, and this theme will emerge frequently as we explore large-scale diversity patterns.

> In **Chapter 10** we examine in detail the implications of equilibrium versus non-equilibrium diversity dynamics for the way we explain global gradients in species richness.

It was in this context that macroecology emerged during the 1980s as a distinct approach to studying biodiversity. The focus of macroecology was explicitly on the description and explanation of the emergent patterns of biodiversity that arise when we consider the collective attributes of species in large assemblages. Some of these patterns could be described in single dimensions, such as the frequency distribution of species' body sizes. Other patterns were described in two dimensions, such as the relationship between body size and average population density across species.

In the seminal macroecology paper published in the journal *Science* in 1989, macroecology was described as 'The division of food and space among species on continents.'[9] This description revealed the community-ecology roots of macroecology. Although large-scale processes, such as speciation

and extinction, were acknowledged as key drivers of large-scale ecological patterns, these patterns, even on continental scales, were still considered to be predominantly due to ecological sorting at the species or lineage levels. Biogeographers and evolutionary biologists, on the other hand, continued to pursue the role of history and chance as the key drivers of present-day diversity, particularly as the analysis of molecular phylogenies (evolutionary trees) became more common.

Figure 1.8 Some biodiversity hotspots are a macroecological and macroevolutionary puzzle. This highly diverse heathland in Australia's south-west lacks many of the features often associated with high plant diversity: it is well outside the tropics, the climate is dry, and the soils are nutrient poor. Yet the local-scale species richness of this heathland is very high, and the region it lies within supports over 3500 endemic plant species. The causes of this exceptional plant diversity remain a mystery, but hypotheses include a long period of comparatively stable climate, a rapid rate of speciation, and a complex mosaic of soil types that promotes ecological specialization.

© Marcel Cardillo

The equilibrium, niche-based perspective of macroecology, and the more historical perspective of biogeography and macroevolution, are still not fully reconciled. At the same time, the boundaries between macroecology, macroevolution, and biogeography have become increasingly blurred, and most biodiversity scientists today would probably adopt a liberal definition of macroecology which includes, for example, large-scale species richness gradients, geographic patterns of phylogenetic diversity, or the effects of climate change on species distributions (**Figure 1.8**). Macroecology has also begun to move beyond its early emphasis on describing and explaining patterns, towards the testing of theoretical models grounded in evolutionary and ecological processes. This has further increased the overlap between macroecology and macroevolution.

> **Chapter 11** asks how a macroecological perspective might help us describe and understand the current extinction crisis, and asks whether macroecology offers tools to help conserve biodiversity.

Key points

- Interest in large-scale ecological patterns and their explanation began in the early nineteenth century as naturalists started travelling widely, and describing the floras and faunas of different regions.

- Many debates in the study of diversity concern the roles of history versus environment, local-scale versus regional-scale processes, and equilibrium versus non-equilibrium dynamics in shaping present-day patterns of biodiversity.

- Macroecology as a distinct discipline emerged in the 1980s, building on equilibrium-focused community ecology, combined with elements of the more historical, non-equilibrium traditions of biogeography and macroevolution.

Testing hypotheses in macroevolution

❝ *Every active mind must form opinions without direct evidence, else the evidence too often would never be collected. Impartiality and scientific discipline come into action effectively in submitting the opinions formed to as much relevant evidence as can be made available.* ❞
R.A. Fisher (1929). *The Genetical Theory of Natural Selection*, Oxford University Press.

It may seem odd that most science textbooks do not contain instructions on how to do science. But the process of scientific investigation is harder to characterize than it would first appear. There is no single approach that can be followed like a recipe in all areas of science. Instead, scientists use a wide variety of approaches in their pursuit of satisfying explanations for natural phenomena. So the best way to learn how to do science is by observing the kind of strategies scientists employ when they are seeking answers to the kinds of questions you are interested in.

There are many important and interesting questions in macroevolution and macroecology that we would like to answer, and in most cases there are several alternative explanations for each pattern. Many of these questions concern events that happened in the past, or processes that operate over such long timescales that no human can expect to witness them directly. How is it possible for a scientist to investigate the cause of an evolutionary event that they have no way of observing directly? How can we judge the plausibility of mechanisms that are too slow to be observed in operation? To address this concern, we need to consider how evolutionary biologists ask and answer questions about the natural world (see Case Study 1).

Testing claims about the natural world lies at the heart of a scientific investigation, but there are many possible ways to conduct a scientific test. Perhaps the most obvious is a manipulative experiment. An experiment usually consists of taking many different subjects, which are randomly assigned to a 'treatment' (a regime where one particular factor of interest is manipulated) or a 'control' (the same regime without the manipulated factor). A robust experiment should include enough subjects in both treatment and control groups to allow the effect of chance outcomes to be discounted (for example, so that the accidental death of a subject does not skew the survival scores in either the control or treatment group). Ideally, only the factor of interest should vary consistently between treatment and control subjects, so any difference between them can only be due to the factor and not chance or bias (for example, by randomly assigning individuals to treatment or control).

Manipulative experiments have a long and noble history in evolution and ecology, from early experiments which demonstrated that slowly raising incubation temperature resulted in a heritable increase in thermal tolerance in bacteria, to more recent 'ecotron' controlled-environment experiments on the impacts of species coexistence and competition on ecosystem productivity. But there are unavoidable limitations to the use of classical manipulative experiments to study macroevolution and macroecology. We can't prove that the processes we witness in the laboratory or the field are indeed responsible for the generation and maintenance of biodiversity in these systems. The amount of change we can directly measure is generally not of the same magnitude as that which separates most species in nature. In particular, manipulative experiments might tell us what *can* happen under particular circumstances, but can't

tell us what actually *did* happen in the past to generate the actual diversity we see today in the real world. Furthermore, many of the processes we wish to study in macroevolution and macroecology are simply not amenable to experimental manipulation. For example, if we want to know whether small-bodied mammalian lineages tend to have higher rates of diversification than larger-bodied mammals, we are not going to be able to set up replicate experimental populations of different-sized mammals that can be left to evolve different numbers of species; it would take tens of millions of years for the experiment to run.

However, the principles of experimental design can be used even where manipulation is not possible. Consider the question: 'Does smoking cause lung cancer?' We can't test this with a manipulative experiment; no-one is going to take a group of young people, randomly assign them to 'treatment' groups who smoke and 'control' groups who don't smoke, and then see which groups have the highest lung cancer rates over the ensuing decades. Yet most people would regard the link between smoking and lung cancer not only as a testable hypothesis but also as proved beyond reasonable doubt, because studies show that people who smoke are more likely to have lung cancer than people who don't, and the more people smoke the greater their risk of lung cancer. These observational studies rely just as heavily on the principles of good experimental design as manipulative experiments.

Similarly, careful design of scientific investigations is as critical in macroevolution and macroecology as it is in all other fields of biology. But the 'treatments' (what happens when …) and 'controls' (what happens in the absence of …) will be produced by careful selection and comparison of naturally occurring phenomena, not by manipulation. So if we want to know whether small-sized mammals have a higher rate of diversification, we can compare lineages of small-bodied mammals to their large-bodied relatives and ask whether the small-bodied lineages have accumulated more species over time. To draw any firm conclusions from this comparison, we need lots of different examples, they need to be statistically independent, and we need to take into account any additional sources of variation that may confound the results. Much of this book will explore how scientists frame hypotheses in macroevolution and macroecology, and how they use a range of analytical approaches to test these hypotheses.

Seeing science in action

Science textbooks should not just teach the products of science (what we know), but also the process of science (how we find out). This is particularly true in fast-moving fields like macroevolution and macroecology. These fields are moving forward with such rapidity that anything we report has the chance of being out of date by the time it goes to print. But by focusing on the process of science we can give you a future-proof insight into the way that ideas in macroevolution and macroecology are generated and tested. You can then approach new ideas and new data armed with the diverse intellectual toolkit of modern biology.

As we have seen in the previous historical account, ideas about the generation and maintenance of biodiversity are constantly changing, and at any given point in time there may be several opposing viewpoints. Therefore we think that the right way to approach major ideas in macroevolution and macroecology is to present them as debates between opposing hypotheses. You will see how scientists frame hypotheses, make predictions, and seek evidence that will support one hypothesis or another. Sometimes we will demonstrate how the evidence seems to support one idea more than another, but we will rarely be in a position to definitively report one hypothesis as fact and the alternatives as false. As time passes, the weight of evidence may swing more behind one idea or another. But in the meantime, you will be asked to think for yourself, form your own opinions based on available evidence, and focus on what data or analyses you would need to support one hypothesis over another.

Because of the emphasis on testing ideas rather than learning facts, this book will not take an exhaustive 'history of life' approach to understanding biodiversity, nor will it cover all major biomes or lineages. Instead, we will focus on a handful of 'case studies,' each asking a different question about the origins of biodiversity. Each case study concerns an area of current debate in biology, and each will highlight different kinds of data and analytical tools being used today to explore these ideas. We can't aim to give you a comprehensive guide to contemporary themes and research tools in these fields—not only is there too much to cover in one book, but new ideas are continuously arising and new methods are always being developed. The best we can do is to offer a selection of examples that illustrate some of the important questions and some interesting ways of seeking to answer them.

Conclusion

❝ *The field of macroevolution embraces the excitement of seeking an understanding of the breadth of life.... We return to this perspective from many quarters of biology and paleontology after many decades of asking far more restrictive questions that tended to put the processes of evolution under a microscope. But now we are stepping back, to take in the broader view.* ❞
Jeffrey S. Levinton (2001). *Genetics, Paleontology and Macroevolution* (2nd edn). Cambridge University Press.

History, like evolution, does not tend to follow neat linear paths, even if we tend to tell stories as if it did. This account of the development of macroevolution and macroecology has selectively focused on opposing points of view that have featured, in various guises, throughout the history of these fields, because these viewpoints provide a starting point for thinking about different possible explanations for the evolution of biodiversity. One view of the evolution of biodiversity is that it is driven by the continuous production and gradual accumulation of small heritable variants in natural populations. Another view is that most evolutionary change occurs when new lineages are created, potentially through large changes to morphology. One view of species distribution is that species are primarily governed at the local scale through availability of, and competition for, resources. Another view focuses on the role of history in shaping current species assemblages. In combination with these different views, there have been debates about the relative importance of selection and chance effects. It would be a mistake, however, to expect that all biologists (past or present) fall neatly into one camp or the other, or to think that all debates about macroevolution and macroecology can be framed in these terms.

Macroevolution and macroecology aim to provide explanations of patterns of biodiversity over space, time, and lineages. These patterns are generated by mechanisms that typically cannot be directly witnessed or manipulated, because they are shaped by events that happened in the past, or by processes that occur over long evolutionary timescales, or arise from the interplay of forces acting on a global scale. While experiments can be used to test the plausibility of different mechanisms and processes, macroevolution and macroecology are typically studied by making predictions about what patterns we would expect to see under a particular hypothesis, and then testing those predictions against observations from the natural world.

In this book, we will use a series of case studies to explore some key themes in macroevolution and macroecology, and to illustrate some of the methods being used to weigh up competing hypotheses. Many of the themes we will explore have been debated for centuries, such as the cause of high biodiversity in the tropics or the role of developmental changes in the evolution of novelty, and so a historical view of the development of these disciplines can help to place these debates in context. While these ideas have been discussed for a very long time, there are many debates still unresolved. This makes it an exciting field for study. But it also means that rather than presenting you with a codified set of answers, this book will instead furnish you with some rather interesting questions and introduce you to some of the tools you might use to work out the answers for yourselves.

◯ Points for discussion

1. How do we decide on an appropriate 'level' of explanation? Should we be aiming to reduce all explanation of biological phenomena to the lowest possible level, such as chemical interactions or physical processes? If we seek an explanation at a 'higher' level, does that mean we are saying lower-level processes are inadequate to explain a particular phenomenon?

2. If we can only directly observe present phenomena, and cannot know for sure what happened in the past, then why shouldn't we invoke different evolutionary mechanisms that only operated in the past and are not observable today?

3. Can manipulative experiments tell us anything about macroevolution and macroecology?

✱ References

1. Mayor A (2000) *The First Fossil Hunters: Paleontology in Greek and Roman Times.* Princeton University Press, Princeton, NJ.

2. Lyell C (1830) *Principles of geology, being an attempt to explain the former changes of the Earth's surface, by reference to causes now in operation.* John Murray, London.

3. Wallace AR (1858) On the tendency of varieties to depart indefinitely from the original type. *Journal of the Proceedings of the Linnean Society of London.* 3(9): 45–62.

4. Huxley TH (1860) The origin of species. In: *Darwiniana: Collected Essays*, Vol 2, pp. 71–9.

5. Darwin C (1859) *On the origin of species by means of natural selection, or the preservation of favoured races in the struggle for life* (1st edn). John Murray, London.

6. Chetverikov SS (1929) On certain aspects of the evolutionary process from the standpoint of modern genetics (English translation published 1961 by M Parker, IM Lerner). *Proceedings of the American Philosophical Society* 105: 167–95.

7. Huxley J (1942) *Evolution: The Modern Synthesis.* Nature Publishing Group, London.

8. Dawkins R (1979) *The Selfish Gene.* Oxford University Press.

9. Brown JH, Maurer BA (1989) Macroecology: the division of food and space among species on continents. *Science* 243(4895): 1145–50.

10. von Humboldt A (1839) Letter to Charles Darwin, 18 September 1839.

11. Willis J, De Vries H, Guppy H, Reid E, Small J (1922) *Age and Area: A Study in Geographical Distribution and Origin of Species.* Cambridge University Press, Cambridge.

12. Huxley TH (1859) The Darwinian hypothesis. In: *Darwiniana: Collected Essays.* Macmillan, London.

13. Lyell C (1863) Letter to Charles Darwin, 15 March 1863. In *Life, Letters and Journals of Sir Charles Lyell* (ed. KM Lyell), 1881.

14. Owen DF, Clarke CA (1993) The medionigra polymorphism in the moth, *Panaxia dominula* (Lepidoptera: Arctiidae): a critical re-assessment. *Oikos* 67: 393-402.

15. Hamilton WD (1964) The genetical evolution of social behaviour. I. *Journal of Theoretical Biology* 7(1): 1–16.

Case Study 1
Testing hypotheses about past events: uncovering the evolutionary origins of a new virus

While scientific training might involve the acquisition of technical skills, such as those associated with laboratory work, computer analysis, or field studies, the most important sets of skills a scientist needs are those associated with asking good questions, framing hypotheses, and designing effective tests. These skills include being able to identify patterns that require an explanation, generate plausible alternative explanations for those patterns, design tests that will reject or support these explanations, gather appropriate data to carry out those tests, and critically examine the findings in light of possible alternative explanations. In this case study, we will use one particular example to illustrate some of the approaches to framing and answering questions in evolutionary biology. No single example could possibly illustrate all the important facets of scientific investigation, but some of the key points we want to cover are as follows.

1. How do we come up with hypotheses?
2. How do we use those hypotheses to generate testable predictions?
3. How do we test those predictions against observations?
4. What confidence can we have that the hypothesis has been supported, disproved, or neither?

There is no single correct approach to generating and testing hypotheses. In fact, questions in macroevolution and macroecology are often profitably examined by taking many different approaches. Typically, no single test will provide a definitive answer.[1] Instead, most scientific investigations in these fields circle around the core issues, with each test providing some degree of support for one explanation over another. In this way, complex processes are gradually illuminated. We will use one well-known example of a past evolutionary event—the emergence of a new human virus—to illustrate some of the strategies that can be used to generate and test hypotheses about phenomena we cannot directly observe in action.

Where did HIV come from?

We start with the biological phenomenon we want to explain. In the 1980s, a perplexing new disease syndrome was reported, which was given the name acquired immune deficiency syndrome (AIDS) and was eventually found to be caused by a retrovirus named human immunodeficiency virus (HIV). AIDS became a global pandemic causing tens of millions of deaths, with millions of people infected every year. We want to answer the question: Why did AIDS suddenly appear in the 1980s? As this is a question about a past event, we cannot make direct observations on its occurrence, so we will have to piece together an explanation from evidence available today. First, we need to narrow it down to a series of more specific questions that we can answer with specific tests. One question we can answer with genome analysis is: What is the evolutionary origin of HIV? To answer this, we start by asking what other kinds of viruses are most similar to HIV.

By comparing the genome sequence of HIV with genome sequences from other viruses, the closest known relatives of the human virus can be identified. Simian immunodeficiency viruses (SIVs) have been isolated from over 40 species of African primates.[2] Comparing the same genes in sequences from different virus genomes allows researchers to use the patterns of similarities and differences between DNA sequences to construct a phylogeny (evolutionary tree)—a hypothesis about the evolutionary relationships between organisms. The branches connecting the sequences (labelled at the tips of the tree) represent lines of descent, and the connections between the branches (nodes) represent common ancestors that gave rise to two or more descendant lineages. We can use this pattern of connections to identify the common stock from which different viruses arose. The phylogeny in **Figure 1.9** suggests that the closest relatives of the human virus HIV-1 (the main pandemic strain causing the majority of HIV infections worldwide) are SIVs from chimpanzees. HIV-2, a strain of the virus predominantly found in West Africa, is most closely related to SIV strains found in sooty mangabeys.

This phylogeny doesn't contain all possible SIV sequences. So on this evidence we can't rule out the possibility that there might be another unknown virus out there that is an even closer relative of HIV. But what we can say is that the human virus HIV is related to simian viruses, and that different strains of HIV appear to be most similar to different kinds of simian viruses. This suggests that SIV viruses have moved from monkeys and apes into humans on multiple occasions. We can gain more detail on this history by adding in more sequences, sampled from a wider range of people and primates (**Figure 1.10**). Since few phylogenies contain all possible sequences, the story we tell depends on which sequences we include.

This phylogeny suggests at least four independent evolutionary origins of HIV because the HIV sequences do not all cluster together in the tree: one of the HIV-1 sequences clusters with a gorilla virus, and others are related to different strains of chimpanzee virus. This suggests that these HIV sequences don't all trace back to a single common ancestral virus which happened to infected a human. But can we be sure that this pattern reveals a biological phenomenon, or could it just be due to an uncertain signal in the data? We can evaluate the reliability of the phylogenetic evidence in both formal and informal ways. Formal methods involve some kind of measure of

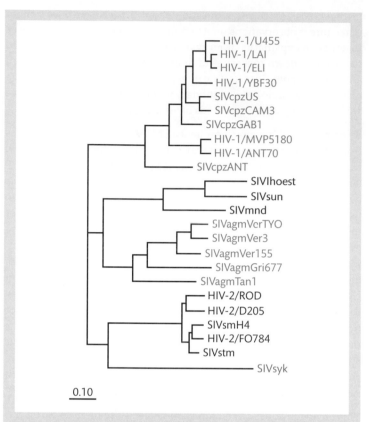

Figure 1.9 This phylogeny of human immunodeficiency virus (HIV) and simian immunodeficiency viruses (SIVs) from a range of primate species (e.g. cpz = chimpanzee, sm = sooty mangabey, agm = African green monkey) was a key piece of evidence that showed not only that HIV was derived from a primate virus, but also that it had crossed from multiple primate species into humans on several different occasions[3]. This phylogeny shows that HIV-1 (the most common form of the virus) is most closely related to the chimpanzee viruses, but HIV-2 is a distinct lineage derived from sooty mangabey viruses. The scale bar indicates the inferred number of amino acid replacements per site.

From Hahn BH, Shaw GM, De Cock KM, Sharp PM (2000), 'AIDS as a zoonosis: scientific and public health implications', *Science* 287(5453):607–14. Reprinted with permission from AAAS.

statistical support. For example, you might explore how well your data support a particular hypothesis by randomly subsampling the data many times, and asking whether these slightly different datasets give you the same phylogeny (a procedure referred to as 'bootstrapping'). Informal methods involve comparing the results of this analysis with other analyses that differ in data or assumptions. We can see examples of both of these approaches in the phylogenetic investigation of a new strain of HIV-1, designated group P (Figure 1.10).[3]

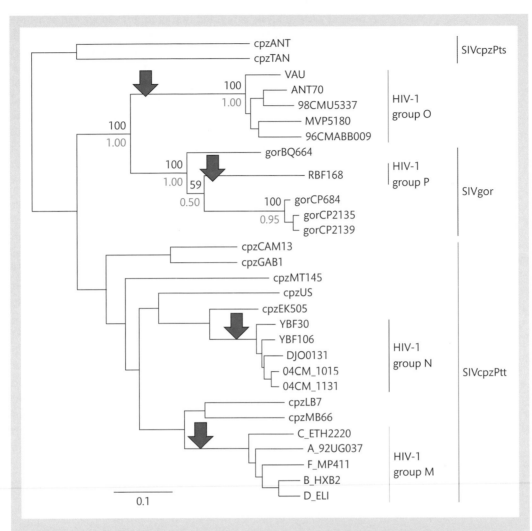

Figure 1.10 Like Figure 1.9, this phylogeny also compares HIV-1 with SIVs, but it samples a different range of viruses—more strains of HIV-1, more SIV from chimpanzees (SIVcpz), and now added gorilla virus sequences (SIVgor).[4] Possible movements from apes to humans are indicated by arrows. The numbers above and below branches are measures of statistical support (see text for details).

Adapted by permission from Macmillan Publishers Ltd: Plantier J-C, Leoz M, Dickerson JE, et al (2009), 'A new human immunodeficiency virus derived from gorillas'. *Nature Medicine* 15(8), 871–2. [10.1038/nm.2016] 15(8):871-2.

Two different measures of statistical support are given on the branches that group HIV-1-P with the gorilla viruses (SIVgor). The numbers in black are bootstrap percentage—branches marked with 100 are those that were supported by 100% of resampled datasets. The numbers below in green are support values derived from a Bayesian analysis—1.00 indicates that all of the most credible phylogenetic solutions found by the analysis contained that grouping. So we can conclude that this grouping is strongly supported by this dataset. Not all parts of the phylogeny are as well supported by the data: you can see that the support for the position of the HIV-1-P sequence within the SIVgor group varied among phylogenetic solutions (being found in 50% of the credible set of Bayesian phylogenies or 59% of the resampled datasets). However, while the exact position of the HIV-1-P sequence within the group is not certain, in all phylogenies it fell somewhere within the gorilla virus clade. So we might not know for sure which lineage of SIVgor the new P-type HIV-1 is most closely related to, but we can be confident that it was derived from a SIVgor virus.

The researchers tested the robustness of this conclusion in a number of other ways. They analysed different types of genetic data, for example using different genes from the virus genome, or analysing protein sequences rather than DNA, and always got the same result. But what if some odd quirk in the data means that the genome always gives a misleading phylogenetic placement? The researchers backed up their conclusion with other lines of evidence. The new P-type virus is clearly different from the main HIV-1 groups, M and O, because it cannot be detected with M-specific assays (so it is sufficiently different from M that it is not recognized by antibodies that bind to M-group HIV), and could not be amplified with group O primers (so its genome is different from O-group HIV). Yet the high level of virus in the patient's blood is compatible with P-type virus being adapted to replicate in human tissues, which suggests that it was not simply a one-off infection of an unlucky human by a gorilla-specific virus. This is consistent with the patient's history—she had never had any contact with gorillas so must have picked up her virus from another human. HIV-1-P has a characteristic amino acid signature of the SIVgor strain of the virus. These observations all add support to the hypothesis that HIV-P is a distinct HIV strain, related to gorilla SIV.

In fact, all phylogenies produced for HIV and SIV strains show that HIV arises from several different SIV lineages, suggesting around a dozen transitions from chimpanzees, gorillas, and monkeys into humans. This raises an interesting question. If SIV has been able to jump from non-human primates to humans on multiple occasions, then why have there not been epidemics of HIV throughout human history? Why did AIDS only appear on the world stage suddenly in the 1980s? One hypothesis that was put forward to explain the relatively recent origin of the HIV pandemic is that it was driven by a particular historical event. It has been proposed that a polio vaccine that was administered to around a million people in the Democratic Republic of the Congo, Rwanda, and Burundi in the late 1950s was contaminated with SIV.[5] This hypothesis makes several testable predictions. Since HIV-1 is most closely related to chimp SIV, then this hypothesis requires that the polio vaccine was produced using tissue infected with the chimp virus (SIVcpz). But analysis of the vaccine shows that it was grown on macaque tissue, and there is no evidence of any SIV, HIV, or chimp DNA in these vaccines.[6]

Another prediction of the polio vaccination hypothesis for the origin of AIDS is that the date of origin of the pandemic should be coincident with the polio vaccination programme. If this is true, then when we compare sequences from HIV-1 viruses taken from around the world, we should see differences between them consistent with a period of less than 60 years of evolutionary change since their last common ancestor. Researchers inferred the rate of molecular evolution in the main pandemic strain, the HIV-1 M-group, by comparing DNA sequences sampled over the last four decades. Using this rate of change, they inferred that the amount of differences between global samples of viruses of the M-strain of HIV-1 suggest a date of origin of this group of viruses between 1915 and 1941, rejecting an origin of the HIV pandemic in the 1960s[7] (**Figure 1.11**).

This example illustrates the difference between precision (how much uncertainty there is in a measurement) and accuracy (how close to the true value it is). Molecular dates are never very precise, so they should be expressed as a range of likely values (usually derived from some statistical measure of confidence from the analysis). The precision of this estimate means that the results cannot be used to say definitively whether the date of origin was 1915, 1941, or any date in between. But if these results are accurate—if this range contains the true date— then they can be used to reject a date earlier than 1915 or later than 1941. Of course, we never know for sure if our results are accurate, as the range of values is dependent on the assumptions of the analysis. So additional exploration and testing are always desirable.

Looking at this graph (Figure 1.11), you can see that the researchers made an assumption about the way that HIV evolves over time: that the rate of change in the well-sampled period from the 1980s onwards was a good reflection of the rate of change throughout the history of the virus. But how do we know if this is a fair assumption? What if the rate of molecular evolution has changed over the history of the pandemic so that more changes accumulated in the past? If the rate of change was faster in the earlier period of the pandemic, then this analysis might have overestimated the date of origin.

One way of checking this is to use the same analysis to estimate a known date, and see if it gets the right answer. The researchers used the same analysis to estimate the age of the oldest known HIV sample, a sequence obtained from a blood sample that had been collected in 1959 in Kinshasa (in the country now known as the Democratic Republic of the Congo). Importantly, this old sequence (referred to as ZR59) was not included in the analysis that estimated the rates, otherwise it might have biased the results. Their estimate of the age of

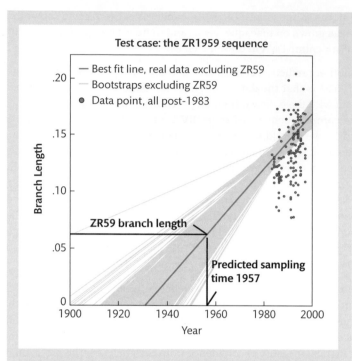

Figure 1.11 Each of the black points represents an HIV sequence of known sample date (all after 1983) from which the amount of evolutionary change (phylogenetic branch length) has been calculated. The central line represents the linear relationship between time and genetic change estimated from these known data points. The point at which the line crosses the *x*-axis is the predicted age of the common ancestor of all of the sequences. Given this relationship, the branch length of the earliest confirmed HIV sequence (ZR59) is compatible with its known sample date.[7]

From Korber B, Muldoon M, Theiler J, et al. (2000), 'Timing the ancestor of the HIV-1 pandemic strains', *Science* 288(5472): 1789–96.

this known sample was 1957, with confidence intervals of 1934–1962, so this estimate is accurate within the reported levels of precision.

Newer analyses, which use more samples, longer DNA sequences, and more sophisticated methods that allow for variation in the rate of genome evolution, have produced very similar results to the earlier molecular dating studies, placing the origins of the global M group of HIV in the 1920s. These newer studies also accurately estimated the age of the oldest known HIV-1 sequence (ZR59) using their method, and obtained a date estimate of 1958 with confidence intervals of 1946–1970.[8] The origin of the HIV-1 pandemic in this region of Africa is also supported by the observation that this area has the greatest genetic variation in HIV-1 M sequences. This is a pattern typical of the centre of origin of a biological lineage, because the area where the lineage has existed the longest has had the most time to accumulate genetic variants within the population.

So phylogenetic and epidemiological analyses suggest that SIV has moved into humans on many different occasions, and many of these cross-species transfers predate the beginning of the global pandemic. This raises a new question: if HIV was already present in human populations in the early twentieth century, why did it only become a globally distributed pandemic in the 1980s? Why not before? Two broad explanations have been offered for the timing of the pandemic.[8] One is that the opportunity for pandemic spread increased with urbanization and transport networks, but by chance only one lineage of HIV-1, group M, expanded to a global distribution while the others remained predominantly distributed in Africa. The other hypothesis is that a genetic change specific to HIV-1 M made it capable of pandemic spread, unlike other SIV strains that had infected humans. Here we see two themes that we will meet again in many other hypotheses for past events: contrasting the role of chance (the luck of being in the right place at the right time) with the influence of adaptation (being the best suited to thrive). How can we test these hypotheses?

Analysis of gene sequences suggests that M-type HIV is most closely related to SIVcpz virus isolated from a particular subspecies of chimpanzee that is found in southeast Cameroon. Spatial analysis of genetic diversity of HIV-1 M in the Congo River basin led to the conclusion that HIV-1 M spread from its origin in chimps in the forest of Cameroon along river transport networks to Kinshasa in the 1920s, and from there by river and railroad to other major population centres, probably reaching Brazzaville in the 1930s and Kisangani in the 1950s.[8] Why did the M-type HIV-1 go on to cause a global pandemic when other strains did not? Researchers used the genetic data to infer population growth rates over time, suggesting that while both O-type and M-type HIV-1 increased in the Congo basin in the first half of the twentieth century, in the 1960s the M type underwent a dramatic increase in infection and spread. This peak in transmission rates was interpreted as being far greater than would be expected on the basis of demographic expansion alone (**Figure 1.12**).

This jump in the transmission rates of M-group HIV1 in the 1960s suggests that the increase in M-group viruses was not due to a special feature of its ancestral virus that made it more infectious than the ancestor of group O HIV-1. The M group was already genetically diverse by the 1960s, so it seems unlikely that a particular genetic change could have caused this increase in transmission because it would have needed to occur in many different viral lineages independently and more or less simultaneously. Instead, researchers combined their genetic analysis with historical information on migration and social

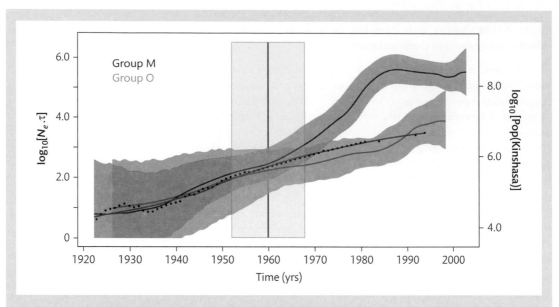

Figure 1.12 The observed differences between genetic variants circulating in the virus population can be used to model the changes in population size over time.[8] This analysis suggests that while both M and O strains of HIV-1 have been evolving within human populations for nearly a century, the M group seems to have undergone an increase in population size from the 1960s to the 1980s, corresponding to the period when M-type HIV-1 became a global pandemic.

From Faria NR, Rambaut A, Suchard MA, et al. (2014), 'The early spread and epidemic ignition of HIV-1 in human populations.' *Science* 346(6205): 56–61

change to suggest that the M lineage became established in high-risk groups such as sex workers, and then spread rapidly through both sexual contact and unsterilized injections. This conclusion is supported by a similar increase in hepatitis C from this region at the same time. This virus lineage spread throughout Africa and to other countries, establishing the family of viruses that are the predominant strain found in Europe, North and South America, and Australia. The phylogeny suggests that HIV-1 was also carried to Haiti in the 1960s where it established the lineage now known as subtype B, the most common strain found in Southern and Eastern Africa and in India.

Review of techniques used

This example illustrates some important points that we will revisit in later parts of this book.

- **Framing a testable question** One of the skills required for scientific investigation is to go from a general area of interest to a more specific testable question. For example, instead of asking 'When did AIDS emerge?', a more specific testable question is: 'What is the age of the common ancestor of all of the M-type strains of HIV-1?' Often these more specific questions seem narrower and less interesting than the broad question. But to go forward you need a specific question that can be answered by a well-designed test or the collection of new data. In many cases, a general area of interest can generate many different specific questions that will each be answered using different data or analyses. In this way, many small bricks (narrow questions) are put together to build a large wall (explaining a broader area of interest).

- **Testing explanations for past events** We cannot go back in time and observe the beginnings of the AIDS pandemic directly. Nor can we do manipulative experiments. Even if we were to set up an unethical experiment involving injection of contaminated polio vaccine into people, it could only tell us whether such a transfer could happen, not whether it did so in the past. Instead, we frame our scientific tests in terms of: 'If this hypothesis is true, then what would we expect to observe?' If it was true that the polio vaccination caused the HIV-1 pandemic, we would expect that the HIV pandemic began in the late 1950s to early 1960s, that the pandemic strains were derived from primate tissue used to make the vaccine, and that the polio vaccine should contain evidence of SIV/HIV contamination. These predictions of the consequences of a past event can be tested using evidence available today, such as using DNA sequences from current HIV strains to infer their evolutionary history, and examining archived polio vaccines for evidence of contamination.

- **Accuracy, precision, and confidence** Estimates, measurements, or predictions are often presented as a single number which represents the best guess of the true value, given the data and the assumptions of the analysis. The precision of the estimate can be represented by confidence intervals, which define the range of plausible values consistent with the data and analysis. An accurate estimate is one where the confidence limits contain the true value. From the molecular phylogenetic analyses, the best estimate of the age of the last common ancestor of the M-type HIV-1 viruses is 1920, but the researchers cannot rule out any date between 1909 and 1930. Even if the estimate is not precise, if it is accurate—if the true date of origin does fall between these limits—it rejects an origin of HIV in the 1960s as being inconsistent with the data and analysis. However, imperfect data or inappropriate analysis could lead to estimates that are inaccurate (confidence intervals do not contain the true date).

- **Assumptions and conclusions** All analyses are based on assumptions. Sometimes these assumptions are obvious; sometimes they are less clear. For example, the first estimates of the date of origin of the HIV pandemic relied upon the assumption that rates of evolution of the HIV genome had remained roughly constant over time. If this assumption was incorrect and rates had changed substantially over time, then the estimates could be inaccurate. There are several ways of testing whether assumptions may be producing misleading estimates. One is to conduct investigations into the reliability of the assumption: for example, using HIV samples of known dates to measure the rate of change over time. Another is to conduct a similar analysis that does not depend on that assumption and see if you get the same answer, for example an analysis that allows for varying rates of change over time. If different analyses all point to the same conclusion, you can be more confident in your results. Note, however, there are some assumptions that all analyses make, because without these we cannot step beyond the things we know to infer the things we don't know. So all of these molecular dating studies rely on the assumption that the rate of change in the earliest part of the history of the M lineage is predictable from what we know about the way M genomes have evolved in the past 50 years (from the first known sample to the present day). We can't rule out that HIV evolution was somehow dramatically more rapid in the pre-1959 period, but we can say that we currently have no good reason to suppose that it was (this is the uniformitarian principle of using known processes to explain past events). Additional information from lines of evidence that do not require these assumptions, such as historical data on migration patterns, can be used as an independent test of the reasonableness of the date estimates.

- **Confirmation and refutation** There is no single model for hypothesis testing in evolutionary biology. Sometimes scientific tests are devised that result in the rejection of one particular hypothesis. The polio vaccine hypothesis for the origin of HIV has been rejected because its predictions are not confirmed by the evidence. But the rejection of one hypothesis does not automatically lend support to an alternative hypothesis. In this case, we cannot conclude that the 'bush meat' hypothesis (that HIV originated through humans hunting and butchering primates infected with SIV) has gained support just because the polio vaccine hypothesis has been rejected. Hypotheses gain support as more observations are made that are consistent with that particular explanation, but it is difficult to rule out the possibility that another untested hypothesis could also explain the data just as well, if not better.

Conclusion

After all these investigations, can we give a definitive answer to the question 'Where did HIV come from?' Yes and no. We can't give a precise and definitive explanation of when, where, how, or why HIV first entered humans. But we do know more than we did when studies of HIV emergence were first conducted several decades ago; we can say with some certainty that HIV has entered human populations on multiple occasions from a range of primate sources of SIV. And we can confidently reject the hypothesis that the HIV pandemic originated with a polio vaccination campaign, because many different observations are inconsistent with that explanation. We have not exhaustively tested all hypotheses, but we have got a little closer to knowing the answer to the question we started with, with growing evidence that the global pandemic of HIV started in Cameroon, underwent a jump in transmission rates in the mid-twentieth century, and then spread throughout the world. And that is how most scientific investigations proceed—using a variety of tests, each of a specific question, to circle around a larger area of interest, ruling out some explanations as inconsistent with observation but supporting others whose predictions match the data.

Questions to ponder

1. We have seen how a range of corroborating evidence can be used to support or refute hypotheses. What if a new virus emerges and we only have DNA sequences, no other historical samples or records. Can we trust phylogenetic estimates of history on their own?

2. How can we test whether viral genomes evolve with a predictable rate of change?

3. Are there limits to reconstructing past events? Are there some parts of the HIV story we will never know?

Further investigation

Ancient DNA derived from human remains is increasingly being used to trace the origins of major epidemics, such as plague and tuberculosis. What are the advantages of using these 'molecular fossils?' What might be the dangers? Can these old sequences be analysed in the same way as the modern sequences? How can researchers be sure their samples are not contaminated with modern viruses? What puzzles might we be able to solve if we could find appropriate samples? Which diseases will be most likely to benefit from this approach, and which ones are unlikely to have appropriate ancient samples?

Note

Molecular phylogenies form the basis of many of the macroevolutionary and macroecological analyses described in this book. If you want to learn more about how molecular phylogenies are produced, tested, and interpreted, you can find a from-the-ground-up description in the companion volume to this book: *An Introduction to Molecular Evolution and Phylogenetics* (L Bromham, Oxford University Press, 2016).

References

1. Bromham L (2016) Testing hypotheses in macroevolution. *Studies in History and Philosophy of Science Part A* 55: 47–59.

2. Sharp PM, Hahn BH (2010) The evolution of HIV-1 and the origin of AIDS. *Philosophical Transactions of the Royal Society B* 365: 2487–94.

3. Hahn BH, Shaw GM, De Cock KM, Sharp PM (2000) AIDS as a zoonosis: scientific and public health implications. *Science* 287(5453): 607–14.

4. Plantier J-C, Leoz M, Dickerson JE, De Oliveira F, Cordonnier F, Lemée V, Damond F, Robertson DL, Simon F (2009) A new human immunodeficiency virus derived from gorillas. *Nature Medicine* 15(8): 871.

5. Hooper E (1999) *The River: A Journey to the Source of HIV and AIDS.* Penguin, London.

6. Poinar H, Kuch M, Paabo S (2001) Molecular analyses of oral polio vaccine samples. *Science* 292: 743–4.

7. Korber B, Muldoon M, Theiler J, Gao F, Gupta R, Lapedes A, Hahn BH, Wolinsky S, Bhattacharya T (2000) Timing the ancestor of the HIV-1 pandemic strains. *Science* 288(5472): 1789–96.

8. Faria NR, Rambaut A, Suchard MA, Baele G, Bedford T, Ward MJ, Tatem AJ, Sousa JD, Arinaminpathy N, Pépin J (2014) The early spread and epidemic ignition of HIV-1 in human populations. *Science* 346(6205): 56–61.

How did evolution get started?

2

Roadmap

Why study the origin of life?

To explain how evolution started on earth, we need to be able to describe how a population of replicators could emerge from naturally occurring molecules. Thinking about the origin of life is a good way to get to grips with the concept of evolution by natural selection (origins of microevolution). It also provides a platform for considering ways that the complexity of information coded for and passed between generations can increase over evolutionary time through a series of major transitions in evolutionary complexity (origins of macroevolution). This chapter explores how we generate hypotheses, building explanations for events in the distant past that we cannot ever witness directly. Because we can't directly observe the origin of life, we need to use a range of scientific tools to study it, so the origin of life provides an opportunity for us to think about the roles of experiments, models, and comparative tests in macroevolution.

What are the main points?

- For evolution to occur, there must be a population of replicators able to copy themselves with near-perfect accuracy, with occasional variations that provide the potential for natural selection to improve copying ability.
- The capacity for heritable information can be increased by combining previously independent replicators into coordinated systems that rely on each other for replication.
- The major transitions in evolution are macroevolutionary events that led to increases in informational capacity, permitting the evolution of diversity and complexity over time.

Photograph: Gorgonian coral. © Jolanta Wojcicka/Shutterstock.com.

What techniques are covered?

- **Experiments:** simple systems manipulated to explore outcomes of particular conditions and processes (such as Miller's spark experiment and Spiegelman's RNA 'monster').
- **Models:** idealized representations for testing the implications of processes and conditions (such as the 'prebiotic pizza' and hypercycles).
- **Comparative analyses:** comparing species to reconstruct ancestral states and histories of change (such as the reconstruction of the last common ancestor, LUCA).

What case studies will be included?

- Using homology to reconstruct history: the common ancestor of all living species (LUCA).

"In crossing a heath, suppose I pitched my foot against a stone, and were asked how the stone came to be there; I might possibly answer, that, for any thing I knew to the contrary, it had lain there for ever ... But suppose I had found a watch upon the ground, and it should be inquired how the watch happened to be in that place; I should hardly think of the answer which I had before given, that, for any thing I knew, the watch might have always been there. . . . . when we come to inspect the watch, we perceive (what we could not discover in the stone) that its several parts are framed and put together for a purpose."

William Paley (1809). *Natural Theology: or, Evidences of the Existence and Attributes of the Deity* (12th edn). Printed for J. Faulder, London.

Natural selection

William Paley (1743–1805), an English academic and philosopher, wrote a number of clearly argued books which were used as standard texts in many colleges and universities in the early nineteenth century, including at Cambridge where Darwin studied Paley as a key part of his degree. Paley begins his *Natural Theology* by pointing out that we seek different kinds of explanation for the existence of a rock and a watch. Unlike the rock, the watch is clearly made for a purpose, and could not function to that purpose without all of the complex parts being exactly so, and working perfectly together. He goes on to point out that this conclusion does not rely on knowing how the watch was made, nor exactly how all the parts work together; even in ignorance of the watch's origin, manufacture and mechanism, we can be sure it is designed for a purpose, not a random object thrown up by chance in the natural world.

Paley made a similar argument comparing the eye to a telescope, both complex instruments designed for seeing, both 'constructed upon strict optical principles'. Darwin, who admired the clarity of Paley's arguments, also used the eye as an exemplar, but as an illustration of how natural selection could produce complex biological features through the accumulation of slight chance variations. Pointing out that the telescope had been developed over time by building on successive improvements, Darwin imagined a similar situation for the development of the eye.[1] Start, he said, with a layer of transparent tissue with a nerve sensitive to light beneath, and then imagine that the structure was constantly being altered in small ways, for example in the density of the transparent material, or the formation of layers. Any such change that made the image more distinct would benefit the individual, increasing the chances of survival, and thus those beneficial changes to the eye's structure would be preserved and multiplied down the generations. Darwin suggested that if this process of multiplication of any slight improvement was allowed to run for millions of years, it could produce an instrument for seeing

as fine as any telescope. This model was formalized over a century later by biologists who found that a simulation test resulted in the evolution of a complex eye in a surprisingly short length of evolutionary time (**Figure 2.1**).[2]

In this chapter, we are going to use the origin of life to explore some fundamental concepts in the study of macroevolution. Most importantly, this is a prime case study for considering how we build plausible hypotheses to explain an event, and a process, that we cannot witness. Unlike other case studies we will consider in this book, we have little direct evidence about the first living things, so we will have to rely on building plausible models, conducting informative experiments, and comparing the long separated descendants of that first ancestor in order to build hypotheses that describe possible paths from the inorganic world to the living biosphere (**Figure 2.2**). The origin of life is also a great example to test your grasp of some of the most fundamental concepts in evolution: replication, variation, selection, chance, inheritance, and phylogeny. From a macroevolutionary perspective, we will need to develop explanations for how we can get from a simple system, born from abiotic chemistry, to one that is capable of increasing in specificity and complexity. This trajectory from simple replicators to complex replicating systems is relevant not only to the origin of life, but also to many of the transitions in complexity throughout the evolutionary history of life on earth. And, after all, how can we study the origins of biodiversity without considering how biological diversity originated in the first place?

Ingredients for evolution

If you have a population of entities that can reproduce with near-perfect heritability, you have the ingredients for evolution by natural selection. Near-perfect heritability means that the entities can make copies of themselves, and those copies will resemble their parents almost exactly. But copying is not 100% accurate, so there will be occasional errors that

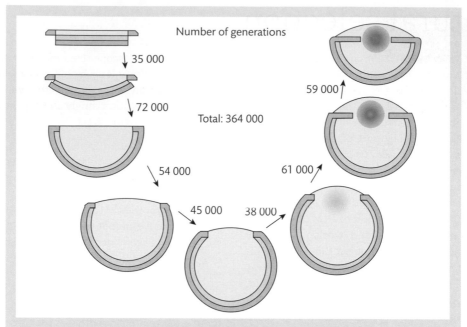

Figure 2.1 Model of eye evolution that allowed researchers to estimate how many steps it would take to get from a simple light-sensitive patch to a complex camera-type eye if each step introduced a 1% change in one of the structures. Using a simple model of heredity, they estimated that the whole series would take less than 400,000 generations to evolve.[2] Like all models, this one simplifies living systems, for example by ignoring the neural equipment needed to make a functional eye, and ignoring the population dynamics that influence rate of evolution. But, like any model, it is designed to illuminate the possible, given a set of simple assumptions about the process.

From Land and Nilsson, *Animal Eyes* (2012), Fig.1.6. By permission of Oxford University Press.

make a copy slightly different from its parent, generating variation in the population of replicators. It is possible that some of the individual differences between replicating entities could influence their ability to make copies of themselves. Naturally, any variants that reduce the chances of that copy surviving and reproducing will tend to be passed on less often, and those that increase the chances of surviving and reproducing will rise in frequency until, in a finite population, they replace all other variants.

Stop and think about systems that might conform to these properties of being a population of copiers with near-perfect heritability. DNA-based replication is a clear example; a cell copies its genome before it divides with a degree of error that can be as low as one in a billion bases copied, but, for a genome of several billion nucleotides of DNA, this generates several new mutations per replication which may be passed on to offspring. But the same argument has been applied to ideas (memes), which may be copied, albeit imperfectly, from one person to another, and some modifications to the ideas might make them more likely to be passed on again. We can also consider that biological lineages have the properties needed for evolution, because lineages reproduce at speciation, and these newly created lineages will share most of their features with their parent lineages. Any lineages that have inherited features from the ancestral lineage that make them more likely to persist and diversify will have a greater representation in the biota than those that are prone to extinction or have a low rate of speciation (see Chapter 7).

Figure 2.2 Darwin famously imagined a possible origin of life: 'But if (and oh! what a big if!) we could conceive in some warm little pond, with all sorts of ammonia and phosphoric salts, light, heat, electricity, &c., present, that a proteine compound was chemically formed ready to undergo still more complex changes, at the present day such matter would be instantly devoured or absorbed, which would not have been the case before living creatures were formed'. But he also recognized the difficulties of reaching so far back with so little evidence to work from: 'It is mere rubbish, thinking at present of the origin of life; one might as well think of the origin of matter'.[34]

We can even construct a minimal evolutionary system in a test tube and observe evolution happening before our eyes. In the 1960s, molecular biologists discovered that they could isolate the replication enzyme from a particular kind of RNA virus, a bacteriophage called Qβ (Q-Beta), and use it to copy RNA molecules in the laboratory. The discovery of this remarkably stable replicase enzyme paved the way for in vitro experiments on replication and evolution in a simple biochemical system.[3] The experiment was started by mixing together RNA molecules and replicase enzymes from the Qβ phage along with the building blocks of RNA (nucleoside triphosphates). After giving the mixture time to incubate, to allow the replicase to make copies of the RNA template, a small sample of the incubation mixture was transferred to a fresh medium. The process was repeated—copy and sample, copy and sample—75 times. Through these repeated cycles of copying, the system evolved. The RNA molecules changed.

The replicase enzyme copied the RNA templates accurately but not without error. Replication introduced occasional mutations, such that the RNA molecules in the solution varied in length and in their nucleotide sequences. Any RNA molecules with variations that caused them to be copied at a faster rate would produce more copies of themselves in any given time period, resulting in more copies of those fast copiers being present in the mixture when it was sampled. Each 'generation', the fastest replicators were likely to have a proportionally greater representation in the sample transferred to the new medium. After 75 generations, the copying rate of RNA molecules in the mixture was more than double that of the original, and the RNA molecules had become shortened to only a sixth of the length of the originals.

In subsequent experiments, researchers went on to show that changing the incubation conditions could change the outcome of the experiment. They could even identify mutations that arose as adaptations to particular environments. For example, a particular three-nucleotide change allowed the RNA molecules to be copied even when incubated with ethidium bromide, a replication inhibitor.[4] In fact, under certain conditions, simply adding the Qβ replicase to a solution containing the RNA building blocks could result in the formation of a population of short RNA templates that could be copied quickly and reliably.[5] These minimal evolution systems are sometimes referred to as Spiegelman experiments, after one of the authors of the original study, and the shortest replicable RNA template is called, somewhat unflatteringly, a Spiegelman monster.

What do you think?

These Spiegelman experiments require three types of biomolecule, nucleic acids (RNA), proteins (replicase enzymes), and building blocks (nucleotide triphosphates). During the experiments, one of these types of molecule evolves and the others don't. Which is which, and why?

All of these evolving systems—RNA molecules, DNA-based genomes, human-generated ideas—consist of entities that rely on the existence of complex machinery for their replication. Genes can't be passed on without a vast suite of proteins and RNAs that copy, correct, and maintain DNA. Ideas can't be passed on without populations of intelligent individuals with appropriate senses, a shared language, and a social structure that facilitates communication. Even the Spiegelman system requires the presence of a sophisticated protein that can copy those particular RNA templates. So, while we can understand how populations of near-perfect copiers can evolve by natural selection, we are going to have to think harder to understand how natural selection could lead to the invention of those complex systems of copying in the first place.

Key points

- Natural selection can occur in any population of copiers if heritable variation can influence the rate of successful copying.

- While we are familiar with natural selection operating on DNA-based organisms through the inheritance of genetic mutations, evolution by natural selection could also occur in simpler systems as long as there is a reliable means of replication and the occasional introduction of heritable variants which can influence success and reproduction.

- Qβ experiments (where an enzyme copies a short RNA template in solution) demonstrate natural selection in a test tube: the fastest copiers produce more copies, so shorter templates become more common because they replicate faster.

Experiments and models

We will face some challenges in considering the origin of life. No one has ever directly observed the generation of a living organism from inorganic matter. Even if we were to be able to create life in a test tube, we wouldn't know if it was the same conditions that created life in the dark distant past, over three billion years ago on the young earth. We can't expect to find direct evidence of the earliest stages of evolution. Rocks as old as the origin of life are rare on earth, and it is unlikely that organisms as small and simple as the first evolving entities would have left fossils that revealed their form. Instead we rely, as did Darwin and Paley, on considering the possible outcomes of different processes operating under plausible conditions, and on inference from observation of the contemporary world.

What form does this inference take? We will discuss several broad approaches to investigating the origin of life (though do not consider these categories absolute or exhaustive, as some investigations may span these boundaries or take different approaches). Experiments are investigations that allow the consequences of particular conditions to be realized. Models are representations of the world that are used as tools for exploring process and outcome. Comparative studies use the patterns of occurrence of traits observed in species to reconstruct their history. Actually, despite a good deal of debate, there are no agreed definitions of what is an 'experiment' or a 'model', and these forms of scientific investigation overlap a great deal, to the extent where some investigations might variously be deemed to be experiments, or models, or comparative studies, or all three at once.

In this chapter, we are going to use the example of the origin of life as a means of exploring different ways of investigating past phenomena that we can't directly witness. The point is not to give you an up-to-the-minute survey of all the latest theories on origin of life, nor to furnish you with convincing biochemical models, but to use these ideas to examine the way that we ask questions in evolutionary biology. We will consider how hypotheses are constructed in macroevolution, building explanations of long-distant events based on what we can observe today. We will also take the opportunity to examine some key ideas in evolutionary biology, such as natural selection, homology, descent, and the importance of considering both chance and necessity in shaping features of the evolved world.

Experiments

The concept of a manipulative experiment has a prominent place in the popular image of science. If you were to ask people 'What do scientists actually do all day?', there is a fair chance they would tell you that they do experiments. A very broad view of experimentation might be to deliberately bring together some specific components of a system under a particular set of conditions in order to see what happens.

Consider one of the most famous experiments in biology. Stanley Miller was doing his PhD in chemistry when he saw Nobel laureate Harold Urey give a lecture on the possibility of organic molecules forming in the oxygen-free hydrogen-rich atmosphere of the early earth. This fired Miller's imagination so much that he persuaded Urey to let him work on a project investigating the effect of electrical discharge on gases. He constructed a sealed system containing gases thought to be present on the prebiotic earth, including hydrogen (H_2), methane (CH_4) and ammonia (NH_3), plus liquid water (H_2O). Heating the water caused it to evaporate, and then sparks from an electrode were discharged through the mix of water vapour and gases (**Figure 2.3**). After a week of running this experiment, the solution that collected as the vapour cooled was found to contain many different amino acids, which are the basic building blocks of protein.[6] Subsequent experiments by others

showed that some of the bases that form the 'letters' of DNA and RNA, could also form spontaneously in such conditions.[7]

Miller's experiment takes the form 'What would happen if . . .?', asking what kind of outcome is possible given a set of starting conditions and a particular process. His experimental set-up looks nothing like a real prebiotic planet, and contains only a few simple components. But it showed convincingly that it is possible for some of the building blocks of key biomolecules to form spontaneously in abiotic conditions. Miller then went on to investigate whether he would get a different outcome if the conditions were slightly different. For example, he made several modifications to his experiment to test the possible influence of volcanoes, such as adding steam and gas to the spark to represent volcano-stimulated lightning, and varying the composition of the experimental atmosphere to include hydrogen sulfide. Both of these modifications produced a greater variety of amino acids.[8] Other researchers have tweaked the components of the Miller spark experiment as ideas have changed about the likely composition of the early earth's atmosphere.

In this sense, the word 'experiment' has an everyday meaning of seeing what happens when you realize a particular set of conditions. For example, you might say that you are experimenting in the kitchen by varying the ingredients of your favourite recipe. But if you asked a scientist what an experiment is, they might consider it is more than just a 'let's see what happens when we do this' kind of approach. A scientist might tell you that a good scientific experiment has a number of key features. A good experiment, one that can provide a robust answer to a given question, needs to be repeated so that we know that the particular outcome really was a result of the particular circumstances and not just a chance event that happened to occur during the experiment. For example, the Qβ replicase experiment will tell us little if it was only conducted once—perhaps the shorter sequences were due to an accidental loss of RNA at an early stage and nothing to do with selection for rapid copiers. But if we run the experiment again and again,

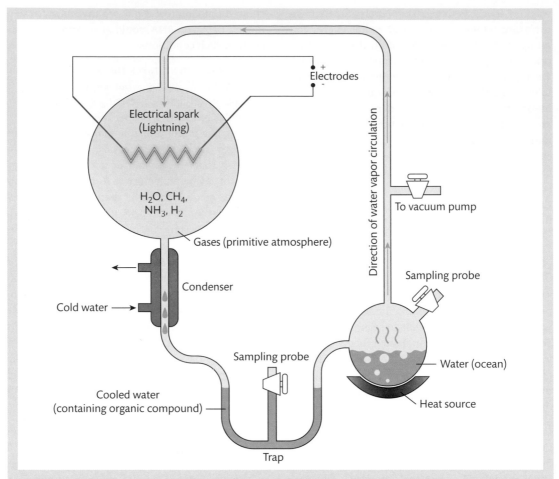

Figure 2.3 Diagram of one of the most famous experimental apparatuses ever constructed: Stanley Miller's closed system of gases circulating through laboratory glassware, being heated and zapped, then allowed to cool and condense. Simple as it is, this experiment proved that it is possible to produce some of the building blocks of life, such as amino acids, in a simple abiotic system. Miller used paper chromatography to identify the amino acids in the reaction mixture. Fifty years later, a former student of Miller's found some of the original laboratory equipment and reanalysed the traces of chemicals in the glassware, and found that Miller had actually produced a wider range of amino acids than he had been able to detect at the time. In an interview in 1996, Miller said 'The fact that the experiment is so simple that a high school student can almost reproduce it is not a negative at all. That fact that it works and is so simple is what is so great about it. If you have to use very special conditions with a very complicated apparatus there is a question of whether it can be a geological process.' (interview published at accessexcellence.org).

Image redrawn from GYassineMrabetTalk via Wikimedia Commons (CC 3.0 license).

starting with the ordinary Qβ RNA template but always ending up with a dramatically shortened 'Spiegelman monster', then we can be confident that it is a predictable outcome of evolving Qβ RNA under conditions of repeated sampling.

Often, experimental repetition takes the form of running many replicate experiments at the same time, which also allows another feature of good experiments: comparing control and treatment. If we want to know if a particular experimental manipulation is causing a particular outcome, then we can run replicates of the experiment in parallel, keeping everything the same except that some replicates are assigned to the 'treatment' (which have the manipulation of interest) and some are assigned to the 'control' (which do not). The concept of a control is familiar from medical trials. If we want to know whether a particular drug is effective, we need to give the drug to some patients, and not to others, and ask whether the patients with the drug have a better outcome than those without.

This strategy of varying just one aspect between treatments and controls is an attempt to isolate particular causal relationships. If we allow only one feature to vary between replicate trials, and the trials with that feature tend to have a different outcome to the ones without it, then we have good reason to think that there is a causal connection between the feature and the outcome. This is why the manipulative experiment is considered by many to have particularly strong explanatory power in science. In practice, it is not simple to design an experiment where the intervention influences one, and only one, feature of the system, complicating the interpretation of causal relationships.[9]

Experiments are often described in terms of hypothesis testing. Initial observations are used to form an idea about how the world works, and this idea is used to generate predictions about the likely outcomes of a particular set of circumstances, which can then be tested by experiment. Some experiments clearly do conform to this description (which is sometimes referred to as a 'hypothetico-deductive' approach). But it would be a mistake to think this is a recipe for all successful experiments. Consider the Spiegelman experiments; the value of these experiments doesn't depend on whether the researchers started with a particular hypothesis they wanted to test, or whether they just chucked the ingredients together to see what happened. Furthermore, it will often be stated that a true experiment can only falsify a hypothesis; it cannot provide confirmation. This is not, in practice, how most science is conducted. While evidence that contradicts a particular hypothesis may be used to argue against the explanatory power of that idea,

results that support an alternative hypothesis are routinely used by scientists to strengthen the argument in favour of that idea. Even if it had failed to produce any biomolecules, Miller's experiment could not have been used to falsify abiotic origins of biomolecules. Yet the fact that the experiment did produce key biomolecular building blocks in a simple inorganic system did provide supporting evidence that helped to build a case for abiogenesis (origin of life from non-living matter).

Scientists may dream of the 'killer experiment'— a single elegantly designed experiment that can comprehensively answer a key question, with no room left for doubt. Such experiments do exist, but more typically experiments are suggestive, rather than decisive, for two reasons. Firstly, experimental systems aim to isolate key causal conditions, but there is no guarantee that those conditions really do adequately represent the real-world systems they are designed to mimic. For example, the conditions may be missing some key ingredient, or there may be unseen connections between the features of the system that the experimenter is not aware of. Secondly, the outcomes of experiments are usually assessed statistically, reporting the results in terms of the probability of achieving that outcome by chance. This means that there is often room for interpretation concerning the degree of support the results of a replicated experiment give to a particular hypothesis. We use statistical conventions to help us express this uncertainty and guide our interpretation.

One such convention is the use of agreed, but arbitrary, probability thresholds to decide whether or not a particular result provides support for a particular hypothesis. One of the most common such thresholds is that a result should not be considered to support a hypothesis if there is a greater than 5% chance that we could have got that result just by chance, even if the hypothesis is not true. This cut-off value for statistical significance has the effect of converting a continuous measure of probability into a binary measure (either yes it supports the hypothesis or no it doesn't).

Imagine we run 20 replicates of a Spiegelman style of experiment under particular incubation conditions. If there was no causal connection between the experimental conditions and the resulting length of the RNA, we would expect the outcome to be random, with templates just as likely to increase or decrease in length. But if we find that in 15 of the replicates the RNA templates get shorter than the corresponding controls, but in the other five replicates they either stay the same or get longer, then that does not look like a random result. In fact, there is a less than 5% chance that we would get 15 out of 20 trials with shorter RNA templates if the experiment did not make template reductions more likely to occur, so we conclude that there is significant evidence that the experimental conditions result in the evolution of shorter templates. But notice that this is an argument from probability, comparing expected outcomes with observations. Imagine running the experiment again with 20 fresh replicates, but this time there were 14 trials out of 20 in which the template ended up shorter. While this result is very close to the original trial, under common scientific conventions it would not be considered to provide significant evidence for selection for shorter templates, because there is a greater than 5% chance that we could have got this result by chance, even if template length is not under selection. Many statisticians and scientists feel that the 5% cut-off is a meaningless way to evaluate the significance of experimental results, and that *p*-values can be a misleading form of statistical inference. Nevertheless, $p<0.05$ remains the accepted standard for most published scientific research.

Thus the interpretation of experiments is not always black and white but a matter of weighing probabilities, using statistical analysis in order to judge how often an experiment needs to conform to predicted outcomes in order to support or reject a particular hypothesis. You can see how it might be possible for different experiments, all designed to test the same question, to lead to different conclusions. Indeed, there is a growing trend towards 'meta-analyses' which take the results of many independent experiments, some of which may have strongly supported a hypothesis, but others less so or not at all, and look for an overall trend.

Meta-analyses would not be necessary if every experiment provided an unambiguous answer to a clearly defined question.

The two classic experiments we have considered give us very important information—that the building blocks of life could have formed spontaneously on the early earth given particular ingredients and conditions (Miller's spark experiment), and that once you have a replicating system you can expect natural selection to result in improved copyability, as the most successful copiers will come to dominate (Spiegelman's RNA experiment). These experiments, and others like them, show that particular steps in the origin of life are plausible, because we can make those steps happen in the laboratory. But there are practical limits to the subjects we can examine in physical experiments. For example, we may lack the time to run an experiment for millions of generations, or have no way of reproducing particular conditions in the laboratory. But there is no limit to our imagination, which can entertain any kind of 'what if' scenario. And this is where models come in.

Models

A model resembles an experiment in the sense that it is an attempt to set up a simplified version of reality, in which the consequences of a particular condition or process can be examined. But unlike experiments, most models are tools of the imagination, explored in the mind, on paper, or with a computer. Models are tools for systematically exploring the consequences of our ideas.[10]

Broadly speaking, models are a way of asking: 'If I had these conditions, could I get that outcome given this particular process?' A model might be an informal 'thought experiment', considering the possible outcomes of a set of conditions. For example, one of the earliest models of the origin of life was developed by J.B.S. Haldane in 1929, by considering the kinds of conditions that might have existed on the early prebiotic earth once it had cooled sufficiently for water to condense. He reasoned that there would have been sources of carbon (carbon dioxide) and nitrogen (ammonia) but little free oxygen, and that the sun's rays would have reached the surface without being filtered through the atmosphere, allowing ultraviolet light to prompt the formation of simple organic substances like sugars. With no organisms to eat them up, these small molecules could have accumulated 'till the primitive oceans resembled hot dilute soup'.[11]

Haldane's primordial soup model is expressed in words, but many models are expressed in statistical or mathematical terms. For example, the astronomer Fred Hoyle considered the possibility that the chemical building blocks of life could spontaneously come together to form a functional protein simply through the random diffusion of chemicals in solution, with reactions occurring when molecules chance to collide. Even the simplest cells have a level of complexity commensurate with that of a passenger plane, with a vast number of intricate parts that must work together if the cell is to function. Hoyle concluded that expecting life to evolve by chance chemical interactions was like expecting a tornado to blow through a junkyard and spontaneously assemble a Boeing 747, and therefore he rejected the hypothesis, preferring instead a hypothesis (known as panspermia) that life originated elsewhere in the universe before colonizing the earth in a relatively advanced state of organization.

When a model fails to support a hypothesis, we can either reject the hypothesis or reject the model. While Hoyle used his random diffusion model to reject the hypothesis of abiogenesis of life on earth, biologists reject Hoyle's random diffusion model on the grounds that it is unlikely to provide a fair description of the origin of life because it is unrealistic to expect that the world went from soup to cells in one single happy accident where all the relevant chemicals bumped into each other in the right way at the same time. Instead, biologists expect that even the simplest life forms are the result of a long process of stepwise modification—just as the passenger aircraft was developed over many decades by a gradual series of improvements to materials, manufacture, and design.

The failure of a model is not a setback, it is a necessary step on the way to improved understanding. We can take existing models, such as Haldane's primordial soup and Hoyle's random diffusion model, and build on them. For example, the assumed starting conditions on the early earth have been debated, influencing the 'ingredients' of Haldane's soup model. Changing the starting conditions (such as different compounds that might have been present in the early atmosphere) can change the outcomes of the model (such as producing a different predicted array of organic molecules). Models that don't rely on random diffusion of molecules in the soup can increase the chance of the right molecules coming together in the right ways at the right times by adding factors that attract and concentrate relevant building blocks. For example, attracting molecules onto a surface could serve to concentrate molecules (increasing the likelihood of stable reactions), orientate them (increasing the chance of them forming coherent structures), and even catalyse reactions (helping molecules join together to create more complex molecules). The notion that early biomolecules could have formed on a solid surface has led to calls for the primordial soup model to be replaced by a model of a primitive pizza.[12]

Several primordial pizza crust materials have been suggested. Fool's gold (iron pyrite, FeS_2) has a positive surface charge that could potentially attract molecules, and it has been proposed that the reaction of pyrite with hydrogen sulfide (H_2S), which is present in abundance in deep sea vents, could provide a reducing environment which might conceivably power a chain of reactions that could ultimately result in the formation of biomolecules such as short peptides (chains of several linked amino acids).[13] Montmorillonite clay, which may have formed on the early earth from volcanic ash, has been shown to catalyse a number of organic reactions, including the polymerization (joining together) of nucleotides to make a short RNA molecule.[14] In this way, rock surfaces could have provided the first steps in the origins of metabolism by stimulating the formation of more complex molecules from simpler building blocks available in the environment. For example, serpentization, a reaction between rock and seawater, could have produced hydrogen gas (H_2) that could drive the formation of hydrogen sulfide and methane, and that methane could then feed a string of reactions (present in some living bacteria and archaebacteria) which can both fix carbon and provide energy.[15]

These different models are all descriptions of possible paths to the origin of life, describing starting conditions and sets of reactions that might represent plausible solutions for how to get from an abiotic earth to interacting biochemical systems. Whether expressed in words, equations, or diagrams, a model is idealized, capturing just a few key elements rather than aiming for an exhaustive representation of conditions in the real world. This simplicity is what makes models useful for developing and testing explanations. But it also means that we should not mistake a model for an accurate representation of the real world. Models are an exploration of what could possibly happen, given a particular set of circumstances, not a description of what actually has happened in evolution.

The trick with model-building is to develop a model that is simple enough to have explanatory power, without missing the important features of the system. There is always a trade-off between simplicity and generality on the on hand, and complexity and realism on the other. Why shouldn't we make our models as complex as the systems we are trying to explain? The more parameters we add to our models—the more features that can take variable values—the easier it is to fit the model to any given set of data, because the variable parameters can be adjusted to match the variation in the data. If we overparameterize our models, we gain a closer fit to each specific case at the expense of general explanatory power. Overfitted models that have too many parameters have poor predictive power, because the results will not be able to be generalized to other datasets. We want the simplest model that can explain a significant proportion of the variation in our data, and we only want to add more features to the model if it significantly increases our general explanatory power.

Steps to an evolvable system

Before there can be heredity, there must be reproduction, and before that there must be growth.
John Maynard Smith and Eörs Szathmary (1999). *The Origins of Life: From the Birth of Life to the Origins of Language.* Oxford University Press.

Metabolism: getting molecules working together

The Miller spark experiments and the prebiotic soup and pizza models describe possible ways that simple molecules, which would have formed spontaneously in early earth conditions, could come together to make more complex molecules. These complex molecules, such as peptides and sugars, must have formed in abiotic conditions in order for there to be raw materials for building living systems. Bringing molecules together to make these components is a critical step, but it is only one step that must be solved. Life is more than a collection of interesting chemicals. Living systems are characterized by the organized interaction between reactions that allow complexity and specificity to develop. So a convincing model for the origin of life can't just explain the origin of biomolecules, it must also get the molecules working together. Not only that, these systems have to be capable

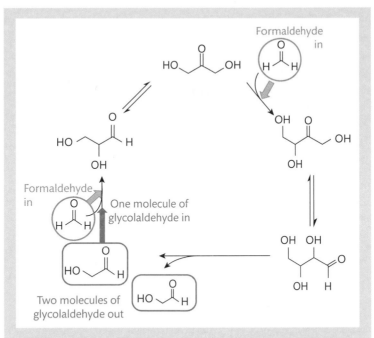

Figure 2.4 The formose reaction is an observable example of a simple autocatalytic reaction, because each turn of the cycle increases the amount of the molecule that stimulates the reaction. One molecule of glycolaldehyde is combined with formaldehyde to produce two molecules of glycolaldehyde. Glycolaldehyde can go on to form simple sugars. This is a 'one-pot reaction' that can occur without biological input. However, changing the environment and conditions of the reaction can result in different types of sugar molecules.

of growth; they have to be able to take in simple molecules from the environment and synthesize more complex structures. In other words, how do we kick-start metabolism?

Some promising models describe sets of chemical reactions, each of which is fed by the products of one reaction and contributes its products to another reaction, forming a cycle of interconnected chemical processes. When we think about the origins of metabolism and growth, autocatalytic reactions present exciting possibilities. A chain of reactions is called autocatalytic if, given sufficient resources to provide the necessary 'food' for the reactions, the products of the set of reactions feed the reaction set itself, creating a self-sustaining production of molecules which resembles a very simple metabolism. For example, glycolaldehyde

is a simple sugar-like organic molecule, which may have been a precursor of ribose, the sugar in RNA.[16] Glycolaldehyde could have been formed in the conditions of the early earth through the formose autocatalytic cycle, in which one molecule of glycolaldehyde is combined with two formaldehyde molecules (which are likely to have been abundant on the early earth) to synthesize two molecules of glycolaldehyde (**Figure 2.4**). So every turn of the formose cycle increases the amount of glycolaldehyde.

While the formose reaction can be observed in the laboratory, some proposed autocatalytic reaction sets are purely hypothetical. These hypothetical cycles describe possible series of interlinked chemical reactions which could theoretically produce more complex molecules than they feed on, thus

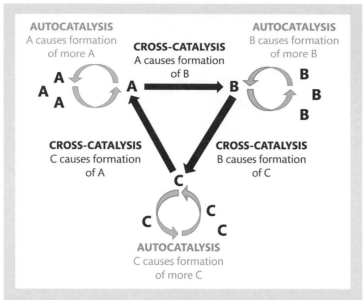

Figure 2.5 A hypercycle is a model of an autocatalytic reaction where self-replicating molecules (A, B, and C) each catalyse the copying activity of another molecule in the set (A catalyses B, B catalyses C, and C catalyses A). Mathematically speaking, hypercycles resemble the equations used in ecology to model the dynamics of any system where the abundance of one element depends on another in some kind of feedback loop (Lotka–Volterra equations). Predator–prey dynamics can be modelled in this way, because as the number of prey animals increases, the predator population grows, reducing the number of prey and thus reducing the predator population, which allows an increase in the number of prey, and so on.

increasing the degree of order in the system. Like all models, these models of chemical production cycles don't have to resemble reality in all its complexity in order to be useful. Instead, they illustrate the possibility of cycles of reactions that could increase the chemical diversity and complexity of the prebiotic environment. But remember that for evolution to work, we need replication—copying with near-perfect heritability that generates a population of individuals that vary in their ability to use the resources of the environment to grow and copy. How can these models be extended to add features necessary for evolution? Let's imagine that not only is the cycle itself autocatalytic (the products of the reaction cause more of the same products to form), but some of the actual

molecules in the cycle are autocatalytic in their own right (the molecule causes more molecules like itself to form).

Picture an autocatalytic set in which the reactions that form the steps in the cycle are actually replication of molecules. Now the steps in the cycle consist of a molecule causing a copy of itself to form, and the formation of each new molecule catalyses the creation of the next molecule in the chain (Figure 2.5). If molecule A makes copies of itself (autocatalytic replication), and those copies stimulate not only the formation of more A but also catalyse the replication of B (cross-catalytic replication), and if the copies of B not only form templates for more copies of B but also catalyse the replication of C, which can not only self-replicate but also catalyse formation of A, then we have a hypercycle where the replicating molecules A, B, and C aid each other's replication.

Given an autocatalytic and cross-catalytic cycle of replicating molecules, we can see how they could increase the number of copies of those molecules. So we have a simple model that provides a hypothetical path to the origin of growth and metabolism. But could these chemical sets evolve by natural selection? Imagine two hypercycles that are similar, but not identical, so that although they draw on the same raw materials, some steps in the cycle differ such that the rate of catalysis is affected, or the sequences of the replicating molecules differ slightly impacting their copy fidelity. If raw materials are limited, the more efficient hypercycle would grow at the expense of the other.

Key points

- Biomolecules can form by chance in an abiotic system, given the right conditions.

- Living systems are characterized by organized series of chemical reactions that build specificity and complexity (metabolism).

- Chemical cycles that result in the formation of molecules that then feed the cycle and make more copies of the molecule (autocatalytic) provide a model for how a simple system can 'grow' by making more biomolecules.

- Growth is not enough for evolution; the system must be able to copy itself.

- A hypercycle is a model where the components of a chemical system make more copies of themselves and other components of the system, allowing the whole system to be copied.

- If these simple chemical systems vary in their ability to make copies of themselves, they can evolve.

But a problem arises: what if a change in one of the replicating molecules in the cycle made it more efficient at copying itself (strong autocatalysis), but less efficient at stimulating the replication of the next molecule in the chain (reduced cross-catalysis)? This variant replicator will increase in copy number, a clear case of selection for better copiers. But consider the effect of this selection on an individual molecule on the system as a whole: it will decrease the output of the cycle as a whole by limiting the next step in the chain. Just as more efficient hypercycles might grow at the expense of others, so more efficient replicators within a hypercycle may copy at the expense of the other replicators, leading to the breakdown of the hypercycle.[17] What could be introduced to the model to overcome the destructive effect of selfish replicators? While the following discussion is about simple molecules, actually the problems it describes—how components can cooperate to form a complex system—are a common theme in macroevolution. We will meet the same problems again later in the chapter when we think about evolution of complex organisms and social systems, and again in Chapter 7 when we consider sexual reproduction.

Compartmentalization

We can generate a hypothesis for the increase in complexity of chemically autocatalytic systems if we can imagine a way of getting molecules to work together to make more copies of each other. But any system that relies on cooperation is vulnerable to exploitation. In a system where molecules cooperate in each other's replication, as in the hypothetical hypercycle described above, a cheater is one that is copied but does not contribute as much to copying others. More copies are made of the cheater, but as a result, fewer copies are made of the other molecules (because it limits the next reaction step), so the cheater grows at the expense of the cooperative elements. Because they favour copying themselves over copying other molecules in the system, a system full of cheaters is one with low replication ability. Individual advantage to cheaters leads to overall disadvantage for the system.

This does not seem like a promising start for the evolution of complexity, if these potential auto-catalytic cycles are prone to being overcome by selfish replication. What can we add to the model to overcome the natural advantage to cheaters? We have to think of a way in which systems of cooperators have a selective advantage over selfish self-copying molecules. Suppose that a system of cooperators—all of them contributing to each other's replication—is more likely to persist and grow than systems that contain cheaters—where each is replicating at the expense of the system. This might mean that a cooperative player has, in the end, a higher rate of being copied because it is part of a successful network where they all copy each other at a higher rate than each could as a selfish individual (see also Chapter 1, Figure 1.6). Now we have to find some way of turning that into an evolutionary advantage, where the cooperators can rise in frequency by natural selection.

One way to change the evolutionary outcome of the hypercycle model is to build an imaginary wall around it. Whether this 'wall' is made of a physical feature of the environment, or constructed from biomolecules coded by the replicators themselves, or simply arises from uneven spatial distribution of molecules into clumps, its effect is to create groups of molecules that share an evolutionary fate. When a group succeeds, its component molecules go on to form the next generation. If it fails, they do not. Now the players in the cycle have a shared evolutionary fate—they all win together, or they all fail. Individual players can't do better by going it alone. Higher persistence means that these cooperative cycles will be more common in the environment, as the cycles with selfish elements self-destruct.

Persistence is all very well, but for a truly evolvable system we need not just metabolism and growth, but also replication. Let's imagine a hypothetical system in which the hypercycle can build up products of reactions (copies of the various replicators) until it 'replicates'—for example the molecules are washed into the environment to reassemble in another location, or the group reaches some

threshold size and splits into two. Now groups can give rise to 'offspring', each consisting of a sample of copies of the component molecules surrounded by a wall. Unsuccessful groups, such as those that contained too many cheaters to grow efficiently, will contribute fewer sequences to the next generation. Successful groups, such as those that gained a growth advantage from containing specialized molecules for replication, wall-building, or uptake of raw materials, could increase in representation as they grow and replicate.

This compartmentalized model introduces the potential for cooperating sets of molecules to evolve greater efficiency and complexity, because now sets of replicators that work well together, each stimulating the others to copy more, could grow faster, budding off more new walled communities of replicators. In this model, sets of cooperative replicators might outcompete those containing more selfish replicators. Binding the individual players together in some kind of container so that they all replicate together as a whole provides a way for good cooperators to gain a selective advantage and hence a greater representation in the environment. In these bound sets of replicators, which replicate as one entity, there is now the potential to develop complementary roles, some specializing in formation of new templates, others specialized to other roles such as forming the wall or providing energy, as long as they are all inherited together and have a shared investment in their evolutionary future. Replication doesn't have to be perfect. It may be that when an existing reaction set divides, some 'offspring', by chance, fail to receive all the component molecules. These incomplete sets might fail, but as long as there are some 'offspring' that have a full complement and are at least as good at growth and replication as their parent, then there is capacity for selection for improved metabolism and reproduction. This 'stochastic corrector' model resembles sociobiological models used to describe how selection may act at the level of groups, by favouring groups that cooperate in ways that enhance their ability to survive and reproduce.[12]

What do you think?

Social groups, where individuals cooperate for a common good (such as bees in a hive or humans in a community), are vulnerable to exploitation by cheaters who gain benefits from the group without contributing fully. In what ways are the potential solutions similar or different to those that maintain co-operation of replicators in a simple evolving system?

Building walls around replicators

This model gives us a plausible framework for imagining the evolution of complex self-replicating chemical sets. But it's all very well to imagine building convenient walls around possible hypercycles. Is it a realistic description of what could actually have happened on the early earth? That would depend on what the components—the replicators and their wall—could have been made of. Modern cell walls are made of lipids which form a protective waterproof layer around the cells, requiring a complicated system of protein-based transportation devices to get molecules in and out of the cell. This system is too complicated to form spontaneously in prebiotic conditions.

However, it is possible for simpler compounds to form which, like lipids, have a water-loving (hydrophilic) head and a water-repelling (hydrophobic) tail. Under certain conditions, such compounds can spontaneously form into simple membranes, organizing themselves by tucking their water-repelling tails in and sticking their water-loving heads out.[18] It has been suggested that the negatively charged surface of montmorillonite clay could, in addition to catalysing the formation of small RNA molecules, aid the formation of small vesicles of these compounds with both hydrophobic and hydrophilic ends. If these vesicles formed around a clay particle, they could carry within them a small catalytic workbench, potentially also hosting self-replicating RNA molecules.[19] But a wall becomes a coffin if it does not allow the ingress of molecules needed for metabolism, growth, and replication. Our early replicating system needs a wall that is selectively permeable, protecting the contents from destruction or dispersal, but letting necessary supplies in.

Perhaps the earliest life forms could have used pre-existing walls. What if the prebiotic pizza was actually a calzone, with the reactive mineral surface turned in on itself to form a container in which biomolecules could form and replicate? Consider deep sea vents, which not only provide energy and a reactive environment in which biomolecules might form, but also mineral surfaces such as clays and pyrites that might act as workbenches for the formation of early biomolecules.[20] If these surfaces contained small pores, they could form micro-chambers for the containment of hypercycles. But while a mineral wall might help keep the products of reactions together, there has to be some means of chemical exchange with the environment (or there is no capacity for growth) and also a way of assembling molecules to form new sets in their own walls (or there is no capacity for replication). One suggested solution to this problem is that metal sulfides, such as iron pyrite, precipitate continuously on active submarine vents to produce a fresh supply of micro-compartments, so that there are always empty apartments for the offspring of successful hypercycles to move into when they leave their parent's home.[21]

Key points

- Simple replicators could have greater capacity for evolution if they are grouped together and depend on each other for replication.
- Compartmentalization could have aided the formation of replicators by controlling the local concentration of molecules, and defining groups that could be selected for efficient cooperative growth and reproduction.

These models of autocatalytic cycles are theoretical constructs. They are plausible stories we tell that describe a hypothetical path by which chemical systems could, without any direction, give rise to simple systems that metabolize, grow, and replicate. But, on the whole, they don't have a clear association with anything we can observe in nature today. They are thought experiments. Now it is time to try to build a hypothesis of what the first replicators could actually have been made of.

First replicators

What might the first replicators have been made of? Simple inorganic replicators could form from a molecule which has a series of units that can be organized in different patterns, if it stimulates the formation of another molecule with the same pattern. In fact, a physicist, Roger Penrose, collaborated with his father, the geneticist Lionel Penrose, to build a simple self-replicating device out of wooden blocks cut into different shaped units, A and B, which could form two alternative two-block structures—either AB or BA. If the box was 'seeded' with one kind of two-block structure and then was shaken so that the units randomly bumped into each other, it would cause more copies of that particular two-block structure to be formed. Thus the AB structures could 'self-replicate' by causing more ABs to form, as could BA. Penrose block replicators also serve to illustrate the diversity of forms of models. While we have described models as being thought-experiments, diagrams, or computer programs, some models are constructed out of real-world materials—think, for example, of Watson and Crick's tin-plate model of the DNA double helix which served to illustrate how the different components of the molecule could fit together into a beautiful stable structure capable of template reproduction. The aim of all models, whether in thought, code, or material, is the same: to illuminate the possible.

What kinds of molecule might be capable of this simple kind of self-replication? It has been suggested that clay crystals might be capable of a simple form of copying, since they can direct the formation of a complementary layer of clay crystal, including the duplication of particular patterns of molecules. The content and structure of clays give

them different properties that might influence their capacity to persist in an environment (e.g. resist erosion) or replicate (e.g. attract materials). Any variations or imperfections in the crystal structure might be passed on to their copies, and these might act as a template for the formation of other molecules.[22] But these simple replicators have limited evolutionary potential; they can make copies very similar to themselves, but the range of heritable variation they can produce is rather restricted, so there is no obvious avenue for increasing in complexity. We need a system where there is more potential for variation that can influence copying potential. In other words, we need a system with a greater capacity for carrying hereditary information that can influence fitness: a truly evolvable system.

What do you think?

If clay crystals or other inorganic materials can stimulate the formation of simple self-replicating molecules, why do we not see multiple origins of life occurring frequently in the natural environment? Or do they occur frequently but we fail to recognize them?

The 'model organism' that has shaped people's imagination of the earliest evolving system is the ribozyme—an RNA molecule that can catalyse biochemical reactions (**Figure 2.6**). Most importantly, some ribozymes have the ability to catalyse the formation of new RNA molecules from an existing template, and so could potentially be partners in an autocatalytic set of replicators. So far, no truly self-replicating RNA molecule has been created in the laboratory. On the one hand, there are RNA templates, like the Spiegelman monster, that can be copied if you add a replicase enzyme. On the other hand, there are RNA molecules with enzyme activity that can perform useful functions like binding together RNA subunits.[23] But, at this point, there are no examples of RNA molecules which can act as both template and enzyme, which can make complete copies of themselves using only

Figure 2.6 A ribozyme is a molecule of RNA that is capable of acting like an enzyme to catalyse specific biochemical reactions. Ribozymes come in a variety of forms and functions—for example, the hammerhead ribozyme (centre) can cut and rejoin RNA strands at specific sites. But unlike protein enzymes, ribozymes could in theory be copied directly by template reproduction. So RNA has been considered a likely form for the first self-replicating molecule.

Image from Lucasharr via Wikimedia Commons (CC4.0 license).

the information in their own sequence. Maybe one will be discovered (or manufactured) soon. In the meantime, there is a gap in the story.

We have interesting and plausible models for chemical systems that can 'metabolize' (build more complex molecules from simpler raw materials) and 'grow' (increase the amount of those molecules) in abiotic conditions. And we can imagine that such a system could split and give rise to offspring that are like their parents. But this is a long way from a molecule like RNA which replicates by template copying, can catalyse other reactions, and has a rich capacity for variation that shapes the chemical functions of the molecule and influences replication. For now, we are going to jump over this gap and take up the story once nucleic-acid-based replicators have evolved. But we encourage you to contemplate the possible steps that connect hypercycles to RNA.

Evolving coding capacity

Once we have a population of replicators that are copied with near-perfect accuracy, we have the capacity for evolution by natural selection. For example, we saw in the Spiegelman experiment that speed of replication affected fitness; faster replicators produced more copies of themselves. What if speed is achieved at the expense of accuracy, so that the fast replicators reproduce with lower accuracy? Raising the error rate increases the chance that the copies made will contain one or more errors, mutations that make the offspring different from their parents in some respect. It is possible that a random copy error might just happen to increase the ability of that copy to make more copies of itself, and this happy accident can lead to more copies made with the fortuitous change. But most random changes to the sequence will either have no effect or make the copy less able to copy itself. A high error rate is likely to lead to more defective copies that are less efficient at copying themselves than their parent molecule was.

In this way, fast copiers may end up leaving fewer viable descendants than slower but more accurate copiers. Therefore, in this simple system, selection may act to balance the advantages of faster copying against the disadvantages of accidentally destroying information needed for being an effective copier. In the Spiegelman experiment, reducing the sequence length led to faster copying. But if the sequence was shortened too much it could not function, because of loss of effective binding sites for the replicase enzyme. A 'Spiegelman monster' represents a compromise: the shortest possible sequence (for faster copying) without losing essential information (for reliable replication).

To explain the origin of life, it is not enough to have a model that gets us to replicating molecules that can evolve by natural selection. We need the model to have the capacity to increase complexity—otherwise it does not provide a good explanation for how we came to have a diverse biosphere full of complex living things. And this is where we run into a paradox. Think about the consequences of increasing the length of the molecule being copied by adding more subunits to the sequence. You now have more coding capacity. But if the chance of making a copy error is roughly the same for every subunit copied, then the longer the sequence, the more likely there will be at least one mistake in each of the copies. For any given copy error rate, the longer the sequence, the less likely it is to produce copies sufficiently like itself to leave a chain of evolvable descendants. The copy error rate places a limit on the amount of information that can be reliably transferred from one generation to the next, and this places a limit on evolvability. At high rates of error per unit copied, the fitness gains of one generation are likely to be erased by mutation, and thus be unable to be transferred to the next generation. We can identify a copy error rate that defines the limit of reliable replication; beyond this 'error threshold' the copies made are too different from their parent for the parent's own copying ability to be reliably transmitted, so the system cannot efficiently evolve adaptations by natural selection. If you want to evolve a longer sequence, you are going to need a lower copy error rate.

The best way to increase reliability of replication is to use specialized tools which actively direct the copying process, rather than relying on passive formation of matching molecules against a template. But making a useful piece of kit like a replication enzyme requires a lot of information. In the Qβ virus genome, the instructions for making replicase are written in over a thousand nucleotide bases. At only a few hundred bases long, a Spiegelman monster could never code for the equipment needed to copy itself (which is why it can only be replicated if incubated with existing enzymes). Here we have a paradox: if you want to evolve a bigger replicator with more coding capacity, you will first need to evolve error-correction tools, but you can't evolve error-correction tools until you have more coding capacity.

How can we break free of this chicken and egg problem (formally known as Eigen's paradox)? If it is impossible to overcome the error threshold by simply increasing the length of a given replicator, then perhaps it might be possible to get many small replicators to cooperate in each other's replication. If each replicator makes enough copies of itself, then the bad copies that can neither copy themselves nor catalyse others' replication will be useless by-products, and the good copies can go on to catalyse the replication of the next replicator in the chain (**Figure 2.7**). Now you can technically have a greater total coding capacity, summed over many shorter replicators, and selection provides an error-filtering stage at each stage of the hypercycle because only good copies contribute to the continuation of the cycle.

Key points

- A molecule that causes copies of itself to form can evolve by natural selection if some heritable variations in structure confer advantages in persistence or copyability.

- However, its evolutionary capacity is limited unless there is the potential for increasing in complexity by increasing information storage and capacity for variation.

Models show that there is a limit to how big the autocatalytic set of replicators can become without becoming unstable, and, just like any co-dependent system, the stability of the hypercycle would be vulnerable to individual replicators which copy

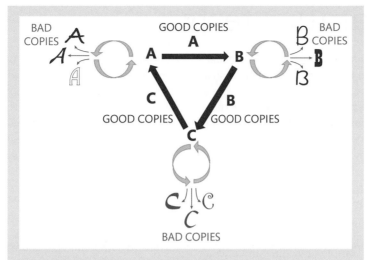

Figure 2.7 Hypercycles have been proposed as a potential way of jumping over the error threshold, because even if the self-replicating molecules (A, B, and C) have high error rates, the occasional good copy (which retains autocatalytic and cross-catalytic activity) can drive the cycle, while the copies with harmful errors (which harm replication) do not take part in the cycle. Note that this hypercycle is hypothetical; it allows us to imagine how a self-replicating system could evolve, but does not supply the details of exactly what chemicals could have taken part in such a reaction. Like many simple models, it is an illustration of the possible.

themselves at the expense of the whole. But if the replicator network was contained in some kind of compartment, and if sets of cooperating replicators were more successful at reproducing themselves than those containing rogue elements, selection could favour the evolution of cooperative replicating systems with the potential to increase their coding capacity above the error threshold.

A hypercycle of replicators is just one possible model that could provide a potential stepping-stone between the abiotic world and the living, evolving world. A plausible model is an important explanatory tool. But it is only a tool, not an answer. Constructing an abstract notion of a self-replicating, autocatalytic chemical cycle tells us that it might be possible for life to have started in this way, but it doesn't tell us whether it actually did or not. Currently there are no known examples of replicative hypercycles beyond those constructed

in people's imaginations. The closest experimental system is a network of RNA fragments of a self-splicing intron derived from a bacterium, modified to only recognize another fragment's sequence as the target of their ribozyme activity, so that they could help the assembly of another RNA molecule but not self-assemble. Unable to replicate themselves, the modified RNAs are therefore forced to cooperate to aid each other's assembly.[24] Ingenious as it is, this experimental system is still some way from a stable collaborative network of self-replicating molecules.

What do you think?

Is the hypercycle model valuable even if we don't have any experimental or observational evidence that they exist in the real world?

Comparative tests

Models are important tools in scientific investigation, allowing us to construct and explore plausible explanations for events we may not be able to directly observe or manipulate. Models are tools, not images of the world; they do not need to exactly resemble the real world, but only represent key aspects of a system that allows us to explore the possible consequences of particular conditions. But of course models will only tell us about the real world to the extent that they capture its salient features. For example, William Paley used the argument from analogy to suggest that the similarities between a complex machine and a living organism are evidence that both were designed for a purpose. The philosopher David Hume criticized the argument from analogy on the grounds that some degree of similarity between two objects or processes does not necessarily tell us that the underlying mechanism that produced them is actually the same. Without a perfect correspondence between the systems being compared, 'the utmost you can here pretend to is a guess, a conjecture, a presumption concerning similar cause'.[25] Models are about what possibly could happen. In order to work out what actually does happen, we need to engage with the real world.

One way to engage with reality is through experiments that use real-world materials. Some people would argue that material experiments have more explanatory power than theoretical models because they are constrained to the possibilities afforded by real systems, and because they can also have more complex components that may behave in surprising ways that the researcher didn't expect. Actually, models and experiments are similar in many ways, as both are simplified systems that aim to reduce the number of confounding variables that could influence the outcome. Often the same tests can be carried out with either material experiments or computer simulations. For example, the demonstration that RNA molecules can be made to form a cooperative network to aid each other's assembly was run both 'wet' (with chemicals in the laboratory) and 'dry' (as a computer model).[24] Even Miller

spark experiments have been run in two very different silicon-based systems—glassware and computers. Material experiments on the origin of life suffer the same limitations as computational models: they are only good explanatory tools inasmuch as they accurately represent the conditions on the early prebiotic earth and the processes that could have occurred there.

While models and experiments are general tools employed in many different scientific fields, there is another method for testing ideas that is particular to evolution. This is broadly referred to as the comparative method (although this is unhelpfully similar to terms used to describe different approaches in linguistics and political science). The evolutionary comparative method makes particular use of descent with modification to leverage causal explanations for biological phenomena. The comparative method can be used to construct 'natural experiments' that test the causal connections between evolutionary phenomena. It uses comparisons between the end products of evolution (which we can observe) to make inferences about evolutionary process and history (which we can't observe).

For example, we have seen how models suggest an error threshold that links the tolerable mutation rate to sequence length; to evolve a larger replicating sequence, you need a lower per-base error rate. Are the predictions of this model realized in the real world? One way to test this is to look at real species and treat each one as an experimental trial of the hypothesis that mutation rate puts a limit on genome size (or alternatively that larger genomes require lower mutation rates). If this hypothesis is true, then we ought to see that species which evolve large genome sizes must also have lower mutation rates. To test this prediction, let us compare mutation rate and genome size in different living microbes (**Figure 2.8**). We can see that the longer the genome, the lower the per-base mutation rate. So species with longer genomes copy more bases, but have a lower error rate per base, which should even out the chances of mutation per

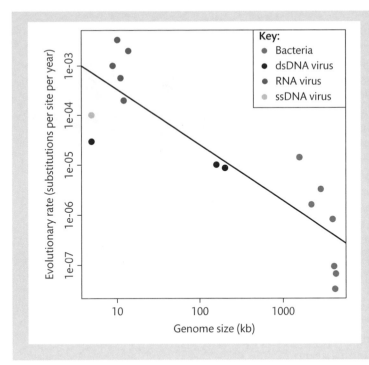

Figure 2.8 A comparison across different living microbes (bacteria, archaebacteria, DNA viruses, and RNA viruses) reveals a correlation between the mutation rate per base and the length of the genome.[35] Patterns such as these have been used to argue that selection shapes mutation rates to a level that allows genomes to be copied without accumulating too many mutations per copy. Clusters of related lineages suggest that these observations are not statistically independent.

From Lynch M. (2010), Evolution of the mutation rate, *Trends in Genetics* 26, 345–52, with permission from Elsevier.

whole genome copy. In fact, figures such as these have been used to suggest that the error rate per genome copy is roughly the same for all microbes (viruses and single-celled organisms), a pattern known as 'Drake's rule'.[26]

RNA viruses might provide a real-world example of the error threshold. The genomes of RNA viruses are copied by an enzyme that has poor error correction, so it has a high per-base error rate; it makes about a million times more errors than replication enzymes in mammalian cells. Because RNA genomes are very short (most are less than 15,000 bases of RNA), this high per-base copy error rate results in roughly one mutation for every new genome copy. This places RNA viruses close to the predicted error threshold. They persist because they produce so many copies of themselves that even if the majority are ruined by mutation, there will be enough copies with no harmful mutations, and potentially some with beneficial mutations, to allow the continuation of the lineage. Their high error rates may give RNA viruses the capacity for rapid response to their host's immune system, but it may come at the cost of limitations

on the evolution of complexity if the high error rate does not permit increases in genome size which could allow the viruses to invent new evolutionary tricks.[27] For comparison, while the mammalian genome is copied with a million times greater precision, it is also a million times larger than the largest RNA viral genome.

Homology

There is another way that comparisons between organisms can shed light on evolution. We can use the patterns of shared heritable traits to reconstruct evolutionary history. This is the argument from homology: if species share a particular feature that we think is unlikely to have evolved more than once, we can use that as evidence that they both inherited that feature from a shared common ancestor. For example, the three main kinds of cellular life currently recognized—bacteria, archaebacteria, and eukaryotes—are very different from each other, but they also have some striking similarities. They all have DNA-based genomes, and they all express the information in the DNA genome in the same basic way, making RNA copies that are translated into proteins using equipment built from RNA and protein. How can we be sure that they didn't just all independently invent similar ways of managing heredity and development? After all, we know that convergent adaptations can end up looking the same even though they have independent evolutionary origins; consider how similar tuna, dolphins, and ichthyosaurs look even though they all evolved that sleek aquatic shape independently from very different ancestors (fish, mammals, and reptiles). Is it possible that there were multiple origins of life that all independently converged on the same working set of molecules—DNA for information, RNA for genome management, protein for building and working?

Even if we thought that nucleotide-based genomes and protein-based metabolism could have been invented more than once, particular features of the system found in all living organisms seem too improbable to have been invented multiple times. For example, many carbon-based molecules, like amino acids and sugars, can occur in different forms which are mirror images of each other. These alternative mirror forms, called enantiomers, have identical chemical components and structures and exactly the same chemical properties. You can think of enantiomers as being like your left and right hands—both the same shape and form but non-identical in relative orientation of the components (**Figure 2.9**). In a Miller spark experiment, both left-handed (L-form) and right-handed (D-form) amino acids are produced in roughly equal numbers. But if you look at a living organism, you will only find L-form amino acids (with very few exceptions). In fact, the chirality of biomolecules was discovered by Louis Pasteur when he realized that crystals derived from living material (specifically, tartaric acid from the dregs of wine) bent light in a specific direction, whereas the same crystals synthesized in the laboratory did not. The living material contained only one enantiomer, so all the molecules bent light in the same direction, but the synthetic material was a mix of both enantiomers, so the light was scattered.

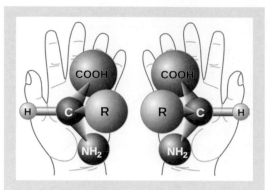

Figure 2.9 A generalized amino acid structure showing left- and right-handed forms of the same molecule. Although both forms are generated spontaneously in abiotic conditions, only left-handed amino acids are used in living cells (with a very small number of exceptions—for example, short peptides from cone snail venom have been found to contain some D-amino acids, as have some antibacterial proteins expressed on toad skin).

Because they have to be able to fit together, many biomolecules will only function if they are made of the building blocks that all shared the same handedness (homochirality). So it is unsurprising that organisms need to use a consistent chirality. But, given that their chemical properties are the same, it shouldn't matter whether a cell uses all L-form or all D-form, as long as it always uses the same one. Why does every single living organism only use left-handed amino acids? And why does every organism only use right-handed sugars?

Imagine an autocatalytic reaction that produces copies of a biomolecule, with each copy having the same chirality as the parent molecule.[28] Whether this molecule is D-form or L-form is unimportant, but to be copied successfully, the molecule must be all D or all L. Incorporating a mix of subunits would lead to breakdown of the molecule because its subunits could not bind together properly. Any productive autocatalytic set that consistently used only one 'hand' of molecules would leave a dominant-handed form in the environment. Notice that we do not need to imply that D-forms or L-forms were any better at forming biomolecules, or that one form could be copied more efficiently than the other. The two forms could be completely equivalent in all respects. But if an autocatalytic cycle has only one enantiomer, and if it outcompetes the others, then its particular handedness will be inherited by all its descendants. So the dominance of one enantiomer could be due to a chance event which is then magnified by inheritance.

As noted by one of the pioneers of molecular evolution, Jacques Monod, evolution is a mix of chance and necessity.[29] It is necessary for all molecules within an organism to have the same chirality in order to produce functional biomolecules. But it might have been just chance which particular chiral form was adopted. But, once that chirality was adopted, it could be inherited by all the descendants of that system. So the fact that all living organisms share the same chirality seems to be one piece of evidence that supports a single origin of all life on earth. If there were many different origins, we might expect some lineages to have originated from an ancestor that, by chance, had a different chirality. In other words, all life on earth can be traced back to a common ancestor. This common ancestor of us all, uniting bacteria, fungi, animals, plants, and even viruses, into a single family, is given the rather attractive name of LUCA (last universal common ancestor).

Last universal common ancestor

What can we know about LUCA? There are very few fossils of early cellular life, and we can't be sure that the fossils found are our LUCA and not a later, or earlier, organism. But we can use a comparative approach to make some inferences about LUCA's characteristics. Using the argument from homology, we can propose that a complex adaptation that is found in many of LUCA's different descendants is unlikely to have arisen by chance in different lineages independently. So a shared complex adaptation found in many different lineages might be taken as evidence that it was also present in their last common ancestor.

As an example of a complex adaptation that is unlikely to have arisen more than once, consider the genetic code, which specifies how the base sequence in the DNA helix is translated to make proteins. DNA consists of four alternative subunits, represented by the letters A, C, T, and G. There are far more subunits that make up proteins—at least 20 amino acids are commonly used in cells. So a one-to-one correspondence between the DNA letters and amino acids is obviously not going to work. Two-letter codes also won't work, because that gives only 16 unique 'words'. So a triplet code is the most sensible solution to using the four-letter DNA alphabet to code for 20 or more amino acids. Each possible combination of three bases—known as a codon—codes for one amino acid. Actually, a three-letter code provides 64 different codons, which leaves redundancy, so some three-base combinations code for the same amino acid (**Table 2.1**). Which brings us to the next question: we can see that a triplet code is the most sensible solution to the coding problem, but why do particular triplets code for the amino acids that they do? In other words, why does CAT always code for histidine in every sequence in every organism, but CAG always codes for glutamine? Why don't some organisms have it the other way around? Wouldn't that work just as well?

The genetic code is translated from DNA to protein via transfer RNA molecules that have a recognition sequence at one end, matching three bases in the DNA sequence, and a specific amino acid at the other. So we can imagine it would be possible to produce a completely different code by changing the amino acid that is matched to each codon. As long as an organism uses a consistent genetic code, so that a given three-base word always translates to a particular amino acid, it shouldn't really matter which codon is assigned to which amino acid. Yet all organisms use essentially the same language to translate from DNA to protein.

One explanation for the universality of the genetic code is that it is, as Francis Crick suggested, a 'frozen accident'.[30] The assignment of codons to amino acids—of words to meanings—could, in the first instance, have been by chance, but once it was set, it was fixed in all descendants. To change the meaning of a codon would cause the placement

Table 2.1 The universal genetic code													
	2nd												
1st	**T**			**C**			**A**			**G**			**3rd**
T	TTT	F	Phenylalanine	TCT	S	Serine	TAT	Y	Tyrosine	TGT	C	Cysteine	T
	TTC	F		TCC	S		TAC	Y		TGC	C		C
	TTA	L	Leucine	TCA	S		TAA	–	Stop Q_6	TGA	–	Stop $M_{1,2,3,4,5}$	A
	TTG	L		TCG	S		TAG	–	Stop Q_6	TGG	W	Tryptophan	G
C	CTT	L	T_5	CCT	P	Proline	CAT	H	Histidine	CGT	R	Arginine	T
	CTC	L		CCC	P		CAC	H		CGC	R		C
	CTA	L		CCA	P		CAA	Q	Glutamine	CGA	R		A
	CTG	L	$T_5 S_7$	CCG	P		CAG	Q		CGG	R		G
A	ATT	I	Isoleucine	ACT	T	Threonine	AAT	N	Asparagine	AGT	S	Serine	T
	ATC	I		ACC	T		AAC	N		AGC	S		C
	ATA	I	$M_{2,3,4,5}$	ACA	T		AAA	K	Lysine N_1	AGA	R	Arginine	A
	ATG	M	Methionine	ACG	T		AAG	K		AGG	R	$S_{1,2} G_3 Stop_4$	G
G	GTT	V	Valine	GCT	A	Alanine	GAT	D	Aspartic acid	GGT	G	Glycine	T
	GTC	V		GCC	A		GAC	D		GGC	G		C
	GTA	V		GCA	A		GAA	E	Glutamic acid	GGA	G		A
	GTG	V		GCG	A		GCG	E		GGG	G		G

All organisms use essentially the same genetic code, but there are small variations in some lineages. Seven codons have variants found in particular lineages—these are listed with the alternative amino acid with the lineage given as a subscript: 1, mitochondrial code of the echinoderms Asterozoa and Echinozoa, and the flatworms Rhabditophora (four differences, two stop codons); 2, mitochondrial code of Nematoda, Mollusca, Crustacea, and Insecta (four differences, two stop codons); 3, mitochondrial code of ascidians (sea squirts) (four differences, two stop codons); 4, mitochondrial code of Vertebrata (four differences, four stop codons); 5, mitochondrial code of the yeasts *Saccharomyces cerevisiae*, *Candida glabrata*, *Hansenula saturnus*, and *Kluyveromyces thermotolerans* (six differences, two stop codons); 6, nuclear code of Ciliata, Dasycladaceae, and Diplomonadida (two differences, one stop codon); 7, nuclear code of the yeast *Candida albicans* (one difference, three stop codons). The chemical properties of the amino acids are indicated by colour: small nonpolar (G, A, S, T; orange), hydrophobic (C, V, I, L, P, F, Y, M, W; green), polar (N, Q, H; magenta), negatively charged (D, E; red), positively charged (K, R; blue).

of the wrong amino acid in every protein made in that cell, which would spell certain disaster. Just as we considered homochirality a necessity (needed for biomolecules to work) but L-form proteins to be chance (could just as well have been D-form), we can consider a triplet code to arise from necessity (required for efficient coding) but the specific matching of triplets to amino acids might be a product of chance (could have been coded differently).

However, it is important to examine such assumptions carefully. Can we be sure that the code is a product of pure chance, a frozen accident? There are two good reasons to think that the code is not wholly a product of chance. One is that the redundancy of the code is used as a buffer against

errors. Because there are 64 codons, but only 20 amino acids are commonly used in proteins, multiple codons can code for the same amino acid (Table 2.1). These are clearly not random assignments; similar codons code for the same amino acid. Since this would not be expected to occur by chance, it is likely to be an adaptation for error minimization.[31] In particular, changing the third base of a codon often does not change the amino acid it codes for, so protein-coding genes are relatively robust to errors in the third position of the codon (appropriately referred to as the 'wobble' position).

The other reason to think that the genetic code is not simply a frozen accident is that it is not entirely frozen. While the code is almost exactly the same in all organisms, there are differences in coding that demonstrate that it can change over time.[32] For example, in most organisms CTG codes for the amino acid leucine, but in the nuclear genome of some yeasts it codes for serine, whereas in the mitochondrial genome of some other yeast species it codes for threonine. Rare modifications of the genetic code show that although it is highly resistant to change, evolution of the code is not impossible.

We have discussed several possible models for the formation of simple replicating systems from non-living matter. This raises an interesting question: if it is possible for evolving systems to emerge from common chemicals, why do we not see new life originating on a regular basis?

Consider the possibility of multiple origins of evolving systems on earth. A successful autocatalytic cycle could have generated an evolving system on one side of the planet, while another kicked off somewhere else, both evolving and gaining complexity over time. What similarities would we expect to see between these independent evolving systems, and what differences?

Perhaps we would not be surprised if different kinds of evolving systems adopted a nucleic-acid-based information storage molecule, because DNA and RNA have the potential to form large molecules with multiple subunits that give them unlimited coding capacity and the ability to be copied. And if they did, we might expect them to adopt a triplet code in order to efficiently translate nucleic acid sequences into polypeptide chains. But even if two different nucleic-acid-based life forms evolved independently, we would be surprised if they both invented exactly the same code for translating those nucleotide letters into amino acids, because we can imagine that an alternative code would also work. Similarly, we might not be surprised that multiple forms of life all adopted proteins as their main working molecules, because the availability of many different amino acids which can be joined together allows great functional variety. But we might not expect all kinds of life to all adopt L-form amino acids, because some independent life forms could have started with D-form amino acids, which presumably would work just as well.

Just as we conclude that the exact translation of the genetic code would be unlikely to evolve independently in different life forms, so we can assume that particular gene sequences within the DNA genome would be unlikely to evolve independently. Even if two organisms were to each independently evolve a protein to fulfil some basic function, we might expect some functional similarities, but not to the extent of having nearly identical amino acid sequences specified by very similar DNA sequences. So when we see the same gene in different organisms we can be confident that both copies were ultimately copied from the same ancestral sequence. This assumption has been used to infer some of the genes that LUCA must have had. And once you know the genes it had, you can begin to guess what kind of capabilities it had (see Case Study 2).

Key points

- Comparative tests use observed data on species to reconstruct evolutionary history, for example by using patterns of homology to infer traits that must have been present in the last common ancestor, or to investigate evolutionary processes, for example by looking for consistent associations between particular traits.

- Many comparative tests resemble experiments in that they represent repeated 'trials' of the association between particular conditions and outcomes in many different biological lineages.

- Homologies between living organisms—such as the genetic code and the consistent chirality of biomolecules—suggest that all identified lifeforms on earth are descended from a single ancestral lineage, the last universal common ancestor (LUCA).

The striking similarities between all living organisms tell us that they all share a common ancestor, so that we can reconstruct the features of that ancestor by considering homologies—shared features of living organisms. This is a little odd, when you think about it. If evolution could start from simple autocatalytic reactions bringing together commonly occurring building blocks in conditions likely to have occurred on the early earth, then it's hard to imagine it could have happened only once. Nothing we have said about LUCA precludes multiple origins of life. But it does suggest that any other origins did not give rise to living descendants. Somehow, even if evolution got started multiple times, just one evolving lineage gave rise to all current members of the biosphere. This brings us to several limitations of the comparative approach to understanding the origin of life on earth.

One is that we cannot go back any further than LUCA by comparing shared features of living organisms. By definition, the last common ancestor of all living organisms was a complex life form with all of the systems that we see across the range of life today: DNA-based genomes, RNA-based genome

management and translation, protein-based construction and machinery, lipid membranes, and so forth. There is a substantial gap between simple replicating molecules, capable of natural selection, and complex cellular life forms represented by the common ancestor of bacteria, archaebacteria, and eukaryotes. While LUCA may be an organism that seems familiar to us, with features that we share with all life on earth, the pre-LUCA replicators might be less recognizable. In fact, Graham Cairns-Smith, an organic chemist, suggested that the first replicators capable of evolution could have been so unlike living organisms that we might not recognize them at all.[22] What if the first replicators were actually mineral crystals, like clay, capable of creating template copies of themselves? If life evolves by stepwise increases in complexity, the earliest forms that provided the stepping stones to complex life may be lost to us. Cairns Smith said that searching for the origin of life by comparing modern organisms might be like trying to understand how a stone

Figure 2.10 If you see a completed brick arch (a), you might wonder how it could have been made, because the structure is not stable unless all the parts are in place. But if you see a brick arch being built (b), you can see that scaffolding is used to hold all the pieces in place until the structure is complete. Graham Cairns-Smith used the construction of an arch as a metaphor for the origin of life: the origin of the complex interdependent structure we have today might appear miraculous and mysterious, but perhaps there was a scaffold on which the complex structure was built which is no longer needed.[22] Cairns-Smith suggested that mineral clays might have provided the scaffolding for the formation of nucleic-acid-based replicators, which later took over as the hereditary material, leaving no contemporary trace of the early construction process.
© Lindell Bromham.

arch was constructed without recognizing that the builders placed the stones over the top of scaffolding that was then removed (**Figure 2.10**).

The other challenge in connecting simple replicating systems to LUCA is to explain how the informational capacity of the replicators increased to allow the evolution of complex life forms. We have already discussed the problem of the error threshold: to allow more complex life forms you need to be able to code for more information, but to code for more information you need an error-correction mechanism—for which you need to be able to code for more information. We have also seen that one hypothetical solution to this problem of vaulting the error threshold has been suggested. If many replicators band together in an autocatalytic cycle, they may be able to increase their net coding capacity while at the same time having an error-correction mechanism through discarding faulty copies. The key to making this kind of system work is to bind those replicators together so that they have a shared fate, and then cooperative systems that work well together might have a competitive advantage. In fact, this has been proposed as a general solution to evolving increasingly complicated organisms. It has been suggested that many of the major transitions in complexity in the history of life consist of changes that leap over informational limits by combining previously independent replicators. In this way, the evolvability of the system evolves through a series of steps that increase the complexity of reproducing entities.

Major transitions

66 *There is no theoretical reason to expect evolutionary lineages to increase in complexity with time, and no empirical evidence that they do so. Nevertheless, eukaryotic cells are more complex than prokaryotic ones, animals and plants are more complex than protists, and so on. This increase in complexity may have been achieved as a result of a series of major evolutionary transitions. These involved changes in the way information is stored and transmitted.* 99
Eörs Szathmáry and John Maynard Smith (1995). The major evolutionary transitions. *Nature* 374, 227–32.

Once we have a population of simple replicators with near-perfect heredity, we have a system capable of evolution by natural selection. But this system may have limited evolutionary potential: if the capacity for variation is limited, and if the total amount of different 'units' that can be coded is small, then selection may simply result in slight variations on a theme, producing copies similar to their parents but not able to build variety and complexity. So it's not enough to be able to explain the origin of a system that can evolve by natural selection; we must also explain how that system can evolve increased complexity. John Maynard Smith and Eörs Szathmáry suggested that biological complexity could evolve by a series of steps where previously independent replicators come together to replicate as a coherent unit. We have already seen one proposed example: the transition from independent self-replicating molecules to communities of cooperating replicators contained within some kind of structure.

Evolution can't aim for complexity. But evolution rewards efficient copiers. Complexity can evolve when it results in more efficient copying. We have seen one way that increased complexity could increase copy efficiency, by allowing the development of specialized replication machinery that could not be encoded by a simpler copier. Other steps might likewise increase copying efficiency by allowing the development of task specialization. This in turn may increase the number of different actions that a replicating entity is capable of performing. However, it is important to remember that just because natural selection can favour complexity if it increases replication efficiency, it does not always do so. Nor is complexity always the sign of selection at work; complexity can evolve even in the absence of a direct selective advantage (see Chapter 3).

The proposed major transitions in evolution are listed in **Table 2.2**. These transitions are not uniform in mechanism or effect. In many cases, they describe the coming together of previously separate replicators into combined units that reproduce by cooperation between their members. This has the result of increasing the capacity for passing heritable information from one generation to the next. For example, separate 'naked' replicators

Table 2.2 Major transitions in evolution		
Replicating molecules	→	Populations of molecules in compartments
Unlinked replicators	→	Chromosomes
Combined gene/enzyme (RNA)	→	Separate gene (DNA) and enzyme (protein)
Prokaryotes	→	Eukaryotes
Asexual clones	→	Sexual populations
Single-celled	→	Multicellular
Solitary individuals	→	Social colonies
Animal societies	→	Human societies (language)

Created from information in Maynard Smith and Szathmary (1997).[12]

might have first joined together into compartmentalized communities of molecules, and then merged further from single genes into multi-gene chromosomes. Once genes join together they could specialize to different tasks. Single-molecule replicators, such as ribozymes, have to be capable of acting as gene and enzyme, being both the template and copying equipment in one molecule. But if you can join several genes together into one cooperating unit, then some genes can make the enzymes that copy the genome, other genes can specialize to making the cell wall or machinery for metabolism, and all genes can be copied by enzymes made by the components of the system.

But this joint enterprise only works if every part of the system is working for the good of the system as a whole. We have seen that 'cheaters' that replicate themselves at the expense of the system as a whole can lead to failure of cooperative replicating entities (we will return to this point in Chapter 7). So another feature of many of the major transitions is that the previously independent entities are now entirely dependent on each other for reproduction, and are no longer capable of separate reproduction. For example, the eukaryotic cell was formed through the union of two kinds of independent cells, probably an early member of the archaebacterial lineage (which contributed the cell structure and nuclear genes) and a bacterium (which formed the mitochondrion). Once this eukaryotic cell was established, the two previously independent cells could no longer divide and go their separate ways. While some eukaryotes have lost their mitochondria, mitochondria cannot revert to free-living existence like their bacterial ancestors. Most genes

from their ancestral bacterial genome were lost or transferred to the nuclear genome, so mitochondria no longer have all the information they need to reproduce independently. Mitochondria are now irreversibly committed to the partnership, their fate joined to that of the cell they inhabit.

The loss of individual identity and independence may seem extreme, but this kind of symbiotic partnership is not uncommon in evolution, where two previously independent organisms fuse into a single unit with a common reproductive outcome. For example, some ants have bacterial symbionts that are housed in special cells in the guts and gonads, and contribute to the metabolism and growth of their ant hosts. The symbionts are passed vertically from mother to offspring, just as mitochondria are. Like mitochondria, the genomes of these bacterial symbionts have lost many of their own genes.[33] At what point do we stop referring to these bacteria as symbionts and start referring to them as subcellular organelles? Unlike mitochondria, these bacterial symbionts are not found in all of the ant's cells. However, the distinction between recognizing two species in the same body (ant plus bacteria) or just one species with two genetic components (nuclear cell with organelles) is far from clear.

Not all of the major transitions involve bringing together separate evolutionary lineages to create a new blended organism. Some involve joining together related individuals into a new cooperating unit. For example, multicellular animals didn't form by fusing different single-celled lineages, but by the close cooperation of individual cells, all descended from a single cell. We can see some similarities between the transition from single-celled to multicellular organisms and the transition from solitary individuals to cooperating societies. Consider an individual bee, living in a large beehive. The bee is made up of individual cells that all cooperate to provide the functions needed to operate an animal—nutrition, respiration, locomotion, and so forth. The hive is made up of individual animals that cooperate to provide all the functions necessary to run the society—food gathering, defence, care of young, and so forth. The individuals (cells or bees) each carry a complete copy of the DNA genome, so could potentially reproduce themselves. But in each case, most individuals do not pass on their genomes to the next generation; only the germ cells in the gonads of the bees copy their genomes into gametes that will form a new bee, and only the queen bee and drones produce larvae that will become the new colony members. All other individuals (body cells or worker bees) will not replicate themselves, but instead will work towards the successful reproduction of the system by providing nutrition, defence, and other necessities to the germ cells (in the bee) or to the queen and drones (in the hive).

Given that the individual cells in a bee, and the individual bees in a hive, have the capacity to copy their own genomes, shouldn't we see an evolutionary advantage to any individual that favours its own reproduction over helping others? Any genome that gains a mutation that causes it to make copies of itself at expense of the others will, ipso facto, end up making more copies of itself. Shouldn't we see that mutations that trigger individual replication have a selective advantage and increase in frequency? The answer is probably the same as for the cooperating hypercycle of replicators: if the fate of the system as a whole is shared by all the components of the system and if systems containing only good cooperators are better at copying themselves than systems containing some selfish replicators, then we should see the evolution of mechanisms that enforce cooperation. Occasionally, a body cell in a multicellular organism does stop cooperating and starts reproducing itself at the expense of the body as a whole; this is cancer, and, if unchecked, it rarely ends well for the system as a whole. Occasionally, worker bees will stop feeding the queen's offspring and start laying their own eggs, so colonies must have policing behaviour that enforces reproductive division of labour. Multicellular organisms whose genomes code for defences against cancer are more likely to make more copies of their own, just as hives that have mechanisms to prevent worker reproduction are more likely to persist and thrive.

One of the major transitions seems to stand out from the others as being of a different kind. The transition from animal societies to human societies does not involve a novel combination of previously independent replicators (Table 2.2). What gives this event the signature of a major transition is the massive increase in information capacity afforded by the invention of language. Like DNA, human language has a massive capacity for information storage and is capable of variation without practical limits. Languages, like genes, are passed from one generation to the next as near-perfect copies, but the introduction of variation every generation allows evolution and divergence. Because of this, language diversification can sometimes be represented as a phylogenetic tree, just as biological evolution can (though languages, like genomes, are also prone to horizontal transfer of words and forms from other languages).

Human language provides a vehicle for cultural and intellectual evolution. Because of language, complex ideas can be copied and passed between generations either by direct instruction or through the medium of recorded language (writing and other forms of transmission). In fact, it is this transition of knowledge between individuals that makes science possible. The language with which we convey ideas allows for their accumulation over many generations, building a body of knowledge by combining many different concepts and observations, gaining in complexity by the gradual accretion of new ideas. The work described in this chapter was carried out by many different people working in distant parts of the world in different time periods, yet our language allows all of those ideas to be combined and transmitted faithfully—and to build a platform for future scientific explorations of the origin of life. These ideas will continue to be copied, modified, selected, and passed on, and thus continue to evolve by descent with modification.

Conclusion

To kick-start evolution, you need a population of copiers that replicate themselves with near-perfect accuracy. If they copy themselves perfectly, then they are all the same, and without variation between individuals there is no capacity for evolutionary change. If they have a high rate of copy error, the copies will not be sufficiently like their parents to inherit their parents' advantages. So evolution requires faithful copying with rare changes, some of which have a consistent influence on the copyability of the sequence. Once copiers with near-perfect replication appear, they can evolve by natural selection. The world fills with copies of entities that are good at copying themselves. This can be seen in experiments that allow RNA templates to evolve in the laboratory, adapting to the conditions in which they find themselves: becoming shorter if selection favours quicker copying, or fixing mutations that allow copying to persist in the presence of various chemicals. Such a system is evolvable, but has limits on the degree to which variation can be generated that allows for an increase in complexity.

The generation of molecules capable of independent self-replication has been investigated using models that allow scientists to explore the consequences of particular conditions and processes. Models such as hypercycles describe a possible mechanism for simple replicating molecules to come together in cooperative systems that allow the evolution of more hereditary information and greater capacity for variation. The aim of these models is to provide plausible scenarios linking the abiotic world to living evolving systems in a series of achievable steps. Models and experiments allow us to explore what could possibly happen, but do not tell us what actually did happen. Comparative tests aim to use observations about the living world to construct links between evolutionary processes and outcomes. Shared patterns between living species can be used to infer the presence of traits in a common ancestor, or the operation of general rules that shape the evolution of different species traits.

There are currently large gaps in the chain of plausible steps between the abiotic world and the living world. How did the earliest replicators cross the error threshold, increasing coding capacity to the point where they could encode specialized copying machinery? What allowed the transition from all-in-one copiers (like RNA ribozymes) to task-specialized genomes where information storage (DNA) is separated from information expression (RNA and proteins)? Some proposed solutions may be particular to specific steps on the evolutionary path to living systems. But, potentially, other models may provide more general solutions. For example, many of the major transitions in evolutionary complexity and evolvability seem to involve the joining together of previously independent replicators into a cooperative system. But these models must incorporate mechanisms whereby the replication of individuals is suppressed if it conflicts with the success of the system as a whole.

The origin of life occurred in the far distant past beyond our powers of direct observation, but we have a diverse scientific toolkit for investigating how it could possibly have occurred. The combination of these approaches will lead to a stepwise construction of a plausible story of the journey from the non-living world to the evolving biosphere.

◯ Points for discussion

1. Some people have proposed that life may have evolved on another planet and then colonized earth, an idea sometimes referred to as panspermia. Does this provide a plausible answer to questions surrounding the origin of life? How would you test the panspermia hypothesis?

2. Should increases in the coding capacity of replicators (for example, increase in genome size) slow the rate of evolution in order to achieve an acceptable rate of replication errors, or increase the rate of evolution because the increased coding capacity allows more potential for variations to arise?

3. John Maynard Smith and Eörs Szathmáry conclude their book *The Major Transitions in Evolution* by suggesting we are currently in the midst of another major transition, with unpredictable (and potentially serious) consequences. Are we living through a transition in the way information is transmitted and inherited? Is it an evolutionary event? What, if anything, would be the consequences of another major transition in evolution occurring?

✳ References

1. Darwin C (1859) *On the origin of species by means of natural selection, or the preservation of favoured races in the struggle for life* (1st edn). John Murray, London.

2. Nilsson D-E, Pelger S (1994) A pessimistic estimate of the time required for an eye to evolve. *Proceedings of the Royal Society of London B: Biological Sciences* 256(1345): 53–8.

3. Mills D, Peterson R, Spiegelman S (1967) An extracellular Darwinian experiment with a self-duplicating nucleic acid molecule. *Proceedings of the National Academy of Sciences of the USA* 58(1): 217–24.

4. Saffhill R, Schneider-Bernloehr H, Orgel L, Spiegelman S (1970) In vitro selection of bacteriophage Qβ ribonucleic acid variants resistant to ethidium bromide. *Journal of Molecular Biology* 51(3): 531–9.

5. Sumper M, Luce R (1975) Evidence for de novo production of self-replicating and environmentally adapted RNA structures by bacteriophage Qbeta replicase. *Proceedings of the National Academy of Sciences of the USA* 72(1): 162–6.

6. Miller SL (1953) A production of amino acids under possible primitive earth conditions. *Science* 117(3046): 528–9.

7. Oró J, Miller SL, Lazcano A (1990) The origin and early evolution of life on Earth. *Annual Review of Earth and Planetary Sciences* 18: 317–56.

8. Johnson AP, Cleaves HJ, Dworkin JP, Glavin DP, Lazcano A, Bada JL (2008) The Miller volcanic spark discharge experiment. *Science* 322(5900): 404–4.

9. Bromham L (2016) Testing hypotheses in macroevolution. *Studies in History and Philosophy of Science Part A* 55: 47–59.

10. Levy A, Currie A (2014) Model organisms are not (theoretical) models. *British Journal for the Philosophy of Science* 66(2): 327–48.

11. Haldane JBS (1929) The origin of life. *Rationalist Annual* 148: 3–10.

12. Maynard Smith J, Szathmary E (1997) *The Major Transitions in Evolution.* Oxford University Press, Oxford.

13. Wächtershäuser G (1992) Groundworks for an evolutionary biochemistry: the iron–sulphur world. *Progress in Biophysics and Molecular Biology* 58(2): 85–201.

14. Ferris JP (2006) Montmorillonite-catalysed formation of RNA oligomers: the possible role of catalysis in the origins of life. *Philosophical Transactions of the Royal Society B: Biological Sciences* 361(1474): 1777–86.

15. Borrel G, Adam PS, Gribaldo S (2016) Methanogenesis and the Wood–Ljungdahl pathway: an ancient, versatile, and fragile association. *Genome Biology and Evolution* 8(6): 1706–11.

16. Joyce GF (2002) The antiquity of RNA-based evolution. *Nature* 418(6894): 214–21.

17. Maynard Smith J (1979) Hypercycles and the origin of life. *Nature* 280(5722): 445–6.

18. Deamer D, Dworkin JP, Sandford SA, Bernstein MP, Allamandola LJ (2002) The first cell membranes. *Astrobiology* 2(4): 371–81.

19. Hanczyc MM, Fujikawa SM, Szostak JW (2003) Experimental models of primitive cellular compartments: encapsulation, growth, and division. *Science* 302(5645): 618–22.

20. Lane N, Martin WF (2012) The origin of membrane bioenergetics. *Cell* 151(7): 1406–16.

21. Koonin EV, Martin W (2005) On the origin of genomes and cells within inorganic compartments. *Trends in Genetics* 21(12): 647–54.

22. Cairns-Smith AG (1990) *Seven Clues to the Origin of Life: A Scientific Detective Story.* Cambridge University Press, Cambridge.

23. Attwater J, Wochner A, Holliger P (2013) In-ice evolution of RNA polymerase ribozyme activity. *Nature Chemistry* 5(12): 1011–18.

24. Vaidya N, Manapat ML, Chen IA, Xulvi-Brunet R, Hayden EJ, Lehman N (2012) Spontaneous network formation among cooperative RNA replicators. *Nature* 491(7422): 72–7.

25. Hume D (1779) *Dialogues Concerning Natural Religion* (2nd edn). London.

26. Drake JW, Charlesworth B, Charlesworth D, Crow JF (1998) Rates of spontaneous mutation. *Genetics* 148(4): 1667–86.

27. Holmes EC (2003) Error thresholds and the constraints to RNA virus evolution. *Trends in Microbiology* 11(12): 543–6.

28. Blackmond DG (2011) The origin of biological homochirality. *Philosophical Transactions of the Royal Society B: Biological Sciences* 366(1580): 2878–84.

29. Monod J (1972) *Chance and Necessity: An Essay on the Natural Philosophy of Modern Biology.* Collins, London.

30. Crick FH (1968) The origin of the genetic code. *Journal of Molecular Biology* 38(3): 367–79.

31. Ardell DH, Sella G (2002) No accident: genetic codes freeze in error-correcting patterns of the standard genetic code. *Philosophical Transactions of the Royal Society of London. Series B: Biological Sciences* 357(1427): 1625–42.

32. Knight RD, Freeland SJ, Landweber LF (2001) Rewiring the keyboard: evolvability of the genetic code. *Nature Reviews Genetics* 2(1): 49–58.

33. Wernegreen JJ, Lazarus AB, Degnan PH (2002) Small genome of *Candidatus blochmannia*, the bacterial endosymbiont of *Camponotus*, implies irreversible specialization to an intracellular lifestyle. *Microbiology* 148(8): 2551–6.

34. Darwin F (ed.) (1887) *The life and letters of Charles Darwin, including an autobiographical chapter*, Vol 3. John Murray, London.

35. Lynch M (2010) Evolution of the mutation rate. *Trends in Genetics* 26(8): 345–52.

Case Study 2
Using homologies to infer history: what was the last universal common ancestor like?

Comparative methods

There are several ways that comparing traits across living species can help us reconstruct the past. If we are confident that the trait has evolved only once, then we can use its shared presence in many species to infer evolutionary relationships, on the assumption that all species with the trait must have inherited it from the same common ancestor. For example, while birds and dinosaurs may be very different in form and habits, the similarity in their skeletons betrays a common ancestry. As Thomas Henry Huxley noted in 1870, 'If the whole hind quarters, from the ilium to the toes, of a half-hatched chicken could be suddenly enlarged, ossified, and fossilized as they are . . . there would be nothing in their characters to prevent us from referring them to the Dinosauria'[1] (Figure 2.11). These shared features of the skeleton are homologies—similar traits found in different species due to inheritance from a common ancestor.

We can also use the identification of homologies to infer what an unknown ancestor was like. All living birds have feathers, no teeth, four-chambered hearts, and are endothermic (metabolically 'warm-blooded'), so it's a fair bet that the last shared ancestor of all birds also had feathers, no teeth, a four-chambered heart and was warm-blooded. Note that this is an argument from probability. It would seem too improbable to suppose that different lineages of birds all independently evolved traits as complex as feathers, with all their shared structural and developmental features. But we can imagine that different lineages could all lose a trait (such as teeth) and end up looking similar. This is because it is unlikely that independently derived complex traits would be identical in all respects, but independent losses could all lead to the same phenotype: absence of the complex trait. So we might end up with the same observable phenotype whether the last common ancestor was toothless, or whether the descendant lineages of a toothed ancestor all lost their teeth. The ancestral explanation might be more probable, but we can't actually rule out the multiple loss scenario.

Some complex traits do evolve multiple times. The four-chambered heart has evolved separately in mammals and birds. Both are endotherms, so we might conclude that if you evolve endothermy you also need to evolve a four-chambered heart for efficient circulation, keeping oxygenated and deoxygenated blood

Figure 2.11 Thomas Henry Huxley used skeletal homologies between living birds and dinosaur fossils to establish their common ancestry, and used rigorous comparative anatomy to show that humans shared a common ancestor with other living apes and were not very different from Neanderthals (the only fossil hominid known at the time). His work showing the homology between the brains of humans and other apes rejected the contemporary notion that human brains had features fundamentally different from those of other species.

Photo from Wellcome Library, London via Wikimedia Commons (CC4.0 license). Thomas Henry Huxley. Etching by L. Flameng after the Hon. J. Collier.

separate. But, on the other hand, crocodiles have a four-chambered heart, and they are not endothermic. And we don't know for sure whether four-chambered hearts evolved independently in birds and crocodiles or whether they both inherited them from a common ancestor (in which case dinosaurs would also have had four-chambered hearts).

If there are only one or few origins of a particular trait, it is difficult to test for meaningful associations between traits. Since there seems to have been only one origin of flight in birds, it's difficult to test which avian features are directly connected to the evolution of flight and which are incidental (can we say that having no teeth is an adaptation to flight because all flying birds have no teeth?). But if there are enough independent evolutionary origins, we can conduct a formal test of the association between two traits (see Chapter 9). There have been dozens of origins of flightless bird lineages, so we can look for general patterns. Compared with their closest flying relatives, flightless bird lineages tend to evolve towards less investment in forelimbs (reduced pectoral regions) and more investment in hindlimbs (larger pelvis).[2,3] This pattern is repeated in many different lineages: it's like an experiment where you take a range of different kinds of birds, put some in an environment where they need flight (mainland), and some in environments where they don't need flight (predator-free islands), and then look for any consistent outcomes.[3] Given enough examples, we may feel confident that the outcomes are not just due to chance but are an effect of the 'treatment' (evolving flightlessness). We can also ask if there are particular traits that predispose bird lineages to evolving flightlessness, like short wing span.[4] Here we gain explanatory leverage by comparing non-homologous traits that have evolved independently in many lineages rather than being inherited from a common ancestor.

Reconstructing LUCA

The power of using homologies in living organisms to infer the properties of the common ancestor is that we can attempt to reconstruct an ancestral organism without any means of direct observation. This might be to infer traits that don't preserve well in fossils (did dinosaurs have four-chambered hearts?) or to give us clues about organisms for which we have no fossils at all (what was LUCA like?). But we musn't forget that reconstructions of past organisms that we can't directly observe are always statements of relative probability. They are hypotheses about ancestral states based on what we can observe today. Here we will focus on a particular study that used comparisons between living species to look back into deep time, long before the earliest fossil organisms, to picture the last universal common ancestor of us all.

Many comparative analyses have relied on assuming that any genes present in all three domains of life—Archaea, Bacteria, and Eukarya—must have been inherited from the common ancestor of all. More specifically, the split between archaebacteria and bacteria is considered to represent the most basal branching event in the tree of life, so researchers have used the presence of shared genes in both archaebacteria and eubacteria to infer the genomic components of LUCA[5] (**Figure 2.12**). They looked for genes present in multiple archaeal

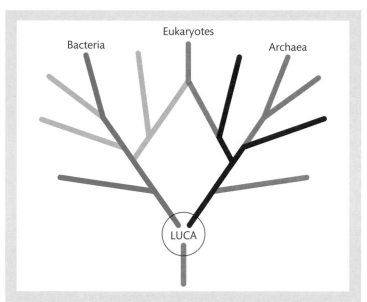

Figure 2.12 Three domains of life. Eukaryotes represent a fusion of genomes from both the archaebacterial lineage (most of the nuclear genes) and the bacterial lineage (which provided the mitochondrial genome). The node that splits the bacterial lineage from the archaebacterial lineages represents the last common ancestor of the three domains, known as LUCA (last universal common ancestor). This diagram illustrates the data selection strategy for a comparative study aiming to identify genes that might have been found in LUCA's genome.[5] Genes that were found in at least two bacterial phyla and two archaeal phyla, and whose phylogeny was consistent with the relationship between the two domains, were considered to have been inherited from LUCA.

From Madeline C. Weiss, Filipa L. Sousa, Natalia Mrnjavac, et al. 'The physiology and habitat of the last universal common ancestor', *Nature Microbiology*, 25 July 2016. © 2016. Reprinted by permission from Springer Nature.

and bacterial lineages, and checked that these showed a pattern of relatedness consistent with inheritance from a common ancestor; the genes had to be found in at least two major groups within each domain, and the sequences of the genes had to recover the expected relationships between the taxa. Accessing the massive public DNA sequence databases, researchers used bioinformatic analysis of over six million protein-coding genes from sequenced prokaryotic genomes to identify 355 protein families that met their criteria for candidate LUCA genes.

Once you have identified homologies that seem likely to have been inherited from a common ancestor, you can start to ask what kind of features that unknown ancestor had. What did these 355 protein-coding genes actually do? The researchers found a number of shared genes associated with chemolithoautotrophic pathways. Breaking this horrendous polysyllabic down, they inferred that these organisms were likely to have been autotrophic, making their own carbon compounds (not heterotrophs that get them by consuming organic compounds made by other organisms); chemotrophic, using chemical energy to fix carbon from carbon dioxide (not by using energy from light like phototrophs do); and lithotrophic, powering these reactions by oxidizing inorganic compounds like iron or hydrogen sulfide (not using organic molecules as organotrophs do).

Here the assumption is that presence in multiple lineages whose connections span the Archaea-Bacteria split are best explained by their inheritance from a shared common ancestor. But there is an alternative explanation. Genes can be both gained and lost over time. Consider the case of reverse gyrase. This gene is found only in thermophiles—organisms that live at temperatures high enough to denature organic molecules in most organisms—and the protein it produces is assumed to be involved in maintaining genome stability in such challenging conditions. In fact, the reverse gyrase gene is apparently so essential to surviving extreme heat that it is ubiquitous in thermophiles. The presence of reverse gyrase in many taxa in the main branches of the tree of life was taken as evidence that they all inherited that gene from an ancestor that likewise needed to survive in hot conditions. So, in addition to inferring that LUCA was an anaerobic (oxygen-free) autotroph, they suggest that it must have lived at high temperatures, probably around deep-sea hydrothermal vents.

But there are two possible explanations of the shared presence of the reverse gyrase gene in the two deepest branches of the tree of life. One is that both lineages inherited it from their heat-loving common ancestor, and the other is that thermophilic taxa in any lineage can only survive if they acquire a copy of reverse gyrase to protect their DNA from denaturing. It has been suggested that several observations about reverse gyrase—such as phylogenetic irregularities, co-location with other archaean genes within bacterial genomes, and presence on a plasmid that can move between cells—suggest that it has been subject to horizontal gene transfer (HGT) between lineages.[6]

In fact, transfer of genes between unrelated lineages appears to have been a relatively common mode of evolution, redistributing key metabolic functions across the tree of life.[7] Some have argued that there has been so much horizontal transfer of genes between lineages that much of the history of life on earth cannot easily be described using the traditional 'evolutionary tree' of diverging branches.[8] This may be especially true of unicellular life, but horizontal gene transfer has also been reported in multicellular species. For example, most animals cannot manufacture carotenoids so must acquire them through food, but the genome of the pea aphid has somehow acquired fungal genes that allow this animal to make its own carotenoids.[9] The frequency of horizontal gene transfer warns that caution is needed whenever we try to infer the presence of genes in a common ancestor from their shared presence in descendants.

In order to use comparisons between species to reconstruct ancestral states, we need to be clear about whether shared traits were inherited vertically from a common ancestor, acquired by horizontal gene transfer, or evolved independently in more than one lineage. How can we tell if the presence of a shared gene indicates shared ancestry? If we suspect that HGT is possible, then presence in two higher taxa within a domain doesn't necessarily tell us whether the gene was inherited from the common ancestor of both domains, whether it was gained by HGT in the ancestor of one (or both) higher taxa, or whether it was gained independently in both lineages within the higher taxon. It's an argument from probability, about which explanation—no HGT events, one HGT event, two HGT events—is most plausible. Without information on rates of HGT, or additional information such as synteny (common genomic location) or shared metabolic pathways, the explanation we find most plausible will depend on our assumptions.

One possible approach is to look for complexity and maintenance of interlocking parts. When we consider feathers, we think it unlikely that they could evolve more than once yet produce exactly the same complex structure, with its many co-adapted parts, in all lineages. So we might consider that presence of a whole metabolic pathway of interconnected genes and regulatory elements might be less plausibly explained by horizontal gene transfer, especially if those genes normally occur in different places in the genome. But again, this is an argument from probability. For example, an interlinked set of 23 genes that form a toxin production pathway has apparently been transferred between fungus lineages,[10] and there is at least one case of the transfer of a whole metabolic pathway from a bacterial host to its giant virus.[11] So it's best to regard such events as improbable but not entirely impossible.

Can we evaluate likely rates of gene loss and gain? Four billion years is a long time and there has been a lot of evolution since LUCA. Consider the search for ancient RNA genes. The ubiquity of RNA, and its potential to act as both template and catalyst for self-replication, has led to the suggestion that it is the original genetic material, predating DNA genomes. Can we compare the RNA sequence across the diversity of life to reconstruct the set of RNA functions found in LUCA, or maybe hints of an even earlier ancestor? When researchers tried this they found that almost all RNAs were specific to a single domain of life—that is, they were found in either Archaea, Bacteria, or Eukarya,

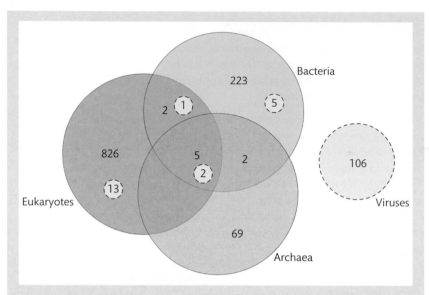

Figure 2.13 Venn diagram of the number of 'families' of related RNA sequences found in the three domains plus viruses. Dashed circles indicate families of viral RNA. This analysis found that 99% of RNA families are specific to only one domain of life.[12] There are only seven families of RNA sequences that are shared by all three domains and viruses. Are these remnants of LUCA?

© 2012 Hoeppner et al. Comparative Analysis of RNA Families Reveals Distinct Repertoires for Each Domain of Life. PLOS Computational Biology 8(11):e1002752.

but not in all three (**Figure 2.13**).[12] So either LUCA had a large set of RNAs, most of which were then lost in two out of three domains, or each lineage has evolved its own characteristic set of RNA tools. Of course, the common ancestor of all three domains must have had useful RNAs, but they have since evolved and produced an array of lineage-specific sequences.

The study we have discussed here has aimed to reconstruct not just the genes present in LUCA, but to use that reconstruction to infer the characteristics of our ancient ancestor. In the case of metabolic genes, this might be fairly uncontroversial. A gene such as reverse gyrase probably does the same job in all taxa. But inference from homology becomes a lot trickier when we have to allow not only for the genes to evolve, but also their functions. We will meet this problem in Chapter 4 when we consider attempts to reconstruct the last common ancestor of all animals by considering which regulatory and developmental genes it must have had. These 'developmental toolkit' genes can be identified in all animals, but they can change function over time. Can we assume that the presence of a shared gene that specifies the development of an eye in all sighted living animals tells us that their last common ancestor could also see?

Questions to ponder

1. Are viruses descendants of LUCA?

2. Homologies are usually identified as complex traits that are unlikely to have arisen more than once. How can you establish homology of genes when all DNA has the same simple components in all organisms, and when mutations to DNA sequences are relatively common over evolutionary time?

3. One bioinformatic approach to detecting incidents of horizontal gene transfer is to estimate the phylogeny of different genes, and look for cases where some trees disagree with the majority. Why is this a test for HGT? How reliable is it? How else might you test for HGT events in the deep distant past?

Further investigation

As humans investigate places beyond our own planet, whether by taking physical samples or analysing observable patterns, how will we recognize living organisms or their products? What features, if any, would you expect off-earth organisms to share with earth life? How similar would they need to be to earth life before you concluded that they must share a common relative on this world or another?

References

1. Huxley TH (1870) Further evidence of the affinity between the dinosaurian reptiles and birds. *Quarterly Journal of the Geological Society* 26(1-2): 12–31.

2. Cubo J, Arthur W (2000) Patterns of correlated character evolution in flightless birds: a phylogenetic approach. *Evolutionary Ecology* 14(8): 693.

3. Wright NA, Steadman DW, Witt CC (2016) Predictable evolution toward flightlessness in volant island birds. *Proceedings of the National Academy of Sciences of the USA* 113(17): 4765–70.

4. McCall RA, Nee S, Harvey PH (1998) The role of wing length in the evolution of avian flightlessness. *Evolutionary Ecology* 12(5): 569.

5. Weiss MC, Sousa FL, Mrnjavac N, Neukirchen S, Roettger M, Nelson-Sathi S, Martin WF (2016) The physiology and habitat of the last universal common ancestor. *Nature Microbiology* 1(9): 16116.

6. Brochier-Armanet C, Forterre P (2006) Widespread distribution of archaeal reverse gyrase in thermophilic bacteria suggests a complex history of vertical inheritance and lateral gene transfers. *Archaea* 2(2): 83–93.

7. Nelson-Sathi S, Sousa FL, Roettger M, Lozada-Chávez N, Thiergart T, Janssen A, Bryant D, Landan G, Schönheit P, Siebers B (2015) Origins of major archaeal clades correspond to gene acquisitions from bacteria. *Nature* 517(7532): 77–80.

8. Doolittle WF (1999) Phylogenetic classification and the universal tree. *Science* 284(5423): 2124–8.

9. Moran NA, Jarvik T (2010) Lateral transfer of genes from fungi underlies carotenoid production in aphids. *Science* 328(5978): 624–7.

10. Slot JC, Rokas A (2011) Horizontal transfer of a large and highly toxic secondary metabolic gene cluster between fungi. *Current Biology* 21(2): 134–9.

11. Monier A, Pagarete A, de Vargas C, Allen MJ, Claverie J-M, Ogata H (2009) Horizontal gene transfer of an entire metabolic pathway between a eukaryotic alga and its DNA virus. *Genome Research* 19(8): 1441–9.

12. Hoeppner MP, Gardner PP, Poole AM (2012) Comparative analysis of RNA families reveals distinct repertoires for each domain of life. *PLOS Computational Biology* 8(11): e1002752.

Does evolution make everything bigger and better?

3

Roadmap

Why study the evolution of size and complexity?

Evolution of life on earth has led to increased size and complexity; the biggest animal that ever lived is alive today, as is the most complex. If we go back in time a billion years, all life would be much smaller and simpler. Does this mean that evolution favours increased complexity over simplicity, and that larger animals have a selective advantage over smaller? Or is it a biased perspective: if you start with the largest and most complex species and trace back to its ancestor, you will inevitably see an increase in size and complexity over time, even if most lineages either stay the same level of complexity (e.g. most bacteria) or undergo simplification (e.g. many parasites). 'Cope's rule' (body size tends to increase over time) and the increase in complexity provide convenient focus points for us to examine how we interpret trends in biodiversity over time.

What are the main points?

- Trends in the fossil record can provide an informative view of evolutionary change over time, but they can often be interpreted in a number of different ways.

- The evolution of horses, which has long been a classic case of an evolutionary trend in size and specialization, provides a convenient illustration of the influence of data incompleteness on views of evolutionary change.

- Proposed explanations for 'Cope's rule' (increase in body size) include microevolutionary advantage to larger individuals, macroevolutionary advantage to lineages with larger size in diversifying or avoiding extinction, and stochastic (undirected) evolution of body size resulting in an overall increase in the maximum size over time.

Photograph: Zebras. © Chantal de Bruijne/Shutterstock.com.

- Complexity is hard to define and measure. There is no clear and simple relationship between genome size, morphological size, and complexity.

What techniques are covered?

- **Palaeontological information:** fossils are patchily distributed in time and space, so patterns need to be interpreted in light of not only the information we have, but also the data we lack.
- **Null models:** before we conclude that a trend is due to selection, we need to ask whether the same pattern could be due to a random walk (undirected change over time).
- **Genomic complexity:** evolutionary patterns in complexity depend on how you define and measure it.

What case studies will be included?

- Accounting for measurement bias: has biodiversity increased over time?

> "I do not think we should embrace scientific theories because they are more hopeful, or more exhilarating. I would like to be able to say that we should embrace them because they are true, but that we can never know. The best we can do is embrace them because they explain a lot of things, are not obviously false, and suggest some interesting questions."

John Maynard Smith, 'Rotteness is all', first published in the *London Review of Books*, reprinted in *Did Darwin get it right? Essays on games, sex and evolution*, Penguin, 1988.

Evolutionary trends

Fossil series linking extinct ancestors to their modern descendants via a chain of intermediate forms have played an important role in shaping ideas about evolution. Early evolutionary biologists used the fossil record as evidence of biological evolution, but in many cases they were troubled by the puzzling gaps. For example, as we will see in Chapter 4, the lack of fossil evidence of the earliest stages of animal evolution was seen as a challenge to Darwin's theory of evolution. But as the fossil record became increasingly well studied, scientists began to piece together sequences of fossils, ordered from oldest to youngest, and they interpreted these fossil series as demonstrating gradual change in form over time.

One of the most celebrated fossil series ever constructed shows the evolution of horses from small forest-dwelling creatures to the noble plains animals of today. This series shows a progression of forms from the small, generalist, and distinctly unhorsey eohippus through a series of extinct species with longer legs, increasingly hoof-like feet, higher-crowned teeth, and longer muzzles (**Figure 3.1**). In other words, this temporal series of fossils gets progressively horsier, culminating in the traits we see in modern horses—long legs, long faces, high-crowned teeth, and one-toed hooves. Similar series from primitive ancestor to advanced descendants have been drawn for many other animals, perhaps most famously for human evolution, which is often represented as a line of figures from the short rounded figure of a chimpanzee walking on its knuckles, through a series of progressively taller, more upright, and less hairy forms, to an erect-postured and nearly hairless man at the end.

Fossil series such as these, setting out a line of proposed ancestors and descendants, have been used for over a century to illustrate the process of evolution. Many are icons of the concept of evolution, such as the chimp-to-human series. These linear series are often used to tell a story of gradual change over time, from a primitive ancestor to a refined descendant, as if each fossil marked a step on the way from simplicity to complexity. Despite their familiarity, we need to think carefully about how we should interpret these linear fossil series. The examples that are included, and the way they are arranged, can shape the way that people think about evolution. For example, some of the pioneering palaeontologists who collected horse fossils used these series to promote the idea of orthogenesis, the hypothesis that evolutionary change tends to occur in a predictable direction from smaller and simpler to larger and more complex (or, more broadly, that biological evolution is progressive). The

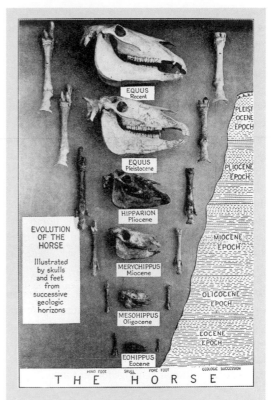

Figure 3.1 A typical museum display (in this case from the American Museum of Natural History) with horse fossils arranged to show a succession of limb bones and skulls from the living *Equus* to the 'dawn horse' eohippus (now named *Hyracotherium*).

Photo from Wellcome Library, London via Wikimedia Commons (CC4.0 license). Evolution of the horse as shown by limbs and skulls recovered from successive horizons in the cenozoic. American Museum of Natural History. Outlines of Historical Geology; 3rd edition Charles Schuchert.

narrative power of the linear series of horse fossils is demonstrated by the surprising observation that many contemporary museum exhibits still display horse evolution as a straight-line series.[1]

There are several reasons to give these progressive depictions a critical appraisal. One is that they often do not represent ancestor–descendant relationships. The popular visual series of figures from chimp to human is misleading, because chimps are our contemporary relatives, not our ancestors. The last shared ancestor of chimps and humans might have looked somewhat chimp-like, but it certainly wasn't a modern chimpanzee (*Pan troglodytes*). The same is true, in a less obvious fashion, for horses; many fossil horses represent extinct relatives of modern horses, rather than their direct ancestors. If we start with the modern horse, we can trace a straight line of forms back 60 million years to the earliest members of the lineage, but this doesn't mean that evolution from the earliest horses to *Equus* followed a straight line. The ancestors of the modern horse also gave rise to a wide range of other taxa, such as the agile forest-dwelling anchitheres and the mighty horned brontotheres (see Chapter 6, Figure 6.1). Considering horse evolution as a process of diversification rather than straight-line evolution changes the way we interpret the patterns in the fossil record.

We are going to use the evolution of horses as an example to examine ideas about evolutionary trends, where evolution appears to be driven in a particular direction. In doing so, we are following in a noble tradition, for horses have been used as a valuable illustration of evolutionary change for centuries.

Evolution of horses

66 *An important reason why horses are so fascinating and why there is so much to say about them is that they are such representative animals. From horses we may learn not only about the horse itself but also about animals in general, indeed about ourselves and about life as a whole, its history and characteristics.* 99
George Gaylord Simpson (1961). *Horses*, American Museum of Natural History, New York.

Among the oldest recognized members of the horse family, Equidae, are the charming 'dawn horses' first described in the 1840s from specimens found in London clay, and given the name *Hyracotherium*. Similar fossils were described from North America in the 1870s and were given the name *Eohippus*. When it was realized these may both be members of a single widespread genus, the scientific name *Hyracotherium* was applied to all, though 'eohippus' was retained as a common name (and the name has since been resurrected for a single North American

species). Eohippus illustrates some of the complications in identifying lines of descent in evolution. The North American populations of eohippus gave rise to the lineage that diversified into the Equidae, whereas the European populations of eohippus gave rise to the palaeothere group, a lineage that eventually produced the magnificent brontotheres—rhino-like animals from the Eocene period (see Chapter 6, Figure 6.1). It has been suggested that the genus *Hyracotherium* has been a 'waste-basket' taxon into which all kinds of early horse taxa were placed, which might actually be recognized as half a dozen different genera, each representing a separate line of descent. This example illustrates the fundamental role that taxonomy plays in understanding evolution. While naming and redefining species, splitting some or combining others, may not be the primary interest of many researchers in macroevolution, species recognition is the fundamental bedrock on which all studies of diversity are ultimately based. Different naming schemes, or changes in taxonomic assignments, can alter our perceptions of the processes that generate biodiversity.

Eohippus were likely to have been forest-dwelling browsers, as were many of their later descendants, nibbling fruits and leaves from trees and shrubs (**Figure 3.2**).[2] But the expansion of grasslands in

Figure 3.2 A classic reconstruction of the dawn horse, which was given two different scientific names: *Hyracotherium* by Richard Owen (based on European specimens), and *Eohippus* by Othniel Marsh (based on American specimens). The name *Eohippus* has now been resurrected for one of the North American species.

the Miocene offered a new opportunity for grazing animals that specialized on eating grass. Grass provides an abundant and consistent food source, but it has lower nutrient content and is more abrasive than leaves and fruit, so grazing specialists tend to show adaptations of teeth and digestion, and larger body size. Grazing horses did not replace the browsing forms, which continued to flourish and diversify in forest environments, or in the mixed environments at the edges of the grasslands. As the global climate changed, open grasslands became an increasingly large proportion of the land area, at the expense of forest cover, and the fortunes of equid lineages reflected these environmental changes. As the world became warmer and drier in the Miocene, the equid lineage underwent its greatest rate of diversification and morphological change. High-crowned teeth (hypsodonty) appeared in many different horse lineages during this period, probably as a response to the wear to teeth caused by silica in grasses and ingestion of sediments from grazing close to the ground. Large body size and changes in leg and foot morphology conferred speed and stamina, perhaps as a defence against predators or to allow a wider geographic range to exploit patchier resources.

We can see these morphological changes to the teeth, skull, legs, and feet reflected in the fossil series shown in Figure 3.1. When arranged like this, with a series of selected fossils arranged from oldest to youngest, the linear series might give the impression that each new form replaced the previous form, for example that large one-toed horses replaced smaller three-toed horses. But the impression of progression from three toes to one, or to higher crowned teeth, or larger size, is largely a result of the great reduction in horse diversity after the Miocene. Before that, there were species with a range of body sizes, browsers as well as grazers, and horses with three toes coexisted with the forms with reduced toe numbers. Considered over the history of the horse lineages, three-toed forms were as successful, in terms of diversity and evolutionary longevity, as the one-toed horses. But only one of the lineages produced in that great Miocene radiation persisted through the last ice age and survived to the present, leaving just the single-toed *Equus*, reduced to a fraction of its previous distribution, persisting only in Africa and Eurasia (**Figure 3.3**). This reduction in diversity appears as a trend in reduction of toes when arrayed as a linear temporal series.

The extinction of horses from their seat of diversity in the Americas reminds us that to understand horse evolution, we need to consider spatial as well as temporal patterns. Eohippus was distributed across North America and Europe in the Eocene but then the continents split, leaving the species divided. In Europe, eohippus gave rise to forms such as the palaeotheres and brontotheres, but that lineage failed to leave modern-day descendants. It was the North American eohippus that gave rise to the equid radiation, and offshoots of this radiation recolonized Europe when land bridges formed between the two continents. Then the North American forms died out, leaving only the living descendants in Afro-Eurasia (American wild horses are all descended from horses brought by Europeans in the 1600s). Yet early ideas of horse evolution were built on fossils from Europe, without knowing that these represented offshoots of the evolutionary diversification of horses in the Americas.

Look at Figure 3.3 and cover the sections representing American fossils, leaving only the 'Old World' taxa visible. Now imagine the impression that having only the European fossils would give you of horse evolution. You would begin with the tiny eohippus in the early Eocene, then a perplexing gap in the record until the appearance of the three-toed browsing *Anchitherium*, which would then be replaced by another browsing taxon, before a sudden late-Miocene takeover of *Hipparion* grazers, which were then replaced by modern *Equus*. The European fossil series is missing the lineages that connect these taxa together because many of the key stages in the evolution of the horse had happened elsewhere. It is more like a series of snapshots, missing the links that join the evolutionary series together. As we shall see in Chapters 5 and 6, geographic biases in fossil sampling can complicate interpretations of evolutionary history.

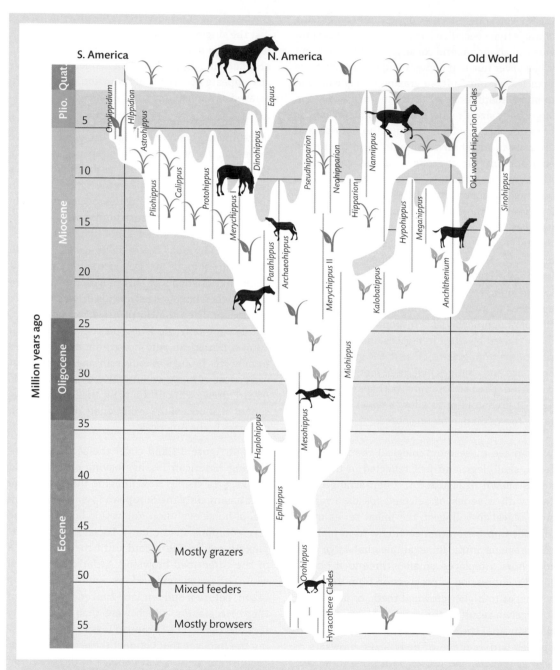

Figure 3.3 The evolutionary history of horses depicted as a branching radiation of forms in time, space, form, and niche. This image reflects classic diagrams drawn by G.G. Simpson.[20] The label 'Old World' is a somewhat archaic label for Afro-Eurasia (the land mass that includes Africa, Europe, and mainland Asia), which can be a little confusing for people who come from other continents that are just as old.

Illustration by Preston Huey/*Science*. From Bruce MacFadden, 2005. 'Fossil horses – evidence for evolution', *Science*, Vol. 307. no. 5716, pp.1728–1730. Reprinted with permission from AAAS.

Key points

- A linear series of fossils from the horse family suggests trends in key characters such as increasing body size, reduction in toe number, and high-crowned teeth. But this line represents only the lineage from the single living horse species to the common ancestor, giving an impression of directed change in characters.

- The diversity of horses throughout their history is better captured by a branching diagram which shows diversification to a range of characteristics in coexisting forms.

Detecting trends

Horse evolution is a classic case study in evolutionary biology because it represents a rich history of diversification in association with changing environment and ecological opportunity. But the example of horse evolution is also important as an illustration of the way a series of fossils can sometimes be interpreted as a linear directional trend even though there may be many other lineages in the same group which do not follow the same trajectory. This is important, because the tendency to describe evolution in terms of directional trends is widespread. One pervasive example of this is in the display of phylogenetic trees.

Phylogenies represent the diversification of descendant lineages from a common ancestral form. Sometimes the tips of the tree represent extinct forms that lived at different times, but more often the labels across the tips are contemporaneous species. Any group of species alive today all have the same length of history back to any given shared ancestor. But, curiously, the forms perceived to be primitive or species-poor are often displayed on the left-hand side of the phylogeny while those considered more advanced or more complex are on the right, subtly implying a kind of progress along the evolutionary lineage from the primitive

to the advanced. It is important to be aware of this kind of bias in order to avoid, as much as possible, thinking that we see evolutionary trends where there are none. So if we want to study trends in the fossil record, the first thing we are going to need to sort out is how to tell when something really is a directional trend and when it is just due to biased perception. The evolution of horses is a poster case for one of the most commonly discussed trends in evolutionary biology—Cope's rule, or the tendency of lineages to increase in size over time.

The term 'law' was once liberally applied in evolutionary biology, from the widely accepted, such as Mendel's laws of inheritance, to the more obscure, such as 'Hunter's law of monstrosity' (large changes in morphology are produced at very early stages of embryo development). Whether due to increasing recognition of the exceptions to proposed laws, or to changing semantic fashion, it became more common to describe generalizations as 'rules'. One such rule is Cope's rule, named for Edward Drinker Cope (1840–1897) (**Figure 3.4**), a pioneering palaeontologist who, amongst many other achievements, found many of the fossils that resulted in horse evolution becoming a paradigm case in evolutionary biology. Cope promoted the idea of orthogenesis—that evolution of new genera followed an ordered series of changes, driven by predictable modifications to developmental trajectories. Oddly enough, Cope did not directly state the 'rule' of increasing body size that now bears his name, so some researchers prefer to refer to it as Depéret's rule, after someone who did (Charles Depéret, geologist, 1854–1929). But, for simplicity, we will stick to the historically inaccurate but widely recognized name of 'Cope's rule' for the hypothesis that evolution tends towards larger body sizes over time.

Consider the way that the classic linear series of horse fossils from eohippus to *Equus* suggests a continuous increase in size over time (Figure 3.1). If we plot body size for members of the horse lineage over time we see that the picture is slightly more complicated (**Figure 3.5**). Like many mammalian lineages, the earliest forms are rather small (a point we will return to in Chapter 6). Almost all species in the horse family were larger than their

Figure 3.4 Edward Drinker Cope (1840–1897) was a pioneering American palaeontologist, with a passion for discovering and collecting fossils that led to 1400 scientific publications in his lifetime (half of which were published in the *American Naturalist*, a magazine he bought and controlled as editor). Cope is nowadays famous for his competition with another leading palaeontologist, Charles Othniel Marsh, that led to the 'Bone Wars,' a race to acquire and name as many dinosaur specimens as possible.

tiny ancestor. Does this imply that larger size was favoured during horse evolution? Or is it simply that if you start small, then the only way is up? The maximum size in Equidae has increased 60-fold since the earliest horses. But, notably, the range of sizes has also increased over time, as horses diversified into a wide range of forms. In fact, apart from the slight increase in the minimum size, you could get the same pattern if body size was evolving by a random walk, sometimes increasing, sometimes decreasing, sometimes staying the same.[3] This does not prove that evolution of body size in horses has been entirely a matter of chance. Larger size may well have been an adaptive advantage in long-extinct horses. But comparing the pattern to

a random walk reminds us that we should not leap to conclusions; just because average size increased in horses over time does not necessarily mean that bigger is always better.

What do you think?

When is it appropriate to describe an observed general phenomenon as a 'law' or 'rule?' Does the identification of examples that don't conform to the general pattern invalidate the law? Do we need to know the underlying mechanism to describe something as a law?

Does body size increase over evolutionary time?

66 *We should not invoke biological principles where statistics suffices.* 99
George C. Williams (1966) *Adaptation and Natural Selection: A Critique of Some Current Evolutionary Thought.* Princeton University Press, Princeton, NJ.

Does Cope's rule generally hold true in biological evolution? Clearly, it is not universally true: some lineages increase in size over time, and some don't. For example, insects reached their largest body sizes in the Carboniferous, when there were bird-sized dragonflies and snake-sized centipedes; they have not continued to increase in size since that time. Very few biological rules don't have any exceptions. So what we want to know is whether there is a general tendency for evolution to produce larger and larger body sizes, such that size increases over time are more common than would be expected by chance. Here we are considering the way body size changes over time, and asking if there are any general directional trends? (In Chapter 8, we will ask a different question about body size evolution: does body size within a species tend to evolve towards an optimum value?)

To test Cope's rule, the first problem we have to solve is how to tell when a lineage has increased

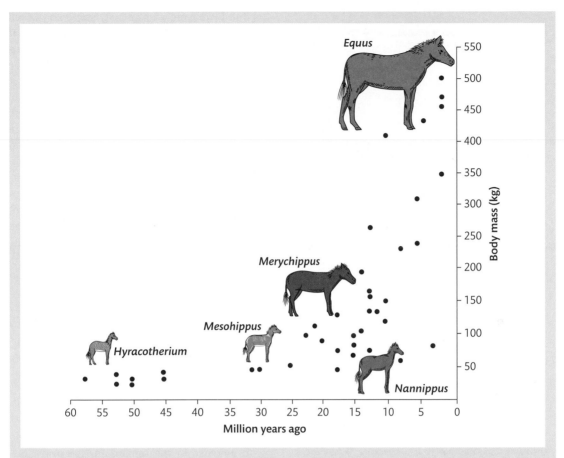

Figure 3.5 Body size of members of the horse family against evolutionary time (millions of years), with mass estimated from fossils. Points indicate different species.[21]

Modified from Bruce J. MacFadden.'Fossil horses from "Eohippus" *(Hyracotherium)* to *Equus*: Scaling, Cope's law, and the evolution of body size', *Paleobiology* Vol. 12, No. 4 (Autumn, 1986), pp. 355–69. Reproduced by permission of Cambridge University Press.

in size over time. This might seem obvious—line up the known species along a timescale and see if they get bigger. But the example of horse evolution illustrated that this is not as easy as it seems. You can line up horse fossils and get an increase in size from eohippus to *Equus* (Figure 3.1). However, if you place the fossils in a phylogenetic context you see that size increase is not a general trend in the horse family, but is largely driven by tracing a line from a small ancestor to a large living species (Figure 3.5). As it happens, the last living species of Equidae is one of the largest members of the group that has ever lived, but this may be a peculiar feature of the horse family. What kind of pattern would we get if we took a much larger group of animals, and looked at the distribution of sizes over the evolutionary history of the animal kingdom?

Marine animals provide a convenient group for studying the evolution of body size over time. They are a diverse group of animals, from tiny invertebrates to large mammals, with a 500 million year history of diversification, and their fossil record has been well studied and documented. If you plot all body sizes against time for all marine animal fossils,[4] you get a clear wedge-shaped pattern

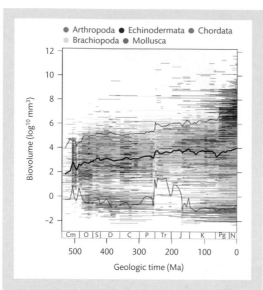

Figure 3.6 Body size evolution across the past 542 million years for genera of marine animals, estimated from fossil evidence.[4] The horizontal lines show the duration of each included genus. The average body size for each stage in geological time (thick black trendline) increases over time (the thin black lines indicate the 5th and 95th percentiles). The geological timescale is given on the horizontal axis: Cm, Cambrian; O, Ordovician; S, Silurian; D, Devonian; C, Carboniferous; P, Permian; Tr, Triassic; J, Jurassic; K, Cretaceous; Pg, Paleogene; N, Neogene (see Appendix).

From Noel A. Heim, Matthew L. Knope, Ellen K. Schaal, et al. 'From Cope's rule in the evolution of marine animals', *Science* 2015. Reprinted with permission from AAAS.

(Figure 3.6), not dissimilar to the pattern seen for horse body sizes over time (Figure 3.5). Unlike horses, the minimum body size of marine animals has decreased slightly over time. But the most notable change in body size distribution is, as for horses, an ever-increasing maximum size and a greater spread in body sizes over time.

On the face of it, this is clear evidence for Cope's rule as a general phenomenon: average body size of marine animals has gone up over time when considered over all marine mammals over half a billion years. Now we need to ask why this pattern occurs. The first thing to notice is that the overall trend in increasing average body size does not tell us what is happening in any particular lineage. If all lineages were being constantly driven towards larger body sizes, transitions to larger size should be more common than transitions to smaller size. But this is not always the case. For example, a study of marine bivalves showed that lineages were no more likely to increase in size over time than they were to decrease.[5] Furthermore, if all lineages tended toward size increases, then we should see smaller-bodied species becoming less common over time, replaced by larger-bodied lineages. This does happen in some cases; for example, the size of pterosaurs increased in the late Cretaceous period, due to both the evolution of larger forms and the disappearance of smaller species (see Case Study 5). But other animal groups show no clear increase in the minimum size. In fact, Figure 3.6 shows that the minimum size of marine animals decreases slightly over time. The minimum bound is largely defined by the size of ostracods, tiny 'seed shrimp' crustaceans which live in marine sediments or as plankton, most of which are only a millimetre or less in size.

The increase in average body size in marine animals over time is not caused by all lineages trending toward larger body sizes over time, but is due to an increase in the maximum size over time. For the first 150 million years of the history of the animal kingdom, the maximum body size is set by arthropods and molluscs, but thereafter it is defined entirely by the upper size in chordates, due to the evolution of large fish, marine reptiles, and cetaceans. So the upward trend in maximum size in marine animals over time predominantly reflects the diversification of vertebrates, rather than any overall trend for all animal lineages to increase in size over evolutionary time. A similar pattern has been noted in particular groups of marine animals; for example, the average size of crabs and lobsters has increased over time due to the appearance of larger-bodied lineages, but the minimum size has not increased.[6] If these patterns are general, it suggests that evolution of ever-larger forms in a few lineages drives Cope's rule, rather than a trend towards larger size in all lineages. If the maximum keeps going up, then the average size will keep going up, even if most animals remain small.

So why does the maximum size have a tendency to increase in many animal lineages? First we will consider explanations that propose a selective advantage to large size. Broadly speaking, two kinds of adaptive explanation have been offered. One is sometimes referred to as a microevolutionary explanation, because it concerns the relative benefits of large size to individuals. If individuals in a population vary in size, and the larger individuals tend to be more successful and have more offspring, then as long as size is at least partly heritable, we might expect to see a gradual increase in average body size in the population over time. This kind of explanation is sometimes referred to as an anagenetic trend. Anagenesis is a gradual directional change in a character over the history of a lineage, such that the descendants are markedly different from their ancestors in some particular character (potentially so different that they are considered to be a separate species by virtue of that change).

Is there any evidence that larger individuals usually have a selective advantage? Several attempts have been made to detect any general microevolutionary trends in body size evolution by gathering together all the published studies on selection acting on populations of different species. Results are mixed, with some studies suggesting that size has fitness advantages in the majority of populations studied,[7] while others claim that there is no consistent pattern of size increase in the population level.[8]

Figure 3.7 The blue whale (*Balaenoptera musculus*) is not only the largest animal alive today, it is the largest animal that has ever lived. The largest whales are nearly twice the estimated weight of the largest dinosaurs. Given that the maximum body size of marine animals has continued to rise for over half a billion years, even larger animals may evolve in future.

For many organisms, even a consistent individual advantage of large size in terms of reproduction must eventually reach a limit beyond which further increases in size are maladaptive. For example, it is possible that the upper size limit for arthropods was reached in the Carboniferous, and that any further increases in size may have been limited by the arthropods' mode of respiration by diffusion of gas through the body surface. For other lineages, there is little evidence that a maximum has been reached. The largest animals that ever lived are alive today (**Figure 3.7**). Will there be even larger animals in the evolutionary future?

In contrast to the within-population microevolutionary explanation, several macroevolutionary explanations for Cope's rule have been put forward. These explanations don't depend on larger individuals having selective advantages within their populations. Instead, these explanations focus on the evolutionary fates of biological lineages with different average body sizes. For example, if large-sized lineages have a greater speciation rate, then whenever a lineage evolves larger size, it will start to produce more descendant species, which will also tend to inherit larger body size. If this is true,

What do you think?

Should Cope's rule only refer to cases where increase in body size over time is driven by some consistent advantage to large body size? Or can Cope's rule also be used to describe passive trends in increasing body size that result from increasing the variance in body size, and therefore the maximum and the average, over time? In other words, should a 'rule' be about pattern or process?

then as time goes on the large-sized lineages will have produced relatively more descendant species, increasing the average size of animals over time. Alternatively, if larger species are less likely to go extinct, when a lineage evolves large size, it will tend to persist for longer. Such explanations are sometimes labelled 'cladogenetic' explanations. Cladogenesis is the splitting of lineages to give multiple descendant lineages. Such explanations have also been referred to as 'species sorting', where trends over time (e.g. increase in average size) are attributed to the differential survival or multiplication of species (e.g. if larger bodied lineages leave more descendant species).

Null models for body size evolution

We have considered two kinds of selective explanation for an upward trend in body size over time: either large individuals have greater reproductive success, or larger-bodied lineages have greater overall rates of diversification. In addition to these directional models, we need to consider a third kind of explanation, which is a random diffusion model. We need to think about the kinds of pattern we would get if body size has no consistent microevolutionary advantage (no predictable fitness advantage for bigger individuals) or any particular macroevolutionary advantage (larger-bodied lineages are neither more likely to speciate nor less likely to go extinct). If body size can change over evolutionary time, but there is no overall tendency for lineages to increase in size either through anagenesis or species sorting, what will happen to patterns of body size over time? That rather depends on where you start.

Imagine that a lineage starts with an intermediate body size and diversifies, with the lineage speciating again and again to give many descendant lineages. And let us suppose that within each new lineage that arises, body size is free to evolve in any direction, and there is no particular selection on increases or decreases. Therefore body size undergoes a random walk along these evolving lineages, so that at any point in time it might stay the same, or go up, or go down. In this situation, over time we would expect to see body sizes spreading out to gradually occupy a greater range of values, even if the average

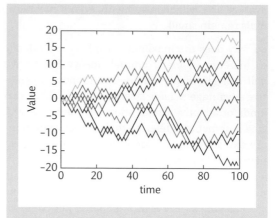

Figure 3.8 A random walk. Starting from a value of zero, each line has an equal chance of increasing or decreasing at each time step (you can replicate the same results by flipping a coin and stepping left for heads, right for tails). This random walk results in diffusion of the maximum and minimum values away from the starting value, and an increase in variance over time, as lines diffuse outwards to occupy a broader range of values.

doesn't change very much (**Figure 3.8**). But this pattern doesn't look much like the distribution of body size over time in horses (Figure 3.5) or marine animals (Figure 3.6). Although we can see an increase in variance (range of sizes) in these examples, the greater size range is mostly due to increases in the upper sizes, rather than decreases in the lower size. The horse lineage never produced a species smaller than the first horses. Ostracods push the minimum boundaries of marine animal size, but the decrease in size over time is much less than the increase in size at the top end. In these examples, and many more, average body sizes go up over time, not down, leaving a wedge-shaped distribution.

You could explain this wedge-shaped distribution of body sizes over time by saying that the tiny ostracods are as small as it's possible for an animal to be, and that you can't make a horse smaller than an eohippus. If you begin at the smallest size, the only way is up. But this only gives us part of an answer. Why is the wedge-shaped pattern so common? Why are there not more examples of lineages that start with a medium-sized ancestor and then diffuse through the size spectrum to increase both the minimum and maximum (Figure 3.8)? Why don't we see more examples of groups that start large then get smaller over time? In other words, why do most lineages start at smaller body sizes?

Some scientists have offered a macroevolutionary explanation for the common wedge-shaped distribution of body sizes over time, suggesting that smaller species are more likely to survive mass extinction events because they have larger populations and shorter generation times, so they are more likely to be able to bounce back from environmental catastrophe and adapt more rapidly to changing conditions. If this is true, extinction events might occasionally reset the size spectrum, resetting clades to begin again from small size. But the plot of body sizes over time in Figure 3.6 doesn't look as if the size increase is being frequently reset back to small again; rather, it seems to show a more or less steady increase in the maximum, and hence the average, size. It could be that the broad sweep of history hides finer-scale

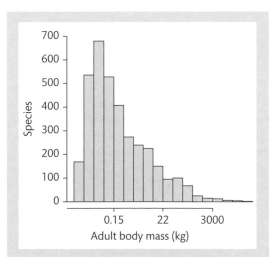

Figure 3.9 An example of a typical body size distribution. This plot is for mammal species, and it shows the number of species in each size class (in kilograms on a log scale). Even though some mammals are very large (>100kg), most mammal species are small (the median size is around 100g), leading to a right skewed distribution (which means that the peak of the distribution is at the smaller sizes on the left while the tail of the distribution stretches out to the larger sizes on the right). We will consider what forces shape this distribution of body sizes in Chapter 8.

patterns of repeated rounds of increase in size and then reset to small, and that the size reset applies to smaller taxonomic groups. But in that case it would not seem to offer a general explanation for the continuous increase in maximum size over time showing the wedge-shaped pattern in so many clades.

There may be a simpler explanation for why most lineages start from ancestors that are below average size. Average size is not actually a good measure of the typical body size of most taxonomic groups of animals. This is because many clades of animals show a skewed body size distribution; there are more small species than large ones (**Figure 3.9**). In such groups, most species are smaller than the average, so there is a good chance that an ancestral species that founds a new lineage will be smaller than the average size. On the assumption that there is a minimum size threshold, a random walk in body size evolution that starts from a smaller than average size will result in an increase in average size over time (**Figure 3.10**).

We have looked at the evolutionary trends within a single lineage, the horse family, and seen that while we can trace paths from descendant to ancestor that suggest a directional change in certain characters over time, we could alternatively interpret the pattern of macroevolution as a multidirectional diversification in response to changing environments. Then we asked whether one of these directional changes—the increase in average body size over time—is a general feature of animal evolution.

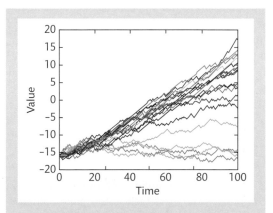

Figure 3.10 A random walk can look like a directional trend if it starts at a minimum value. Variance increases as the maximum goes up over time.

We can see that for many animal groups, at many different scales of resolution, there is a clear and consistent increase in average size over time. But what is driving this? Is it because bigger animals tend to be the fittest members of their populations? Because lineages of larger-bodied animals have a higher speciation or lower extinction rate?

Or is it simply a result of a random evolutionary diversification that results in the gradual increase in the upper size limit, bringing the average size up with it? We are going to meet many of the same kinds of hypotheses as we explore another proposed trend in macroevolution: the tendency for complexity to increase over evolutionary time.

Key points

- Average body size increases over evolutionary time for many groups of animals,

- The increase in average size is typically a result of an increasing maximum size, not an increasing minimum size.

- The increase in maximum size might be due to passive diffusion, or population-level selection for larger individuals, or greater macroevolutionary success of large-bodied species.

- In some cases, the increase in the maximum is due to a trend toward the evolution of larger taxa and the disappearance of smaller-bodied lineages.

- In many cases, the increase in average size appears to be a result of starting from a relatively small ancestor and then increasing the variance in body size over time, resulting in an ever-increasing maximum and average size.

Evolution of complexity

Has there been an overall trend toward increases in complexity over the history of life? The answer would appear to be obvious. If we were to go back in time over a billion and half years ago, all living things would be single-celled organisms, whereas we can look around us now and see towering forest trees, mobile intelligent animals, and massive fungi. But let us examine this idea in more detail. Just as showing that the only living horse species is larger than its early ancestors does not tell us whether the horse lineage has undergone a directional increase in size, so showing that we are more complex than our earliest ancestors does not tell us that evolution has an inbuilt drive to increased complexity.

If we want to examine evolutionary trends in complexity, our first task is to define complexity. Unfortunately this is rather harder than it first appears. There have been long and hearty debates about the best way to define organismal

complexity, and they show no clear sign of being resolved any time soon. One simple measure that has been used to compare the complexity of different organisms is the number of different cell types typically found in the body of an individual of a particular species. For example, humans contain several hundred different types of cells that perform different roles, such as red blood cells to carry oxygen around the body, spindle neurons that transmit electrical signals across parts of the brain, and specialized olfactory sensory cells in the lining of the nose. But not all animals have as many different cell types as humans. Jellyfish typically have less than a dozen different types of cells (though one of these is the formidable stinging nematocyst). Insects tend to have around a quarter as many cell types as mammals. **Figure 3.11** shows a plot of the age of the first appearance of a lineage in the fossil record against the number of different kinds of cells typical for that taxon. There appears to be a clear upward trend—as time goes

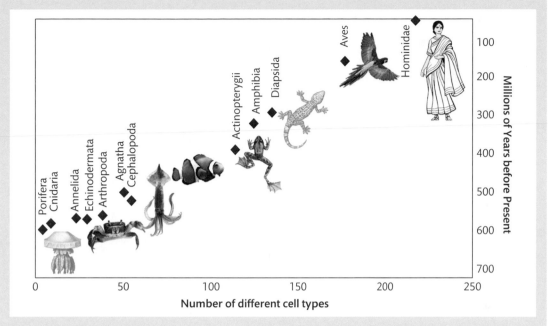

Figure 3.11 Number of different cell types per taxon against the time of first appearance in the fossil record.[22]

on, animals with more and more different types of cells appear. Evolution seems to be driving toward more complex animal life. What could be causing this trend?

One possible explanation is that complex animals have some kind of evolutionary advantage over their simpler relatives. We saw in Chapter 2 that it has been suggested that, in the earliest steps of the evolution of life, any forms that could reproduce more efficiently might have displaced less efficient forms, promoting the evolution of more complex replicators with enzymes, cell walls, and so forth. Does Figure 3.11 bear witness to this process writ large, with animals with more cell types being more efficient at using resources, better at protecting themselves, or more reliably reproducing, and therefore increasing at the expense of simpler forms? We can see this is not the case. The points on the lower part of the graph are not past forms, long replaced by better models. Instead, they represent contemporary species. Porifera (sponges) and Cnidaria (jellyfish) are not our ancestors, they are our relatives. The animals with fewer cell types

have not been outcompeted by more complex animals. In fact if we plot the number of cell types against the number of species in the phylum we would get a very different picture (**Figure 3.12**). There is no obvious trend in evolutionary success, in terms of number of species alive today, and complexity, as measured by the number of cell types.

But you may have noticed that this is not a very fair comparison. The points on the graphs in Figure 3.12 represent different levels of the taxonomic hierarchy. Porifera, Cnidaria, Annelida, Arthropoda, and Echinodermata are phyla, the largest subdivision in the animal kingdom (Chapter 4). Phyla generally represent over half a billion years of evolution, so it is hardly surprising that some of them have accumulated more species than the class Aves (birds), which has had less than 100 million years to accrue species, and the family Hominidae, which is only around 20 million years old. There is another aspect of this graph that does not seem a fair comparison—some of the data points are nested within the others. For example, birds (Aves) are one of the lineages within the Diapsida (birds,

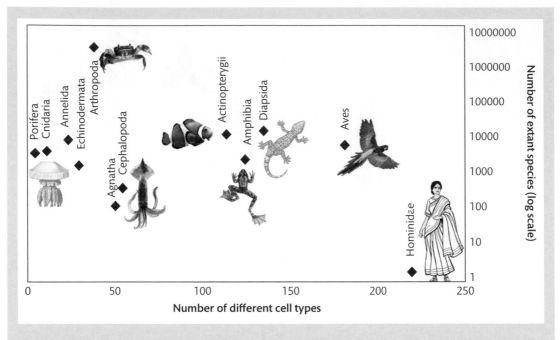

Figure 3.12 Number of different cell types per taxon plotted against the species richness of the group. Note that these groups are different taxonomic ranks. For example, Hominidae is a family (over ten million years old), Aves is a class (over a hundred million years old), whereas Arthropoda, Annelida and Porifera are phyla (over half a billion years old).

lizards, snakes, crocodiles). In other words, some members of Diapsida have more cell types than others. We will meet this problem of nested comparisons again later in this chapter when we consider genomic complexity.

This points us towards another way of interpreting this graph. If we connect these observations on a phylogeny, it looks less like a linear progressive trend, and more like a pattern of diversification (Figure 3.13). The earliest animal forms were relatively simple; more complex forms evolved over time, but the simpler forms didn't disappear. Metazoan lineages spread out in complexity space, just as horses diversified in ecological space. It has been suggested that this pattern of evolution looks more like an undirected search of the space of possible forms. Stephen Jay Gould referred to this process as 'diffusion from the left wall', which is a reference to the drunkard's walk analogy—a simple thought experiment which demonstrates how an undirected process can end up with a predictable outcome.[9]

Imagine that a person leaves a pub in a state of lowered mental clarity and reduced physical coordination and begins to walk home. If she attempts to walk home across a field, taking completely random steps in any direction, we can expect to find her not far from where she started, although her exact position is not predictable. But now imagine that the pub is located in an alleyway. On her left side is the pub wall, and if she staggers into that she bounces back into the alleyway. On the right side of the alleyway is a gutter, and if she happens to stagger into that she falls down and stays there. So now, even though her direction of walking is still entirely random, the outcome is wholly predictable—if we come back in a few hours we will find her in the gutter. Why? Because the lower bound (pub wall) can't be breached (reflecting boundary), and the upper bound (gutter) is an absorbing boundary—if at any point she happens to reach the gutter, the walk ends.

What has this got to do with the evolution of complexity? Unlike the drunk in the field, complexity

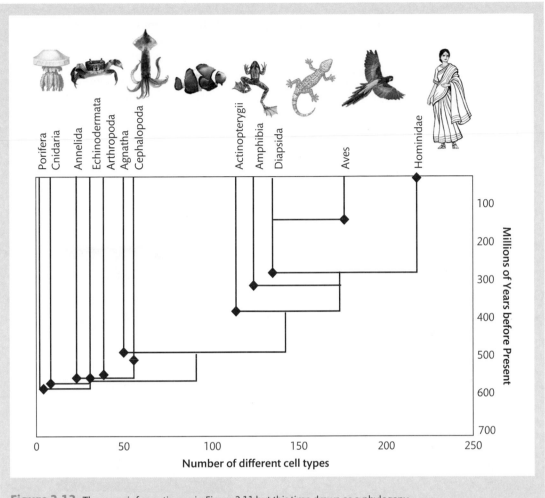

Figure 3.13 The same information as in Figure 3.11 but this time drawn as a phylogeny.

cannot go just as far in any direction. There is a lower bound to complexity of one cell type—a cellular life form can't have zero cell types (nor can it have negative values). Like the drunk in the alley, life starts at the 'left wall' (unicellular life), and then the only way is stay the same or increase (increase in number of cell types over time). What is less clear is whether biological complexity has an upper limit (like the absorbing boundary of the gutter) or whether it can just keep increasing.

One way of looking at the upper limit on complexity is that it has been periodically redefined by the

evolution of major innovations in the way that organisms are put together. For example, it has been suggested that the evolution of the eukaryotic cell, through the merger of previously independently evolving lineages, may have opened up new evolutionary possibilities, allowing the development of multicellular organisms with many different cell types.[10] If this is true, then without the evolution of eukaryotes (which represent a single evolutionary event), the upper limit to complexity might have remained single cells with relatively small genomes. 'Major transitions in evolution', such as the origin of the eukaryotic cell and subsequent evolution of multicellularity, may allow the evolution to hop over the 'right wall' and increase in complexity. A series of such transitions, each building on the last, would result in a directional trend in the maximal complexity, with each transition pushing the upper limit further from the start point (see Chapter 2).

What do you think?

Can complexity continue to increase, or has it reached an upper limit? Could another major transition occur that would allow increased complexity to evolve?

Gain and loss of complexity

66 *There is no reason why evolution by natural selection should lead to an increase in complexity, if that is what we mean by progress. At most, the theory suggests that organisms should be better, or at least no worse, at doing what they are doing right now. But an increase in immediate 'fitness'—that is, expected number of offspring—may be achieved by losing eyes or legs as well as by gaining them.* 99

John Maynard Smith and Eörs Szathmáry (1995) *The Major Transitions in Evolution.* Oxford University Press, New York.

Clearly, some lineages have increased in complexity, and it seems fair to say that, however you measure complexity, the maximum complexity has gone up over evolutionary time. Does this mean there is a driven trend toward complexity, due to some microevolutionary or macroevolutionary advantage? Just as the upper bound in body size in marine animals was defined by the big vertebrates (Figure 3.6), so is the upper bound in complexity (Figure 3.11). As we found when considering horse evolution, the contemporary species we tend to focus on happens to be in the upper right hand of the graph; modern *Equus* are larger than all their ancestors. Similarly, modern humans have more different cell types than their ancestors. So when we

start at human and trace backward, we inevitably see a steady rise in complexity, at least when complexity is defined as the number of cell types. But how typical is this pattern?

Let's pick another contemporary species and trace it back to its ancestors. In honour of Charles Darwin, who spent eight years of his life working on a mighty four-volume taxonomic treatise on both living and fossil barnacles, we will start with a barnacle. Barnacles are a diverse group of arthropods, with over 1200 living species (**Figure 3.14**).

Figure 3.14 Barnacles are a diverse group of sessile marine arthropods. They occupy a range of marine habitats, from living in rockpools in the intertidal zone to travelling attached to sea-going vertebrates to living within the tissues of other arthropods—as in the parasitic barnacle *Sacculina*, pictured centre, emerging from an infected crab (see Figure 4.3). This drawing is from Ernst Haeckel's *Kunstformen der Natur* (*Art Forms of Nature*) (1904), a masterful work combining zoological detail with aesthetic sensibility. Haeckel had a huge influence on both science and art, and played a key role in the public understanding of evolution. We owe some of the key terms in this book to him, such as ecology and phylogeny.

They develop via two distinct larval stages: the first larval stage is adapted for moving, growing, and feeding, but the second larval stage is a non-feeding form primarily directed towards finding a suitable place to settle. When they find one, they cement their head to the surface and transform into the adult stage of the life cycle, which grows a protective armour of hard shell, and spend the rest of their lives glued to the spot, filtering food from the seawater using feathery appendages. Barnacles are clearly complex animals, with a complicated life cycle and many specialized organs that adapt them to their particular lifestyle. But they are arguably simpler than many other arthropods because, due to their sessile life, they have jettisoned some of the complex features of the typical arthropod body plan: adult barnacles do not have compound eyes, they have no heart or gills, they have lost one set of antennae, and their brains have become simplified. You don't need to see or move or think much if you live your life stuck to a rock slurping passing morsels out of the seawater.

Some barnacle species have undergone a more dramatic simplification. The first two larval stages of the barnacle *Sacculina* resemble those of other barnacles, but the second larval form metamorphoses into a specialized juvenile form called a kentrogon. The trigger for metamorphosis is, as for other barnacles, finding an appropriate substrate, but in the case of *Sacculina*, the substrate isn't a rock but a living crab. Nestled inside the kentrogon is the next stage of the *Sacculina* life cycle, the tiny slug-like vermigon. The kentrogon inserts a spike into the crab's carapace and injects the vermigon into the crab's body. Within the crab's body tissues, the vermigon grows into the adult form, a mass of cells with radiating tendrils that resemble fungal hyphae. *Sacculina* then lives as a parasite in the crab's body, not only drawing nutrients from its host, but also changing its host's behaviour. *Sacculina* suppresses the crab's reproduction and prevents moulting, limiting the crab's growth. If it infects a female crab, it triggers her brooding behaviour, directed not towards her own offspring but to nurturing and releasing *Sacculina's* larvae

(see Chapter 4, Figure 4.3). If a *Sacculina* infects a male crab, it causes him to act like a female and then start brooding the parasites.

In some ways, *Sacculina* is more complex than most barnacles: it has specialized life-cycle stages not found in non-parasitic barnacle taxa, and it has a sophisticated chemical weaponry that allows it to take control of the crab's behaviour. But in terms of adult morphology, *Sacculina* is much less complex than most barnacles, with virtually no organs or appendages—essentially just a gonad with a chitinous skin. You don't need sense organs or a brain or limbs or armour or anything else if all you do is absorb nutrients from a slave crab.

The point of this slightly disturbing story of parasitism is that if we constructed a series of ancestor–descendant forms for *Sacculina,* as has been done for horses and humans, we wouldn't get a clear increase in complexity over time. Instead, we would start with a relatively simple metazoan ancestor, which evolved into a more complex arthropod ancestor, and then to a morphologically very simple contemporary species. Evolution of simplification is common in many parasitic taxa as they jettison features of independent living that are no longer needed. If you are tempted to dismiss simplification in parasites as a rare exception, consider that there are almost certainly more parasitic species on earth than host species. For example, it has been estimated that the roughly 45,000 species of vertebrates have at least 76,000 associated species of parasitic trematodes, cestodes, nematodes, and acanthocephalans.[11] Even parasitic barnacles have their own 'hyperparasites,' such as parasitic isopods (the crustacean group that includes slaters, pillbugs, and woodlice). Like their hosts, these isopods are morphologically simple, having lost many of the features typical of other isopods such as legs, feelers, mouthparts, and eyes, yet they also seem to suppress their host's fertility, preventing female parasitic barnacles from producing eggs. While not all parasites are simpler than their free-living relatives, we can recognize that evolutionary trajectories towards increasing simplicity must be reasonably common.

Key points

- If you start with a complex organism and trace its evolutionary history back far enough, you will see an increase in complexity over time. But many evolutionary lineages maintain the same level of complexity over time (e.g. bacteria) and some reduce in complexity (e.g. many parasites).

- Complex organisms have not replaced simpler ones, but because some lineages evolve increased complexity, maximum complexity must go up over evolutionary time.

- The increase in maximum complexity over time could be explained through selection; for example, if occasional evolutionary innovations permit one or few lineages to give rise to more complex forms, which then multiply or diversify. Or it could be explained by a random walk; as lineages evolve, occasionally one may increase in complexity, and if it persists it may give rise to even more complex lineages, whether or not they have an evolutionary advantage.

Genomic complexity

Thus far we have focused primarily on just one measure of organismal complexity: the number of different cell types. But this might not capture all elements of complexity. For example, adult *Sacculina* have relatively few cell types, but they have a complex life cycle with five morphologically distinct stages, and they are able to chemically control the behaviour and physiology of a complex host. Would we get a different view of the evolution of complexity if we were to consider measures of complexity other than number of cell types? Attention is increasingly being turned to the complexity of the genome. Given that major transitions in the evolution of complexity have been defined in terms of heritable information (see Chapter 2), we might expect to see a clear pattern of increase in genomic complexity over the

history of life. But how are we to define complexity at the genome level?

Genome size might provide a simple measure of genome complexity, on the grounds that the more DNA you have, the greater your informational capacity. Simple unicellular organisms, such as bacteria and archaebacteria, tend to have relatively small genomes (between 0.5 and 10 million bases of DNA), while large multicellular organisms with many different cell types tend to have much bigger genomes (from 20 million to 130 billion bases). This makes intuitive sense: to build an organism with many different tissue types and complex behaviour, you will need a lot of different genes and some sophisticated gene regulation mechanisms. But surprisingly, when looked at more closely, there is very little correlation between genome size and morphological or behavioural complexity. For example, we might expect different species of fish to have roughly equivalent morphological complexity, but while a pufferfish (*Takifugu rubripes*) has a genome of 390 million bases, the lungfish (*Protopterus aethiopicus*) has a mighty 130 billion bases of DNA per cell. The lungfish genome is over 300 times bigger than the pufferfish and over 40 times bigger than the human genome. Multicellular plants show a similarly bewildering range of genome sizes, from the Brazilian carnivorous plant *Genlisea* with 64 million bases to *Paris japonica,* a Japanese sub-alpine plant with a whopping 150 billion bases of DNA per cell. Genome size, as measured by picograms of DNA per cell, is traditionally referred to as the C-value of a species, so the striking lack of correlation between organismal complexity and genome size is known as the 'C-value paradox'.

One possible resolution to the C-value paradox is that it's not the amount of DNA in the genome that determines complexity, but the number of genes it contains. The total amount of DNA in the genome is a poor predictor of number of genes, because the amount of non-gene DNA varies so much between species. For example, humans and pufferfish share a similar level of morphological complexity, each having over a hundred different cell types. But the pufferfish genome is an eighth of the size of the human genome. The difference between the 3.2 billion base human genome and the 0.39 billion base fugu genome is not in the amount of genes—actually the pufferfish probably has around the same number of genes as the human. Instead, the difference is in the amount of non-genic DNA, the sequences between and within genes that are not transcribed into functional RNA molecules or translated into proteins. The pufferfish genome has much less non-translated sequence within coding sequences (introns), and less non-coding DNA between genes, than the human genome. Amount of DNA per cell might place a theoretical limit on information capacity. But in practice, at least for multicellular organisms, the actual amount of coding DNA is a tiny fraction of the genome, and is more or less unrelated to the total genome size.

Now the question is: is the amount of coding DNA a reflection of organismal complexity, and if it is, does it increase over evolutionary time? The predicted number of genes from whole genome sequences does not give the impression of a clear correspondence between number of genes and number of different cell types.[12] For example, the tiny transparent pond flea *Daphnia* probably has nearly as many genes as the the human genome (Figure 3.15). The tiny flatworm *Caenorhabditis elegans*, studied for its simplicity, has only a thousand cells in total in its adult body. Although it has only a sixth of the number of cell types as a human, it has nearly as many genes (19,000). So number of genes doesn't seem to predict morphological complexity.

Perhaps it's not how many different genes you have, but what you do with them that counts. Differential gene expression allows the same genome to be 'read' in different ways to produce different phenotypic outcomes. For example, in most species, males and females receive identical or near-identical copies of the genome, but can end up with distinctly different morphologies and behaviour. Sex differentiation occurs through the initiation of different developmental programs that express the genetic information in the genome in different ways (see the Case Study 4). Similarly, development can respond to environmental influences, switching on or off certain genes to produce a phenotype appropriate to conditions. For example, the level of CO_2 in the atmosphere influences the number and spacing of stomata (openings for gas exchange) on leaves, due

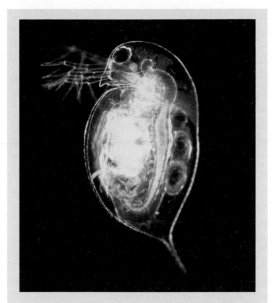

Figure 3.15 The tiny pond flea *Daphnia* is only a few millimetres in length. Initial estimates suggested over 30,000 genes in the Daphnia genome.[23], but recent estimates are much lower.[24]

Photo by Paul Hebert via Wikimedia Commons (CC2.5 license) from Gewin V (2005) Functional Genomics Thickens the Biological Plot. *PLoS Biology* 3(6): e219.

to genetic signalling cascades that detect CO_2 levels and adjust gene expression accordingly. If these signalling cascades are faulty, the plant can end up with a suboptimal number of stomata, and an inefficient metabolism as a result. Genetic regulation, controlling which genes are turned on, where, and when, can generate morphological, developmental, and behavioural complexity.

Gene expression allows different combinations of genes to be expressed, allowing multiple phenotypes to be generated from the same genome. In some cases, a single gene can be read in different ways to give more than one possible gene product. Genes, coded in DNA in the genome, can only be expressed if an RNA copy is made of the gene (transcription), and that transcript either forms a working RNA molecule or acts as a template for the production of a protein (translation). Some genes can give rise to many different transcripts, for example by starting transcription at alternative places, or including different parts of the gene in the transcript (see Chapter

4, Figure 4.11). For example, the human GNAS gene can produce many different gene products by including different combinations of its 17 blocks of protein-coding sequences (exons) in the gene transcript by alternative splicing. In addition to the mixing and matching of different protein sequences, GNAS can produce two completely different proteins by starting translation of the mRNA transcript at alternative places, so that the sequence is read in two different reading frames, resulting in different amino acid sequences.[13] If there are alternative ways of transcribing or translating many genes, then just because two species have the same number of genes, we can't necessarily infer that they can produce the same number of different gene products.

So it may be that the reason that gene number doesn't correlate with morphological complexity is that gene number is only loosely correlated with protein diversity (or number of functional RNA molecules). If many genes can produce multiple products, for example through alternative splicing or multiple start codons, then morphological complexity might depend more on the regulation of gene transcription than on the baseline number of genes. Since changes to the timing or the spatial organization of gene expression can influence phenotype, regulatory sequences can have a large impact on an organism's morphology. It has been suggested that it is the 'non-coding DNA' in the genome that determines complexity through the operation of a sophisticated genome control system, by encoding regulatory factors that govern gene expression. A survey of the biochemical activity and sequence content of the genome came to the surprising conclusion that the majority of the non-gene sequences in the genome have some kind of function, despite a lack of sequence conservation between species.[14] This claim that most of the vast wildness of intergenic DNA, which in humans makes up over 98% of the genome, is actually involved in gene regulation and the construction of complex phenotypes has been greeted with delight by some researchers and derision by others. There is an ongoing, and very vigorous, debate about how much of the human genome sequence is functional, and how we can tell when non-gene DNA has a functional role in influencing phenotype.

Many years of investigation and debate lie ahead of us before we can come close to an answer on the perplexing relationships (or rather lack of relationships) between genome size, gene number, gene expression, and organismal complexity. But it is interesting to note that many of the graphics used in these debates bear some resemblance to the progression diagrams that have been used for over a century to support evolutionary hypotheses. Often, the species considered the most 'advanced' (usually *Homo sapiens*) are in the top right-hand corner (in addition, you may notice that these figures tend to display only a very biased representation of human diversity). Sometimes there is a subtle implication of progress by ordering values from smallest to largest, left to right, even when the values are not directly comparable (for example, some are averaged over large groups, and others for subsets of those groups). And, for all genomic data, the comparisons do not represent ancestor–descendant relationships, or changes in genome size and architecture over time, but can only include contemporary species, all existing at the same point in time with the same amount of evolutionary history behind them. We need to be careful to evaluate the evidence with an open mind and not project onto the data preconceived ideas about progressive evolution of biological complexity.

What do you think?

How could you test the hypothesis that lineages with more complex genome regulation are more adaptable (in the population-based microevolutionary sense) or have higher evolvability (in the macroevolutionary sense of the ability of lineages to adapt and diversify)? How would you distinguish the predicted patterns from stochastic processes?

Complexity by drift

66 *Where, then, is the direct supporting evidence for the assumption that complexity is entirely rooted in adaptive processes? No existing observations support such a claim, and, given the massive global dominance of unicellular*

species over multicellular eukaryotes, in terms of both species richness and numbers of individuals, if there is an advantage of organismal complexity, one can only marvel at the inability of natural selection to promote it. 99
Michael Lynch (2007) *The Origins of Genome Complexity.* Sinauer Associates, Sunderland, MA.

Many hypotheses concerning the evolution of complexity rest on the assumption that greater complexity is advantageous in some way, such that increases in complexity will be favoured by selection. Stop and ask yourself whether the evidence we have seen so far supports this basic assumption. Simple organisms not only continue to coexist with more complex organisms, they thrive. Many lineages decrease in complexity. Furthermore, even if complexity has advantages, it might also come with costs, such as the burden of maintaining a larger genome, or the metabolic costs of operating a large body or sustaining a complex way of life. More parts may also bring more risk if there is more that can go wrong, as a mutation in an essential gene or regulatory sequence might cause loss of function or expression of genes in the wrong place at the wrong time or in the wrong amount. Even mutations in non-essential DNA can be harmful if they cause a disruptive 'gain of function'. A spare copy of a gene produced by duplication might acquire a mutation that ruins the protein product and makes it toxic, causing it to build up in the cell and inhibit normal function. Excess non-coding DNA might collect a chance mutation that creates an additional regulatory signal, causing neighbouring genes to be turned on or off inappropriately.

One way to explain the increase in genome size and complexity, despite apparent costs, is to say that the benefits must outweigh the costs, and so the fitness advantages of having more DNA overcomes the disadvantages of increased metabolic burden and greater mutational risk. But we should also consider an alternative possibility that, rather than being driven by innate advantage, the evolution of complexity may be the result of a failure to maintain simplicity. We have seen that genome size and gene number seem to bear little obvious relationship to organismal complexity. What if complex

genomes evolve not as a result of selection for morphological complexity, but as the result of a series of accidental changes to genome architecture that make gene function and expression more convoluted. If these changes are not removed by selection they may accumulate over time.

If it is difficult to imagine how complexity could increase in the absence of any advantage, and even in the face of costs, then consider Parkinson's law. Although this phenomenon relates to the growth of human bureaucratic institutions, it helps to illustrate how complexity can increase without any actual benefit, and even in the face of high overall costs. Reflecting on his experiences in the civil service, Cyril Northcote Parkinson wrote an insightful and amusing account of the way that organizations can increase in size and complexity without actually increasing in output or efficiency.[15] Parkinson starts with the observation that work expands to fill the time available for its completion. You may have noticed this with assignment deadlines; if the essay is due tomorrow it will by definition take less than 24 hours to write, but if it is not due for another week it may take much longer to complete the same task. So an employee will usually find that the work allotted to them takes up all of their available time. This will inevitably lead to the feeling that really they have too much work to do given the time they have to do it, and thus they will argue that another employee is needed to help complete the work. But as soon as you have another employee, their available time will rapidly become filled with tasks, and so they too will feel that they have more work than they can do.

Thus the job that was initially done by one person expands to fill the days of two people, then three, and so on. Parkinson supported his hypothesis with observations from real organizations, such as the British Colonial Office which continued to expand even as the number of colonies it administered decreased. He presented data on the British Navy which showed that in a period when the number of ships was reduced by more than half, the number of sailors decreased by only a fifth, and the number of administrators and officials increased by more than a third. Each step towards increasing bureaucracy may come at a relatively modest cost, in the overall scheme of things, but when summed over the whole organization over many years, the result can be a substantial increase in running cost with little increase in productivity.

If you are wondering what the British civil service has to do with genome evolution, a similar process has been suggested to result in increases in genomic size and complexity, where small increases accumulate even if they bring no benefits, and even if they have slight costs. Accidents in genome management can result in an increase in the amount of DNA. For example, slippage during replication can result in the creation of repeats of the same sequence, and uneven cross-over between chromosomes can result in extra copies of genes being created. What will happen to this extra DNA? If it harms the individual, reducing their chances of survival and reproduction, then it will tend to disappear from the population as fewer carriers leave descendants with the extra copies. But what if this extra DNA comes at little or no cost, such that an individual with an extra copy of a sequence is at no major disadvantage, and carrying a slightly larger genome makes little difference to chances of having healthy offspring? Then the fate of the increase in DNA will be due to chance. An individual with an extra copy of a gene might survive, reproduce, and, by chance, have lots of descendants that also have that extra sequence. And some of those descendants might also have offspring with slight increases in DNA that they also pass on with relatively little added cost.

Eventually it may happen that, by chance, all members of the population are descendants of this extra-DNA line, because the lack of benefit and the slight cost has not been enough to stop them reproducing. The larger genome has now become a feature of that population. Fixation of a slightly costly mutation is more likely to occur in a small population, because it only takes a relatively small number of generations where the extra-DNA individuals just happen to reproduce more than the less-DNA individuals before all living members of the population carry the larger genome. So species with small populations might be more prone

to accidental increases in genome size and complexity than those with larger populations, where selection can more effectively remove slightly costly mutations. And, as it happens, species that are already large and complex often have smaller population sizes than those that are smaller and simpler, potentially generating a 'trend' in increasing genome size and complexity in those lineages.[16]

Instead of being viewed as a consequence of natural selection for adaptation, the evolution of genome complexity can be considered in light of the nearly neutral theory.[17] This theory states that a significant proportion of random mutations will be slightly deleterious (nearly neutral), and that the fate of these mutations is governed by the interaction between selection, drift, and population size. In a large population, selection efficiently removes these deleterious mutations, but in a small population there is more chance that they can accidentally become fixed in the population due to the random sampling effects that occur every generation, where alleles may increase in frequency just by chance. So slightly deleterious mutations can accumulate in small populations. This means that, if all species are prone to occasional accidental increases in amount of DNA, genomes will increase the most in size and complexity in species with small populations. Since large-bodied species tend to have smaller population sizes, they may be most vulnerable to increases in genome size and complexity, as extra copies of functional sequences or expansion of non-coding DNA will not be efficiently removed. Of course, not all species with small populations will undergo increase in genome size and complexity. Some species may be under strong selection pressure to reduce genome complexity. Just as parasitic taxa can evolve a lower level of morphological complexity as they shed features no longer needed, so the genomes of parasitic species can slim down as they lose genes associated with independent life. Selection for rapid reproduction or small cell size could conceivably also drive genome size downwards.

When we look at the increase in body size over time, we can consider that the increasing maximum size over time, which drags the average size ever upwards, might be the result of a random walk in morphology space. Now we must consider whether the increase in organismal complexity over time might also be the consequence not of selection for superior organisms with a higher degree of organization, but as a random walk in complexity space which, in small populations, may be unchecked by the cleansing power of natural selection. Despite the revelation that we do not have the largest genome, nor the most genes, there has been a persistent tendency to assume that features of the human genome are intimately connected with morphological and behavioural complexity. If this link to complexity cannot be found in the genes themselves, then, many assume, it must be in the way that gene expression and development are regulated by a complex web of genome regulatory elements scattered throughout the intergenic regions of the genome. Furthermore, it is usually assumed that features such as complex gene regulation or multiple gene products from single genes must be the result of selection for complexity.

Just as we have considered apparently less complex species whose genomes are bigger or more gene-rich than humans, so we should consider species whose complex genome architecture and regulation are more intricate than for humans. For example, unlike humans whose mitochondrial genome is on a single circular chromosome, the mitochondrial genome of the single-celled eukaryote *Diplonema papillatum* is carried on a series of mini-circles of DNA, none of which carry a whole gene.[18] Some genes are split across a dozen of these little 'chromosomes'. RNA transcripts from these partial genes have to be matched and spliced together before they can be translated into gene products. Note that this complicated post-transcriptional splicing does not result in a more complex phenotype, because the end result is the same products that most organisms produce by simpler means.

Other single-celled eukaryotes also display bewildering complexity in genomic regulation without any apparent connection to more complex phenotypes. For example, for mitochondrial RNA genes in trypanosomes to be expressed, they must first

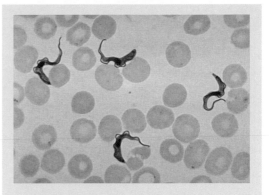

Figure 3.16 Trypanosomes (dark pink), with human blood cells (light pink). Trypanosomes are single-celled eukaryotes that illustrate the co-occurrence of simplicity and complexity—while they are morphologically simple, they can have a complex life cycle consisting of a dozen different forms occupying several different hosts, and they have some insanely complex genome management features. Mitochondrial sequences are divided among many mini-circles, and mRNA is extensively edited to add and subtract bases before being translated.[19]

be copied base for base into an RNA transcript, and then the transcripts undergo RNA editing, adding or subtracting bases from the sequence (**Figure 3.16**). This post-transcriptional editing requires the construction, maintenance, and operation of specific cellular equipment, involving hundreds of specific reaction steps.[19] Yet this complex processing has no direct effect on phenotypic outcome of gene expression, because the same RNA sequences could have been coded directly into the gene without the need for editing (as they are in most eukaryotes). Why have a complex RNA editing step in genome regulation, rather than just coding the genes with the required sequence? As in all such cases, we need to consider several possible explanations for the evolution of a complex trait: it was selected for because it conveyed a fitness advantage; it persisted by chance because it did not have any costs; or it evolved despite costs because natural selection had insufficient power to remove it from the population.

Key points

- The lack of a clear relationship between genome size, number of genes, number of gene products, and phenotypic complexity has complicated the search for connections between genomic complexity and phenotypic or behavioural complexity.

- Evolution of complexity in the genome may be driven by selective advantage of complex features, or be due to the gradual accumulation of excess DNA or more complex forms of expression that are insufficiently costly to be removed by selection. In this sense, genetic complexity might be regarded as a failure to maintain simplicity.

Conclusions

The fossil record can be used to identify evolutionary trends where a biological feature undergoes a consistent direction of change over long periods of time. One trend seen in many, but not all, animal lineages is Cope's rule, the tendency for body size to increase over evolutionary time. This trend has been attributed to both microevolutionary processes (larger individuals have a fitness advantage within populations) and macroevolutionary processes (larger-bodied lineages persist for longer or leave more descendants). It has also been suggested that the pattern of increase in body size is similar to that expected from a random process. The increase in average size is largely driven by an ever-increasing maximum size. A similar pattern is observed for increase in the morphological complexity of animals, where the increase in number of cell types over evolutionary time reflects the maximum continuing to rise while most lineages maintain or decrease in complexity. It is difficult to relate these patterns to any clear trends in genomic complexity, where there is no simple relationship between genome size, gene number, or phenotypic complexity.

Evolutionary trends are evident in the fossil record, but their interpretation is a matter of debate. If we pick a living species that is at the top end of the range of some trait, such as size or complexity, when we trace back through its ancestors we will tend to see a continuous increase in that trait. But this may provide a false impression of directionality; other lineages descended from the same ancestor may have decreased or stayed the same over time. Similarly, trends will be apparent if the ancestral state is at the low end of a trait spectrum. If the ancestor of a given lineage is small or simple, any evolutionary change over time, whether random or promoted by selection, will tend to drive the average value of that trait up. These observations do not deny the existence or importance of consistent trends in evolution, or directional change in traits over long periods of time. But detecting trends and testing ideas about their causes requires comparison with null expectations and a keen eye for sources of observational bias.

◉ Points for discussion

1. Many of the papers supporting Cope's rule use palaeontological data (include fossil species), while many neontological studies (using only comparisons between living species to infer past evolutionary patterns) fail to find support for a tendency towards size increase. Could this difference be due to different ways of measuring body size, or different timescales, or sampling biases? Or does it reveal an important evolutionary pattern?

2. Given a distribution of species' body sizes over time, how would you test whether the pattern represents a 'passive trend' (diffusion without bias in any particular direction of change) or a 'driven trend' (differential success over time leads to a consistent directional change)?

3. If it is true that the 'non-coding' majority of the human genome is responsible for generating morphological and behavioural complexity, what would you expect to see when you compare the genome sequences of complex species with less phenotypically complex species?

✳ References

1. MacFadden BJ, Oviedo LH, Seymour GM, Ellis S (2012). Fossil horses, orthogenesis, and communicating evolution in museums. *Evolution: Education and Outreach* 5(1): 29–37.

2. Semprebon GM, Rivals F, Solounias N, Hulbert RC (2016) Paleodietary reconstruction of fossil horses from the Eocene through Pleistocene of North America. *Palaeogeography, Palaeoclimatology, Palaeoecology* 442: 110–27.

3. Shoemaker L, Clauset A, Arita H (2014) Body mass evolution and diversification within horses (family Equidae). *Ecology Letters* 17(2): 211–20.

4. Heim NA, Knope ML, Schaal EK, Wang SC, Payne JL (2015) Cope's rule in the evolution of marine animals. *Science* 347(6224): 867–70.

5. Jablonski D (1997) Body-size evolution in Cretaceous molluscs and the status of Cope's rule. *Nature* 385(6613): 250.

6. Klompmaker AA, Schweitzer CE, Feldmann RM, Kowalewski M (2015) Environmental and scale-dependent evolutionary trends in the body size of crustaceans. *Proceedings of the Royal Society B: Biological Sciences* 282(1811): 20150440.

7. Kingsolver JG, Pfennig DW (2004) Individual-level selection as a cause of Cope's rule of phyletic size increase. *Evolution* 58(7): 1608–12.

8. Gotanda KM, Correa C, Turcotte MM, Rolshausen G, Hendry AP (2015) Linking macrotrends and microrates: Re-evaluating microevolutionary support for Cope's rule. *Evolution* 69(5): 1345–54.

9. Gould SJ (1996) *Full House: The Spread of Excellence from Plato to Darwin.* Harmony Press, New York.

10. Lane N, Martin W (2010) The energetics of genome complexity. *Nature* 467: 929.

11. Dobson A, Lafferty KD, Kuris AM, Hechinger RF, Jetz W (2008) Homage to Linnaeus: How many parasites? How many hosts? *Proceedings of the National Academy of Sciences of the USA* 105(Suppl 1):11482–9.

12. Vogel C, Chothia C (2006) Protein family expansions and biological complexity. *PLOS Computational Biology* 2(5): e48.

13. Nekrutenko A, Wadhawan S, Goetting-Minesky P, Makova KD (2005) Oscillating evolution of a mammalian locus with overlapping reading frames: an XLαs/ALEX relay. *PLoS Genetics* 1(2): e18.

14. Encode Project Consortium (2012) An Integrated encyclopedia of DNA elements in the human genome. *Nature* 489(7414): 57–74.

15. Parkinson CN (1957) *Parkinson's Law, or the Pursuit of Progress.* John Murray, London.

16. Lynch M (2007) *The Origins of Genome Architecture.* Sinauer Associates, Sunderland, MA.

17. Ohta T (1992) The nearly neutral theory of molecular evolution. *Annual Review of Ecology, Evolution, and Systematics* 23: 263–86.

18. Vlcek C, Marande W, Teijeiro S, Lukeš J, Burger G (2011) Systematically fragmented genes in a multipartite mitochondrial genome. *Nucleic Acids Research* 39(3): 979–88.

19. Stuart KD, Schnaufer A, Ernst NL, Panigrahi AK (2005) Complex management: RNA editing in trypanosomes. *Trends in Biochemical Sciences* 30(2): 97–105.

20. MacFadden BJ (2005) Fossil horses—evidence for evolution. *Science* 307(5716): 1728–30.

21. MacFadden BJ (1986) Fossil horses from 'Eohippus' (*Hyracotherium*) to *Equus*: scaling, Cope's Law, and the evolution of body size. *Paleobiology* 12(4): 355–69.

22. Valentine JW, Collins AG, Meyer CP (1994) Morphological complexity increases in metazoans. *Paleobiology* 20(2): 131–42.

23. Colbourne JK, and 67 other authors (2011) The ecoresponsive genome of *Daphnia pulex*. *Science* 331(6017): 555–61.

24. Ye Z, Xu S, Spitze K, Asselman J, Jiang X, Ackerman MS, Lopez J, Harker B, Raborn RT, Thomas WK (2017) A new reference genome assembly for the microcrustacean *Daphnia pulex*. *G3: Genes, Genomes, Genetics* 7(5): 1405–16.

Case Study 3
Accounting for measurement bias: has biodiversity increased over time?

At first glance, it looks as if the question 'Has biodiversity increased over time?' is easy to answer: simply count all species from different time periods, and see if the numbers go up over time. But estimating past biodiversity is very difficult, and is subject to a number of limitations and biases. We are going to use this basic question about increasing biodiversity over time as a way of exploring some of the ways that people have attempted to overcome measurement biases.

One of the earliest attempts to systematically investigate patterns of biodiversity over time was made by palaeontologist John Phillips, a contemporary of Darwin, who compiled lists of fossils from British strata, revealing an increase in diversity over time (**Figure 3.17**).[1] While

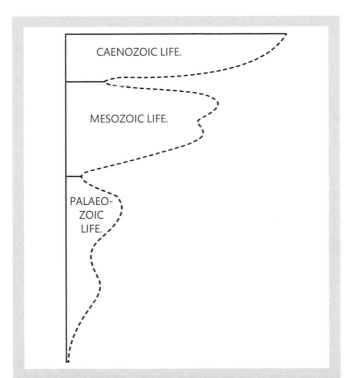

Figure 3.17 John Phillips published one of the first broad-scale quantitative evaluations of biodiversity over time in 1860.[1] Phillips grew up under the guardianship of his uncle William 'Strata' Smith, and helped him to gather data for his groundbreaking stratigraphic maps. Phillips likewise contributed to the formalization and generalization of stratigraphy by publishing the first global geological timescale.

he recognized the potential biases in the record and the uneven rates of deposition represented in different strata, Phillips had sufficient confidence in the fossil record to consider that it revealed that each major part of the Phanerozoic had more biodiversity than the last, although diversity had dipped dramatically between these three periods.

But could this increase be an artefact of the incomplete state of knowledge? Perhaps in Phillips's day the older strata had been poorly characterized, making them appear more species poor than they really were. If this was the case, then as more fossils were collected, we would expect the appearance of a biodiversity increase over time to soften or disappear. Increasing palaeontological knowledge might fill in the gaps and even out the species counts per period. A century after Phillips, the number of known fossil taxa had increased approximately 25-fold, yet the overall pattern was similar, showing an uneven increase in the earlier part of the Phanerozoic but a sharp increase towards the present (**Figure 3.18**).[2]

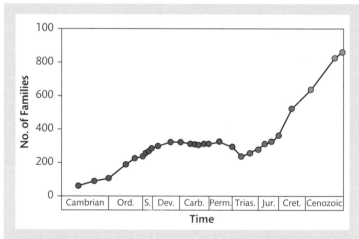

Figure 3.18 Norman Newell considered that the patterns of fossil diversity over time showed that rates of 'evolutionary activity' were not constant over time, and that many different taxa showed congruent patterns of rapid radiation and mass extinction.

From Newell ND (1963) 'Crises in the history of life', *Scientific American* 208: 76–93.

The same basic approach has been repeated many times since. Each successive attempt incorporates a greater number of described fossils and a wider geographic extent, and accounts for taxonomic corrections (where fossils are reinterpreted and assigned to different groups). Furthermore, the analytical sophistication grows with each new analysis. The most famous analysis of biodiversity over time was published by Jack Sepkoski, who spent over a decade compiling and analysing fossil occurrence data (**Figure 3.19**).[3] This iconic image is sometimes referred to as the 'Sepkoski curve'.

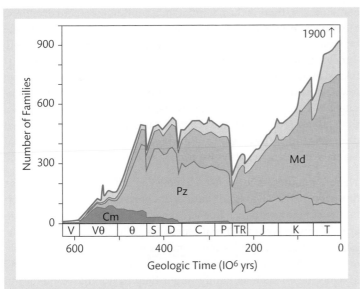

Figure 3.19 Jack Sepkoski was part of a new wave of palaeontologists taking an analytical approach to investigating patterns of diversity over time.[3] Like Newell, he considered that the results led to the conclusion that the earth's biota has suffered periodic mass extinctions. He identified three different evolutionary faunas of well-preserved marine families whose fortunes waxed and waned over time: Cambrian fauna (Cm), such as trilobites and monoplacophorans; Palaeozoic fauna (Pz), such as cephalopods and ostracods; and the Modern fauna (Md), including bivalves and vertebrates. The upper layer (in grey) is the poorly preserved families that were not clearly associated with any of the three faunas.

From Sepkoski JJ (1984) 'A kinetic model of Phanerozoic taxonomic diversity. III: Post-Paleozoic families and mass extinctions', *Paleobiology* 10(2): 246–67. Reproduced with permission of the Paleontological Society.

Subsequent redrawings of this curve, based on improved databases, have largely preserved the overall pattern, which has three main features: (a) an uneven increase in biodiversity over time, rising sharply to the present day; (b) a series of dramatic drops in biodiversity, interpreted as mass extinction events that each temporarily reduced global diversity; and (c) sharp rises in biodiversity after these drops, interpreted as radiation of surviving lineages. Here we are going to focus only on the inference of an overall increase in biodiversity over time, but we will discuss one of the inferred mass extinction events in Chapter 5 (dinosaur extinctions) and the subsequent proposed rapid radiation (mammals) in Chapter 6.

Does the data represent the biosphere?

The classic Sepkoski curve plots the number of families of marine animals. How fairly does this represent biodiversity at any given time? Marine animals are a good choice for a data-rich analysis of the fossil record. In the oceans, unlike on land, deposition is more or less continuous so there is the potential for a continuous record over time. Many marine animals have hard parts that preserve well and provide taxonomically valuable characteristics that allow them to be classified. But of course the record will never be complete. Many small or soft-bodied marine animals will not have been preserved at all or will have few records, making their evolutionary time range hard to define accurately. We expect some places, periods, and taxa to be better known than others. More broadly, we can't be sure how well marine animal diversity reflects the diversity of the biosphere as a whole. Terrestrial animals or aquatic plants or bacteria might show different patterns.

There is another issue we need to consider when interpreting these diversity curves. The unit of data is the presence or absence of each family of marine animals in each geological period. This is because species counts from the fossil record are difficult to estimate accurately,[4] so pooling species into families might give a more reliable overview. Since species assignments are also prone to change, using a higher taxon such as family might help to iron out some of the differences in species recognition between groups. But can we rely on counts of the number of families described, rather than directly counting species, as a meaningful measure of biodiversity? Families can contain very different numbers of species; while some families contain thousands of recognized species, others are 'monotypic', containing only a single species. Not surprisingly, families that have received more taxonomic attention tend to have more recognized species,[5] and the number of species per family is higher for recent taxa than it is for the oldest families (**Figure 3.20**). We also expect families that are more widespread and abundant to have a greater chance of being sampled in the fossil record. So counting the number of families for each period does not necessarily give a clear indication of the number of different species alive at the time. Opinions differ on the appropriate taxonomic level of analysis, but a number of studies suggest that you get the same basic pattern whatever taxonomic level you use. For example, the number of genera of marine bivalves provides a similar curve with increasing biodiversity over time, but with less pronounced drops at the inferred mass extinction events.[6]

Our knowledge of fossil taxa depends wholly on the hard work of the palaeontologists and taxonomists who find and describe the fossils. Sadly, there are not enough working palaeontologists to make sure that all areas of the earth and all strata and taxa within each area receive due attention. Therefore our knowledge of diversity must be to some extent dictated by the places, geological periods, and taxa that have received most attention. In fact, the numbers of palaeontologists working on each period bears a remarkable resemblance to the diversity of the period (**Figure 3.21**).[7]

Is the match between the number of palaeontologists and the diversity of different periods because there are more species (or higher taxa)

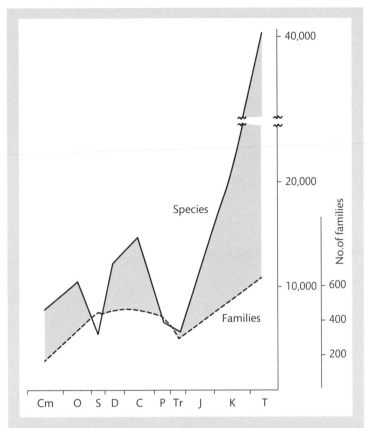

Figure 3.20 The shape of the diversity curve over time depends partly on the taxonomic level of the analysis. The curve for families of marine animals with hard parts (dashed line) is smoother than the curve for species (solid line).[4]

From Flessa KW, Jablonski D (1985) 'Declining Phanerozoic background extinction rates: effect of taxonomic structure?', *Nature* 313(5999): 216–18. © 1985. Reprinted by permission from Springer Nature.

described from periods that have the most professional attention? Or do more palaeontologists work on more biodiverse periods, because there is more to be described?[8] Can we correct the raw biodiversity accounts for some measure of sampling intensity to get a number that reflects the number of known species corrected for the number of palaeontologist-hours spent on obtaining the identifications of those species?

One of the reasons that more palaeontologist-hours are spent on more species-rich periods is that there are more fossil-bearing beds from those periods. Fossil-bearing rocks are relatively rare and unevenly distributed. Plots of amount of fossil-bearing rock by time period often closely match plots of biodiversity over time (**Figure 3.22**).[9]

The amount of available fossil-bearing rock influences measures of biodiversity because it is the fossil-bearing rocks that allow us to 'see'

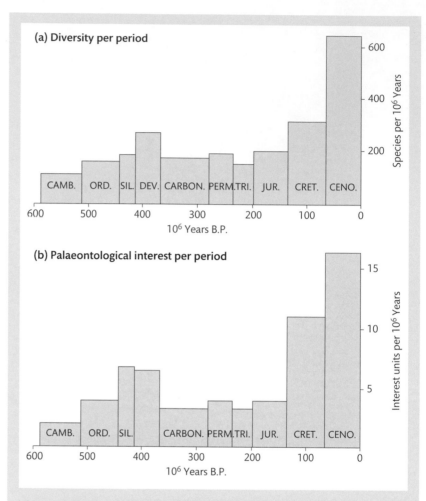

Figure 3.21 In this analysis from 1977, the number of described species per geological period (a) follows the amount of 'palaeontological interest' (b), estimated from a published directory of palaeontologists.[7] Note that the geological periods do not represent equal time durations.

From Sheehan, P. (1977) Species diversity in the Phanerozoic: A reflection of labor by systematists?, *Paleobiology*, 3(3), 325–8. Reproduced by permission of Cambridge University Press.

the species present at any given time period: the more rock, the more fossils, the more known species. Biodiversity for time periods with less fossil-bearing rock available is probably undersampled. Variation in geological processes, such as sedimentation, exposure, and erosion, influences the availability of fossil-bearing rock over time (different periods are represented by differing amounts of rock) and space (rock representing particular periods are not available in all places).[10] For example, sea-level change might influence patterns of marine diversity over time by influencing the amount of sedimentary rock available.[11]

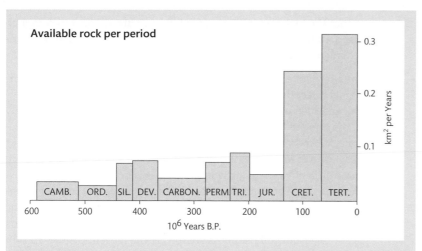

Figure 3.22 From the same analysis as Figure 3.21,[7] the available areas of rock correlate significantly with the number of palaeontologists working on each period and the number of species per period.

From Sheehan, P. (1977) Species diversity in the Phanerozoic: A reflection of labor by systematists?, *Paleobiology*, 3(3), 325–8. Reproduced by permission of Cambridge University Press.

If measures of biodiversity follow the amount of available rock or the degree of palaeontological effort applied, is it possible to detect biological changes in diversity over time from the fossil record? One approach has been to ask whether there is any variation in biodiversity that cannot be accounted for simply by palaeontological effort or rock volume. Researchers can use statistical analyses to ask how much variation in diversity can be explained as a result of sampling effort or rock volume, and then look at the remaining (residual) variation in diversity and ask whether this residual variation in diversity suggests an increase over time. This approach has resulted in some dramatic reshaping of diversity curves. For example, it has been suggested that most variation in dinosaur diversity over time can be attributed to the amount of fossil-bearing formations, and that dinosaur diversity actually declined during the Cretaceous despite an increase in described species[12] (see Chapter 5).

The problem with correcting for the number of fossil-bearing formations is that the amount of available rock might itself be connected to biodiversity, and not just a source of measurement bias (e.g. Case Study 5). For example, sea-level changes will influence not only the rock record, through the deposition or erosion of sedimentary rock, but also the available habitat for both marine and terrestrial taxa. Therefore the geological processes that influence the amount of fossil-bearing rock, such as sea-level change and plate tectonics, could also have influenced biodiversity patterns. So a 'common cause' explanation for the close match between the rock record and biodiversity is that the two are fundamentally linked in life, not just in sampling effort. In this case, correcting for rock availability or formations studied might remove patterns of real changes in biodiversity over time.

An alternative way of overcoming biased sampling effort is to randomly select the same number of observations from each time interval, and then use that standard sample size to estimate the diversity of the interval. The observations might be specimens, or fossil collections, or known fossil-bearing formations. If biodiversity really does goes up over time, then the same number of samples should contain more different species (as long as richness is not so skewed that most individuals are all from the same common species and the rest of biodiversity is very rare). When marine animal diversity is analysed using this standardized sampling procedure, the diversity curve over time is flattened, but still rises (less sharply) to the present (**Figure 3.23**).[13]

What, if anything, does the increase in fossil diversity toward the present day tell us? There are two broad ways to interpret this pattern. One is as a macroevolutionary phenomenon: if it accurately represents patterns of global biodiversity, then the increase towards the present day tells us that the biosphere is not saturated with

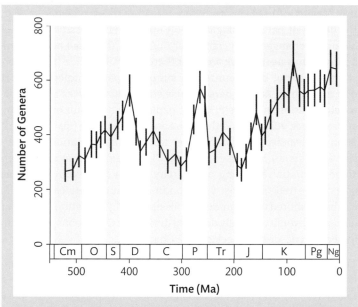

Figure 3.23 Genus-level diversity of both extant and extinct marine invertebrates based on a sampling-standardized analysis of the Paleobiology Database. Diversity for each period is estimated by sampling fossil collections for each time interval until each bin contained the same number of specimens (16,200). This curve represents the average values from repeating this sampling procedure 20 times. The authors interpret this as a 'modest rise in diversity with no clear trend after the mid-Cretaceous'.[13]

From Alroy J, et al. (2008), 'Phanerozoic trends in the global diversity of marine invertebrates', *Science* 321(5885): 97–100. Reprinted with permission from AAAS species, but that more species can and are being constantly added (see Chapter 10). The second interpretation is that the apparent rise in biodiversity a side effect of measurement: an increase in species counts close to the present is what we would expect to see, even if biodiversity is not actually increasing, because recent taxa are more likely to be counted. We have considered some possible biases toward increasing biodiversity counts in recent periods, such as greater rock volume and increasingly well-studied and informative fossils. But there is an additional bias that results from the evolutionary process itself, rather than simply being a sampling artefact.

The number of species, families or other taxonomic levels present in a given time period is typically based on inference from temporal ranges. That is, if a taxon is found in an older stage and also in a younger stage, we can also infer that it must have also been present in the biota in the intervening period even if we don't have any fossils for that stage. The more fossils we have, the better the inference of temporal range will be. If we know that a species is still alive today, we can complete the range from its first appearance to the present, even if there are no known fossils in between. Over time, species are being added to the biota by speciation and lost by extinction. The time to extinction may vary, but we can see that younger species have had less time to go extinct. Given that contemporary species are likely to be better sampled than at any period in the past, and that recent taxa might have a higher chance of being preserved as fossils and being discovered and described, we expect to be able to count more of the recent species than we can for any period in the past. The 'pull of the recent' means that we should expect a strong 'uptick' in the number of lineages recorded for time periods closer to the present day, whether counted from the fossil record or from phylogenies. Just as we need to be able to correct for the influence of rock volume or sampling effort, we need effective ways of accounting for 'pull of the recent' bias if we are to read meaningful signals of changing biodiversity over time.

Questions to ponder

1. We have seen how the diversity curve has changed as the fossil record becomes increasingly well-described. How will we know when we have found enough fossils that the picture of diversity over time is accurate and unlikely to change in future?

2. Can we be sure that the dips in the diversity curve over time are due to macroevolutionary events, and not due to bias in the fossil record such that diversity is undersampled at the boundaries between some major periods? What tests could you devise to evaluate the possibility of temporal bias in the record creating the impression of mass extinctions and radiations?

3. Can we extrapolate from marine animal diversity to terrestrial animal diversity, or the diversity over time for other organisms such as plants, or microbes? Can you use one 'indicator taxon' to predict diversity in other taxa?

Further investigation

Palaeontologists typically recognize five mass extinction events, but some consider there have been more than five mass extinctions, and some less. Why do scientists disagree on something as major as a mass extinction event? Is this disagreement due to disputed definition of 'mass extinction', or the way we measure extinction rates, or both? Do you expect all mass extinction events to share a common cause? Are we in the middle of another mass extinction event? How can we meaningfully compare current rates of species loss to past extinctions from the fossil record?

References

1. Phillips J (1860) *Life on the Earth: Its Origin and Succession*. Macmillan, London.

2. Newell ND (1963) Crises in the history of life. *Scientific American* 208: 76–93.

3. Sepkoski JJ (1984) A kinetic model of Phanerozoic taxonomic diversity. III: Post-Paleozoic families and mass extinctions. *Paleobiology* 10(2): 246–67.

4. Flessa KW, Jablonski D (1985) Declining Phanerozoic background extinction rates: effect of taxonomic structure? *Nature* 313(5999): 216–18.

5. Lloyd GT, Young JR, Smith AB (2011) Taxonomic structure of the fossil record is shaped by sampling bias. *Systematic Biology* 61(1): 80–9.

6. Mondal S, Harries PJ (2016) The effect of taxonomic corrections on Phanerozoic generic richness trends in marine bivalves with a discussion on the clade's overall history. *Paleobiology* 42(1): 157–71.

7. Sheehan PM (1977) Species diversity in the Phanerozoic: a reflection of labor by systematists? *Paleobiology* 3: 325–8.

8. Raup DM (1977) Systematists follow the fossils. *Paleobiology* 3: 328–9.

9. Benton MJ (2015) Palaeodiversity and formation counts: redundancy or bias? *Palaeontology* 58(6): 1003–29.

10. Vilhena DA, Smith AB (2013) Spatial bias in the marine fossil record. *PloS One* 8(10): e74470.

11. Smith AB, McGowan AJ (2005) Cyclicity in the fossil record mirrors rock outcrop area. *Biology Letters* 1(4): 443–5.

12. Barrett PM, McGowan AJ, Page V (2009) Dinosaur diversity and the rock record. *Proceedings of the Royal Society of London B: Biological Sciences* 276(1667): 2667–74.

13. Alroy J, and 34 other authors (2008) Phanerozoic trends in the global diversity of marine invertebrates. *Science* 321(5885): 97–100.

Why did evolution explode in the Cambrian?

4

Roadmap

Why study the Cambrian explosion?

The Cambrian explosion has been called one of the greatest unsolved mysteries in evolutionary biology. Why do so many animal phyla (the highest taxonomic division in the animal kingdom) all appear in the fossil record during a single geological period? This burst of animal diversity has been interpreted by some as marking an unusually inventive period of evolution, with a greater rate of change in fundamental features of body plan than has ever occurred before or since. Many hypotheses have been proposed to explain the near-simultaneous appearance of many animal phyla in the fossil record, including environmental conditions (such as available oxygen) and ecological triggers (such as the rise of predation). In this chapter we are going to examine only one hypothesis: rapid macroevolutionary change was facilitated by changes to a core genetic 'toolkit' of body-patterning genes. The universality, conservation, and diversity of regulatory genes, such as those in the Hox cluster, and their association with body patterning in embryos, has led to suggestions that they have played a key role in the evolution of different animal body plans.

What are the main points?

- The Cambrian explosion presents two challenges to Darwinian gradualism: explaining evolution of novel features for which we have little direct evidence of intermediate forms, and explaining why animal phyla appear in a particular time period and not before or since.

- One proposed solution to both these challenges is that body plans evolved not by a long series of gradual changes but by relatively few changes to developmental genes that allowed new distinct new forms to be generated without intermediates.

- To investigate this hypothesis, we need to know if such 'macromutations' are possible, and if it is plausible that they could give rise to a successful lineage of animals with a novel body plan.

What techniques are covered?

- **Evo-Devo:** the field of evolutionary development considers evolution of form in light of changes in gene regulation.
- **Mutation and substitution:** evolution is driven by heritable variation which must arise, rise in frequency, and become a fixed part of persistent lineages.

What case studies will be included?

- Evo-Devo: Regulatory genes and the development of body plan.

> "And Mr Darwin's position might, we think, have been even stronger than it is if he had not embarrassed himself with the aphorism, 'Natura non facit saltum' [nature does not make jumps], which turns up so often in his pages. We believe that nature does make jumps now and then, and a recognition of the fact is of no small importance in disposing of many minor objections to the doctrine of transmutation."

Thomas Henry Huxley (1863) *Evidence as to Man's Place in Nature and Other Essays*. Williams & Norgate, London.

The Cambrian explosion

Darwin's explanation for the evolution of diversity was that it formed gradually by the accumulation of many small variations. Individuals within populations vary in myriad ways: some slightly bigger, or darker, or more efficient or fecund than others. While some variation may be due to changes acquired as an individual grows and responds to its environment, or due to the random slings and arrows of outrageous fortune, at least some of this variation will be caused by change in the hereditary material, meaning that any offspring of those individuals could inherit the same altered feature. If some such variations cause individuals to be slightly more likely to survive and reproduce, then those traits will tend to be more common in the following generation than variants that conveyed a slight disadvantage. Darwin proposed that if you considered any two living species, you ought to be able to trace each of them back through a series of ancestors, each differing only slightly from the previous generation, until you reach their common ancestor: 'if my theory be true, numberless intermediate varieties, linking most closely all the species of the same group together, must assuredly have existed'.[1]

But the fossil record of animals does not always provide clear evidence of a series of gradual changes. In particular, the beginning of the animal fossil record starts with a bang, in an astounding burst of diversity referred to as the 'Cambrian explosion'. Instead of bearing witness to a gradual diversification, with a stepwise building of forms by incremental change from a common ancestor, the major different types of animals such as arthropods, annelids, and molluscs all appear more or less simultaneously in the fossil record around half a billion years ago (**Figure 4.1**). In Darwin's day, these early animals represented the start of the known fossil record. But the fossil record was so evidently

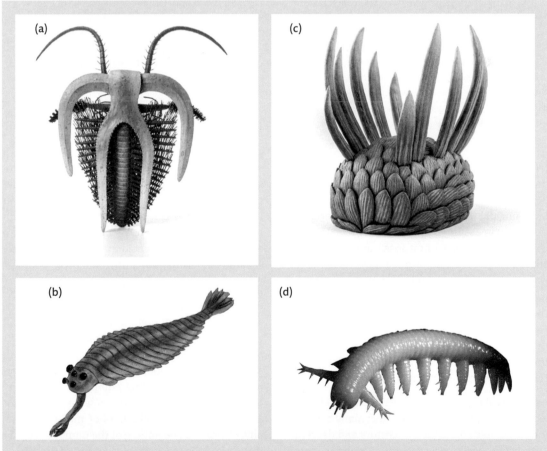

Figure 4.1 The Cambrian explosion marks the first undisputed appearance in the fossil record of members of animal phyla such as arthropods (a) *Marella*; (b) *Opabinia*, molluscs (c) *Wiwaxia*, and onycophorans (d) *Aysheaia*.

(a) Matteo De Stefano/MUSE via Wikimedia Commons (CC3.0 license). (b) Jose Manuel Canete via Wikimedia Commons (CC4.0 license). (c) Matteo De Stefano/MUSE via Wikimedia Commons (CC3.0 license). (d) © Citron via Wikimedia Commons (CC3.0 license).

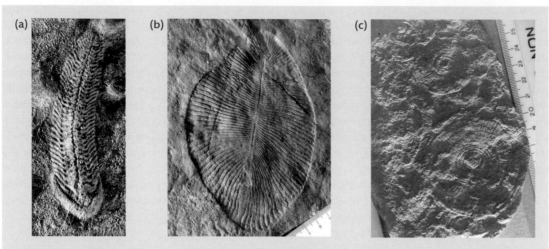

Figure 4.2 The Ediacaran fauna is known from locations all over the world from the latest Precambrian period (579–542 million years ago). Since their discovery, Ediacaran fossils such as (a) *Spriggina*, (b) *Dickinsonia* (b) and (c) *Cyclomedusa* have variously been interpreted as ancestral animals, or modern metazoans, or a sister group to modern animals, or a completely independent experiment in multicellularity. The Ediacaran faunas show a different form of preservation to the later Cambrian animals; they are preserved as impressions, possibly between a microbial mat covering the sea floor and a layer of sediment. It may be that the conditions in the late Proterozoic—particularly the lack of burrowing animals—allowed this preservational window, which then closed with the onset of the Cambrian.

Photos by Verisimilus via Wikimedia Commons (CC2.0 license).

patchy that Darwin was confident that the burst of animal diversity in the Cambrian represented the beginning of the recorded history of the animal kingdom, not the actual beginning of animal evolution. Darwin assumed that the intermediate forms that connected these first fossil animals to their shared common ancestor must have existed, even if they were hidden from our view: 'Consequently, if the theory be true, it is indisputable that before the lowest Cambrian stratum was deposited, long periods elapsed, as long as, or probably far longer than, the whole interval from the Cambrian age to the present day; and that during these vast periods the world swarmed with living creatures'.[1]

As the fossil record became increasingly well studied, many gaps were filled in, and many series of intermediates connecting ancestors and descendants were described. Yet the Cambrian fossil record retained its explosive character, without a clearly defined chain of intermediates between the major types of animals. When older fossils were discovered—the enigmatic Ediacaran assemblage—they served to heighten the mystery rather than solve it (**Figure 4.2**). Debate continues over the affinity of the Ediacarans, with some researchers recognizing in them the ancestors of modern animals, but others considering them an entirely separate experiment in multicellularity.

Whether or not they are on the direct line of descent of the animal kingdom, the Ediacarans appear to be a different grade of organization to the Cambrian animals. The Ediacarans are apparently all soft-bodied, without evidence of hard armour, and, with a few possible exceptions, no sign of directed movement, specialized feeding equipment, or sophisticated sense organs. If there were arthropods, annelids, and other members of modern animal phyla before the beginning of the Cambrian, either they lacked the recognizable features of those body plans—such as legs, eyes, shells, mouthparts, or segmentation—or for some reason they were not preserved alongside the soft-bodied Ediacarans.

Furthermore, the trace fossil record also suggests a remarkable transformation, with the number and complexity of burrows, tracks, and traces increasing over the transition from the Precambrian to the Cambrian period. In fact, the start of the Cambrian period is defined by the appearance in the record of a particular trace fossil which provides the first clear evidence of animals burrowing down into the sediment. The first evidence of hard parts also appeared around the same time, with the 'small shelly' fauna post-dating the Ediacarans but just preceding the appearance of members of modern phyla.[2] While some of these represent the complete shells of tiny organisms, others are likely to be scales or other hard parts from larger animals.

Molecular dates complicate this picture of explosive radiation because they tend to place the origin of the animal phyla long before the Cambrian, suggesting a hidden history of diversification. But, unlike Darwin and his contemporaries, we can no longer comfortably explain away the Cambrian explosion by assuming a long period of gradual change hidden from view by the absence of fossils. The fossil record paints a picture of a transition from simple, sedentary, passive-feeding, soft-bodied creatures to a multilayered ecology of sophisticated animals with directed locomotion and sense organs, and with innovations in structure and behaviour associated with diversification into a wide range of niches. The evolution of hard parts provided implements for those who would eat others, and armour for those who would avoid being eaten. Animals moved up into the water column and down into the sediment. As animals diversified, they exploited a greater range of resources, modified the environment, and constructed webs of interactions between species. Even if the deep lineages go into the Proterozoic, we still need to explain why there is a rapid expansion of diversity and complexity over the Precambrian–Cambrian boundary.

Nothing like it before or since

The evolution of sophisticated animals or complex ecologies is not in itself surprising. We know that complex animals with hard parts evolved because we see them around us today, and they must have evolved from simpler soft-bodied ancestors.

What is surprising is the relative speed. The first undisputed members of modern animal phyla all appear in the early to mid-Cambrian, over a period of approximately 10–20 million years. While this may seem like a long time, the explosion of diversity is striking because there are no obvious parallels at any other time. There are no other geological periods in which so many animal phyla make their first appearance. In fact, almost all animal phyla are reckoned to have their beginnings in, or not long after, the Cambrian explosion. This has been interpreted as a sign that the Cambrian period was characterized by a remarkable evolutionary inventiveness where fundamentally new ways of building animal body plans were established, with all evolution subsequent to the Cambrian being merely modification of these existing body plans.

To explore this remarkable period of change, we need to unpack several aspects of the Cambrian explosion. First, we need to consider what we mean by a body plan, and how it differs between animal phyla. Second, we need to consider the mechanisms that could generate body plan differences between animal lineages. Third, we need to ask why the origins of most animal phyla are dated to the Cambrian explosion, with few additional phyla appearing in the fossil record before or since. It is the second point that will be our main focus in this chapter, because it gives us a chance to think about the role that generation of variation plays in the tempo and mode of evolution. But, to frame that debate, we need to start with the first question: what is an animal phylum?

Phylum and body plan

Phylum is a taxonomic category: it splits the animal kingdom into groups of organisms that are united by a common history and share fundamental features of organization. For example, the phylum Mollusca is characterized by unsegmented bodies with a soft mantle that excretes calcareous substance to form a shell, spicules, or plates (along with other features of the nervous and digestive system)—so these features are assumed to have been present in the most recent common ancestor of all living molluscs and then were inherited by its living descendants. Arthropods, by comparison, have segmented bodies with chitinous hard parts that form a jointed exoskeleton with paired appendages.

The key features that define the shared architecture of members of an animal phylum are commonly referred to as the body plan (sometimes called the bauplan). The notion of a body plan has an instinctive appeal; we can all recognize the fundamental difference between a soft-bodied mollusc leaving a silvery trail with its muscular foot and an arthropod scuttling past on jointed legs. But all animal characteristics evolve: old features are lost, new features form. And if we compare any taxonomic categories, we will find that they differ in a number of traits. So, of all the many differences between animals in form, function, and behaviour, is there something special about the body plan? Is 'phylum' just an arbitrary division of the animal kingdom into higher-level categories, or is there something special about phyla and the way that they evolved?

One way to define body plan is those features that define phyla, being shared by all of the members of a phylum and not present in members of other phyla. There are several problems with this approach. One is that it is somewhat circular: we must first define phyla, then look for any traits that are invariant within the phylum, and then call those invariant traits the phylum-defining body plan.[3] Other fundamental features of animal design won't be counted in the body plan if they vary within phyla. For example, there are many kinds of eyes found within the mollusc phylum, ranging across light-sensitive cells in bivalves, simple ocelli in chitons, stalked eyes in snails and camera-type eyes in cephalopods, so none of these can be assigned to a phylum-invariant body plan. Innovations within a phylum will also not be counted under this definition. We might think of wings as a fundamental innovation in body design, but wings evolve within phyla and do not represent shared common features of any whole phylum (not surprising, if phyla all originated at a time when all animals were aquatic). Another problem with assigning body plan features to phyla is that it is surprisingly difficult to pin down characters that are found

Figure 4.3 Two arthropods, one of which shows classic features of the arthropod body plan (the crab *Carcinus maenas*), with a segmented body, jointed exoskeleton and paired appendages, and one of which has none of these features (the parasitic barnacle *Sacculina carcini*, appearing as a pale blob emerging from the base of the crab's abdomen).

Photo by Auguste Le Rou via Wikimedia Commons (CC4.0 license).

in all members of a single phylum. For example, some molluscs, such as octopuses, lack a shell, and some arthropods, such as parasitic barnacles, lack an exoskeleton and paired appendages (**Figure 4.3**).

Another way to define body plan characters is the set of discrete characters that represent distinct forms of animals, with no intermediate states. Phyla have been regarded by many as representing distinct modes of organization, separated by 'bridgeless gaps.' For example, we may say that body plan characters such as paired appendages, calcareous hard parts, or segmentation are all-or-nothing phenomena—either you have them or you don't. Unlike continuous characters like size, or features with different organizational grades like eyes, there is no graded series of intermediate states between discrete characters like number of legs. This throws the challenging aspects of body plan evolution into sharp relief: how can discrete characters such as these evolve gradually from a common ancestor in a series of small modifications? How can a species have slightly paired appendages or 4.1 legs?

Perhaps the impression of body plan as consisting of discrete states with no intermediates is, at least in some cases, due to failure to identify intermediates, either in the living fauna or in the fossil record. For example, the Cambrian animal *Opabinia* was once considered to have such peculiar body plan features—such as five stalked eyes and a single enormous feeding appendage that resembles a vacuum cleaner hose with a claw attached—that it could not be fitted into any known phylum. It has since been reinterpreted as falling somewhere between the living phyla of Arthropoda and Onycophora (velvet worms), having some but not all of the features of each phylum (Figure 4.1(b)).[4] One way of approaching this issue is to consider body plan evolution in light of phylogeny; if phyla are hierarchically related, so that each shares more features with close relatives than with more distant relatives, this suggests that body plan characters were acquired stepwise along lineages, even if we have no direct evidence of taxa with intermediate states between living phyla.[5]

So we must consider two different views of the nature of animal phyla. One is that they represent distinct body plans, formed by an evolutionary process that generated few recognizable intermediate states. The other view is that phyla are a convenient way of dividing animal diversity into categories, and that the appearance of 'unbridgeable gaps' between phyla is due to a lack of informative fossils or difficulty in identifying close relatives of existing phyla from the known Cambrian fossils. If the first view is accurate, then phyla represent a fundamental level of biological organization, or the products of a special evolutionary process. If we accept the second view, then we would expect that, while some divisions into phyla are clear, in other cases we will have more trouble drawing clear lines between phyla. In fact, Darwin used a similar argument about the arbitrariness of species to support his gradual view of evolution. He cited the fact that taxonomists argue over which groups should be defined as species, and which are only sub-species or races, as evidence that species form gradually, creating a range of intermediate stages that might make clear divisions problematic.

Defining phyla

The examples of phyla we have used so far are distinct and excitingly well equipped, such that we can easily recognize the fundamental differences between them. But many animal phyla are small and soft-bodied, often with a generally 'wormy' appearance.[6] In such cases, the alignment of body plans to phyla can be less than obvious, and some such groups have been moved around the animal kingdom as they are reinterpreted and reclassified. Some types of animals that had been previously considered to be a phylum have been subsumed into existing phyla. For example, the spoon worms (Echiura) and acorn worms (Sipuncula) were demoted from phyla to classes when DNA evidence showed that they nested within the phylum Annelida, suggesting that they had lost the defining annelid body plan characteristic of segmentation.[7,8]

In other cases, animals once grouped within an existing phylum have been promoted to their own phylum. *Xenoturbella* has a simple body plan, with only two cell layers, no brain, no eyes, no gonads, no respiratory system, and not even an anus—in fact no defined organs of any kind (Figure 4.4). DNA evidence originally placed it within the phylum Mollusca, but it became apparent that this placement was the mistaken result of contamination, with those first DNA sequences representing the animal's breakfast, not its body. Then *Xenoturbella* was elevated to its own phylum (Xenoturbellida), allied to deuterostomes such as Echinodermata and Hemichordata. Recently, *Xenoturbella* has been

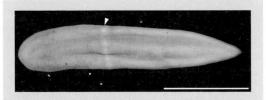

Figure 4.4 *Xenoturbella* is a marine animal with a rather simple body plan. The four currently known species range from a few centimetres to 20cm long. All almost entirely lack any definable body plan features (no eyes, no appendages of any kind, no gut, no anus, no brain, no nervous system, no gonads), and it has been described as resembling a 'crumpled sock'.

Photo from Nakano et al. (2017). 'A new species of *Xenoturbella* from the western Pacific Ocean and the evolution of *Xenoturbella*'. BMC Evolutionary Biology 17: 245.

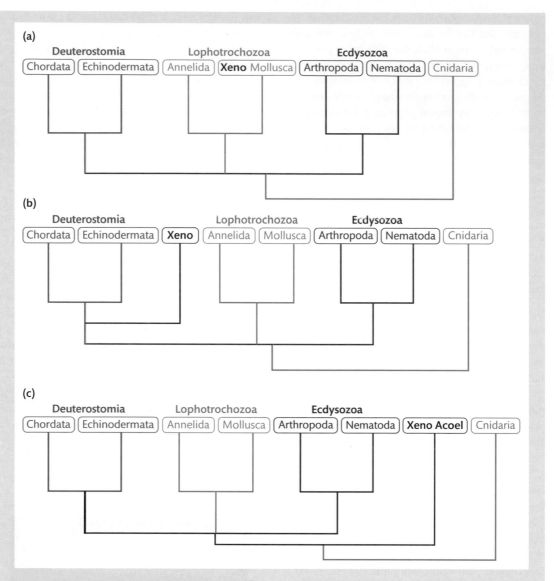

Figure 4.5 Phylum of fortune: the changing place and taxonomic status of *Xenoturbella* in the tree of animal phyla. Only a few key phyla are included here for illustration (phyla marked by boxes). (a) Based on DNA evidence, *Xenoturbella* was first placed within the phylum Mollusca, but this result was later attributed to DNA contamination from its diet. (b) New analyses placed it as a sister group to Deuterostome phyla; this placement elevated *Xenoturbella* to its own phylum, Xenoturbellida. (c) More recently, *Xenoturbella* has been grouped with the Acoelomorpha, and this group has been placed at the base of the tree of other bilaterian phyla—this lineage has been given the phylum name Xenacoelamorpha.

moved to a much deeper position in the metazoan tree, grouped with another bunch of enigmatically simple marine animals (the Acoelomorpha), and is now classified as a sub-phylum (**Figure 4.5**).

The case of *Xenoturbella* highlights the role of molecular phylogenies in reorienting the definition of phyla. If DNA sequence analysis connects a taxon as a sister lineage to another phylum, then convention dictates that it must also be regarded as a phylum, regardless of whether it has obviously distinct body plan characters. To complicate matters further, most of the 'minor phyla', with their small soft bodies, are exactly the kind of animals that rarely leave informative fossils. These are the 'silent majority' of the animal kingdom who may have existed in swarms without leaving any trace in the rocks for us to find.[6] The elevation of these lineages to phylum status increases the explosiveness of the Cambrian, because even if they are not found in Cambrian deposits, the origin of these phyla must date to the same time as their more fossilizable sisters.

There are two important lessons to draw from the changing fortunes of animal phyla. One is that there is a tension between the recognition of phyla as fundamental biological units, consisting of distinct forms generated by a special evolutionary process, and phyla as arbitrary units of taxonomy, convenient categories in which to organize different kinds of animals, or ways of dividing up the phylogenetic tree by slicing it at deeply branching lineages. And the other lesson is that we may find it challenging to consider the evolution of discontinuous traits within a Darwinian gradualist framework if we cannot observe, or even imagine, intermediate steps between fundamentally different body plans.

The origin of animal phyla as a macroevolutionary puzzle

The origin of animal phyla has been considered by many, including Darwin himself, to provide a challenging case against which the universality of Darwinian gradualism must be tested. Darwinian evolution predicts a chain of intermediates between any two forms, each representing a modification of the existing morphology or behaviour, with each new variant being either beneficial to survival and reproduction or at least not very harmful to an individual's chance of passing on its genes. Not only that, Darwinian evolutionary theory is built on the uniformitarian principle of explaining past changes using processes we can witness in action, not invoking special forces

that operated only in the past and are not directly observable today (see Chapter 1).

The Cambrian explosion presents two challenges to this gradualist view. One is the challenge of explaining the evolution of distinct forms for which we have no chain of intermediates from which to build a series of ancestors and descendants, and where we may find it difficult to even imagine possible intermediate states between discrete characters. And the other is that the tempo and mode of evolution in the Cambrian explosion seems so different to that inferred from subsequent periods that some researchers have doubted that we could explain it using processes currently in operation, and must instead invoke special macroevolutionary mechanisms, rejecting uniformitarianism.[9]

There have been many possible solutions proposed to the Cambrian conundrum, including the suggestion that the apparent 'leaps' in animal evolution at this time are due to bias in our reconstruction of evolution.[5] But for the purposes of this chapter, we are going to consider a hypothesis that has been proposed as a single solution to both the apparent rapidity of change in the Cambrian, and the perceived lack of intermediates between body plans. It has been suggested that the explosion of body plans in the Cambrian was driven by radical developmental changes that resulted in large changes in body organization, allowing new body plans to arise in relatively few evolutionary steps. This macromutational hypothesis for the origin of body plans aims to explain both the speed at which phyla arise (if they do so in a few large jumps rather than a long series of small steps), and the apparent discontinuous nature of body plans (jumping to a new body plan creates no intermediate states). But the macromutational hypothesis also faces some serious challenges. If phyla can form by macromutation, why did these macromutational changes affect animals in the Cambrian period and not throughout history? And how could an animal be radically altered in form yet still be viable and able to contribute its novel properties to the next generation? To investigate the plausibility of this hypothesis, we need to first ask if such large changes are possible, and then we must ask ourselves whether it is likely that they drove body plan evolution. To do this, we have to consider the role of mutations in macroevolution: how mutations are generated and how those changes come to characterize new lineages.

> ## Key points
>
> - The sudden origin of animal phyla in the fossil record has been considered a challenge to the Darwinian explanation of the evolution of biodiversity. Instead of a long series of ancestral taxa representing intermediate stages between different kinds of animals, the first recognizable members of most fossilizable animal phyla appear more or less simultaneously.
>
> - The relative morphological, behavioural or ecological simplicity of the Ediacaran fossils, the increase in number and sophistication of trace fossils at the end of the Ediacaran, and the appearance of small shelly fauna suggests a rise in animal complexity over the Precambrian–Cambrian boundary.
>
> - There is no other period in geological history when so many animal phyla appear at once, prompting some researchers to suggest that the Cambrian represents a particularly inventive period in evolutionary history, and that we must invoke special mechanisms to explain why it occurred then and not before or since.
>
> - One proposed mechanism for rapid body plan change is the modification of gene regulatory networks, allowing large changes in body plan from relatively few genetic changes. We are going to examine this particular hypothesis for the Cambrian explosion as a case study in thinking about the role of mutation and variation in the evolution of novelty.

Mutation drives evolutionary change

> ❝ *If Darwin did not solve the problem, he gave us the hope of a solution—that these questions may be profitably addressed.* ❞
>
> **William Bateson** (1894) *Materials for the study of variation treated with especial regard to discontinuity in the origin of species.* Macmillan, London.

The classical Darwinian view of mutation is that it is an ever-turning engine of variability. Because the availability of heritable variation was a core aspect of his theory, Darwin spent decades gathering information on variation, for example by breeding fancy pigeons, cross-pollinating garden plants, and writing to a worldwide network of naturalists, animal breeders, and horticulturalists (see Figure 1.4). As part of his detailed study of the natural world, Darwin spent nearly a decade dissecting and comparing barnacles, the result of which was his declaration that he could find no feature that did not vary between individuals in some slight way.[10] In particular, he noted that any feature that could be found to vary between species would also be found to vary within species. For Darwin, this was evidence of gradual evolution at work: the variation among individuals in populations is the raw material from which the differences between species are constructed.

Mutations—heritable changes to the genome—are constantly generated because no copying process is perfect (see Chapter 2). For most organisms, every reproduction event involves the generation of copy errors in the genome. In addition, DNA is a complex biomolecule that is vulnerable to molecular damage. If damage is repaired imperfectly, the sequence of bases that carries the genetic information may be permanently altered. DNA copy mutations are mistakes, and DNA damage is due to accidents. They are not modifications made deliberately for a particular purpose, so they will be random with respect to fitness. In fact, most mutations will either have no effect on the organism, or will be bad for it. Possibly very bad indeed—many random changes to important sequences will result in non-functioning organisms that cannot live, grow, or breed. These bad mutations won't be passed on. But the ones that have little or no effect can be passed on. And the occasional lucky change that just happens to accidentally make the organism more likely to survive and reproduce has an increased chance of finding its way into the gene pool of the next generation.

The neo-Darwinian view of evolution typically sees mutation as a constant source of trial and error, modifying existing organisms in all possible features in all possible directions. Most such mutations will be discarded when their carriers fail to survive or breed. Random generation of variation has the effect of exploring design space around existing organisms—in what ways can the organism be altered and still survive? If any of these ways are better than what is already there, then that mutation might persist and eventually become a standard part of the genome of that species. By viewing mutation as a constant undirected exploration of possibilities, this perspective on evolution rests on the assumption that if there is a possible change that is one biochemical step from where an organism is now, then, given enough time and a large enough population, it will probably occur at some point. In this sense, mutation is 'backgrounded' in the classical neo-Darwinian view of evolution—a constant source of possibilities that neither limits nor promotes any particular evolutionary direction.

What do you think?

Is randomness of mutation a fundamental part of evolution? All evidence suggests that DNA mutations are undirected with respect to outcome: they are mistakes and damage that happen without being directed at making the organism better. But could an organism evolve that was able to skew mutations towards more positive outcomes, avoiding harmful mutations and promoting advantageous changes? Would such an organism still evolve by Darwinian evolution? Do other evolving systems have random mutation? For example, languages evolve, but language change is sometimes deliberate and directed: does this make language evolution fundamentally different to biological evolution, or is this simply a difference in the supply of variation not in the process of evolution itself?

The Cambrian explosion throws into focus another fundamental challenge to the Darwinian view of the nature of variation and change. Darwin's revolutionary idea was that there was a single evolutionary process that generated all diversity: the variation always present in populations accumulates until there is sufficient difference for the population to be recognized as a distinct species. Like Lamarck, Darwin thought that it was often hard to draw a clear line between varieties, sub-species, species, and genera, which was why taxonomists were in a constant state of dispute over which groups deserved to be recognized as species (Chapter 1, Figure 1.1). The Darwinian theory is a unified view of evolution. The evolutionary process within populations is the same process that generates species: macroevolution is simply the cumulative effect of long periods of microevolution. From this perspective, studying the process of change within natural or laboratory populations gives us insight into the changes we can't usually witness, such as the production of new species.

But, from Darwin's time onward, some biologists have been dissatisfied with the unified view. Instead, some consider that species and higher taxa are often distinct, separated by clear gaps, and that the features that define species and higher taxa are of a different kind than the variation that arises each generation within populations. Are there particular kinds of mutations that trigger the formation of new species, or even new phyla? Could the Cambrian explosion be due to the generation of macromutations that make radical changes to body plan?

Macromutations

❝ *The decisive point is the single change which affects the entire reaction system of the developing organism simultaneously, as opposed to a slow accumulation of small additive changes.* ❞
Richard Goldschmidt (1940) *The Material Basis of Evolution.* Yale University Press, New Haven, CT.

Some biologists have proposed that the differences between species are generally not the same kinds of variants that are constantly arising in populations through random mutation. Instead, they argue that new species, and higher taxa like phyla, might be generated by special kinds of heritable variation. If this is the case, then, counter to the Darwinian position, we might not learn much about the origin of distinct types by studying the fate of small continuous variations that are constantly found in populations. A strong advocate of this 'two-speed' view of evolution was William Bateson (see Chapter 1, page 14), one of the first geneticists (in fact, we might say he actually was the first geneticist since it was he who invented the term 'genetics').

Bateson agreed with Darwin that natural selection could shape traits by selecting the continuous variation that is constantly arising in populations. But he did not feel that this could provide a full explanation for the evolution of species diversity, because differences between species were not always in continuously varying traits, but instead were characterized by discrete forms: 'Species are discontinuous; may not the Variation by which Species are produced be discontinuous too?'[11] To Bateson, there was not, as Darwin had suggested, one single unified evolutionary process, but two: natural selection acted on continuous variation within populations, but new species were produced by discontinuous variation.

The production of continuous variation is observable; in both natural populations and domestic breeds, individuals vary in size, shape, and temperament, and some of this variation is inherited by their offspring. Can we observe the origins of discontinuous variation? Bateson studied 'meristic' variation that caused individuals to have different numbers of existing features (e.g. more body segments) and 'homeotic' mutations that transformed one body part into another (e.g a second row of petals where sepals would normally grow). In doing so, Bateson was one of the first geneticists to study the contribution of macromutations to the formation of new biological types. But he was certainly not the last.

Richard Goldschmidt (1878–1958) also studied developmental genetics, and, like Bateson, he considered that Darwinian evolution was all very well for explaining the variation in continuous traits in populations over time (microevolution) but that it could not provide an adequate explanation of the origin of species or higher taxa (macroevolution). Goldschmidt suggested that single genetic changes, such as alteration of developmental genes or chromosomal rearrangements, could occasionally produce a 'hopeful monster', an individual that radically departed from the usual phenotype yet was sufficiently well integrated to survive and breed. Such changes could be very rare, at a frequency low enough to be effectively unobservable in normal circumstances, and yet still have a profound effect on evolution by starting a distinct new lineage: 'a single mutational step affecting the right process at the right moment can accomplish everything'.[12]

Goldschmidt's ideas were largely rejected by his contemporaries. But investigations into the possible role of developmental changes in body plan evolution underwent an enthusiastic resurgence with the comparative study of developmental genes in different animal phyla, starting in the 1980s and 1990s and continuing to the present day. Just as the work of Bateson, Goldschmidt, and others had shown, these new studies of developmental genes demonstrated that dramatic phenotypic changes could be generated by changing the expression patterns of regulatory genes that controlled developmental processes. Of particular interest were homeotic mutations that swapped the position or identity of body parts, like Goldschmidt's 'Antennapedia' mutant flies that grew legs where their antennae should be, and 'Ultrabithorax'

Figure 4.6 Changing Hox gene expression changes segment identity in fruit flies. While normal flies have one pair of wings and one pair of halteres (a), these mutants have two pairs of wings (b) or two pairs of halteres (c).

Ian Duncan and Geoffrey Montgomery, 'E.B. Lewis and the bithorax complex: Part I', *Genetics* April 1, 2002 vol. 160, no. 4 1265–72. Republished with permission of Genetics Society of America; (parts (b) and (c) courtesy of E.B. Lewis).

mutants that grew two pairs of wings instead of one (**Figure 4.6**). Could mutations such as these give rise to new body plans in a single mutational leap? Might this be the solution to the mystery of the Cambrian explosion of animal phyla?

Regulatory genes in evolution

There is an emerging theme here: many of those who have promoted the idea of macromutations as an evolutionary force have studied development (not just of animals but also of plants). Development is the process by which the organism grows by cell division and differentiation. This is a remarkable process. A single cell, the fertilized egg, contains all the information it needs to produce a complex body with many different tissue types, organs, and structures. At every cell division, each daughter cell receives an entire copy of the genome. So if every cell carries the same genetic information, how do they end up developing different features, with some becoming skin cells, some nerve cells, and so on? A complex network of gene regulation pathways switch different genes on or off in different cells. Without proper regulation, the embryo could not grow into an organized body with specialized tissues for performing different functions. Now consider the practical upshot of this arrangement: every cell has the information it would need to grow any feature. So tweaking the regulatory switches that call upon different genes to act at different times can change cell fate and tissue identity, using the same underlying genetic information in different ways to make different structures.

We do not have to look far to appreciate the role of developmental genes in producing remarkably different phenotypes from the same genome. Sex differences within species generally rely on developmental genes to guide the formation of different morphologies. Both sexes share most of the same genes, but the developing female embryo must call upon a different set of genes to act than the male (Case Study 4). Other examples where different morphologies are produced from the same genome by triggering different developmental pathways include mimicry in butterflies and castes in termites. Could changes to genes that switch developmental pathways on and off lead to the rapid evolution of radically different phenotypes?

The developmental genes that have been most often discussed as possible triggers of rapid body plan evolution in the Cambrian are those that generate homeotic mutations, which affect the identity of whole body parts. We have seen how the pioneers in these fields studied homeotic mutations in plants and animals through their effect on individual phenotype. But with the advent of gene sequencing and precision mapping of expression patterns, studies of homeotic mutations could move beyond their phenotypic expression. Developmental genetic studies in the 1980s and 1990s revealed the rather shocking observation that the same basic set of regulatory genes were involved in specifying aspects of body plans in many diverse animals.

In particular, much attention has focused on the Hox genes, a particular class of regulatory genes which have some intriguing properties. In the geneticists' favourite study animal, the fruit fly *Drosophila*, eight Hox genes, arranged in two clusters along one chromosome, are active in shaping segment identity (**Figure 4.7**). Their expression patterns help to define the boundaries of body segments in the developing embryo and dictate what features those segments will have, such as legs on all three thoracic segments, wings on the second, and halteres on the third. It was loss of function of particular Hox genes that led to the dramatic homeotic mutants that Goldschmidt studied, such as Antennapedia and Ultrabithorax, so, following tradition, these genes were each named for the mutant phenotype that occurred when their normal function was disrupted.

But Hox genes are not just about making flies. The Hox genes in *Drosophila* match those in vertebrates, where they help to define the sections of the vertebral column (Figure 4.7). In fact, all animals have versions of the Hox genes, or related genes. To understand why Hox genes have been considered to play a critical role in animal body plan evolution, we need to look at some key features of Hox genes: their duplication into many copies, the remarkable conservation of a core sequence (the homeobox), and their fascinating combination of conserved functions in some aspects of body plan, and co-option into novel body plan features.

Hox genes: same but different

Since all animal phyla have Hox genes (with the possible exception of sponges, which have related ParaHox genes), the last common ancestor of all animals must also have had Hox genes, so their origin predates the diversification of phyla (see Case Study 2). But Hox genes show a pattern of duplication and loss across the animal kingdom, such that living animal phyla show a great diversity of number and type of Hox genes. This has led to the suggestion that gene duplication of these body-patterning genes could have provided the genetic substrate for the rapid evolution of animal phyla in the Cambrian by allowing the rapid evolution of new structures through rewiring the developmental pathways that specify body plan.

Gene duplication plays an important role in evolution. Making two copies of an essential gene can free one copy to evolve new functions (neofunctionalization), or allow the two copies to each specialize in different roles originally performed by a single ancestral gene (subfunctionalization). For example, the genes that specify different taste receptors in vertebrates have evolved from duplication and divergence of a particular family of signalling genes (*Tas1R* and *Tas2R*), and different species have different sets of genes generated by duplication and loss. Diversification has allowed these taste receptors to specialize to particular tasks, such as sweet, bitter, or salty taste. The fate of many gene copies is to decay into redundancy; if their function can be adequately covered by just one copy of the gene, then any mutation that knocks out the other copy will have no fitness cost to the organism, creating non-functional pseudogenes. For example, there is little fitness advantage in being able to taste sugar if you are exclusively a meat-eater, so the sweet taste gene has decayed into a pseudogene in many different carnivorous species.

Hox genes show a similar pattern of duplication, diversification, and loss. Through duplication of ancestral Hox genes, their many copies have taken

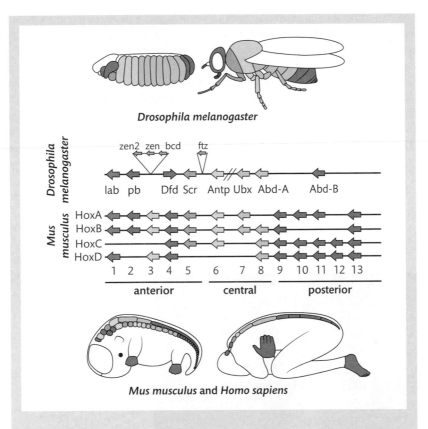

Figure 4.7 Hox genes contain sequences that are sufficiently conserved that they can be matched in distantly related animals such as flies, mice, and humans. Moving along the Hox cluster in *Drosophila*, you find genes associated with the head and mouthparts (Labial, lab; Proboscipedia. Pb; Deformed, Dfd), the first thoracic segment (sex combs reduced, Scr), second thoracic segment (Antennapedia, Antp), third thoracic segment (Ultrabithorax, Ubx), and abdominal segments (Abdominal A and B, Abd-A, Abd-B). You can also see that in vertebrates (here represented by a mouse embryo and an adult human) there are many more Hox genes, resulting from duplication of particular genes, as well as multiple duplications of the whole cluster. However, some duplicate copies have evidently been lost, resulting in gaps in the clusters. Some duplicate Hox genes in *Drosophila* have taken on new roles, such as fushi tarazu (ftz, 'segment deficient' in Japanese), zerknullt (zen, 'crumpled' in German), and bicoid (bcd). These genes act at a much earlier stage of embryo development than most Hox genes, before segment identity is established. Despite their very different body plans, Hox genes in humans show some similarity of expression pattern with those in the fly, being associated with 'segments' of the spine and nervous system. In some lineages, such as in flies and humans, the lineal patterning of Hox genes has been maintained, along with their serial expression patterns. But in other lineages, the collinearity of genes and expression has been broken as the Hox genes have become scattered around the genome.[24]

Redrawn from Hueber SD, Weiller GF, Djordjevic MA, Frickey T (2010) 'Improving Hox Protein Classification across the Major Model Organisms'. *PLoS ONE* 5(5): e10820.

on a range of roles in different animal phyla, such as specifying segment identity, directing limb formation, and the development of internal organs. Sometimes duplications create a copy that can take on a new role in development. For example, fushi tarazu (ftz) in *Drosophila* is a copy of a Hox gene, but it has changed its role, acting to delimit segments in the early embryo rather than specifying segment identity. The pattern of Hox genes in different phyla shows that some copies of Hox genes have either disappeared or changed so much that they are unrecognizable, resulting in 'missing'

Hox in some lineages (Figure 4.7). But while duplication of Hox genes may have contributed to the array of functions of this class of regulatory genes, Hox gene diversification does not always result in body plan evolution. For example, Hox genes seem to have undergone an expansion in the genome of the sea anemone *Nematostella* without any obvious sign of morphological innovation (**Figure 4.8**).[13]

Despite the differences in number and function of Hox genes among animals, some parts of these developmental genes are highly conserved. One

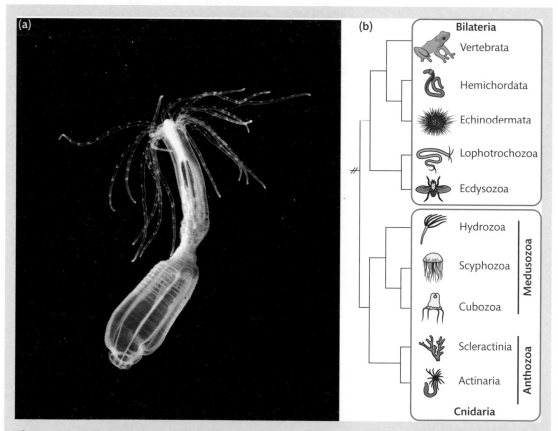

Figure 4.8 The delightfully named, and delightfully apportioned, starlet sea anemone (*Nematostella vectensis*) has as rich an array of Hox genes as the fruit fly.[13]

From Michael J. Layden, Fabian Rentzsch, Eric Röttinger (2016) 'The rise of the starlet sea anemone *Nematostella vectensis* as a model system to investigate development and regeneration', *Wiley Interdisciplinary Reviews. Developmental Biology*, 5: 408-28. 2016. © 2016, John Wiley and Sons; Photo: © Eric Röttinger/Kahi Kai Images.

particular part of the gene sequence, termed the 'homeobox', can be recognized in hundreds of different regulatory genes across the genome, in organisms as widely divergent as animals, plants, and fungi.[14] This small DNA sequence, only 180 bases long, encodes part of the transcription factor protein that binds to DNA. Transcription factors interact with the genome to influence the rate of expression of other genes, turning them on or off (or up or down). These genetic switch sequences can, in some cases, be exchanged between different Hox genes in a cluster with little or no phenotypic effects.[15] In some cases the homeobox sequence can even be transferred from one species to another distantly related species and still function to switch gene expression on and off.

Sequence conservation is not, in and of itself, particularly surprising. There are many genes that are as conserved as homeobox sequences across animal phyla. In fact, the construction of molecular phylogenies of animals relies on selecting sequences that change so slowly that they can be compared between lineages that have been separated for half a billion years or more (Figure 4.5). Genes for fundamental functions that all animals rely on, such as genes associated with basic metabolic pathways or DNA replication, typically have slow rates of change. But in most cases, the genes are performing the same function in each lineage. What is interesting about the Hox genes is not that their sequence is conserved, but that the same genes are used to specify quite distinct body plans.

Conservation and change

> **❝** *One feature of such signalling systems . . . is that they are 'symbolic'. By a 'symbol', semioticians mean a signal whose form is causally unrelated to its meaning . . . The same is true of genetic signals. The small-eye gene in a mouse means 'make an eye here', but, as far as its form is concerned, it could equally well mean 'make a whisker', or 'don't make a toe'.* **❞**
> **John Maynard Smith** (1998) *Shaping Life: Genes, Embryos and Evolution.* Weidenfeld & Nicolson, London

These fascinating observations—the universality of Hox genes in the animal kingdom, their involvement in establishing the body plan of growing embryos, their high degree of conservation yet variable number and remarkable evolutionary lability in phenotypic effects—have led to the suggestion that the evolution of the Hox cluster played a key role in driving body plan evolution in animals. This presents a puzzle. How can a set of genes be on the one hand fundamental to all animals and on the other hand responsible for the key body plan differences between animal phyla? The reason that Hox genes are so evolutionarily labile is that they have a very general function; instead of making particular structures, Hox genes are regulatory genes that act as control switches to modulate the expression of other genes in the genome.

While the Hox cluster has attracted the most attention, there are many other similar sets of regulatory genes. The DLX family of homeobox-containing genes is found throughout the animal kingdom. This gene family is named after the *Drosophila* gene *distal-less* (*Dll*) which is required for the proper formation of limbs, not only the legs and wings but also the antennae and mouthparts. Mystifyingly, *distal-less*-like DLX genes have been found to be important in the development of very different kinds of limbs in a wide range of animals, from the water-filled tube feet of starfish to the wings of chickens and the soft caterpillar-like legs of velvet worms. Some of the features that *distal-less* is involved in can only with a great stretch of the imagination be called 'limbs' at all, such as the ampullae that project from seasquirts to attach them to the rocks and the spines of sea urchins. This array of different 'limbs' does not seem to represent homologous structures, i.e. they are unlikely to all represent modifications of the same ancestral limb-like structures. It seems more likely that these different appendages evolved independently in each lineage. So why do they all use the same homeobox-containing regulatory gene?

Distal-less is expressed in outgrowths of the developing embryo (referred to as the proximodistal axis). Remember that the homeobox sequence is a very general piece of kit—it binds to DNA and can be used in a wide range of different gene products that must interact with DNA to regulate gene expression. So if the homeobox-containing *distal-less* gene was already being expressed in ancestral animals along the margins of embryos or associated with outgrowths of the embryo wall, then as those animals diversified and evolved exciting new body plans, *distal-less* expression might have been co-opted into new features that grew out along the proximodistal axis, such as legs, wings, spikes, and tube feet. DLX genes are also associated with neural development and formation of sensory systems in many animals, so again it could have been co-opted from association with neural development into the formation of useful features like limbs, antennae, and mouthparts. A gene that acts as a switch can be co-opted to switching on any other gene, in the same way that an electrical switch could be adapted to turn on a toaster, a television, or a light.[16]

Other observations are consistent with the idea that *distal-less* genes are handy switches that can be co-opted into a wide range of roles. *Distal-less* is not only used for phylum-level differences in body plan like limbs and sensory apparatus, it has also been implicated in the diversification of the morphology of jaws and anal fins in cichlid fish,[17] in patterning butterfly eyespots, and in the development of defensive chemical weaponry in worker termites.[18] Like other regulatory genes, *distal-less* acts embedded in cascades of gene action, where the expression of gene is influenced by 'upstream' regulators that occur before it in the regulatory cascade, and then it influences the expression of 'downstream' genes that occur after it in the cascade (see Case Study 4). In *Drosophila*, *Dll* is influenced by the upstream regulatory genes that establish the limb primordia (such as the gene *wingless*). *Dll* then influences several downstream genes that control and shape limb growth and elongation (including the gene *dachshund*, and you can probably guess the phenotype of flies in which these particular upstream and downstream genes do not function).

The functioning of regulatory genes like *distal-less* in genetic pathways gives us some insight into an evolutionary puzzle. If the homeobox sequence is so common in the genome, and so general in function that it can be co-opted to different developmental roles, how can it play a critical role in shaping different animal body plans? The answer is that none of these genes acts in isolation. Since every cell in the body contains the same genetic information, the differences between cells and tissues and organs cannot be due to different genes, but must be how those genes are expressed. Developmental cascades ensure that only the genes that are actually needed for that structure are called upon to act.

Homeotic mutants

Hox genes, and other body patterning regulatory genes, act by calling genetic programs consisting of all the 'downstream' genes needed to make a structure. If the regulatory switch is turned on, the whole program will be called, potentially resulting in a complex structure forming out of place, such as the classic fruit fly homeotic mutants Anntenapedia (legs where antennae should be) and Ultrabithorax (wings where halteres should be) (see Case Study 4). A famous example of this kind of change is the production of ectopic (out of place) eyes through manipulation of *Pax6* gene expression.

The discovery that many animal phyla use the same regulatory gene to direct eye formation has been one of the more astonishing results of the field of Evo-Devo (evolutionary development). Technically this gene has different names in different species, such as *small-eye* in mouse and *eyeless* in fruit fly, but we are going to refer to all the related versions of this gene occurring in different animal lineages by the generic name *Pax6*. The *Pax6* gene is yet another homeobox-containing gene; it encodes a transcription factor that can bind to DNA and switch on gene expression. *Pax6* is important in neural development, and in most species is essential for proper eye formation. The astounding thing about this gene is that it seems to have a common role in eye development among widely divergent animals, even though they have very different kinds of eyes, including the camera eyes of cephalopods and vertebrates, the compound eyes of arthropods, and the eye-spots of flatworms.

Even more astounding, the *Pax6* gene of a mouse can stimulate eye formation in *Drosophila*, not only in the usual place but wherever it is expressed on the body. But of course, the mouse version of *Pax6* does not cause a mouse eye to grow on a fly. It causes a fly eye to grow. That is because the role of *Pax6* is to call the eye-making gene regulatory network. In a fly, this consists of the genes needed to make a compound eye. In the mouse, this network contains a different set of genes needed to make a camera-style eye. The ectopic eyes are not correctly wired to the fly's nervous system, and the lack of functional integration with the rest of the body makes them useless as sense organs, (although modification of a different regulatory gene in a beetle has been reported to cause eye-like structures with some functionality).[19]

Key points

- Regulatory genes are genes whose products act by influencing the expression of other genes.

- Many regulatory genes have complex patterns of expression, embedded within gene networks, working in partnership with upstream and downstream genes to regulate the development of specialized tissues and structures in a growing individual.

- Changes to regulatory genes can cause changes of large phenotypic effect by failing to switch on the appropriate gene cascade for a particular part of the developing body, or by causing a structure to form in a different position.

Mutation

At this point you may be wondering why a chapter on the Cambrian explosion has turned into a discussion of the details of developmental genetic pathways, so let's recap how we got to this point and where we hope to go next. The mystifying thing about the Cambrian explosion is that fundamentally different kinds of animals seemed to have evolved in a relatively short evolutionary interval, resulting in a rate of body plan evolution that does not seem to have been equalled in any period before or since. The soft-bodied Ediacaran fauna has become increasingly well documented, yet there are still no unambiguous finds of early members of animal phyla before the Cambrian, and little evidence of gradual build-up of intermediate states of modern animal body plans. Instead, ongoing palaeontological discoveries reinforce the picture of a surprisingly rapid transition from the soft-bodied simplicity of Ediacaran life to the diversification of animal form and ways of life in the Cambrian. Virtually all phyla are considered to

date to the Cambrian period, either due to fossil evidence or inference from phylogeny (i.e. identified as a sister to a Cambrian lineage and therefore at least as old).[20]

The simultaneous appearance of members of so many animal phyla in the Cambrian fossil record, and the lack of origin of new animal phyla in other periods, has led some researchers to suggest that the kind of evolution that occurred in the Cambrian was different from the kind of evolution we see occurring in the animal populations that we live amongst today. Some explain this radical period in animal evolution as the result of a particular set of environmental or ecological conditions that occurred then and only then. But others reach beyond the unique environmental conditions and invoke special macroevolutionary mechanisms that operated in the Cambrian, but not before or since, generating phyla which were then incapable of undergoing such radical modification ever again.

What could be the nature of this proposed special mechanism? A prominent suggestion has been that the early period of animal evolution represented a time when the genetic architecture underlying body plans was forming, allowing large changes in phenotype to be generated by relatively few genetic changes. The identification of body-patterning regulatory genes with conserved elements identifiable in all animals, which have a pattern of duplication and loss across the animal kingdom and which can cause dramatic phenotypic modifications when disrupted, has led to a hypothesis that these regulatory genes represent an evolutionary 'toolkit' for body plan evolution. Specifically, Hox genes, and similar regulatory genes involved in embryo patterning, have been considered by some scientists to play a driving role in body plan modification.

The study of developmental genetics shows that mutations that alter an individual's basic body plan are possible. But production of mutations is not enough in itself to drive evolution of new body plans. Those body-plan-changing mutations have to become fixed features of functioning lineages for us to be able to observe them in the living biota. For macromutations to contribute to the evolution of body plan, they have to be transmitted and amplified over generations. So, to evaluate the evolutionary contribution of developmental changes that radically alter phenotype, we need to take a more general look at the debates concerning the fate of mutations. We have considered how changes to developmental genes can cause macromutations in body plan, but to frame the debate about whether macromutations played a role in the rapid origin of animal body plans in the Cambrian, we now have to consider the process of substitution, whereby a mutation that occurs in a single individual rises in frequency to become characteristic of an entire lineage.

Substitution

A mutation originally occurs in the genome of a single individual, changing a particular DNA sequence. If that individual fails to reproduce, that mutation dies with it. But if it does reproduce, the mutation has the chance to enter the gene pool of the next generation (remember that in a sexually reproducing species, each offspring only gets half of each parent's genetic variants (see Chapter 7)). The mutation will initially be at low frequency in the population, occurring in just that original mutant individual and in some of its children and grandchildren. If misadventure befalls this handful of mutant individuals, the mutation may be lost. But if enough survive, and they are not handicapped by the mutation, they too may reproduce, passing copies of the mutation from generation to generation. If their descendants live and grow and reproduce along with other members of the population, then we say the population is polymorphic for that mutation because it is carried by some but not all members of the population. If individuals carrying this mutation happen to out-reproduce others in the population (whether by virtue of the mutation or just by chance), the mutation will rise in frequency. And, through the re-assortment of alleles that occurs every generation in a sexually reproducing population, there may come a day when all members of the population carry a copy of this mutation, originally formed in a single ancestral individual. At this point, we say that the mutation has become a substitution; the new mutation has been substituted for the previous variants.

We can see that there are steps that every mutation must take on the path from being a heritable change in an individual genome to becoming a genetic variant that characterizes a whole lineage. The individual with the mutation must survive, reproduce, and pass on its novel genetic variant to the next generation. The descendants who carry this variant must do at least as well as others, by fitness or fortune. The mutation must eventually rise in frequency and replace all other variants of that trait in the population. And that population must give rise to a chain of descendants that persists over millions of years as part of a successful evolutionary lineage. Now, to assess the chances of a mutation that causes a large change to body plan becoming a substitution that characterizes a whole lineage, we need to think about the influence of size of mutational effect on the process of substitution.

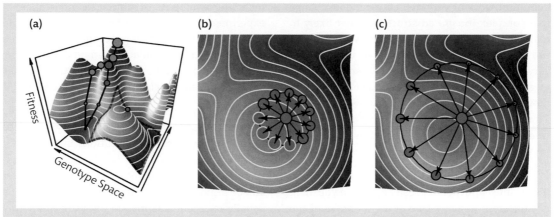

Figure 4.9 We can think of a population as occupying a landscape of fitness, with peaks on the landscape representing high fitness values of a trait in a given environment (a). The path taken to the peak will depend on the mutations that occur (the possible steps), the nature of the population (the efficiency of selection), and the form of the landscape (for example, whether a stable single peak or a rugged, multipeaked, and ever-changing landscape). To consider the influence of size of mutation on the path to adaptation, imagine viewing one of these peaks from above. If an individual has a genotype that is not at optimum fitness, what is the chance that a random mutation will increase the fitness of its offspring? Fisher's geometric model predicts that a very small change to the current state has a nearly 50% chance of moving in the right direction, toward the fitness optimum (b). But a large change has a high chance of producing a genotype that is further from the fitness optimum (c). This model makes a number of assumptions about the nature of the landscape, the way genes interact, and the population genetic process.[25]

© Russell Dinnage.

The fate of macromutations

The classical neo-Darwinian position is that mutations with a wide range of effects on phenotype, small and large, will be constantly generated, but it is the tiny modifications, representing minor tweaks to the existing form, that are more likely to persist and contribute to the evolution of a lineage. If an animal has survived and successfully reproduced, by definition that individual is well fitted to its environment. All of its complex systems are working together to create a functioning individual that can get what it needs from the environment, avoid becoming another animal's breakfast, and (in the case of sexually reproducing animals) persuade another individual to invest its DNA in some genetically shared offspring. Given that any successful parent is already doing very well, a large change is more likely to make its offspring worse, not better.

One way to think about the effect of mutation size on fitness is to visualize individuals on a 'fitness landscape', with peaks representing the best possible value of a particular trait in a given environment and the valleys representing low-fitness values (Figure 4.9). As Darwin and others have shown, every population is full of variation, so every individual differs from others in some way. We can picture these individuals arrayed on a landscape of fitness, with those who have genetic variants that allow them to be a functioning successful individual near the top of the fitness mountain, and those with less propitious gene variants occupying a position away from the fitness summit. Individuals will have offspring that vary from their parent due to mutation, shifting them away from their parents on the adaptive landscape.

Mutation is undirected with respect to fitness. A mutation might make the offspring better than its parents (toward the fitness peak). But it's more likely to move it away from the fitness peak. Why? Because any individual that becomes a parent is, by definition, working very well, and a random change to a complex integrated structure is not likely to make that organism work better. Now think about the effect of size of mutation. If an individual is very close to the adaptive peak, a small change might inch it closer to the adaptive peak, or it might move it across the slope or slightly downwards. But a large change will almost certainly end up further from the optimum than its parents, as it will either jump downslope or overshoot the peak and head down the valley on the other side (Figure 4.9).

If the landscape model doesn't appeal to you, here is a different analogy. One of the founders of the population genetic view of evolution, R.A. Fisher, likened the effect of mutation size on fitness to the art of focusing a microscope. If you are nearly in focus, small random adjustments have an even chance of making the image clearer or worse, but a large change will almost certainly overshoot and swing the microscope way out of focus. The chances of achieving focus from a random move are also reduced if the microscope has many different knobs that can all be adjusted to improve the focus; getting the right combination of moves has an even lower probability than a random single move. This model reflects the common assumption that the more complex a device is, with more parts that must interact to achieve an outcome, the less likely that random adjustments will improve performance.

Living organisms are very complex indeed. Not only do they have thousands of genes that must be turned on and off at the appropriate times for sensible development of the growing organism, but, as we have seen, genes do not act alone but as players in complex networks of interactions. Fisher's model predicts that if a population must adapt to a new circumstance—say an environmental change in resource availability, or the opportunity to exploit a new niche, or the need to escape from a novel predator—then if selection is acting on genes embedded in interacting networks, the path to adaptation is far more likely to occur through a series of small steps than by a few large changes.[21] There are many empirical examples of the long path to adaptation by many small steps,

including evolution of antibiotic resistance in experimental populations of bacteria: while single mutations can occur that generate resistance in a single jump, most populations adapt to increasing antibiotic concentration by accumulating a series of mutations that each contribute to resistance.[22]

Does Fisher's microscope rule out evolution by large changes? No, it suggests that large changes are unlikely to become substitutions, but not that it is impossible. But, as Goldschmidt argued, successful macromutations might be exceedingly rare yet still be important in evolution. Even if populations usually take long adaptive journeys of many tiny steps, this does not tell us whether there might be occasions when evolution leaps to a new state by macromutation. In fact, any large changes that are advantageous will have faster rates of fixation than positive changes of small effect. Changes with a large beneficial effect are more likely to escape the accidental loss that befalls most mutations when they are at low frequency in the population, because they will put their lucky carriers at substantial advantage in the survival and reproduction stakes. So we have something of a paradox. 'Hopeful monsters' are expected to be exceedingly rare because large change is almost always ruinous. But if a large change happens that is advantageous, it has more chance of becoming a substitution than a small positive change. It might even leap to a new part of the adaptive landscape, and be able to exploit a new niche.

It has been suggested that these two forces acting in opposition—small mutations are less likely to be harmful but large beneficial changes are more likely to be fixed—will result in success in the middle ground, with positive changes of medium impact most likely to contribute to adaptation.[23] Furthermore, the size of mutations contributing to adaptation might change as evolution proceeds. For a population adapting to environmental change or the opening up of a new ecological opportunity, it's possible that their journey across the adaptive landscape will be characterized by large steps at the beginning of the journey, when there is much to be gained from significant changes in phenotype, followed by smaller changes as the population nears the adaptive peak, where fine-tuning changes are more likely to be beneficial.

Evolving developmental genes

The plausibility of a macromutational model of rapid body plan evolution doesn't rest on whether such mutations can occur (they can), or whether or not many of them will be disastrous (they will), but on how likely it is that a 'hopeful monster' will arise that is better than the others in the population. As we have seen, there is ongoing debate about how unlikely it is that a large change will be beneficial with respect to the fitness advantage over others in the population. To some extent this will depend on the form of the 'fitness landscape'. But some developmental biologists have offered an alternative way to increase the chance that large changes will be advantageous, or at least avoid being disastrous. If development is structured so as to buffer the potential harm of large changes, there may be a lower average cost to mutations.

The classic Fisherian view assumes that the more gene interactions there are, the less likely it is that any given random change will be positive, because it will influence so many traits. But an alternative viewpoint has been offered. What if more complex systems, with many interconnecting gene regulation networks, are more robust to change? This idea was championed by C.H. Waddington, a developmental biologist (and noted skeptic of neo-Darwinism). Waddington felt that the complexity of developmental pathways would buffer the developing embryo against perturbation by genetic mutations. Like a ball rolling down a hill, if mutation blocked one developmental pathway, the embryo would find another path to the same endpoint. Waddington highlighted the possibility that development could result in the same phenotype from different genotypes.

So changes to developmental genes may be less harmful if their negative effects are buffered against disrupting development (Waddington's canalization). And the predicted costs of epistatic interactions—the more genes act together, the less chance a random change will be favourable—may not have to be paid if they switch on integrated genetic programs that act harmoniously to generate a whole new structure (homeotic or meristic mutations). But does this provide a coherent account of the evolution of novelty by developmental macromutation? While switching on and off existing regulatory paths may make large changes less disastrous, it is most likely to result in a novel arrangement of existing structures, rather than generating a novel structure.

What do you think?

Could modification of regulatory genes cause an entirely new feature to grow, or can new features only evolve through changes to the downstream target genes that are used to build body structures?

Evolutionary persistence

Mechanisms that reduce the probability of a macromutation being harmful might increase the chance that such changes can contribute to the evolution of body plan, but they won't solve all the hopeful monster's problems. If a macromutation occurs and happens to produce a well-integrated individual who can exploit a new niche, is that sufficient to start a new lineage? Not quite. Remember that mutations are at the greatest risk of being lost when they are rare. Our monster must make a way for itself, either in its ancestral environment or by exploiting a new niche, and it must produce more of its kind in order to start a new and distinct lineage. This might be straightforward if the mutant individual can reproduce asexually. It's more of a problem if our monster, like most animals, relies on biparental reproduction. Even if that monster can persuade another individual in the population to combine gametes with it, the resulting combination has to produce a well-functioning offspring. If the macromutation is very rare, then it is unlikely that two parents will both carry the same mutation, so the offspring will have to function with one copy of its genome from its monster parent and one from an ordinary member of the population. This may be why the earlier research into macromutations was predominantly focused on plants; more plants than animals are capable of self-reproduction.

Indeed, it is possible for new and distinct lineages of plants to form in a single step by macromutation. In particular, changes in chromosome number can not only influence phenotype but might also make the mutant individual incapable of mating with 'normal' members of the population. For example, in a population of diploids (each individual has two copies of every chromosome), a triploid individual which has, by accident of birth, three copies of each chromosome may be perfectly functional. Because of the difficulties of pairing three chromosomes at meiosis, the triploid may be unable to produce viable pollen or ovules. But if it can reproduce clonally it can give rise to a population of triploid individuals which cannot reproduce with members of the parent population, thus forming a new genetically isolated species in a single macromutational jump. For example, the Australian plant species *Grevillea renwickiana* probably formed from a triploid, and now consists of several populations of clonally reproducing plants (new plants form by suckering from the roots of existing plants) (**Figure 4.10**). Because it is restricted to a relatively small area, and shows little genetic diversity, the species is considered endangered and at risk from habitat loss or climate change. While chromosomal speciation is more common in plants, it can also occur in animals. For example, a number of families of stick insects contain triploid and tetraploid species, offshoots of sexually reproducing lineages that are now obligate parthenogens (with uniparental reproduction). However, while these new species might represent genetic macromutations, they have not, in these cases, caused major perturbations of body plan.

The association between gene expression patterns, or number of gene duplicates, and change in body plan over time is not in itself sufficient to tell us about the nature of the causal link between developmental genes and body plan evolution. It's much

Figure 4.10 New species can arise by a single 'macromutation'—at least in some lineages. *Grevillea renwickiana* probably arose by a genome duplication, making a triploid lineage that is incapable of sexual reproduction but can reproduce clonally by suckering from the roots. The species probably arose from only one or a few mutant invididuals. Should we recognize such species as successful hopeful monsters, because a single genomic change in one individual can establish a new lineage, or as evolutionary dead-ends, a clonal population with little chance of giving rise to a long line of varied descendant?

Photo from *Atlas of Living Australia*.

trickier to go from pattern to process. We might observe the same pattern if changes to Hox genes caused changes in body plan, or if changes in body plan generated by other genetic changes brought about a change in Hox gene expression pattern, or if both were incidentally associated.

Consider another example of this kind of dilemma. Many genes in metazoans have introns. In some genes, differential splicing at introns produces different forms of proteins. The gene *doublesex* is a good example; by splicing together different combinations of exons, different regulatory proteins can be produced which can then trigger different developmental pathways, resulting in dramatically different morphologies, such as males and females, or different mimic patterns (see Case Study 4). So there is a clear pattern that introns are associated with alternate splicing that produces different protein products. But what was the evolutionary process that led to this association? Did introns evolve in order to allow differential splicing so that different morphologies could be produced from the same genes? Or did the presence of introns in genes allow alternate splicing, stimulating those genes to be co-opted into multiple roles?

Contemplating the role of body-patterning genes in evolution faces the same challenges of separating cause and effect. In some cases the causal arrow is clear. Genes that are essential to proper wing formation in flies, such as *wingless* and *distal-less*, are found across the animal kingdom. Their roles in animal development long predate the evolution of wings or flies (see Case Study 4). So these genes must have had pre-existing roles in embryo development in insects, and were then co-opted into wing formation in some lineages. In other cases, untangling cause and effect is more challenging. What was the role of *Pax6* in the ancestor of all animals? Did it specify eye formation, and if it did, does that tell us that all the wild diversity of eyes in the animal kingdom are modifications of an original *Pax6*-triggered ancestral eye structure? Or have eyes evolved independently in many different animal phyla, yet in each case adopted the same available developmental trigger?

Why was the Cambrian explosion unique?

The macromutational model of body plan evolution throws into focus the flipside of the question 'What caused the Cambrian to explode?'. Why were there no such explosions at other times? If changes to developmental genes can generate new body plans, and were responsible for the rapid evolution of the majority of animal phyla, then why didn't similar macromutations generate new animal phyla in subsequent periods? Why do we not see new body plans evolving in present-day populations or laboratory stocks? We can consider several general explanations.

One possible explanation for the uniqueness of the Cambrian explosion is that the Cambrian represents an unprecedented confluence of environmental conditions. Evidence of glaciation has led to the proposal of several severe 'snowball earth' events prior to the Cambrian. Life must have persisted through any such periods of global glaciation, but it seems unlikely to have provided conditions conducive to the evolution of multicellular life, which might then have been delayed until after the last thaw in the late Proterozoic. It has also been argued that oxygen concentrations rose after the last glaciation, providing a necessary resource for the growth of complex multicellular animals. Changes in ocean chemistry may have made key minerals available, making the evolution of mineralized hard parts possible and contributing to the development of sophisticated body plans. But even if conditions prior to the Cambrian reached a threshold point that allowed the evolution of complex body plans, that does not tell us why there have been no new phyla since that time, given the sufficiency of oxygen, biominerals, and so forth for complex animal life.

Alternatively, the uniqueness of the Cambrian explosion may be due to unparalleled ecological opportunities. Animals were diversifying to take advantage of new, previously unexploited opportunities, such as sediment burrowing or active predation. These new ways of being may have increased the chance that large changes in phenotype would find a successful home. As all of those new 'openings' became progressively filled by the evolution of new forms of animals, the pay-off for large changes might have diminished. In other words, the nature of the fitness landscape might have changed during the course of the Cambrian explosion.

The prediction that large changes will usually have a disastrous effect on fitness is predicated on the assumption that the parent individual is near the adaptive peak for that population, in which case large mutations will overshoot the peak and land in a maladaptive valley. But what if the adaptive landscape is so wide open that large mutations jump a lineage to a whole new region of ecospace, allowing it to exploit a new resource? Then once those niches are filled, the landscape might change from a wild untamed frontier to a series of crowded hills, and the only changes that can increase fitness are tiny tweaks. If empty ecospace allowed the formation of new phyla, then we should expect to see a burst of body plan evolution when new ecological opportunities opened up, for example the movement onto land. Indeed, the colonization of the land did prompt new innovations in body plan such as wings. But the move to the terrestrial sphere did not lead to lineages that are recognized as new phyla.

An alternative explanation has been offered for the lack of evolutionary inventiveness at other times. It has been suggested that once the complex gene regulatory networks were set in place during the phylum-level radiation of animals in the Cambrian, they became resistant to further change. Once gene networks are embedded into complex interacting webs, it might be hard to make any major changes without causing the whole network to crash. However, this can't be universally true, because there are examples of changes to core genetic networks.

For example, one of the most fundamental aspects of body plan is segmentation, because it occurs early in embryo formation and provides the foundation on which many other body plan features are built (for example, many homeotic mutations alter segment identity, such as Ultrabithorax and Antennapedia). Segmentation was once considered such a fundamental aspect of animal body plan that it was used as the basis of classification, not only for defining the body plan of a phylum but also for grouping related phyla together. However, molecular phylogenies show that segmentation has a surprisingly labile evolutionary history. Segmentation appears to have evolved independently several times, and then been lost in some lineages (such as the demoted annelid sub-phyla Sipuncula and Echiura, see page 129). More intriguingly, the genetic basis of segmentation has been entirely rewired in at least one lineage: while most arthropods use *Notch* genes to make segments, the lineage leading to flies and beetles evolved an entirely new method using a different set of 'pair-rule genes' to define segment boundaries (one of these genes, *fushi tarazu*,

was derived from a Hox gene, but has evolved a new role). This is a case where the developmental basis of a trait (pair-rule genes) has completely changed even though the phenotypic outcome (segmentation) remains the same.

It is important to emphasize that, given the number of genes involved in the construction of any body plan feature, developmental macromutation is not the only way to generate discrete animal phyla. If enough genes change, each contributing in some ways to a distinct phenotype, a taxon can be separated from its closest relatives by major aspects of phenotype. Gradual evolution can produce evolutionary novelties with no living intermediates, as extinction of intermediate states creates 'unbridgeable gaps' between the descendants. Consider the baleen of large whales, the sting of the scorpion, the wings of a bat, or barnacle cement. None have intermediate forms found in close relatives, yet all can be explained as extreme modifications of existing structures, as the end of a chain of Darwin's imagined series of intermediate taxa.

The longer lineages evolve separately, the more they will differ from each other. This suggests an alternative view of phyla, as the amount of difference you get when you allow animal lineages to accumulate changes for half a billion years. Under this view, generation of variation drives the evolution of higher taxa, but it is the constant accumulation of many tiny changes rather than the occasional macromutation. So although the suddenness of appearance of animal phyla may be a challenge to Darwinian gradualism, the gradualist view does provide a possible explanation for why no new phyla have appeared since the Cambrian: if phylum-level differences take half a billion years to accrue, then there can be no young phyla. But this view also presents two challenges. Firstly, it suggests that if phyla were already distinct in the Cambrian, then they must have had a lengthy prior period of accumulation of changes. Secondly, it also suggests that if we come back in another 500 million years, maybe the *Drosophila* lineage will have given rise to a whole new animal body plan.

Key points: Why no new phyla since the Cambrian?

- The Cambrian may have represented a unique opportunity when the right combination of environmental factors—such as sufficient oxygen, available minerals, and reasonable temperature—allowed the diversification of large complex animals (but why did other cases of radiation into empty niches, such as the colonization of the land and sky, not stimulate as much change?).

- The evolution of complex gene regulatory networks in the Cambrian may have 'locked in' animal body plans which were then too inflexible to change (yet morphological novelties have evolved since then, and co-opted conserved regulatory genes into their development, such as wings).

- The biosphere may have become saturated with body plans in the Cambrian and there was no ecological opportunity for high-level innovation after that point (but animals that have colonized novel environments, such as deep caves or remote islands, have not produced new phyla even if few other competing phyla are present).

- If a phylum is a phylogenetic category that reflects lineages of Cambrian age, then any new form that evolves after the Cambrian, however much it departs in form from its ancestors, will not be counted as a new phylum.

Conclusion

Do extraordinary periods of change require extraordinary explanations? The debate over the Cambrian explosion has a central role to play in macroevolution. In particular, the Cambrian explosion has been used to challenge the most fundamental tenet of Darwinism, that the processes that we can observe in populations today—mutation, selection, drift, divergence—are those that produce all evolutionary change. But what if things were different in the past? Can we be sure that there were not additional evolutionary processes in some periods? No one denies that the population-level processes identified by Darwin—now referred to as microevolution—can generate change in the frequency of variations over generations. But some have questioned whether Darwinism fails to capture all the critical features of evolutionary change. The Cambrian explosion has frequently been drawn upon as an example of an evolutionary event that does not seem to fit the mould of gradual change by the accumulation of many tiny changes.

In this chapter, we have not considered the possible environmental triggers for the Cambrian explosion, such as the cessation of glaciation, increase in oxygen, or available biominerals. Nor have we examined ecological arguments, such as diversification into empty niches, or the cascading effects of the rise of mobile predators. Instead, we have focused entirely on the idea that the kinds of variations that could arise and be fixed in evolutionary lineages were in some way different in the Cambrian, generating a much greater rate of fundamental change in body plan in a relatively short amount of time. These ideas have been discussed throughout the history of evolutionary biology, but they have received renewed impetus through the study of the action of genes that are key players in the development of body plan in animals. It has been suggested that it was the establishment of regulatory gene networks that triggered the Cambrian explosion by providing the means for rapid modification of body plan.

The issue at the heart of this debate is variation. The Darwinian model proposes that everyday minor variations between individuals are the fuel of evolution, but some people doubt they are sufficient to account for the origin of discontinuous higher taxa such as phyla, so have suggested that special kinds of 'macromutations' can create new animal types in one or a few large changes to body plan. In fact, micromutations, the variation we see between all individuals in a population, and macromutations, causing rare 'monsters' with dramatically different forms, can be considered as occurring at the ends of a long spectrum of mutations of varying effects. Small mutations of tiny effect might have a greater chance of becoming fixed in a population, particularly when the population is well fitted to the environment, and larger changes may have a much smaller chance of leading to a new type unless they have the phenomenally good luck to hit upon a successful phenotype, in which case they will be promoted by selection. Both micromutations and macromutations may contribute to evolution.

But no mutations can bypass the normal Darwinian processes. If a mutation results in an individual with a lowered chance of survival and reproduction, it has less chance of being passed on than those changes that increase the chances of success. So the key issue at the heart of this debate is not whether large mutations can occur, but the likelihood of large changes helping, or not harming, chances of survival and reproduction. The debate has perhaps been unnecessarily polarized; it is possible for both sides to be true, or at least for neither to be entirely false. Perhaps much of the debate tells us more about the relative emphasis on aspects of the evolutionary process: generation and expression of variation is emphasized by those who study body-patterning genes and development, heritability and population processes by those who study microevolution, and the patterns of origination of novelty and higher taxa by those who study macroevolution.

Earlier in the chapter, we saw that mutation is 'backgrounded' in the neo-Darwinian view of evolution—it is presumed to provide a constant but undirected exploration of design space around existing organisms. Since, by definition, a living organism that reproduces must be well adapted to its environment, minor changes to its phenotype are more likely to produce a working successful offspring than a wild modification to the existing plan. Biologists who focus on origins of major body plan changes or developmental evolution are sometimes frustrated by the apparent indifference of neo-Darwinians to the generation of novelties. But, on the other hand, researchers who concentrate on the developmental mechanisms of body plan have tended to 'background' the population genetic processes that fix novelties in populations. A mutation in a single individual cannot contribute to evolution if the new feature is not sufficiently integrated with existing physiology, behaviour, and niche to allow survival and reproduction. And unless the new type can compete against the existing form, or find sufficient space, ways of life, and mates to start a new lineage free from competition, it is unlikely to rise in frequency and become established as a new lineage.

Clearly, evolution requires the generation of new heritable types able to survive, reproduce, and out-reproduce alternative forms. Different biologists may focus on different parts of the evolutionary spectrum from mutation to development to population genetics to macroevolution. Evolution requires all these steps, even if most biologists tend to focus on one particular part of the story. Find the part of the story you find most fascinating and see what you think is the best explanation of diversity you can find, but don't forget that the others are important parts of the narrative as well.

Points for discussion

1. Which kind of genes have a greater influence on the evolution of variation in animal form and function: regulatory genes, that turn other genes on and off, or structural genes, that specify different aspects of form and behaviour?

2. Can you design an experiment that would test whether mutations can arise that are more likely to be advantageous than would be expected from random changes? How would you control for the fact that beneficial mutations are more likely to lead to surviving (and therefore observable) offspring?

3. Will new animal phyla evolve in the future? Or are there now as many animal phyla as there will ever be?

References

1. Darwin C (1869) *The origin of species by means of natural selection, or the preservation of favoured races in the struggle for life* (5th edn). John Murray, London

2. Budd GE, Jensen S (2017) The origin of the animals and a 'Savannah' hypothesis for early bilaterian evolution. *Biological Reviews* 92(1): 446–73.

3. Hejnol A, Dunn CW (2016) Animal evolution: are phyla real? *Current Biology* 26(10): R424–6.

4. Zhang X, Briggs DE (2007) The nature and significance of the appendages of *Opabinia* from the Middle Cambrian Burgess Shale. *Lethaia* 40(2): 161–73.

5. Budd GE (2013) At the origin of animals: the revolutionary Cambrian fossil record. *Current Genomics* 14(6): 344–54.

6. Sperling EA (2013) Tackling the 99%: Can we begin to understand the paleoecology of the small and soft-bodied animal majority? *Ecosystem Paleobiology and Geobiology* 19: 77–86.

7. Struck TH, Schult N, Kusen T, Hickman E, Bleidorn C, McHugh D, Halanych KM (2007) Annelid phylogeny and the status of Sipuncula and Echiura. *BMC Evolutionary Biology* 7(1): 57.

8. Dordel J, Fisse F, Purschke G, Struck TH (2010) Phylogenetic position of Sipuncula derived from multi-gene and phylogenomic data and its implication for the evolution of segmentation. *Journal of Zoological Systematics and Evolutionary Research* 48(3): 197–207.

9. Erwin DH (2011) Evolutionary uniformitarianism. *Developmental Biology* 357(1): 27–34.

10. Darwin C (1854) *A monograph on the sub-class Cirripedia*. Ray Society, London.

11. Bateson W (1894) *Materials for the study of variation treated with especial regard to discontinuity in the origin of species*. Macmillan, London.

12. Goldschmidt R (1940) *The Material Basis of Evolution*. Yale University Press, New Haven, CT.

13. Ryan JF, Burton PM, Mazza ME, Kwong GK, Mullikin JC, Finnerty (2006) The cnidarian–bilaterian ancestor possessed at least 56 homeoboxes: evidence from the starlet sea anemone, *Nematostella vectensis*. *Genome Biology* 7(7): R64.

14. Holland P (2015) Did homeobox gene duplications contribute to the Cambrian explosion? *Zoological Letters* 1(1): 1.

15. Ruff JS, Saffarini RB, Ramoz LL, Morrison LC, Baker S, Laverty SM, Tvrdik P, Capecchi MR, Potts WK (2017) Mouse fitness measures reveal incomplete functional redundancy of Hox paralogous group 1 proteins. *PLoS One* 12(4): e0174975.

16. Maynard Smith J (1999) *Shaping Life: Genes, Embryos and Evolution*. Yale University Press, New Haven, CT.

17. Diepeveen ET, Kim FD, Salzburger W (2013) Sequence analyses of the *distal-less* homeobox gene family in East African cichlid fishes reveal signatures of positive selection. *BMC Evolutionary Biology* 13(1): 153.

18. Toga K, Hojo M, Miura T, Maekawa K (2012) Expression and function of a limb-patterning gene *Distal-less* in the soldier-specific morphogenesis in the nasute termite *Nasutitermes takasagoensis*. *Evolution & Development* 14(3): 286–95.

19. Zattara EE, Macagno ALM, Busey HA, Moczek AP (2017) Development of functional ectopic compound eyes in scarabaeid beetles by knockdown of orthodenticle. *Proceedings of the National Academy of Sciences of the USA* 14(45): 12021–6.

20. Valentine JW (2004) *On the Origin of Phyla*. University of Chicago Press, Chicago, IL.

21. Orr HA (2005) The genetic theory of adaptation: a brief history. *Nature Reviews Genetics* 6: 119.

22. Lindsey HA, Gallie J, Taylor S, Kerr B (2013) Evolutionary rescue from extinction is contingent on a lower rate of environmental change. *Nature* 494(7438): 463–7.

23. Kimura M (1983) *The Neutral Theory of Molecular Evolution*. Cambridge University Press, Cambridge.

24. Schiemann SM, Martín-Durán JM, Børve A, Vellutini BC, Passamaneck YJ, Hejnol A (2017) Clustered brachiopod Hox genes are not expressed collinearly and are associated with lophotrochozoan novelties. *Proceedings of the National Academy of Sciences of the USA* 114(10): E1913–22.

25. Waxman D, Welch JJ (2005) Fisher's microscope and Haldane's ellipse. *American Naturalist* 166(4): 447–57.

Evo-Devo: regulatory genes and the development of body plan

Evo-Devo (evolutionary developmental biology) is an exciting addition to the macroevolutionary toolkit. The aim is to use a detailed understanding of development (how a complex multicellular adult grows from an undifferentiated cell), combined with both experimental genetics (what happens to development when you alter key genes) and comparative genetics (similarities and differences in key genes between species), to generate hypotheses about how major evolutionary innovations arise.

In order to understand the role of developmental genes in the evolution of body plan, it is helpful to consider some examples that illustrate the way regulatory genes can switch developmental programs on or off. Here we consider two examples of developmental genes. One example is a gene that influences the whole morphology of the individual (*doublesex*), and the other specifies the identity of particular body parts (*ultrabithorax*). Both of these examples focus on *Drosophila* (fruit fly) development. If we are concerned about the diversification of animal phyla, why spend so much time talking just about fruit flies? Because there is a long tradition of intense study of developmental genetics using *Drosophila* as the model organism; findings in fruit flies have formed the basis for extending our understanding to other animals.

doublesex and sex determination

Male and female fruit flies carry essentially the same genetic information, but they have different sets of sex chromosomes (XX in females, XY in males). However, male and female embryos have different developmental pathways; not only is their reproductive equipment different, but other aspects of their morphology and behaviour must also align with their sex. The genes that specify these differences are not all on the sex chromosomes; many are on the autosomal chromosomes shared by both sexes.

If a *Drosophila* embryo has two X chromosomes (and two of each autosome), the developing fruit fly has sufficient gene product of the *sex-lethal* gene (*sxl*) to trigger expression of the *transformer* genes (*Tra* and *Tra-2*) (**Figure 4.11**). Their gene products, the proteins TRA and TRA2, influence the expression pattern of the *doublesex* (*dsx*) gene, which is a key player in triggering sex-specific gene cascades. How can one gene trigger different gene regulatory pathways in males and females? Although it is only one gene, it can produce two different gene products through alternative splicing. Remember that for a protein-coding gene to be expressed, the information in the DNA (in the cell nucleus) must be transcribed into RNA (the messenger that takes the information to the cytoplasm), and then the information in the

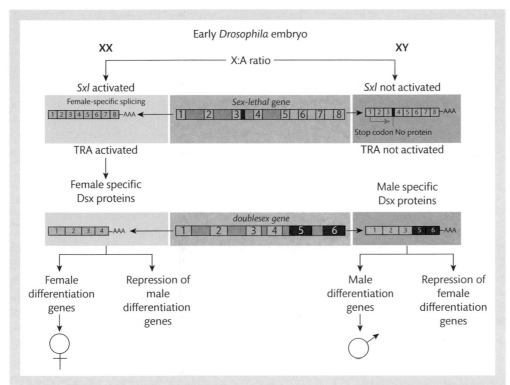

Figure 4.11 The gene regulatory network of sex determination in *Drosophila*. The cascade is triggered by the ratio of X chromosomes to autosomes. If there are as many X chromosomes as autosomes, the expression of the *sex lethal* gene is triggered. The spliced mRNA of *sex lethal* (*sxl*) in females triggers expression of *transformer* (*tra*) which then promotes female-specific splicing of *doublesex* (*dsx*), which calls on female-specific developmental pathways and suppresses expression of male-specific genes.

RNA is translated into an amino acid sequence (in the case of protein coding genes). The *dsx* gene is split into six separate blocks of protein-coding sequence (called exons) separated by untranslated sequences (introns). Once this gene has been transcribed into a messenger RNA, the parts of the sequence that are not needed to create the protein are cut out to produce the final transcript from which the amino acid sequence of the protein will be read (a process known as splicing). In a female embryo, the TRA and TRA2 proteins bind to the *doublesex* transcript at the fourth exon, so only the first four coding blocks of the gene will be translated to make the DSX protein. So the female version of the *doublesex* gene product, DSXF, incorporates exons 1-2-3-4. In a male embryo, where there are few or no TRA or TRA2 proteins present, all of the *dsx* gene apart from the fourth exon is translated. So the male form of the protein, DSXM, has exons 1-2-3-5-6. These different proteins influence other regulatory genes (including the Hox genes), triggering cascades of gene action affecting hundreds of genes needed for the proper development of female, or male, morphology and behaviour.

You might personally not be particularly interested in what creates girl or boy fruit flies. But this example illustrates some key features of regulatory gene networks. A very large number of genes are involved in generating different phenotypes from the same genotype. Some of these are structural genes needed to make parts to build the different bits of the animal. But many other genes have products that are not incorporated into structures. Instead, many regulatory genes make products whose role is to govern how other genes are expressed. Disrupting the normal function of these regulatory genes can result in dramatically different phenotypes as the right genes fail to turn on at the right times, or different genes switch on when they shouldn't. For example, the action of the sex-determining genes in flies was origially investigated by studying mutants in which normal expression was disrupted; that is why these genes have names like 'sex-lethal' and 'doublesex,' after the phenotypes of the mutant flies. Without proper expression of *sex-lethal* and the *transformer* genes, female developmental pathways can't be triggered, so even XX embryos will become males, even though they have all the genetic information they need to become females. Since *doublesex* expression is needed to trigger both male and female development, individuals without proper *dsx* expression become neither male nor female, but turn into intersex adults.

doublesex, sex-lethal, and *transformer* are regulatory genes that influence the expression of other genes. Recognition sequences, or active sites, determine the targets of the regulatory genes; some produce transcription factors that bind to DNA, turning gene transcription on and off, and some make proteins or RNA molecules that bind to gene transcripts (mRNA) to influence their translation into protein. So, just as structural genes can evolve by tweaks to the sequences that change the product they build, regulatory genes can evolve through tweaks to the sequences that change which other proteins or nucleotide sequences they interact with (or the sequence of target genes could change to alter their interaction with different regulatory genes). This means that regulatory genes can change function by changes in their response to different regulatory triggers, or in the way they interact with different genetic targets.

We can illustrate this by looking at the way the gene *doublesex* has taken on varying roles in different animals. The targets of *dsx* can evolve quite rapidly, leading to diversification of sex-associated structures like sex combs in fruit flies, which can vary between different species.[1] *Doublesex* can also lead to evolution of other distinct phenotypes. As in fruit flies, *doublesex* triggers the development of sex differences in butterflies, but in one particular type of butterfly *dsx* has taken on an additional role that illustrates the power of regulatory

Figure 4.12 Multiple phenotypes from the same genome. The swallowtail butterfly *Papilio polytes* grows from an egg (a) to a caterpillar (b), and then becomes a pupa (c) and undergoes metamorphosis into one of four alternative adult morphs: males are all *cyrus* morph (d), as are some of the females. But other females mimic different species: the *stichius* form (e) mimics unpalatable *Pachliopta aristolochiae* and the *romulus* morph (f) mimics *Pachliopta hector*. In other parts of its range, *Papilio polytes* mimics other species.

genes to shape phenotype. In the swallowtail butterfly *Papilio polytes* females and males have different phenotypes, not just their sexual organs but also their wing patterns (**Figure 4.12**). But not all females are the same. There are four very distinct forms of females, differing in wing shape, colour, and pattern, three of which closely mimic other butterfly species. These four female forms are all members of the same interbreeding population, so they share the same genetic background. *doublesex* is responsible for switching on the gene pathways need to create the different mimetic forms.[2]

Curiously, *doublesex* in these mimetic swallowtail butterflies maintains its other regulatory functions such as sex differentiation, as it does in dungbeetles (*Onthophagus taurus*) in which *dsx* has been co-opted into the growth of horns in males (and the suppression of horn growth in females). Interfering with *dsx* expression in *O.taurus* disrupts the growth of horns in males, but causes females to grow horns. In a closely related species, *Onthophagus sagittarius*, *dsx* expression is linked to a bizarre sex reversal, where the females grow large horns on their heads and bodies (and males have only small head horns).[3] This is not because the female *O.saggitarius* dungbeetles are expressing the male form of the DSX protein, but because in this species the female form of the *doublesex* transcription factor, DSXF, triggers horn formation. This illustrates how regulatory genes can evolve to trigger different gene pathways.

ultrabithorax and haltere formation

Most winged insects have two pairs of wings. Flies, including *Drosophila*, have one set of wings and one set of wing-like flight organs called halteres. The Hox gene *ultrabithorax* (*Ubx*) is required for proper development of halteres in *Drosophila*. As Goldschmidt and others showed, if you knock out *Ubx* function, the fly can develop a second set of wings where its halteres should be (Figure 4.7). And if *Ubx* expression is turned on in the embryo's wing discs, it can cause them to develop as halteres.

But *Ubx* does not act in isolation. Haltere formation in *Drosophila*, like wing formation, is triggered by upstream regulatory genes such as *wingless* (*wg*) which prompt the formation of a wing primordium, an area of tissue from which the growing wing or haltere will develop. Within this primordium, *Ubx* acts to modify the expression of hundreds of downstream genes that contribute to the formation of wing structures such as veins, margins, cuticles, and bristles, upregulating or downregulating genes to produce the characteristic morphology of the haltere.[4] So we can see that there will be many ways to change a body structure associated with a particular Hox gene: alter the upstream triggers so that the Hox gene is turned on in a different time or place, alter the expression pattern of the Hox gene itself so that it responds to different triggers or interacts with different downstream genes, or alter any of the downstream genes that contribute to building or regulating the formation of the structure.

Switching on *Ubx* can cause a normally two-winged fly to grow four wings. But would the reverse be true: could modification of the *Ubx* expression in the four-winged ancestor of flies have resulted in evolution of halteres from wings? To make halteres, *Ubx* modifies the actions of hundreds of genes. Could change to the regulatory switch alone cause a new feature like halteres, or would it also require modifying all the structural genes that make halteres different from wings? Having the necessary genetic switches is not enough. After all, *ultrabithorax* and *wingless* predate wings; related genes are found in many animal species that do not have wings, including humans and flatworms. So the ancestral metazoan must have had a *Ubx* gene which was not associated with wing development. Then, when wings evolved, *Ubx* was drafted in to play a role in the regulation of wing development. The downstream targets of *Ubx* differ between species, allowing diversification of wing morphology.[5] As two-winged flies evolved from their four-winged ancestors, the role of *Ubx* changed again, to act in the suppression of the second set of wings and the growth of halteres in their place.[6,7] In fact, regulatory genes are such handy switches that they are co-opted into a wide range of different roles. In flies, *Ubx* is involved not only in haltere formation, but also in the development of the heart, trachea, gut, and muscles.

Questions to ponder

1. How can a regulatory gene like *doublesex* take on a new role (such as specifying the mimic phenotype) while maintaining its ancestral role (sex determination)?

2. Which came first: modification of downstream genes that specify the structural aspects of the mimic phenotype (affecting wing shape and coloration) or the upstream genes that regulate the switching of developmental programs between mimic phenotypes?

3. If *Ubx* expression predates the evolution of wings, why do all insects use *Ubx* to modify wing development? Why haven't some species co-opted a different regulatory gene?

Further investigation

There is growing interest in regulatory genes whose products are not proteins but small RNA molecules that can influence the expression of genes, particularly by preventing gene transcripts from being translated. What challenges might the study of the evolution and function of these small RNAs have for evolutionary biologists and geneticists? Would you expect the same basic patterns and processes that characterize regulatory genes that make transcription factor proteins to also apply to regulatory RNAs?

References

1. Tanaka K, Barmina O, Sanders LE, Arbeitman MN, Kopp A (2011) Evolution of sex-specific traits through changes in HOX-dependent *doublesex* expression. *PLOS Biology* 9(8): e1001131.

2. Kunte K, Zhang W, Tenger-Trolander A, Palmer D, Martin A, Reed R, Mullen S, Kronforst M (2014) *doublesex* is a mimicry supergene. *Nature* 507(7491): 229–32.

3. Kijimoto T, Moczek AP, Andrews J (2012) Diversification of *doublesex* function underlies morph-, sex-, and species-specific development of beetle horns. *Proceedings of the National Academy of Sciences of the USA* 109(50): 20526–31.

4. Pavlopoulos A, Akam M (2011) Hox gene *Ultrabithorax* regulates distinct sets of target genes at successive stages of *Drosophila* haltere morphogenesis. *Proceedings of the National Academy of Sciences of the USA* 108(7):2855–60.

5. Prasad N, Tarikere S, Khanale D, Habib F, Shashidhara LS (2016) A comparative genomic analysis of targets of Hox protein *Ultrabithorax* amongst distant insect species. *Scientific Reports* 6: 27885.

6. Carroll SB, Weatherbee SD, Langeland JA (1995) Homeotic genes and the regulation and evolution of insect wing number. *Nature* 375: 58.

7. Medved V, Marden JH, Fescemyer HW, Der JP, Liu J, Mahfooz N, Popadić A (2015) Origin and diversification of wings: Insights from a neopteran insect. *Proceedings of the National Academy of Sciences of the USA* 112(52): 15946–51.

Were dinosaurs evolutionary failures?

5

Roadmap

Why study dinosaurs?

Despite the great affection for dinosaurs from scientists and the general public alike, they rarely feature in evolution textbooks. But dinosaurs make an ideal case study for examining the macroevolutionary processes of adaptive radiation (diversification of a lineage into many different forms and ways of life) and mass extinction (simultaneous loss of many lineages, far exceeding normal background extinction rates). Considering the rise of dinosaurs allows us to examine the role of both key adaptations and chance in diversification. And considering their fall prompts us to examine our ability to resolve events and determine cause and effect in deep time.

What are the main points?

- The study of fossils gives us so much more than size and form of extinct species; it also sheds light on physiology, locomotion, behaviour, and ecology.

- The diversification of dinosaurs may have been driven by key adaptations, such as upright gait and fast metabolism, which allowed them to exploit a wide range of niches, or it may have been triggered by opportunity through the extinction of other reptile groups.

- Given that vertebrate fossils are rare, and biased in terms of location, taxa, and time period, there is some uncertainty over the timing and nature of dinosaur extinctions.

Photograph: Dinosaur footprint. © Celiafoto/Shutterstock.com.

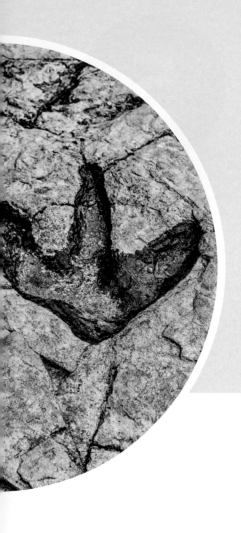

What techniques are covered?

- **Temporal resolution:** qualitative and quantitative assessment of the effect of incomplete sampling on our ability to assess timing and rate of evolutionary events.
- **Hypothesis testing in deep time:** making predictions based on proposed mechanisms and weighing up evidence for and against particular causes.

What case studies will be included?

- Testing evolutionary hypotheses with fossil evidence: Did birds out-compete pterosaurs?

"The public image of dinosaurs is tainted by extinction. It's hard to accept dinosaurs as a success when they are all dead. But the fact of ultimate extinction should not make us overlook the absolutely unsurpassed role dinosaurs played in the history of life."

Robert Bakker (1986) The Dinosaur Heresies: *New Theories Unlocking the Mystery of the Dinosaurs and Their Extinction.* Longman Scientific & Technical Press, Harlow, UK.

Adaptive radiation

The study of dinosaurs is a rare topic that can unite professional biologists, students, and members of the public in joint enthusiasm. They are also an underexploited case study in macroevolutionary biology for several key reasons. Firstly, even a cursory overview of modern studies of dinosaurs reveals the amazing breadth of understanding that can be gained from studies of fossils. Secondly, dinosaurs are a fine illustration of an adaptive radiation, the diversification of a lineage into a variety of forms which exploit a wide range of ecological niches. Thirdly, debate over the patterns and causes of dinosaur extinction provides a challenging area for considering the process of hypothesis testing in macroevolution. How can we reconstruct past events and test ideas about the causes of phenomena that cannot be directly witnessed? Fourthly, the patchy distribution of dinosaur fossils in space and time affects our ability to resolve the timing and nature of specific events. Do we have to accept limits to our ability to discriminate different hypotheses, or can evidence always be found to weigh the explanatory power of different hypotheses?

In this chapter, we will briefly address these four issues of reconstruction, radiation, historical inference, and resolution. Throughout the chapter, we will critically examine the nature of the information we can derive from the fossil record, and how we can use that information in a hypothesis testing framework, taking both qualitative and quantitative approaches to the evidence. We are going to begin by considering the impact that the discovery of dinosaurs had on people's ideas about earth history and biological change over long time periods.

Fossils reveal a hidden history

To understand the importance of fossils in shaping our understanding of biodiversity, try to imagine what the discovery of dinosaur bones would mean to people who did not know about the evolutionary history of life on earth. Dinosaurs, so strikingly different from anything alive today, were shocking evidence that life on earth had changed over time. Remains of long extinct creatures provided a connection to a past time entirely different from the world we know, a glimpse of a world we could not have imagined without fossil evidence.

Dinosaur bones must have been occasionally uncovered throughout human history, and may even be the basis of some stories of mythical beasts. For example, it has been suggested that ceratopsid bones exposed in the Gobi desert gave rise to tales of gryphons, with their stout legs, tough 'beaks', and bony head adornments interpreted as lion-footed eagle-headed winged beasts (see Chapter 1, Figure 1.2). One of the first scientific descriptions of a dinosaur was made by Gideon Mantell, a doctor from Lewes, England, and a keen collector of fossils. In 1822, he and his wife Mary Ann collected a number of unusual teeth in East Sussex (**Figure 5.1**). By comparing these teeth with those in natural history collections, Mantell inferred that they were from a large iguana-like animal. Further finds of vertebrae and limb bones led him to reconstruct *Iguanodon*

as an arboreal lizard, but because the bones were disarticulated (jumbled together) he made some mistakes in reconstruction, most famously adding a horn to the nose which later discoveries revealed was actually a thumb spike. He later realized that the forelimbs were relatively light. Subsequent reconstructions portrayed *Iguanodon* as bipedal, sometimes resting on its tail like a kangaroo. However, this bipedal stance has been challenged by studies which suggest that although a juvenile *Iguanodon* may have been capable of standing on two feet, as it grew larger and heavier, it must have increasingly moved on all fours (Figure 5.1).

The changing form of *Iguanodon*, from arboreal lizard to bipedal dinosaur, illustrates an important point: as species that we can never observe directly, our understanding of dinosaurs changes markedly with new discoveries and reanalyses of existing evidence. For example, sauropods, the largest terrestrial animals to have ever lived, were once commonly portrayed as laconic swamp dwellers, too large to support themselves on land, and, with small heads and peg-like teeth, incapable of eating anything tougher than waterweed. But ongoing palaeontological research has completely overturned this image. Footprints have shown that sauropods not only walked on land, but travelled in large herds. Biomechanical modelling suggests that some sauropods had sufficient bite force to process hard foliage, and their digestive capacity could have been increased if tough material was processed in a stone-filled crop (as in birds) or fermented in the gut (as in ruminants).[1,2] Biomechanical models also suggest that the long tail counterbalanced the long neck, which may have allowed a wide reach in high or low browsing to reduce the cost of moving such a large body to forage.[3]

In the colloquial sense, the word 'dinosaur' is often used to refer to any of the large extinct reptiles that may be found as reconstructed skeletons in museums, as cheerfully coloured monsters in children's books, or rampaging through adventure movies. But Dinosauria is a specific taxonomic category, describing a super-ordinal clade that contains some animals you might not think of as dinosaurs, such as chickens, and excludes some very dinosaur-like animals, such as sail-back lizards

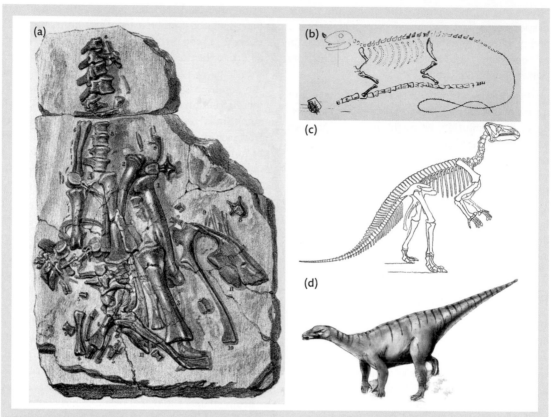

Figure 5.1 The changing stance of *Iguanodon*. (a) This rock found in Maidstone, England, known as as the 'Mantell-piece', played an important role in one of the first dinosaur reconstructions by Gideon Mantell in the 1820s. Disarticulated fossil remains such as these present challenges to reconstruction. Mantell's initial reconstructions were of a quadrupedal arboreal lizard (b), but later reconstructions, such as this one by Othniel Marsh in 1896 (c), portrayed *Iguanodon* as capable of bipedal stance. Nowadays, *Iguanodon* is often reconstructed as moving on all fours (d) due to the interpretation of the wrist bones as being weight-bearing and doubt that the tail could have supported the kangaroo-style tripodal stance.

(**Figure 5.2**). Like any taxon, membership of the Dinosauria can be defined in two different ways. One is by the possession of key characteristics; for example, Dinosauria has distinctive traits associated with the pelvic girdle and upright gait. An alternative definition is the position of the taxon in a phylogeny; in the case of Dinosauria, this includes all descendants, living and dead, of the most recent common ancestor shared by a tricera tops and a sparrow. Today, there are around 10,000 extant species of Dinosauria, better known as birds. Approximately 1000 extinct non-avian dinosaur species have been described in over 300 genera. Hereafter, we shall follow popular convention and refer only to the extinct non-avian members of Dinosauria as 'dinosaurs'. What follows is not a comprehensive guide to dinosaur evolution, but a selective account that picks out a few interesting points as food for thought for developing a macroevolutionary perspective. Interested readers are encouraged to delve deeper into the scientific and popular literature for more information.

Figure 5.2 Not a dinosaur: not everything that fits the image of a dinosaur is actually a member of Dinosauria. *Dimetrodon*, a sail-back pelycosaur reptile from the Permian (found from around 280 to 265 million years ago), is more closely related to mammals than it is to dinosaurs. The sail on its back has commonly been considered to function for thermoregulation by pumping blood through the sail to warm it in the sun or dissipate excess heat to the air. However, this hypothesis has been questioned, because the sail first evolved in small-bodied pelycosaurs which, according to metabolic modelling, would not have benefited from thermoregulation organs. Instead, the authors of the study suggest that the sails may have been sexually selected ornaments.[42]

Photo by D'Arcy Norman (CC3.0 license).

Triggers of adaptive radiation

The rise of the dinosaurs can be viewed as part of the colonization of the land by vertebrates. The vertebrate lineage can be traced back to the Cambrian, but was entirely aquatic until the late Devonian or early Carboniferous period. The earliest terrestrial vertebrates probably resembled amphibians that lived on the margins of the marine environment or in shallow freshwater. They had lungs to breathe air and bony limbs to support themselves on land, but they were still dependent on water for reproduction and were vulnerable to desiccation, because of loss of moisture through their wet skin.

Reptiles descended from these first land-dwellers, acquiring additional adaptations to life of land. They had tough waterproof skin which reduced desiccation, modifications to their excretory systems which helped them to conserve water, limbs that conferred a striding gait for greater speed and agility, stronger skulls and teeth to bite and chew hard terrestrial food, and, importantly, the amniote egg, with its desiccation-resistant shell, which allowed reptiles to break the dependence on returning to water to reproduce. These adaptations to life on land allowed the reptiles to diversify into a wide range of niches, including large and small predators, herbivores, and scavengers. Several groups of reptiles underwent parallel adaptive radiations, each producing a diverse range of forms. This makes dinosaurs, and other reptiles, an excellent case study in adaptive radiation, a topic of much interest in macroevolutionary biology.

There are many different definitions of adaptive radiation, but the key feature is the relatively rapid diversification of a single ancestral lineage into a wide variety of niches (ways of living). An adaptive radiation produces many descendant species (diversity) which have many different modes of life (disparity). In theory, the process of adaptive radiation could be expected to resemble the following basic pattern.[4] First, a lineage colonizes a new resource for which there are few existing competitors. This allows the population to thrive and expand. Then, there may be selection, either on standing variation or new mutations, to allow different parts of the resource spectrum to be utilized, providing the driving force for the evolution of different adaptations in different populations. Once different groups have specific adaptations to particular resources, there may be selection against mating between populations if hybrids between differently adapted populations are less well suited to exploiting either resource. This could drive the evolution of reproductive isolation between species, further allowing ecological specialization. Adaptive radiation can be viewed as diversification in response to opportunity, rather than being driven by the division of populations into isolated groups by geographical barriers or distance.

What makes a particular lineage produce an unusually high diversity and disparity of descendants at a particular point in time? There are two commonly discussed triggers of adaptive radiation, both of which focus on the opening of opportunities to expand into new ways of life.

Empty niche space

An adaptive radiation may be triggered when a change in environment makes available new, relatively unexploited ecological opportunities, so that resources can be utilized without much competition from other species. Such opportunities are commonly referred to as 'empty niche space' (although it has been objected that the niche is a property of a species rather than a property of the environment). The availability of these underexploited resources can trigger both a rapid rate of evolutionary change, as species adapt to take advantage of the resources, and a high rate of speciation, as lineages diversify to specialize in different parts of the resource spectrum.

Lineages may encounter empty niche space when they move into a previously unexploited area. Island radiations, such as Darwin's finches on the Galapagos, provide classic examples of radiation into empty niche space. A change in the environment may also create new opportunities. For example, the expansion of grasslands in the Miocene provided the opportunity for the radiation of grazing mammals (see Chapter 3). Alternatively, opportunities may open up through the extinction of other species which vacate previously occupied niches for exploitation by new species. For example, the radiation of mammals into a wide variety of niches in the early Palaeogene has commonly been interpreted as filling niches left vacant by the extinction of the dinosaurs (see Chapter 6).

Key adaptation

The other commonly discussed trigger of an adaptive radiation is key adaptation—evolutionary innovations that allow a lineage to move into a new niche. For example, the amniote egg can be considered a key adaptation to life on land because it opened up opportunities for reptiles to move into fully terrestrial niches. But the identification of key adaptations is not as simple as it might seem. Not all traits that allow exploitation of a new niche prompt a radiation. Some innovations provide the basis of specialization to a particular way of life. For example, several mammal lineages have developed adaptations that allow them to eat ants, an abundant resource that most mammals cannot eat. But although these adaptations allow access to a new resource that is relatively competition free, ant-eating lineages have not shown elevated diversification rates. And just because a trait is found in a lineage that has undergone adaptive radiation does not imply that that particular trait was the key adaptation that allowed the diversification to happen. For example, while many ancient bird lineages had teeth, none of the living descendants of the adaptive radiation of modern birds have teeth. Can we infer from this that toothlessness was a key adaptation that permitted the diversification of modern birds?

This illustrates a point we will meet several times in this chapter. Unique events present challenges in testing causal hypotheses. Compared with their living reptilian relatives, living birds have many distinct characteristics, including bones with air chambers, a beak, no teeth, and feathers. Which of these traits were intrinsic to their adaptive radiation, and which incidental? We might be tempted to say that feathers are a key adaptation that allowed the evolution of powered flight in the theropod lineage, thus opening up the aerial niche and triggering an adaptive radiation that has resulted in the tens of thousands of bird species alive today. But feathers evolved in non-flying theropod dinosaurs, and may have originally functioned in thermoregulation or display. They were co-opted into flight in only a single lineage. So the origin of feathers didn't cause flight to evolve, but was co-opted into serving an essential role in the flight machinery. Can we consider a single feature (e.g. feathers) to be a key adaptation in some lineages (e.g. birds), but not in others (e.g. theropods)?

Key points

- Adaptive radiations mark the diversification of a lineage into a wider range of niches than it previously occupied.

- The increase in evolutionary diversity (number of lineages) and disparity (adaptation to different niches) could be triggered by the availability of empty niche space (through the advent of a new resource, colonization of an unexploited resource, or extinction of incumbent species) or through evolution of a key adaptation, evolutionary innovations that allow a lineage to exploit a new resource or environment.

- We must be cautious in attributing a causal role to an apparent key innovation that coincides with an adaptive radiation, as many unique traits will be associated with any given radiating lineage.

Adaptive radiation of dinosaurs

Now that we have considered some of the general features of adaptive radiation, we can turn our attention to the diversification of the dinosaurs. The oldest dinosaur fossils are from the late Triassic, around 230 million years ago, although there may be some older footprints.[5] The first dinosaurs were relatively small bipedal carnivores or omnivores. But in the early Jurassic, they diversified into a wide range of body sizes and shapes and ways of life, with a global distribution. In fact, many hallmark features of the dinosaur radiation, like large size, quadrupedalism, and herbivory, evolved several times independently in different dinosaur lineages, as the dinosaurs radiated into a wide range of niches.

In addition, dinosaurs weren't the only reptile lineages undergoing adaptive radiations. The pterosaurs diversified into aerial niches, and marine reptiles such as plesiosaurs and ichthyosaurs evolved in the oceans. One of the most fascinating of these parallel adaptive radiations is the Crurotarsi (crocodile family, previously referred to as Pseudosuchia), which included a diverse range of forms including large terrestrial herbivores, carnivores, and piscivores (**Figure 5.3**). In many ways, the crurotarsan radiation mirrors that of the dinosaurs. In particular, like dinosaurs the crurotarsans evolved a range of different locomotion styles, including both quadrupedal and bipedal forms.[6] The convergence of form is so striking that it has occasionally led to misclassification. *Shuvosaurus* is so similar to an ornithomimid dinosaur in morphology that it has only recently been reclassified as a crurotarsan reptile (Figure 5.3).

Given that many different reptile lineages were diversifying in the Mesozoic era, why did the dinosaurs come to dominate the terrestrial niches? Or, to put it another way, why can most people recognize a dinosaur but few outside vertebrate palaeontology have heard of the crurotarsans? Two broad hypotheses have been put forward for the adaptive radiation of dinosaurs, and these fit neatly into the two triggers of adaptive radiation discussed above: key adaptation and empty niche space.

The 'competitive' hypothesis proposes that, due to particular key adaptations, dinosaurs were

Figure 5.3 More non-dinosaurs: the crurotarsan radiation produced many similar forms to the dinosaurs. Since dinosaurs and crurotarsans are separate evolutionary lineages, this is a clear example of convergent evolution.

(a) Dmitry Bogdanov (CC3.0 license). (b), (c), (e) Nobu Tamura (CC3.0 license). (d) Smokeybjb (CC3.0 license).

superior competitors to other reptilian lineages, such as the crurotarsans, and so they replaced them and went on to dominate all the major vertebrate niches. Alternatively, the 'opportunistic' hypothesis suggests that the crurotarsans had all the major terrestrial niches covered in the Triassic, preventing the dinosaurs from radiating into these

same niches. The dinosaurs only diversified following the extinction of competing crurotarsans at the end of the Triassic period, which opened up empty niches for the dinosaurs to colonize. We can think of these two hypotheses as contrasting 'good luck' (dinosaurs were the lucky beneficiaries of empty niche space) with 'good design' (dinosaurs had superior competitive abilities). As with so many ideas in evolutionary biology, these two hypotheses are the extreme ends of a broad spectrum of possibilities. Dinosaur adaptive radiation may have been triggered by key adaptations, empty niche space, or a combination of these and other factors.

Is it possible to use patterns in the fossil record to weigh up the relative merits of these different explanations? The relative abundance of different lineages over time may give some insight into the competitive replacement of one lineage by another (see Case Study 5). Replacement by competition might be expected to produce a gradual loss of crurotarsans and a concomitant rise in dinosaurs. Instead, there is a sharp drop in crurotarsan diversity at the end-Triassic mass extinction event and a rapid increase in dinosaur diversity in the early Jurassic. The dramatic increase in maximum body size of dinosaurs in the early Jurassic has been interpreted as a sign of release from competition following the extinction of other large terrestrial vertebrates. Indeed, some scientists feel that dinosaurs did not have any unique innovations that could be considered to be key adaptations that would give them a special advantage, and that their diversification was more a product of opportunity than of superiority. For example, palaeontologist Steve Brusatte and colleagues stated in 2010 that: 'There was nothing predestined or superior about dinosaurs when they first arose, and without the contingency of various earth-history events during the early Mesozoic, the Age of Dinosaurs might never have happened'.[7]

Key adaptations in dinosaurs

If there was nothing special about dinosaurs, why did they survive the end-Triassic extinction event when the crurotarsans were destroyed? Did they have particular adaptations that allowed them to avoid extinction, or were they just lucky? As discussed in the previous section, it is notoriously difficult to decide whether a particular trait is a key adaptation for a clade, because it is not always easy to confidently assign causal links between specific adaptations and the process of adaptive radiation. Although we can't know for sure what led to the rise of the dinosaurs, we are going to consider just two dinosaur traits that might have contributed to their adaptive radiation: gait and metabolism. These are not necessarily the only, or even the most important, features of dinosaurs, but they provide a convenient focus for thinking about the kind of information we can get from the palaeontological record. Both of these traits show how we can go beyond using fossils merely to study the shape, size, and location of extinct animals, to reconstruct movement, physiology, and even behaviour.

Walking with dinosaurs: gait

Consider the way that modern lizards walk. They typically have a 'sprawling' stance, with the body slung between legs that are bent at the knee. To walk, they swing their backbone back and forth horizontally, which has the effect of throwing each leg forward in turn to take a step (Figure 5.4(a)). Viewed from above, this gives reptile locomotion a sinuous quality, as the backbone (including the tail) snakes from side to side and the limbs move in the horizontal plane.

There are a number of limitations of the sprawling stance and side-to-side movement of the backbone. The sprawling stance requires active muscle use to maintain the body above the ground and it also limits the length of the limb, putting an upper limit on body size and running speed. In addition, walking by flexing the body sideways distorts the chest with each forward step, compressing the lung on one side, and stretching it on the other. When an animal with a sprawling stance runs, the rapid chest compressions might place limits on efficient breathing (a phenomenon known 'Carrier's constraint', though the degree to which it impedes sustained locomotion is debated). Some reptiles have adaptations that address the limitations of sprawling stance. For example, crocodiles can vary their stance from sprawling to an erect stance, and can use a 'high walk', with straightened legs

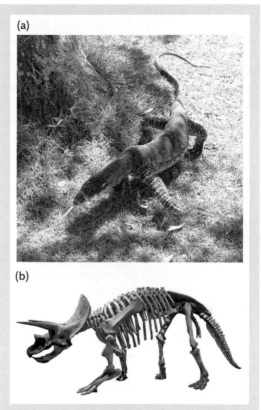

(a)

(b)

Figure 5.4 Standing tall: in the sprawling stance typical of modern lizards and crocodiles, like this goanna (a), the body is suspended between the legs and the spine flexes sideways as the foot is brought forward, unlike dinosaurs, such as this ceratopsid dinosaur (b), with legs held upright beneath the body.

(a) Photo by Quartl (CC3.0 license). (b) Photo by MathKnight (CC3.0 license).

beneath the body, when they need to put on a burst of speed. The largest living lizard, the Komodo dragon, seems to compensate for Carrier's constraint by actively pumping air into the lungs, by "swallowing" air from a throat pouch.

In contrast with the sprawling stance of most modern reptiles, dinosaurs held their legs beneath their body in an erect stance. Modification of the hip joint allowed them to stand with their knees beneath their hips, so that the legs were held vertically underneath the body and the limbs moved in the vertical plane (Figure 5.4(b)). Dinosaurs must have swung their legs beneath their body, just as mammals do today, a mode of walking known as parasagittal gait. Upright stance allowed dinosaurs to support a larger bulk, and may also have enabled faster running speeds. It takes less muscle power to sustain the upright stance over long periods, so it is better suited to continuous activity than a sprawling stance, which is more typical of animals that intersperse periods of activity with inactivity. However, size itself is likely to have set an upper limit on speed; running speed cannot simply be scaled up linearly with size. For a *Tyrannosaurus* to be able to run fast, its leg muscles would need to take up a ludicrously large proportion of its body mass.[8]

So how fast could dinosaurs run? The speed of long-extinct animals can be estimated in several different ways, for example by comparison with living animals with the same size or stride length, or using biomechanical modelling. Dinosaur trackways (preserved series of footprints), combined with their skeletal anatomy, have led to estimates of running speeds of up to 40km/hour for the smaller bipedal dinosaurs, but much less for large dinosaurs.[9] Despite the common depiction of *Tyrannosaurus* charging across prehistoric landscapes, many scientists contend that it probably rarely, if ever, broke into a run, and if it did would not have been faster than the average running speed of a human (around 20km/hour), which is good news if you ever find yourself being pursued by a tyrannosaur.

What do you think?

How could we test whether evolution of a striding gait was directly responsible for the diversification of dinosaurs into a wide range of niches? Can we distinguish whether it was gait itself that promoted diversification, or whether it had an indirect effect by allowing the evolution of gigantic size? Can we consider gait a key to dinosaur success when other similar reptile lineages also developed an upright stance but did not have the long-term success of the dinosaurs?

Running hot: metabolism

We have seen that we cannot understand dinosaur biology by simply scaling up modern lizards and birds to an equivalent size. For example, a bipedal living member of the Dinosauria, like an ostrich, can achieve far higher running speeds than one of the larger bipedal extinct dinosaurs such as a *Tyrannosaurus rex*. The same applies to physiology; we cannot simply extrapolate from observations of living lizards to predict dinosaur metabolism.

Extant reptiles are poikilothermic. This means that their body temperature is not maintained at a steady level but changes with the environmental temperature. When reptiles have a low body temperature, they have slow metabolism and low levels of activity. But when they warm up, for example by basking in the sun, they use the heat gained from the environment to increase activity. In contrast, mammals and birds are homeotherms. They use a substantial part of their energy budget to produce a constant body temperature, regardless of the environmental temperature. This allows mammals and birds to be continuously active, and to fine-tune their physiology and biochemistry to run at a predictable temperature. But maintaining a constant warm temperature comes at metabolic cost, so mammals and birds must maintain a high input of energy. So homeothermy not only allows high levels of activity, but also requires constant levels of activity for its maintenance.

There is an unfortunate tendency to think of 'cold-blooded' as a primitive condition superseded by the evolution of more advanced 'warm-blooded' mammals and birds. But metabolic strategies should be viewed as adaptations to different ways of living, and different strategies have different costs and benefits. Unlike homeotherms, which must maintain a constant high body temperature, poikilotherms can tolerate a range of body temperatures, but like homeotherms they need a relatively high body temperature for full activity. The advantage of poikilothermy is that it is not as metabolically costly as maintaining homeothermy, and does not require constant activity to supply enough energy. So poikilothermy allows animals to stay inactive for long periods, but it also requires that

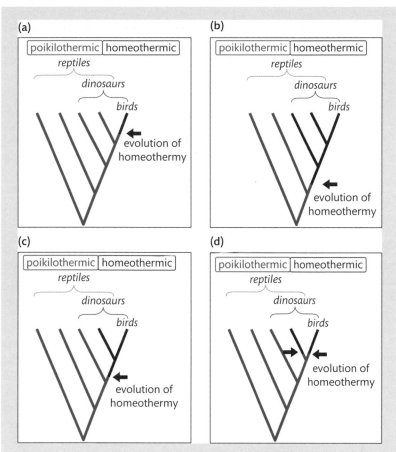

Figure 5.5 Alternative phylogenetic explanations for the origins of homeothermy in Dinosauria. Living dinosaurs (birds) are all homeothermic. From phylogenetic relationships alone, we can't tell whether homeothermy evolved in the bird lineage (a), or in the shared ancestor of all dinosaurs including birds (b), or within the Dinosauria (c). Alternatively, homeothermy may have evolved independently in some dinosaurs and also in birds (d). After all, we know that homeothermy has evolved independently at least twice (in birds and in mammals). In fact, poikilothermy has also evolved multiple times, including within mammals (naked mole rats are effectively poikilothermic, at least under some conditions).

they stay inactive when their body temperature is low. Typically, modern reptiles tend to have periods of inactivity punctuated by bursts of movement.

But dinosaurs, with their large body size, erect stance, and running gait, don't look like the sort of reptiles that spend most of their day doing nothing.

How could they grow so large, and be so active, with a sluggish 'cold-blooded' metabolism? Over the last few decades, there has been a growing debate about dinosaur physiology. Where once it was taken for granted that dinosaurs, being reptiles, must have been cold-blooded, there are now many different lines of evidence that suggest that at least some dinosaurs were warm-blooded. But how can we tell anything about operating temperatures when all we have is bones? What kind of information can we leverage to understand dinosaur physiology?

One argument for warm-blooded dinosaurs is derived from phylogeny; all living members of Dinosauria are homeothermic, so it's possible that extinct dinosaurs shared this trait. But given that homeothermic birds must have evolved from poikilothermic ancestors, we can't infer from phylogeny alone when the transition happened (Figure 5.5). Homeothermy might have evolved only in the avian lineage, as a special adaptation to the metabolically expensive activity of flight. However, evidence that many non-flying dinosaurs had downy feathers could suggest that homeothermy evolved before the origin of birds, given that these feathers would be useless for flight but might provide valuable insulation to prevent heat loss (like hair on mammals).

Another observation that has been interpreted as evidence for homeothermy is that the bones of some dinosaurs were highly vascular. This kind of bone structure, similar to modern mammals (but unlike modern reptiles), is associated with rapid growth, suggesting a high sustained metabolism. Studies of bone growth in dinosaurs reveal that they grew much faster than modern reptiles. For example, a sauropod that hatched at 10kg could grow to 10 tonnes in as little as ten years, a growth rate more like a whale than any living reptile.[10] And although patterns of seasonal growth in dinosaur bones have been interpreted as signs of dependence on environmental temperature, typical of poikilothermic animals, similar features have also been found in modern large-bodied mammals.[11] However, the interpretation of bone structures and growth rates in dinosaurs is, like most areas of historical inference, subject to debate and discussion.[12]

While some observations suggest dinosaurs may have had a high metabolism, that does not prove that they used metabolic energy to actively maintain a constant temperature (homeothermy), as modern birds and mammals do. It may be that some of the larger dinosaurs achieved high operating temperatures without the metabolic cost of actively maintaining homeothermy. Even in a 'cold-blooded' animal, activity generates heat within body tissues, which dissipates through the animal's skin. A large animal with a relatively low surface-to-volume ratio could, through normal activity, generate a large amount of heat, yet lose relatively little heat to the surroundings. Therefore large active animals can achieve a high operating temperature without really trying, a phenomenon known as gigantothermy. In fact, metabolic scaling analyses combined with isotopic analyses of bones suggest that gigantothermy could potentially have generated such high temperatures in the largest dinosaurs that they might have had mechanisms for reducing their body temperature to keep it at levels similar to that of modern mammals.[13] For example, it has been suggested that the long necks of sauropods may have helped to dissipate excess body heat.

Not all dinosaurs were large enough to benefit from gigantothermy. Yet some dinosaurs were able to live in polar regions which, although warmer then than they are today, would have been cool temperate regions.[14,15] Diverse dinosaur remains have been found in both the northern and southern polar regions, in places where no other ectotherms, such as lizards or amphibians, have been found.[14,16] While some polar dinosaurs might have undergone seasonal migration, others are likely to have overwintered.[17] Given that polar dinosaurs have been found in areas where there is no evidence of other poikilotherms, this could suggest that the dinosaurs had a different way of coping with low temperatures.

Do these observations prove that all dinosaurs had an active warm-blooded metabolism, like their modern avian relatives? Given how diverse the non-avian dinosaurs were in size and lifestyle, they may have varied widely in metabolism between species and even in different life stages. Operating at constant high temperature may even have evolved more than once in dinosaurs, as different lineages adopted different metabolic strategies. Like their upright stance, the physiology, ecology, growth patterns, and activity of dinosaurs show that they were not simply scaled-up lizards.

Key points

- Fossils of extinct taxa not only provide information about size and shape, but also allow investigation of
  - physiology, e.g. using bone structure to infer growth rates
  - behaviour, e.g. using trackways to study herding behaviour

- metabolism, e.g. feathers suggest maintenance of body heat.
- Palaeontology draws on a wide range of analytical tools, including biomechanical modelling, comparative studies, phylogeny, and population biology, to reconstruct dinosaur niches.

Mass extinction

We have had a brief look at the evolutionary success of dinosaurs, considering some of the traits that may have enabled them to exploit a wide range of niches and diversify into a variety of forms, habitats, and ways of life. But one of the most remarkable facts about dinosaurs, from the time of their discovery to the present day, is also one of the most obvious: they are no longer around. The prior existence of massive reptiles, now conspicuously absent, forced natural historians to confront the reality of extinction (see Chapter 1). Just as we have used dinosaurs as a case study of diversification, now we are going to use them as a case study of mass extinction. In particular, we are going to use the final disappearance of the dinosaurs at the end of the Cretaceous period to consider how we can investigate events that happened in the deep past. How can we test hypotheses about events that we cannot witness directly, replay, or experiment on? (see Case Study 1).

Extinction is the irreversible loss of all of the unique genetic information that defines a species. So what does it mean to say 'the extinction of the dinosaurs?' Particular dinosaur species arose and went extinct throughout the Mesozoic; like any major biological group, there was gain and loss of species over time. For example, there are no known species of *Brachiosaurus* (one of the largest kinds of sauropod dinosaurs) after the end of the Jurassic period, around 145 million years ago. And the dinosaur lineage is still with us today, with approximately 10,000 living species (better known as birds). But there is no record of any non-avian dinosaurs living after the end of the Cretaceous period. Why not?

If you ask most biologists today what killed the dinosaurs, they would probably say that the dinosaurs met their doom due to a very unlucky event, the chance collision between the earth and a large object from space. Yet the impact extinction hypothesis is only one of a large number of hypotheses that have been put forward to explain dinosaur extinction. How can we be certain that it is the right explanation? Is it possible to prove any hypothesis about dinosaur extinctions either true or false?

The impact extinction hypothesis

66 *It is not fantastic to consider the likelihood that an extra large impact affected this planet some sixty million years ago. No human being was here to see it, or study it with radar. But just such extinctions occurred, and just such survivals occurred, at the end of the Cretaceous, as would be expected to occur as the result of impact from a planetesimal or an extra large shower of meteorites. May we not take this into consideration as possibly having been the doom of the dinosaurs?* 99

M.W. De Laubenfel (1956) Dinosaur extinction: one more hypothesis. *Journal of Paleontology* 30: 207–18.

One of the great challenges of reading evolutionary history from the palaeontological record is relating the record in the rocks to absolute time. Because the amount of deposition varies over time, the evolutionary timescale cannot be simply read from depth in the strata. For any given site, there may be periods when a large amount of sediment accumulated in a short time, and other sections where little or no sediment collected, representing a hiatus in the record. A hiatus is like pressing 'pause' on the recording of history. If there is no accumulation of sediment, fossil evidence might not be preserved, so we will be missing information on the species present at that particular time and place.

This raises an important question about how we assess the passage of time from geological evidence. It is usually difficult to date fossils directly (carbon dating is only useful for recent material, not fossils that are millions of years old). Radioactive isotopes of uranium and potassium can be used to date igneous rock (formed from volcanic activity), but this is not the rock that contains fossils, which form in sedimentary rock. Therefore the age of fossils can only be bracketed by considering the date of igneous layers above and below, but this rarely gives fine temporal resolution for biotic changes. So when there is an obvious disjunction in the fossil record, marking the sudden disappearance of many species, how can we tell when it reflects a sudden change in the biosphere of the time, or when it is a failure to record the biotic changes over a more extended period?

In the 1970s, a multidisciplinary team, consisting of a geologist, a physicist, and two nuclear chemists, set out to test if rare metals might provide a record of passing time, independent of the amount of deposition in the rock record (**Figure 5.6**).[18] Platinum group metals, such as platinum and iridium, are rare in the earth's crust but more common in meteorites, so their presence in sediments is commonly attributed to meteorite dust raining down continuously on the earth. This constant rain of space dust might provide a continuous clock, as iridium would continue to accumulate even when sedimentation stopped; a thin geological section with a high amount of iridium might indicate a long period of low deposition, while a thick section

Figure 5.6 Interdisciplinarity at work. Although an extraterrestrial impact had been previously proposed as a possible cause of end-Cretaceous extinctions, it was the work of this research team that put the impact theory front and centre of dinosaur extinction theories: chemists Helen Vaughn Michel and Frank Asaro, geologist Walter Alvarez, and physicist Luis Alvarez.[18] Interestingly, and in common with a recent interdisciplinary team that published a comprehensive account of evidence for the impact hypothesis,[19] this research team did not include a vertebrate palaeontologist.[43]

© 2010 The Regents of the University of California, through the Lawrence Berkeley National Laboratory.

with little iridium might indicate a short period of high deposition. The researchers measured iridium levels over one centimetre of clay sediment that provides a sharp boundary between the Cretaceous and Tertiary marine sediments in the Appenine Mountains in Italy. If this clay layer represented a hiatus, then it should have a greater concentration of iridium than surrounding strata.

Indeed, the boundary layer did have a raised level of iridium. But it had so much iridium, 30 times that of the preceding and subsequent sections, that the team felt it could not simply be explained by a pause in the sedimentary record. The iridium spike was also found at other locations around the world at the Cretaceous–Tertiary (KT) boundary (which is now formally referred to as the Cretaceous–Palaeogene or K–Pg). They concluded that the iridium spike provided evidence of a massive influx of extraterrestrial material at the end of the Cretaceous. The authors suggested that the best explanation was that a meteorite measuring 10km across slammed into earth, sending dust into the atmosphere which blocked the sun, halted photosynthesis, and resulted in the collapse of food chains on earth, bringing about catastrophic extinctions. Using information on craters and astronomical observations, they estimated that meteorites of that size have a chance of hitting the earth every two hundred million years, and they employed a comparison with the Krakatoa volcano to argue that dust ejected into the earth's atmosphere could have worldwide effects.

But where was the impact crater? A prime candidate was found 10 years later. The Chicxulub crater on the Yucatán peninsula in Mexico, estimated to be 180km in diameter, dates to the end of the Cretaceous period. Other evidence has been found that supports a large impact at this time, such as shocked quartz that may have been caused by massive shock waves moving through the rock, tiny spherical bodies that may be droplets of molten rock thrown up from the impact, and sediments that may have been deposited by a massive tsunami caused by the impact.

All of these lines of evidence, including the iridium layer, crater, shocked quartz, and ejecta, build a convincing picture of an impact occurring at the end of the Cretaceous period.[19] But note that all these lines of evidence are about a physical event; none of these observations is directly connected to biological extinctions. Many researchers consider that the coincidence of a massive impact with the final disappearance of the dinosaurs is convincing enough proof that one caused the other. Is the evidence of impact all we need to close the case on dinosaur extinctions?

What do you think?

Is the coincidence of two rare events sufficient to establish a connection between the two? Can causes of unique past events ever be convincingly established?

Predictions and observations

In this chapter, we will use the impact extinction hypothesis to examine some important issues in macroevolution. Is it possible to scientifically test a hypothesis about a single event that happened millions of years ago? We can't witness the event, we can't rerun it, and we can't directly test the effect on the earth's biota. But hypothesis testing does not need to involve direct experimental manipulation. Instead, we make predictions about what we should expect to see if that hypothesis is true (see Case Study 1). So here we need to ask: if it is true that a massive extraterrestrial impact caused dinosaur extinctions, what patterns should we expect to see in the fossil record? The first step is to formulate what would be the expected outcome of an impact of this magnitude. We are focusing only on dinosaurs, so this will not be a comprehensive investigation of all K–Pg extinctions (in fact, much of the debate has concerned the record of marine plankton). However, we can draw on information from other taxa to evaluate the hypothesized 'killing mechanisms' for dinosaur extinctions.

An impact big enough to form the gigantic Chicxulub crater would have caused an unimaginable disaster, vaporizing everything in the immediate vicinity, sending shockwaves through the surrounding regions, potentially triggering a destructive tsunami, and possibly starting massive wildfires. But, even though they would cause broadscale devastation, these are regional effects, largely confined to one continent or one ocean. Dinosaurs were distributed across all continents, from the poles to the tropics. To explain the global disappearance of dinosaurs, we are less interested in the local consequences of a meteorite impact. Instead, we need to focus on any predicted worldwide effects.

The explosive force of a large impact would be expected to throw a large amount of material high into the atmosphere, where it could move away from the site of the impact and potentially affect the entire globe. There are several major predicted effects, some immediate and some more gradual. The blast could have generated an immediate thermal pulse, potentially global in extent, as the red-hot debris from the blast re-entered the atmosphere. The injection of fine dust into the atmosphere could have created a barrier to sunlight, bringing darkness and an 'impact winter'. Sulfur released by the impact might have caused acid rain. If these modelled effects of a massive impact are broadly correct, then we should expect to see a significant effect on the earth's biota at the time of the impact. Can we predict what the effects would have been?

Burning skies: thermal radiation and acid rain

It has been proposed that a massive thermal pulse caused by the impact would have killed all exposed individuals.[20] Large terrestrial animals and plants would have had little chance of shelter from the instant baking heat, but small animals and plants (including seeds and eggs) that were sheltered by rocks, soil, or water may have been able to survive. Consistent with this hypothesis, there is no evidence of large-bodied terrestrial vertebrates surviving from the end of the Cretaceous period into the early Palaeogene. But we might expect that animals capable of sheltering from the blast of heat, or whose habitat provided thermal protection, might have survived the initial impact. Yet the marine reptiles, including mosasaurs and plesiosaurs, also disappear at the K–Pg boundary, even though the oceans should have provided protection from the thermal pulse. Large-bodied crocodiles also went extinct, even though they could potentially shelter in water or underground dens like modern crocodiles. Evidence is emerging of burrowing dinosaurs,[21] so perhaps some dinosaur species had the capacity to shelter from a firestorm. If any dinosaurs buried their eggs to incubate them underground, as many modern turtles do and as has been suggested for some pterosaurs, their offspring might have survived the initial blast.

Acid rain, triggered by the release of sulfur into the atmosphere, might have had more prolonged effects. Acid could have had a direct negative impact on vulnerable organisms, such as those with calcium carbonate shells. While the oceans might have been buffered against a significant change in acidity, smaller bodies of water might have been more strongly affected. However, species often considered to be most at risk from acid

rain today, such as freshwater fish and amphibians, do not seem to have suffered a high level of extinctions at the end of the Cretaceous.[22,23] Interestingly, acidification could possibly affect fossilization, for example eroding the shells of calcareous plankton or destroying bones.[24,25] This is a reminder that the fossil record is influenced by both geological and biological processes, so the record is not wholly independent of changes occurring in the biosphere.

The long night: impact winter

If any dinosaurs survived the initial effects of the impact, what kind of environmental effects would they have had to endure? An 'impact winter', caused by the dust from the impact site blocking the sun, could have reduced light levels and dropped global temperatures. While larger particles might have settled out of the atmosphere within days, smaller dust particles might have affected light levels for months. If global darkness was prolonged, the reduction in photosynthesis would have had catastrophic effects on the food chain, reducing food availability for herbivores, which in turn would reduce prey for carnivores. There is evidence of a major perturbation of plant communities in several regions of the world, particularly North America, although other areas seem to show fewer signs of disturbance and most of the world did not suffer a high rate of plant extinction.[26,27] What about photosynthesis in the oceans? Like the terrestrial plant record, there is evidence of high phytoplankton extinction rates in North America, but less so further from the impact site,[28] and no clear signal of a global collapse of photosynthesis-dependent marine food webs.[29]

Can prolonged darkness explain the dinosaur extinctions? Clearly, a drop in photosynthesis activity would have massive ecological effects, as nearly all species on earth are ultimately dependent on energy captured from the sun by primary producers, either directly or via food webs. So the really puzzling question is how an impact winter could cause mass extinction in some groups (such as dinosaurs), while other groups (such as fish and amphibians) survived without suffering high levels of extinction. It has often been assumed that poikilothermic dinosaurs would have been particularly vulnerable to global temperature change during an impact winter because of a dependence on environmental heat. But the existence of a diverse community of polar dinosaurs suggests that at least some dinosaurs were adapted to survive extended periods of low light. While some polar dinosaurs may have migrated seasonally, it seems likely that some must have overwintered, so must have been capable of surviving prolonged periods of cold and dark.[17] Environmental temperature dependence does not provide a clear distinction between those taxa that suffered mass extinction at the K–Pg boundary, such as lizards, snakes, and dinosaurs,[30] and those that did not, such as crocodiles and amphibians.[31]

Could impact winter have killed off the dinosaurs by skewing their sex ratio? Many modern reptiles have temperature-dependent sex determination, where the sex of hatchlings is determined by their incubation temperature. If the sex of offspring depends on temperature, then rapid climate change might result in a reduction in numbers or loss of one sex, bringing about a sharp decrease in reproductive output. If dinosaurs had temperature-dependent sex determination, could a rapid change in global temperature have brought an end to dinosaur reproduction?[32] We can't directly test this hypothesis for dinosaurs, because their sex determination system is unknown, but some researchers have asked whether other reptile lineages that are likely to have had temperature-dependent sex determination fared poorly at the end of the Cretaceous. So far, there is no evidence that lineages that used temperature to determine sex of offspring suffered higher levels of extinction at the K–Pg boundary.[33]

What do you think?

The proposed killing mechanisms of the Chicxulub impact don't always seem to draw a clean line between the survivors and the extinct lineages. Does this mean that we can reject the impact extinction hypothesis? Or does the balance of evidence weigh on the side of catastrophic impact, even if the predictions are not perfectly met? How do we judge when conflicting evidence is sufficient to reject a hypothesis?

Timing of dinosaur extinctions

66 *This is the story of one terrible day in the history of the Earth.* 99
Walter Alvarez (1997) *T. rex and the Crater of Doom.* Princeton University Press, Princeton, NJ.

One of the fascinating aspects of the impact extinction hypothesis is that an extraterrestrial impact is a catastrophe that happens suddenly at a specific point in time. If the Chicxulub impact was the primary cause of dinosaur extinctions, then all of the extinctions should be sudden and simultaneous, occurring at, or just after, the impact. Looking back 66 million years into the past, can we distinguish an instantaneous event from one that occurs over a protracted time period?

A mass extinction event is typically defined as one in which a significant proportion of the world's biota disappears in a 'geologically insignificant time period', with respect to the resolution of the record. To consider why resolution of the record is pertinent to testing the impact extinction hypothesis, we must keep in mind that fossilization is rare. For a dinosaur to become fossilized, its body must have become buried in sediment soon after death, for example in river mud, volcanic ash, or estuarine silt, in conditions that prevent it rotting away to indistinct compost. The sediment must have remained undisturbed, not eroded by a river nor washed away by rain nor churned up by burrowers, and eventually turned into rock. The rock must have escaped being crushed, heated, or eroded for tens or hundreds of millions of years, yet be present at or near the surface in the present day, and in an accessible area. And finally, the fossil must have been exposed and detected by someone who was motivated to carefully extract the specimen and identify it. Consider that many dinosaur species, although they must have been represented by many tens of thousands of living animals over the life of the species, are known from only one or a few incomplete specimens. Given that new dinosaur species are still being discovered, there must be many dinosaur species from which we currently have no fossils at all. Furthermore, there are very strong biases in the species that we know about. For example, we are more likely to have

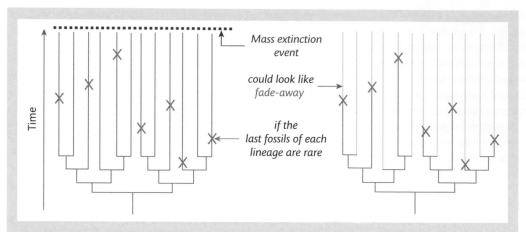

Figure 5.7 The Signor–Lipps effect describes how limits to temporal resolution arise from the stochastic nature of the fossil record. Since fossilization is rare, the chance of the last living member of a species being recorded is very low. Instead, the last fossil will be from some date before the final extinction. Even if all species went extinct on the same day, the final fossils will occur over a range of times before that catastrophic event. Therefore it might be difficult to distinguish a gradual pattern of extinctions from a sudden mass extinction.

species from environments that provide opportunities for fossilization, such as river margins. Large animals might also be more likely than small ones to find their way into palaeontological collections.

The rarity of fossils introduces limits to the resolution of temporal changes in biodiversity. Even if all the individuals of a particular dinosaur species died on the same day, the chances of one of those last surviving representatives of a species being in our current palaeontological collections is vanishingly small. Instead, we would probably find that the last known fossil preceded the date of the final species extinction by many millions of years. So even if many different species all went extinct at the same time, the final fossil of each species would be found over a range of dates before the extinction event (Figure 5.7). This limit of temporal resolution means that, for taxa like dinosaurs with relatively sparse fossil records, it may not be possible to tell whether species all went extinct at the same time or whether the extinctions were spread out over many millions of years.

This problem is illustrated by the debate over the timing of the dinosaur extinctions at the end of the Cretaceous. Dinosaur species arose and went extinct throughout the Mesozoic, so many species were extinct long before the end-Cretaceous events. For example, the iconic genus *Stegosaurus* is known only from the Jurassic period, between 155 and 150 million years ago. It is surprising difficult to answer the question: Did the dinosaur species that were alive in the last part of the Cretaceous all go extinct at the same time or not?

With a whimper or a bang?

❝ *Sampling biases can generate patterns in the fossil record which appear to reveal biologically significant phenomena where, in fact, none exist.* ❞
Phillip W. Signor, Jere H. Lipps (1982) Sampling bias, gradual extinction patterns and catastrophes in the fossil record. *Geological Society of America Special Paper* 190: 291–6.

There are many fossil sites around the world that have dinosaur fossils from the Cretaceous period. But to ask whether dinosaur extinctions coincide with the Chicxulub impact, we need to examine sites that have a continuous terrestrial vertebrate fossil record running from the latest parts

Figure 5.8 Some of the dinosaurs found in the Hell Creek formation, Montana, USA, which spans the late Cretaceous and early Palaeogene: *Ankylosaurus, Tyrannosaurus, Quetzalcoatlus, Triceratops, Struthiomimus, Pachycephalosaurus, Acheroraptor,* and *Anzu*.

of the Cretaceous through to the beginning of the Palaeogene. In other words, we need a series of fossils that records changes in the fauna before, during, and after the K–Pg event. Currently, such sites are restricted to the western interior of North America (though some Southwest European dinosaur sites have fossils from just below the K–Pg[34]). So while we have a rich record of dinosaur diversity from around the globe, our picture of changes to biodiversity over the K–Pg boundary is currently focused on one particular region. We must be careful to consider whether the information we have on late Cretaceous dinosaur extinctions is sufficient to paint a picture of global patterns of extinctions, or whether it largely provides insight into the ecological changes in a particular region (we saw a similar problem in Chapter 3 with the effect of geographic sampling on understanding the evolutionary history of the horse family).

The best-known site for the latest Cretaceous dinosaurs is the Hell Creek formation in Montana, USA (Figure 5.8). Although Hell Creek has been studied intensively, there is still a lively debate about whether the pattern of fossil occurrences in these strata constitutes evidence that the end-Cretaceous dinosaur extinctions were gradual or sudden. Some researchers interpret the lack of fossils in the sediments immediately preceding the K–Pg boundary as indicating that, in this area at least, dinosaurs were gone long before the Chicxulub impact, so their extinction must be due to other causes. Others consider that the lack of fossils just below the K–Pg layer is a sampling artefact that could arise even if dinosaur diversity was not declining, so cannot be used as evidence against the impact extinction hypothesis.

As a further complication, whether the diversity of dinosaurs declined or not during the late Cretaceous depends on how you estimate diversity. For example, straight counts of dinosaur species offer no clear evidence of decline in dinosaur diversity. But some statistical corrections for sampling bias, taking into account the number of dinosaur-bearing formations available over time, suggest that dinosaur taxonomic diversity was declining throughout the end of the Cretaceous period (see Case Study 3).[35,36] A study that used dinosaur phylogenies, scaled by fossil evidence of first and last appearances, suggested that extinction rates of dinosaurs were greater than speciation rates throughout the Late Cretaceous, leading to a long-term decline in dinosaur diversity.[37] Of course, different kinds of dinosaurs might show different patterns of diversity over time. A study that used measures of anatomical variability between species over time suggested that morphological disparity was declining toward the end of the Cretaceous period for some large herbivores, particularly ceratopsids and hadrosauroids, but was increasing for the sauropods.[38]

Key points

- The impact hypothesis makes a number of testable predictions concerning the timing, duration, and selectivity of extinctions.

- A thermal pulse following the blast may have killed unsheltered organisms, although burrowing, dormant, or marine species may have had some protection.

- Acid rain from the explosive release of sulfur may have altered shallow aquatic environments, though freshwater fish and amphibians do not seem to have suffered mass extinction.

- An impact winter may have caused global cooling, perturbing plant and planktonic communities, though there is no direct evidence that primary productivity was severely disrupted at all locations.

- Resolution of the timing of dinosaur extinction is currently disputed; some interpret fossil series spanning the K–Pg as evidence for sudden extinction at the boundary, others as evidence for most extinctions occurring prior to the impact.

Alternative hypotheses

We have seen that some evidence seems to fit the impact extinction hypothesis, but other observations are harder to explain. In addition, the limits to temporal and spatial resolution in the dinosaur fossil record leave room for disagreement between palaeontologists over the interpretation of the evidence. It may be that future research will reveal patterns that clearly support the impact extinction hypothesis. Alternatively, new studies might reveal evidence that speaks strongly against a catastrophic impact-induced extinction. Probably, studies offering evidence and analysis supporting the impact extinction scenario, as well as studies casting doubt on the match between observation and one or more of the predictions of the hypothesis, will continue to be published. This is one of the reasons that dinosaur extinction is such an important case study for appreciating both the delights and the challenges of macroevolutionary biology. We have to learn to deal with the data we have, and keep a weather eye open for uncertainty and bias.

What are we to make of studies that suggest a lack of fit between the predictions of the impact extinction hypothesis and observations or analyses? Should we regard the impact extinction hypothesis as the best explanation we have? Do we need a viable alternative hypothesis before questioning the most widely accepted explanation? There have been dozens of theories for dinosaur extinctions, but here we will briefly consider only three alternative explanations.

Fire from below: volcanoes

The Deccan Plateau in central India bears witness to one of the most extraordinary volcanic events in earth history. Solidified lava, up to several kilometres deep, covers an area of 500,000km^2, though the original lava flows may have been many times bigger (**Figure 5.9**). The eruptions that produced this mighty volume of lava occurred over a period of around 750,000 years, beginning a quarter of a million years before the K–Pg boundary. The first phase of the eruption started just before the end of the Cretaceous, the second and largest phase occurred around the time of the K–Pg boundary, and the

Figure 5.9 The Deccan Traps in India represent one of earth's most impressive volcanic events, with a series of eruptions occurring over several million years in the latest Cretaceous. The greatest pulse occurred just before the K–Pg boundary. Lava flows may have covered up to 1.5 million square kilometres. The series of basalt flows gives rise to the stepped pattern of rock ('traps' is derived from a Swedish word for 'stairs').

Photo by Nicholas (Nichalp) (CC2.5 license).

third phase in the earliest part of the Palaeogene. Such massive eruptions are likely to have been extremely destructive to the surrounding region. However, as with the impact hypothesis, it is not the immediate catastrophic effects of the eruptions on the local region that concern us here, but the likely effect on the global atmosphere and climate. Given that more modest recent volcanoes, such as Tambora (1816), Krakatoa (1883), and Pinatubo (1991), have caused a tangible drop in global temperature and sunlight for a year or more after the eruption, it is easy to imagine that such massive and sustained volcanic activity would have had a substantial effect on the whole biosphere.

Many of the predicted effects of the Deccan eruptions are similar those of the impact hypothesis, including the release of gases that could affect global climate, injection of particulate matter into the atmosphere causing a prolonged 'winter', ejected sulfur precipitating acid rain, and surface deposition of rare earth elements (including iridium).[39] The major difference is the timescale. Whereas an extraterrestrial impact occurs at a particular instant in time, the Deccan Traps were formed over hundreds of thousands of years. Interpretations of fossil distributions as indicating gradual extinction of species leading up to the K–Pg boundary have been used to support the volcanic hypothesis for end-Cretaceous mass extinctions.[40] Since the proposed killing mechanisms are similar, some of the same objections can be raised against the volcanism hypothesis as against the impact hypothesis, such as why all dinosaurs suffered mass extinction when other lineages such as frogs and crocodiles survived, and why polar dinosaurs, which must have been adapted to periods of dark and cold, could not have persisted through a volcanic winter.

Left high and dry: sea-level changes

Marine regression (drop in sea level) in the late Cretaceous would have reduced the amount of shallow seas on continental shelves, resulting in a reduction of habitat for many marine species, and a shift in distribution for others. For example, sharks and rays disappear from Hell Creek deposits at the K–Pg boundary, but obviously this does not signify a global extinction event since sharks survived into the Palaeogene. As the inland sea withdrew, the sharks went with it, returning later as the continental seas briefly returned.

What effect would marine regression have had on land-dwelling dinosaurs? The loss of the inland sea in North America would have reduced the coastal plains that had been home to many abundant dinosaur species. On a global scale, a drop in global sea levels is associated with climate change, as more of the earth's water becomes locked in ice caps as the world cools. Marine regression also affects the amount of fossil data. Recall that fossils can only be preserved where it is possible for them to be covered in sediment and then lie undisturbed, so many terrestrial fossils are from shallow lakes, river beds, and marine sediments. If marine regression reduces the margins where the water meets the land, this might reduce the opportunity for fossilization (see Case Study 3). However, while sea-level changes could have had a severe effect on some dinosaur communities, some scientists doubt that marine regression could be sufficient to explain global loss of all dinosaur species, rather than having a localized impact on survival and distribution.

Business as usual: nothing in particular caused dinosaur extinctions

Throughout this book we will emphasize the importance of considering the null hypothesis when weighing up competing explanations. While meteorites, super-volcanoes, and changing sea level may be much more exciting, we must also consider the alternative explanation that nothing in particular caused dinosaur extinctions. Dinosaur species arose and went extinct throughout the Mesozoic, and some researchers think that dinosaur diversity declined steadily during the late Cretaceous. Furthermore, it is important to remember that the dinosaur lineage survived the end of the Cretaceous and flourished throughout the Tertiary, with thousands of extant species. Under this hypothesis, the non-avian dinosaurs fail to make it into the Tertiary simply because there were none left at the end of the Cretaceous, or because the dinosaur fauna was already so depauperate and unstable that the remaining species did not survive any of the environmental perturbations that may have occurred in the latest stage of the Cretaceous. Dinosaurs, like many other lineages before them, may have gone out with a whimper, not a bang.

Which explanation is the most convincing?

The patterns of vertebrate survival and extinction from the Cretaceous to the Tertiary are considered by some researchers to closely match the expected effects of massive bolide impact at the end of the Cretaceous, with the effects on the biota most notable in North American sites, not far from the hypothesized impact. But other researchers point out puzzling exceptions that don't seem to fit a global catastrophe, such as the survival of amphibian and fish species, yet extinction of marine and polar dinosaurs who might be expected to have been less affected by an impact. In addition to considering which species survived and which did not, we can consider the timing of extinctions. If an impact was responsible for the K–Pg mass extinction, then the extinctions and the impact event should be simultaneous. It may seem simple to ask the question: Did the dinosaur extinctions occur at the same time as the impact? But the resolution of the fossil record typically does not allow recovery of instantaneous events.

Another way to examine the link between impact and mass extinction is to try to find other examples to establish a general pattern. Some researchers claim that most, if not all, mass extinction events in earth's history can be linked to extraterrestrial impacts, but the evidence of impacts associated with other mass extinctions is thus far less convincing than for the end-Cretaceous. Others suggest that, given that the number of impacts exceeds the number of mass extinction events, there is likely to be coincidence of impacts with at least some extinctions purely by chance. However, even if no other mass extinction events are associated with impacts, that does not disprove the K–Pg impact extinction hypothesis; there is no a priori reason why any of the 'big five' mass extinctions should share a common cause.

However, if the K–Pg impact extinction hypothesis is true, we should expect that other massive impacts should also have had a significant impact on global biodiversity. Yet there are other large impact craters that do not seem to be associated with global extinctions, such as the 'comet shower' which produced several very large craters around 36 million years ago. While Chicxulub is one of the biggest impact craters on earth, the Kara crater in Russia may have been two-thirds the size of Chicxulub or more, and is dated to just five million years before the K–Pg boundary. Several smaller impact craters, around half the size of Chicxulub, are known from the Mesozoic. Why did the dinosaurs survive these impacts but not the Chicxulub disaster?

It is important to acknowledge that in scientific debates, many (if not most) scientists develop strongly held convictions about which explanation is likely to be correct. It is not uncommon to find that proponents of one hypothesis will tend to always find evidence in favour of that hypothesis, while supporters of an opposing hypothesis will always find evidence that supports their explanation. Indeed, a blind test of patterns of extinction of planktonic foraminifera from sections spanning the K–Pg boundary was interpreted by different scientists as evidence both for and against the impact extinction hypothesis.[39,41] While this may reveal a strong bias amongst researchers, it would be overly simplistic to jump to the conclusion that the science is flawed. Instead, the debate can be viewed as something like a court case, in which the prosecution and the defence each put forward arguments that support their own position, but the ultimate aim is to reveal the truth of what really happened. Strong belief in one hypothesis or another may provide the driving force that spurs on further scientific research. The most important thing is to be honest in reporting and appraising evidence, and to be prepared to change ideas when new data emerges. Keep in mind Charles Darwin's advice to himself:

66 *I have steadily endeavoured to keep my mind free so as to give up any hypothesis, however much beloved (and I cannot resist forming one on every subject), as soon as facts are shown to be opposed to it.* 99
Charles Darwin, (1892) Charles Darwin: his life told in an autobiographical chapter, and in a selected series of his published letters (ed. F. Darwin). John Murray, London.

Conclusions

The word 'dinosaur' is commonly used to mean something old-fashioned and outmoded, destined to be replaced by superior models. But the image of giant, lumbering, slow-witted reptiles has been replaced by a view of dinosaurs as diverse, active, and supremely well adapted to a wide range of niches. Dinosaurs evolved and diversified throughout the Mesozoic, and are happily still with us today. Yet we are missing some of the astounding species of the past, so different from anything we know today that they seem somehow unbelievable. Now, rather than viewing dinosaur extinctions as examples of evolutionary failures, many scientists consider the dinosaurs to be victims of bad luck. If the Chicxulub meteor had, by chance, whistled past the earth instead of colliding with the Yucatán Peninsula, would dinosaurs still be roaming the earth today?

The extinction of the dinosaurs is an informative case study for examining the way that we generate and evaluate hypotheses about past events. The resolution of the terrestrial vertebrate palaeontological record makes fine-scale analysis of patterns of extinction challenging, so it is currently difficult to determine if dinosaur extinctions happened simultaneously due to a single global catastrophe or more gradually over thousands or millions of years. Will the record gradually improve in resolution as more fossils are discovered, allowing these hypotheses to be discriminated? Or are there fundamental limitations to resolution that will not be overcome?

Dinosaur extinction theories also highlight the criteria we use to accept or reject particular hypotheses. Given that the Chicxulub impact and the Deccan volcanoes must have had a huge effect on the global climate, it would seem ridiculous to recognize these events as coincident with the dinosaur extinctions without according them a causal role. But is temporal co-occurrence sufficient evidence of causality, convincing enough to override any conflicting observations? Many scientists feel that the weight of evidence is with the impact hypothesis, and that conflicting evidence is overridden by observations that do provide a good fit. Other scientists feel that rather than upweighting confirmatory evidence, we should pay more attention to the observations that seem not to implicate impact winter in dinosaur extinctions.

What if all proposed hypotheses have some flaws, or some key predictions where they do not match observations—do we reject all hypotheses as inadequate or do we accept the least-worst as the best available explanation? As with all such debates, there are scientists who have entrenched opinions on the matter. But if there is one thing that the study of dinosaurs has demonstrated, from their discovery to the present day, it is that new data prompt new ideas, and new ideas spur the search for new data. So, whichever explanation for dinosaur extinctions you find most convincing, prepare to have your convictions challenged as new evidence emerges.

Points for discussion

1. Palaeontological advances have revealed more about dinosaur biology than would have once been thought possible, such as appearance (from impressions of skin and feathers), behaviour (e.g. from nests and footprints), and development (such as using multiple individuals to reconstruct growth patterns). What more might we find out in the near future? Is there any information about dinosaurs that is beyond the reach of scientific investigation?

2. Is there a 'killer test' or definitive piece of evidence that could be found that would unambiguously either confirm or reject the impact extinction hypothesis?

3. Will the resolution of the end-Cretaceous dinosaur fossil record improve until the timing of the extinctions is clear, or could it be that the record is unavoidably too coarse to allow discrimination of an instantaneous event from a gradual reduction in diversity?

References

1. Hummel J, Clauss M (2011) *Sauropod Feeding and Digestive Physiology.* Indiana University Press, Bloomington, IL, p 11.

2. Button DJ, Rayfield EJ, Barrett PM (2014) Cranial biomechanics underpins high sauropod diversity in resource-poor environments. *Proceedings of the Royal Society B. Biological Sciences* 281(1795): pii 20142114.

3. Ruxton GD, Wilkinson DM (2011) The energetics of low browsing in sauropods. *Biology Letters* 7(5), 779–81.

4. Losos JB (2010) Adaptive radiation, ecological opportunity, and evolutionary determinism. *American Naturalist* 175(6): 623–39.

5. Brusatte SL, Niedzwiedzki G, Butler RJ (2010) Footprints pull origin and diversification of dinosaur stem lineage deep into Early Triassic. *Proceedings of the Royal Society of London B. Biological Sciences* 278(1708): 1107–13.

6. Bates K, Schachner E (2012) Disparity and convergence in bipedal archosaur locomotion. *Journal of the Royal Society Interface* 9(71): 1339–53.

7. Brusatte SL, Nesbitt S, Irmis RB, Butler RJ, Benton MJ, Norell MA (2010) The origin and early radiation of dinosaurs. *Earth-Science Reviews* 101: 68–100.

8. Hutchinson JR, Garcia M (2002) *Tyrannosaurus* was not a fast runner. *Nature* 415(6875): 1018–21.

9. Thulborn RA (1982) Speeds and gaits of dinosaurs. *Palaeogeography, Palaeoclimatology, Palaeoecology* 38(3–4): 227–56.

10. Benton MJ (2009) Dinosaurs. *Current Biology* 19(8): R318–23.

11. Köhler M, Marín-Moratalla N, Jordana X, Aanes R (2012) Seasonal bone growth and physiology in endotherms shed light on dinosaur physiology. *Nature* 487(7407): 358–61.

12. Brusatte SL (2012) *Dinosaur Paleobiology.* John Wiley, Chichester.

13. Eagle RA, Tütken T, Martin TS, Tripati AK, Fricke HC, Connely M, Cifelli RL, Eiler JM (2011) Dinosaur body temperatures determined from isotopic (^{13}C-^{18}O) ordering in fossil biominerals. *Science* 333(6041): 443–5.

14. Godefroit P, Golovneva L, Shchepetov S, Garcia G, Alekseev P (2009) The last polar dinosaurs: high diversity of latest Cretaceous arctic dinosaurs in Russia. *Naturwissenschaften* 96(4): 495–501.

15. Spicer RA, Herman AB (2010) The Late Cretaceous environment of the Arctic: a quantitative reassessment based on plant fossils. *Palaeogeography, Palaeoclimatology, Palaeoecology* 295(3): 423–42.

16. Rich TH, Vickers-Rich P, Gangloff RA (2002) Polar dinosaurs. *Science* 295(5557): 979–80.

17. Bell PR, Snively E (2008) Polar dinosaurs on parade: a review of dinosaur migration. *Alcheringa* 32(3): 271–84.

18. Alvarez LW, Alvarez W, Asaro F, Michel HV (1980) Extraterrestrial cause for the Cretaceous–Tertiary extinction. *Science* 208: 1095–1108.

19. Schulte P, and 40 other authors (2010) The Chicxulub asteroid impact and mass extinction at the Cretaceous–Paleogene boundary. *Science* 327(5970): 1214–18.

20. Robertson DS, McKenna MC, Toon OB, Hope S, Lillegraven JA (2004) Survival in the first hours of the Cenozoic. *Geological Society of America Bulletin* 116(5-6): 760–8.

21. Varricchio DJ, Martin AJ, Katsura Y (2007) First trace and body fossil evidence of a burrowing, denning dinosaur. *Proceedings of the Royal Society B. Biological Sciences* 274(1616): 1361–8.

22. MacLeod N, Rawson P, Forey P, Banner F, Boudagher-Fadel M, Bown P, Burnett J, Chambers P, Culver S, Evans S (1997) The Cretaceous–Tertiary biotic transition. *Journal of the Geological Society* 154(2): 265–92.

23. Friedman M, Sallan LC (2012) Five hundred million years of extinction and recovery: a phanerozoic survey of large-scale diversity patterns in fishes. *Palaeontology* 55(4): 707–42.

24. Retallack G (2004) End-cretaceous acid rain as a selective extinction mechanism between birds and dinosaurs. In Currie PJ, Koppelhus EB, Shugar MA (eds), *Feathered Dragons: Studies on the Transition from Dinosaurs to Birds.* Indiana University Press, Bloomington, IN.

25. Premovic PI (2011) Distal 'impact' layers and global acidification of ocean water at the Cretaceous–Paleogene boundary (KPB). *Geochemistry International* 49(1): 55–65.

26. Wappler T, Currano ED, Wilf P, Rust J, Labandeira CC (2009) No post-Cretaceous ecosystem depression in European forests? Rich insect-feeding damage on diverse middle Palaeocene plants, Menat, France. *Proceedings of the Royal Society B. Biological Sciences* 276(1677): 4271–7.

27. Barreda VD, Cúneo NR, Wilf P, Currano ED, Scasso RA, Brinkhuis H (2012) Cretaceous/Paleogene floral turnover in Patagonia: drop in diversity, low extinction, and a *Classopollis* spike. *PLoS One* 7(12): e52455.

28. Jiang S, Bralower TJ, Patzkowsky ME, Kump LR, Schueth JD (2010) Geographic controls on nannoplankton extinction across the Cretaceous/Palaeogene boundary. *Nature Geoscience* 3(4): 280.

29. Alegret L, Thomas E, Lohmann KC (2012) End-Cretaceous marine mass extinction not caused by productivity collapse. *Proceedings of the National Academy of Sciences of the USA* 109(3): 728–32.

30. Longrich NR, Bhullar B-AS, Gauthier JA (2012) Mass extinction of lizards and snakes at the Cretaceous–Paleogene boundary. *Proceedings of the National Academy of Sciences of the USA* 109(52): 21396–401.

31. Mannion PD, Benson RBJ, Carrano MT, Tennant JP, Judd J, Butler RJ (2015) Climate constrains the evolutionary history and biodiversity of crocodylians. *Nature Communications* 6: 8438.

32. Miller D, Summers J, Silber S (2004) Environmental versus genetic sex determination: a possible factor in dinosaur extinction? *Fertility and Sterility* 81(4): 954–64.

33. Escobedo-Galvan AH, Gonzalez-Salazar C (2012) Survival and extinction of sex-determining mechanisms in Cretaceous tetrapods. *Cretaceous Research* 36:116–18.

34. Vila B, Galobart Ä, Canudo J, Le Loeuff J, Dinarès-Turell J, Riera V, Oms O, Tortosa T, Gaete R (2012) The diversity of sauropod dinosaurs and their first taxonomic succession from the latest Cretaceous of southwestern Europe: clues to demise and extinction. *Palaeogeography, Palaeoclimatology, Palaeoecology* 350–352: 19–38

35. Barrett PM, McGowan AJ, Page V (2009) Dinosaur diversity and the rock record. *Proceedings of the Royal Society B. Biological Sciences* 276(1667): 2667–74.

36. Lloyd GT (2012) A refined modelling approach to assess the influence of sampling on palaeobiodiversity curves: new support for declining Cretaceous dinosaur richness. *Biology Letters* 8(1): 123–6.

37. Sakamoto M, Benton MJ, Venditti C (2016) Dinosaurs in decline tens of millions of years before their final extinction. *Proceedings of the National Academy of Sciences of the USA* 113(18): 5036–40.

38. Brusatte SL, Butler RJ, Prieto-Marquez A, Norell MA (2012) Dinosaur morphological diversity and the end-Cretaceous extinction. *Nature Communications* 3: 804.

39. Courtillot V (2002) *Evolutionary Catastrophes: The Science of Mass Extinction.* Cambridge University Press, Cambridge.

40. Schoene B, Samperton KM, Eddy MP, Keller G, Adatte T, Bowring SA, Khadri SF, Gertsch B (2015) U–Pb geochronology of the Deccan Traps and relation to the end-Cretaceous mass extinction. *Science* 347(6218): 182–4.

41. Arenillas I, Arz JA, Molina E, Dupuis C (2000) An independent test of planktic foraminiferal turnover across the Cretaceous/Paleogene (K/P) boundary at El Kef, Tunisia: catastrophic mass extinction and possible survivorship. *Micropaleontology* 46(1): 31–49.

42. Tomkins JL, LeBas NR, Witton MP, Martill DM, Humphries S (2010) Positive allometry and the prehistory of sexual selection. *American Naturalist* 176(2): 141–8.

43. Archibald JD, and 28 other authors (2010) Cretaceous extinctions: multiple causes. *Science* 328(5981): 973.

Case Study 5
Testing hypotheses with fossil evidence: did birds outcompete pterosaurs?

Pterosaurs were a lineage of winged reptiles that coexisted with dinosaurs from the late Triassic to the end of the Cretaceous period. Over a hundred species of pterosaur have been described. One of the largest known pterosaurs, *Quetzalcoatlus*, was the size of a small aircraft, with an estimated wingspan of over 10 metres. Since the largest flying animals today have wingspans only a third as long, there has been much debate about whether the largest pterosaurs could have undertaken powered flapping flight, whether they mostly relied on soaring or gliding, or whether they were flightless quadrupedal walkers. Pterosaur biology is an active and exciting area of research, with studies as diverse as biomechanical modelling of feeding behaviour, trackway analysis of individual movement and group behaviour, studies of growth rates based on large samples of individuals from the same populations, dietary analysis from microwear on teeth, and reproductive biology from eggs and nests. In this case study, we will concentrate on one particular hypothesis about pterosaur evolution: did the evolution of birds lead to extinction of pterosaurs by competitive displacement?

If two or more species attempt to exploit the same resource, then any species with a consistent advantage in accessing that resource might succeed at the expense of the others, as it more effectively turns resources into individuals through metabolism and reproduction (see Chapter 2). Growth of the population of the successful competitor might lead to local or global extinction of the other. 'Resources' refers to anything that members of a species need to successfully complete their life cycles, such as food or water, appropriate habitat, access to mates, or a safe place to raise offspring. For example, if two different species use the same nest sites, the species that selects nest sites slightly earlier in the season might have a competitive advantage, as the species that begins nesting later may find insufficient nest sites left to allow the reproductive potential of the population to be realized. To support a hypothesis of competitive displacement, we first need to establish that the species in question relied on the same resource, in the same place, at the same time. Then we need to show that the success of one competitor came at the expense of the other.

So the first question we need to ask is: did pterosaurs and birds coexist in space and time? If they were not present in the same places at the same time, then they could not have directly competed. Pterosaur fossils are found from the Jurassic to the end of the Cretaceous. They had a global distribution, with fossils found on every continent.[1] Birds also first appeared in the Jurassic, and over a hundred bird species are described from the Mesozoic. However, the timing of the diversification of modern birds, Neornithines, is debated, with molecular phylogenetic estimates pushing the radiation back into the Cretaceous, earlier than a consideration of fossil evidence alone would suggest.[2] Given that some deposits contain both bird and pterosaur fossils, and there are fossil trackways with both putative bird and pterosaur prints, it seems very likely that birds and pterosaurs did coexist in some places, though it is possible there were some areas or habitats where they did not overlap. It has been suggested, on the basis of distribution of trackways, that pterosaurs and birds may have predominated in different niches. For example, birds may have dominated freshwater environments, with pterosaurs more common in coastal areas. But assignment of species to habitats on the basis of body or trace fossils may be affected by preservational biases; for example, if birds inhabited rocky coastal areas or offshore islands there may have been little opportunity to leave fossil traces.[3]

Morphological patterns of diversity

We have established that birds and pterosaurs were part of the biota at the same time and overlapped in geographic extent. But to show that they competed, we have to show that they both accessed the same resources in the same places. This is harder for extinct species than it is for living species, because we cannot directly observe which resources they use. Several approaches to looking for overlap have been taken. One is to try to infer habitat and diet of Mesozoic birds and pterosaurs from fossil evidence. Pterosaur diet has been inferred in a number of ways, such as biomechanical inference from the morphology of jaws and 'beaks', analysis of stomach contents, and interpretation of teeth as specialized to particular tasks, such as grabbing fish from the water (although late Cretaceous pterosaurs are all toothless).[4] Comparison of tooth wear patterns in pterosaurs with those of modern species such as birds and crocodiles suggests a wider range of diets than previously supposed, including species that concentrated on terrestrial invertebrates.[5] Pterosaur diets might have changed over time. It has been suggested that flight may have evolved in the early small-bodied pterosaurs so that they could exploit the abundant resource of flying insects.[6] The evolution of lineages with a more mixed diet may then have given rise to specialist piscivores. Reconstruction of the diet and habitat of Mesozoic birds is a matter of debate, but some researchers suggest that many early bird species were small ground or aerial foragers that may have had little ecological overlap with large fish-feeding pterosaurs.[7,8]

Another more general approach to assessing the likelihood of competition between pterosaurs and birds is to compare their relative morphospace, on the assumption that if species are vastly different in size and morphology they are unlikely to use the same resource.[9]

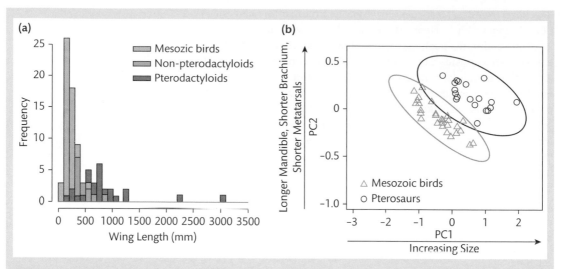

Figure 5.10 Morphometric analysis of size and shape of wings in birds has been used to evaluate how likely it is that pterosaurs and birds competed directly with each other, by comparing Mesozoic birds that co-existed with two major groups of pterosaurs (pterodactyloid and non-pterodactyloid). Morphology can be compared in one dimension, such as wing length (a), or multiple dimensions can be analysed simultaneously. (b) Principal components analysis of many measurements of different wing bones is summarized in the two axes that explain most of the difference between birds and pterosaurs.[11]

Reproduced by permission of The Royal Society.

Plotting the morphospace of the range of limb sizes and jaw length revealed that birds and pterosaurs had distinct morphologies, which has been interpreted as indicating they were unlikely to have directly competed with each other (**Figure 5.10**).[10,11]

But there are two grounds on which we could query this conclusion. The first is on biological grounds: we can observe today that animals of different sizes and shapes can compete for the same resource. For example, grazing animals as diverse as rabbits, wallabies, and cattle can all compete for grass in the same place at the same time, despite several orders of magnitude in size difference. The second is on methodological grounds: can we be sure that the range of morphologies was adequately sampled and that the lack of overlap is not due to missing data?[12]

Temporal patterns of diversity

We can sidestep the problems associated with inferring ecological interaction by asking what pattern we would expect to observe in the record over time if birds did outcompete pterosaurs. If it is true that birds caused pterosaur extinctions, we should see a link between pterosaur decline and bird increase over time. Competitive replacement can be observed in the present day, such as when a native species is

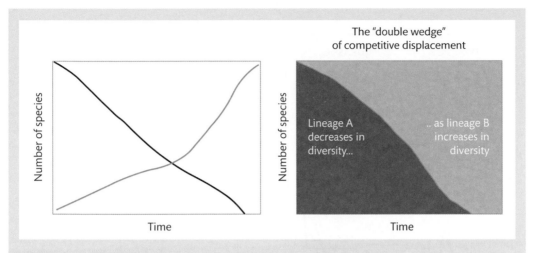

Figure 5.11 When the number of recognized species is compared over time between two groups, the appearance of a 'double wedge'—one group rises in diversity as the other declines—is often taken as evidence in favour of long-term competitive displacement as one lineage diversifies at the expense of the other.

displaced by an invasive species. But even if one species were to take 1000 generations to replace another, this would probably appear as an abrupt and instantaneous change in the fossil record, not a gradual rise in one taxon and a corresponding decline in the other. Clearly birds did not instantly replace pterosaurs, as they coexisted in the global biota for millions of years. So this hypothesis concerns a different scale of competitive replacement, where a clade of related lineages increases in representation in the biota at the expense of the other through differential diversification. This pattern is sometimes referred to as a 'double wedge' (Figure 5.11).

Did pterosaur diversity drop as bird diversity rose? If the number of known species of pterosaurs is plotted against time, the diversity curve rises unevenly through the Mesozoic to reach a high point in the mid-Cretaceous, followed by a sharp decline (Figure 5.12).[13] Is the drop in diversity in the mid-Cretaceous the sign of pterosaurs under assault by Mesozoic birds? While birds do increase in diversity toward the end of the Cretaceous, they also show a dramatic drop in diversity at the same point as the pterosaurs (Figure 5.12).[14] So were birds and pterosaurs both affected by the same global environmental change that caused a catastrophic extinction event? A more prosaic explanation is that both the bird and pterosaur species counts are strongly influenced by the same sampling biases. In fact, the pterosaur and bird diversity closely matches the sampling patterns of all terrestrial tetrapods across that time period, a pattern that might be driven by the availability of fossil-bearing rock (Figure 5.13).[12,14,15]

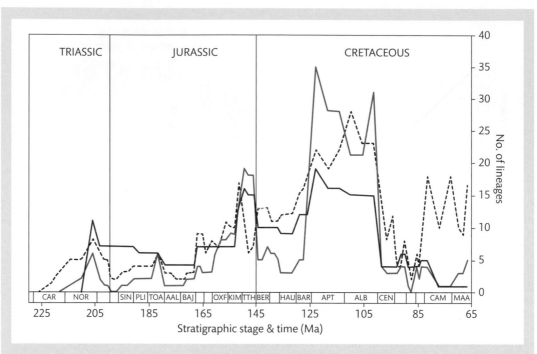

Figure 5.12 Number of recognized pterosaur species over time (blue line) and the phylogenetic diversity of pterosaurs (red line) which includes lineages inferred to be present in a time period using information on their temporal distribution and relationships. Diversity tracks the number of pterosaur-bearing formations identified over time (dashed line).[13]

© The Paleontological Society

Both birds and pterosaurs have features that might reduce their chances of fossilization, such as light hollow bones. Both bird and pterosaur diversity patterns are strongly associated with the occurrence of lagerstätten, sites of exceptional preservation that record fine anatomical detail and soft-bodied organisms. Such sites preserve specimens to a level of detail and completeness that may rarely be found at other sites, and may reveal more delicate and small-bodied taxa (**Figure 5.14**).[16] In fact, pterosaur diversity over time shows a distinct 'spikiness', with the peaks associated with the occurrence of around a dozen lagerstätten.[17] The completeness of specimens of both birds and pterosaurs is also strongly correlated with lagerstätten (unlike more robust taxa such as sauropods).[16] We have to consider the possibility that the drop in diversity of pterosaurs during the mid to late Cretaceous is due to the absence of lagerstätten, rather than a reduction in the number of pterosaur species living at the time. When fossil occurrence is patchy, it is difficult to distinguish a gradual decline in diversity from a sudden extinction event (the Signor–Lipps effect, see Figure 5.7).[18]

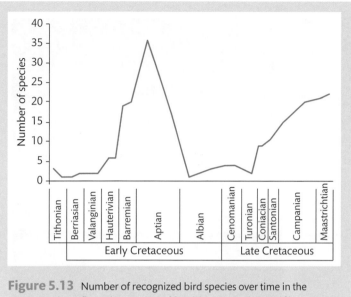

Figure 5.13 Number of recognized bird species over time in the Mesozoic also reflects the number of fossil-bearing collections.[14]

Courtesy of PLoS One.

Should we adjust the record for the number of lagerstätten? When pterosaur diversity is corrected for number of lagerstätten, the apparent extinction events disappear as signals of low sampling rather than low diversity.[16] But we have to be careful here: are we simply correcting a sampling bias, or do we risk removing an important biological signal? If the number of sites of exceptional preservation is a reflection of important changes in the biosphere that also influenced biodiversity patterns, applying a correction for sample size might remove biologically interesting signals in the data.[17] It may be that both the existence of lagerstätten and pterosaur diversity fluctuate through time because they are both connected to particular global conditions, such as sea-level changes. In this case, the amount of fossil-bearing rock might be linked to diversity not just as a sampling artefact but because the two are driven by a 'common cause'.[13]

Competitive displacement

If we are interested in testing the hypothesis that birds outcompeted pterosaurs, there is a range of other ways in which we can explore the data beyond straight diversity counts. Instead of asking how many different bird or pterosaur species there are over time, we can ask what kinds of species there are. If it is true that competition with birds drove pterosaurs extinct, we might expect to see that pattern most clearly in the pterosaur species most likely to compete with birds. There are interesting trends in pterosaur diversity over time that might have some bearing on their interaction with birds.

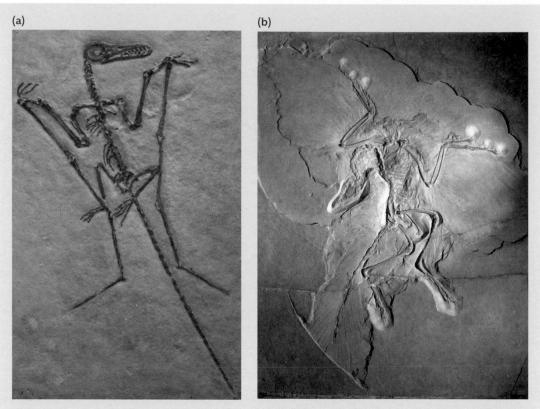

(a) **(b)**

Figure 5.14 (a) A pterosaur (*Rhamphorhynchus,*) and (b) a bird (*Archaeopteryx*) from the Solnhoffen limestone in Germany, a lagerstätte from the Upper Jurassic.

(a) Photo by Kevin Walsh (CC2.0 license). (b) Photo by H. Raab (CC3.0 license).

One is that pterosaurs present one of the most striking cases of Cope's rule—increase in body size over time (see Chapter 3). Unlike many taxa, where the increase in average size over time is driven by increasing variance, extending both the maximum size and the overall size range over time, in pterosaurs the increase in size during the Cretaceous is driven both by increasing size at the high end of the size spectrum and a loss of species at the low end (Figure 5.15).[18] At the same time, birds were also increasing in size, though their pattern of size increase is more like the diffusion model where the maximum and range increase over time.[18] A fossil assemblage from Morocco dated to the end of the Cretaceous suggests a clear size separation between the large-bodied pterosaurs and the smaller birds (Figure 5.16).[19]

Does this prove that the rise of birds pushed pterosaurs into larger-bodied niches, and resulted in the extinction of smaller species of pterosaurs? As with all such patterns, the size shift over time must be interpreted with respect to any possible biases in the record. What if smaller-sized pterosaurs existed in the latest Cretaceous but have not

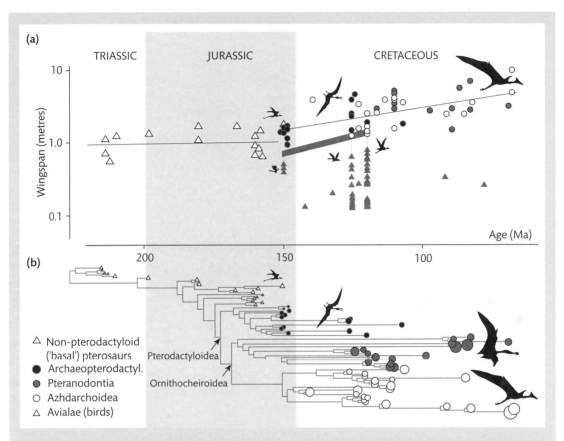

Figure 5.15 (a) Wingspan in pterosaurs and birds over time.[18] Regression lines (black lines) show no significant increase in wingspan in pterosaurs in the Triassic and Jurassic, but a significant rise in body size over time in the Cretaceous. The thick grey line shows an increase in wingspans in birds in the relatively well-sampled period spanning the late Jurassic and early Cretaceous. (b) A phylogeny showing adult wingtip diameters proportional to log10(wingspan).

Courtesy of Nature Communications (CC3.0 license).

been recovered, due to either chance or taphonomic biases against detecting smaller, lighter specimens, or geographic biases that create a false impression of global diversity? Future discoveries of diverse smaller-bodied pterosaurs from the latest Cretaceous would challenge this picture. Better understanding of the timing of the radiation of modern birds would also shed light on this question. Just as for mammals (Chapter 6), molecular dates have suggested that modern birds began diversifying in the Cretaceous,[2] although, again as for mammals, these results are a matter of debate.[20] There are potential biases in both the fossil and molecular records, so more work is needed to clarify the historical signal in the data. Knowledge of both pterosaurs and Mesozoic birds is changing so rapidly, with new data and new analyses constantly being published, that we can expect an exciting and fast-moving debate ahead.

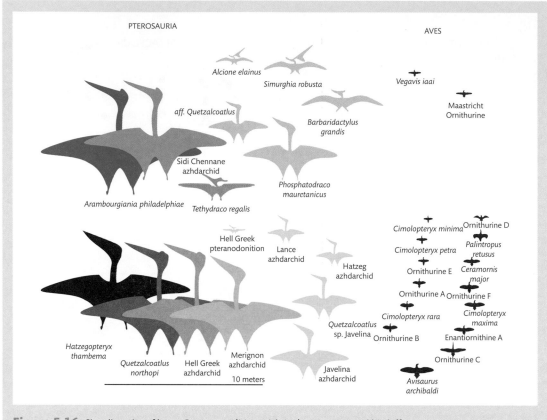

Figure 5.16 Size disparity of latest Cretaceous (Maastrichtian) pterosaurs and birds.[19]
Courtesy of PLoS One (CC4.0 license).

Questions to ponder

1. Do you consider lack of size overlap as convincing evidence for lack of competition between birds and pterosaurs during the Mesozoic? Could flying vertebrates of different size or flight styles have competed directly with each other?

2. What evidence could palaeontologists use to investigate the diets of pterosaurs and Mesozoic birds?

3. Some of the late Cretaceous pterosaurs are so large that questions have been raised about whether it was possible for them to fly, and many images of these species show them as earthbound quadrupeds walking on their elbows. What evidence could be gathered to answer the question of whether large pterosaurs were capable of active flight, or powered take-off?

Further research

How do biologists assess competition between living species? Are tests for competition used in the field or laboratory applicable, or adaptable, to investigating the potential for competition between long-extinct species? A growing number of studies seeks to evaluate the signature of past competition between species using phylogenetic relationships and current geographic distribution. What are the strengths and weaknesses of this approach?

References

1. Barrett PM, Butler RJ, Edwards NP, Milner AR (2008) Pterosaur distribution in time and space: an atlas. *Zitteliana* 28: 61–108.

2. Haddrath O, Baker AJ (2012) Multiple nuclear genes and retroposons support vicariance and dispersal of the palaeognaths, and an Early Cretaceous origin of modern birds. *Proceedings of the Royal Society B: Biological Sciences* 279(1747): 4617–25.

3. Lockley MG, Rainforth EC (2002) The track record of Mesozoic birds and pterosaurs. In Chiappe LM, Witmer LM (eds), *Mesozoic Birds. Above the Heads of Dinosaurs.* University of California Press, Oakland, CA, pp.405–18.

4. Hone D, Henderson DM, Therrien F, Habib MB (2015) A specimen of *Rhamphorhynchus* with soft tissue preservation, stomach contents and a putative coprolite. *PeerJ* 3: e1191.

5. Pickrell J (2018) Tooth scratches reveal new clues to pterosaur diets. *Nature* 553(7687): 138.

6. Ősi A (2011) Feeding-related characters in basal pterosaurs: implications for jaw mechanism, dental function and diet. *Lethaia* 44(2): 136–52.

7. Mitchell JS, Makovicky PJ (2014) Low ecological disparity in Early Cretaceous birds. *Proceedings of the Royal Society of London B: Biological Sciences* 281(1787): 20140608.

8. Brusatte SL, O'Connor JK, Jarvis ED (2015) The origin and diversification of birds. *Current Biology* 25(19): R888–98.

9. Prentice KC, Ruta M, Benton MJ (2011) Evolution of morphological disparity in pterosaurs. *Journal of Systematic Palaeontology* 9(3): 337–53.

10. McGowan AJ, Dyke G (2007) A morphospace-based test for competitive exclusion among flying vertebrates: did birds, bats and pterosaurs get in each other's space? *Journal of Evolutionary Biology* 20(3): 1230–6.

11. Chan NR (2017) Morphospaces of functionally analogous traits show ecological separation between birds and pterosaurs. *Proceedings of the Royal Society B: Biological Sciences* 284(1865): 20171556.

12. Butler RJ, Brusatte SL, Andres B, Benson RB (2012) How do geological sampling biases affect studies of morphological evolution in deep time? A case study of pterosaur (Reptilia: Archosauria) disparity. *Evolution* 66(1): 147–62.

13. Butler RJ, Barrett PM, Nowbath S, Upchurch P (2009) Estimating the effects of sampling biases on pterosaur diversity patterns: implications for hypotheses of bird/pterosaur competitive replacement. *Paleobiology* 35(3): 432–46.

14. Brocklehurst N, Upchurch P, Mannion PD, O'Connor J (2012) The completeness of the fossil record of Mesozoic birds: implications for early avian evolution. *PLoS One* 7(6): e39056.

15. Butler RJ, Benson RBJ, Barrett PM (2013) Pterosaur diversity: Untangling the influence of sampling biases, Lagerstätten, and genuine biodiversity signals. *Palaeogeography, Palaeoclimatology, Palaeoecology* 372:78–87.

16. Dean CD, Mannion PD, Butler RJ (2016) Preservational bias controls the fossil record of pterosaurs. *Palaeontology* 59(2): 225–47.

17. Benton MJ, Dunhill AM, Lloyd GT, Marx FG (2011) Assessing the quality of the fossil record: insights from vertebrates. *Geological Society, London, Special Publications* 358(1): 63–94.

18. Benson RB, Frigot RA, Goswami A, Andres B, Butler RJ (2014) Competition and constraint drove Cope's rule in the evolution of giant flying reptiles. *Nature Communications* 5: 3567.

19. Longrich NR, Martill DM, Andres B (2018) Late Maastrichtian pterosaurs from North Africa and mass extinction of Pterosauria at the Cretaceous–Paleogene boundary. *PLOS Biology* 16(3): e2001663.

20. Jarvis ED, and >70 other authors (2014) Whole-genome analyses resolve early branches in the tree of life of modern birds. *Science* 346(6215): 1320–31.

Was the diversification of mammals due to luck?

6

Roadmap

Why study the mammalian radiation?

The diversification of mammals is often considered a classic example of ecological opportunity prompting evolutionary radiation. The fossil record shows a dramatic increase in the diversity of placental mammals as they evolved to occupy the niches left vacant by the extinction of the dinosaurs. But date estimates based on DNA analysis have been used to question the assumed timing of this radiation, suggesting that the major placental mammal lineages diverged from each other long before the dinosaurs' demise. Given that different molecular dating analyses give very different estimates for the timing of the mammalian radiation, how reliable are these molecular dates? Can we estimate the timing of divergence events from DNA sequences when rates of change vary between lineages?

What are the main points?

- Molecular dates are a powerful tool in macroevolution because they can reveal the timing of diversification events.
- DNA changes accumulate in evolving lineages, so the amount of genetic difference between two lineages can give an indication of how long it has been since they last shared a common ancestor.
- Rates of molecular evolution vary between species, complicating the inference of dates of divergence from molecular data.
- Molecular dates rely on a large number of assumptions about the evolutionary process, and changing the assumptions can influence the date estimates obtained.

Photograph: Red panda. © Jelena Janjetovic/Shutterstock.com.

What techniques are covered?

- **Molecular evolution:** to understand how the genome records evolutionary history, we need to understand the mechanisms of DNA sequence change over time.
- **Molecular dating:** comparisons between DNA sequences from different species can be used to infer an evolutionary timescale.

What case studies will be included?

- Bayesian molecular dating: timing of the radiation of placental mammal orders.

"Inconspicuous synapsids evolved during the Triassic into true mammals, of which the first were 5cm long and nocturnal. Their unobtrusive scuttlings gave little evidence of what was to become the most exciting radiation in vertebrate history when, well over 100 million years later, in the late Cretaceous period, the nonavian dinosaurs lumbered into oblivion."

David W. Macdonald (2006) *The Encyclopedia of Mammals.* Oxford University Press.

Timing and causes of the mammalian radiation

The dinosaurs were a hugely successful lineage of vertebrates that occupied a diverse range of niches for over 150 million years, from gigantic herds of browsers to small darting insectivores, and included pursuit predators, seed-eaters, fish-eaters, herbivores, and omnivores (Chapter 5). Other reptile groups also occupied a range of niches in the Mesozoic era (252–66 million years ago (Mya)), such as aerial piscivores and massive aquatic predators. But all the non-avian dinosaurs,

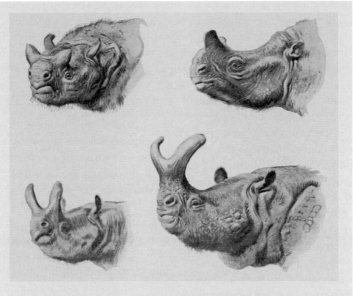

Figure 6.1 Brontotheres (formerly known as titanotheres) were one of the earliest fossil mammal groups described from North America. They are members of the Perissodactyla, the placental order which includes horses and rhinos. They are known from the Eocene period, first appearing 10 million years after the end-Cretaceous extinction of the non-avian dinosaurs.

From Osborn HF (1929) *The titanotheres of ancient Wyoming, Dakota, and Nebraska.* United States Geological Survey., DC.

along with the pterosaurs and large marine reptiles, disappear from the fossil record by the end of the Mesozoic. What happened next?

The palaeontological record shows that within 10 million years of the last dinosaur fossils, mammals occupied many of the niches previously occupied by reptiles. There were wolf-like carnivores, horse-sized herbivores, and semi-aquatic hippo-like mammals. The earliest known fossil bats, around 55 million years old, show animals capable of powered flight with wings as sophisticated as those of modern bats. In some cases, there is a degree of physical resemblance between some placental mammals and Cretaceous reptiles, as mammals independently evolved similar solutions to related niches. Rhinoceroses and brontotheres, with their stout bodies and facial armaments, are not entirely dissimilar to ceratopsids (**Figure 6.1**). Early whales had similar body shapes to the aquatic reptilian mosasaurs.

The explosion of mammalian diversity in the Palaeogene (66–23 Mya) does not mark the origin of the mammals. Mammals evolved from a reptile-like ancestor during the Late Triassic (237–201 Mya), which is also the epoch when the first dinosaurs appeared. Classic mammalian traits such as agile gait, heightened sensory capacity, larger brains, and a fast metabolism seem to have evolved in the Jurassic (201–145 Mya).[1] But undisputed members of modern placental orders, such as primates, rodents, and artiodactyls, do not appear in the fossil record until after the end of the Cretaceous (65 Mya). Because the first members of many different lineages of placental mammals appear not long after the dinosaur extinctions, the mammalian radiation has been interpreted as a classic case of rapid radiation into 'empty niche space' (see Chapter 5). Although mammals had existed for as long as the dinosaurs, the end-Cretaceous extinction event may have freed up ecological opportunities for mammals to exploit, triggering a rapid increase in diversity, disparity, and average size.

This interpretation of mammalian radiation as a response to evolutionary release implies that the radiation of mammals had previously been inhibited by the ecological incumbency of the dinosaurs. But although the earliest known mammals were small and shrew-like, diversification of mammals in the Jurassic produced a wide range of forms including gliders, burrowers, and semi-aquatic beaver-like animals.[2] So if mammals evolved at the beginning of the dinosaur radiation, if they already had traits that characterized their high-energy lifestyle, and if they occupied a range of niches in the Mesozoic, why is mammalian evolution so often described as a post-dinosaur phenomenon? Most fossils of Cretaceous mammals belong to long-extinct lineages often described as 'archaic mammals', such as the multituberculates and haramyadins. While they share key mammalian characteristics, they are typically considered to be lineages with no living descendants, rather than being the direct ancestors of modern mammals.

All modern mammals are descended from a common ancestor which gave rise to the three major lineages of living mammals: monotremes, marsupials, and placentals. The split between these three major branches of modern mammals probably occurred deep in the Mesozoic era. Monotremes (five living species) and marsupials (over 300 living species) are diverse in both appearance and habits. But the placentals are by far the most numerous and diverse group of modern mammals, with over 5000 named species and a global distribution. This chapter is going to concentrate entirely on the placental mammal radiation. The Placentalia include mammal lineages that gestate their offspring in utero until a relatively late stage of fetal development, such as rats, cats, and bats. Because of this, some of the key anatomical features that define Placentalia are in the pelvis and epipubic bones, to allow the gestation and birth of almost fully formed offspring.

While members of the lineage Eutheria (the group that includes placentals and their direct ancestors) have been described from the late Jurassic, fossil representatives of modern placental orders, such as Carnivora, Artiodactyla, and Rodentia, have only been found after the end-Cretaceous extinction of the dinosaurs. Taken at face value, this suggests

that even if the first eutherian lineages arose in the Mesozoic, the prolific radiation of the placental orders waited until the dinosaurs were no longer getting in the way. This pattern of radiation has been described as a 'long-fuse' model, with earliest lineages persisting for a considerable period of time before explosively radiating when the opportunity arose.

This raises a very interesting and fundamental question in evolutionary biology. If the radiation of modern mammals was triggered by the extinction of the dinosaurs, what would have happened had the dinosaurs not gone extinct? Would mammals have diversified anyway? Or would the biosphere look rather different today, with no large mammals—and therefore no humans? This question is particularly intriguing if we consider the possibility that astronomical or geological phenomena played a key role in dinosaur extinction (Chapter 5). Without the chance impact of an extraterrestrial object, or without the eruption of gigantic volcanoes in India, would the dinosaurs have persisted and prevented mammalian adaptive radiation? More generally, is the success and diversity of biological lineages the result of special features that give them an unprecedented advantage over others, or the unpredictable result of random catastrophes or lucky flukes?

These issues of chance and contingency are important to our understanding of the origins of biodiversity. In this chapter, we are going to look at one particular case study (the radiation of placental mammals) and one specific line of evidence (molecular dates) that can help us to evaluate competing hypotheses for the triggers of evolutionary radiations. This focus on molecular dates does not imply that DNA evidence is the most valuable tool we have for investigating the causes of the mammalian radiation. Clearly, there are many possible approaches to understanding the timing, causes, and outcomes of the placental radiation, such as looking at fossil distribution and disparity, and considering biogeographic models in the context of global change. And there are many very interesting aspects of mammalian diversification that we will not be able to cover here. But the mammalian radiation gives us an informative case study for examining the application and reliability of molecular dating in macroevolution and macroecology, for several reasons.

Firstly, the mammalian radiation is a case where knowledge of lineage divergence dates would help us to test hypotheses about the drivers of evolutionary change. If we knew for certain that placental mammals diversified long before the end of the Cretaceous, then we could rule out dinosaur extinction as the primary trigger of the radiation. Secondly, the mammalian radiation is an interesting case study because for many years there was a clear mismatch between the evidence in the fossil record and the story told by molecular dates, which prompts us to consider very carefully what could be causing this discrepancy. In this chapter, we are going to focus on the accuracy and precision of molecular dates, with only a brief discussion of possible sources of bias in the fossil record of mammalian radiation. Thirdly, because patterns and causes of variation in rate of molecular evolution have been well studied for mammals, they make a good case study for considering how rate variation might impact on molecular date estimates in macroevolutionary hypothesis testing. Fourthly, more recent molecular date estimates are commonly considered to be in closer agreement with fossil dates than earlier studies, which has been taken as an encouraging sign that molecular dates are improving. This makes the placental mammal radiation an excellent opportunity to examine whether we can be more confident of recent date estimates produced by sophisticated methods that allow for variation in the rate of molecular evolution. This is a question of much broader significance than understanding whether placental diversification was triggered by dinosaur extinction. Given that the reach of molecular data is ever increasing, we need informative test cases such as this one to investigate the promises and pitfalls of molecular dating analyses in macroevolution and macroecology.

Molecular dates for the mammalian radiation

To put this debate in context, we first need to consider the history of molecular date estimates for the placental mammalian radiation. This history tracks developments in the field, showing how molecular dating analyses have increased in sophistication and computational power since the earliest studies were published half a century ago. Then we will look at the factors that influence the rate of molecular evolution, which might impact on our ability to read time from molecular distance. Then we will consider how newer molecular dating techniques allow for variation in rate of molecular evolution. It is these new molecular dates that are generally considered more complementary to fossil evidence. Finally, we will ask whether molecular date estimates have allowed us to answer the long-standing question of whether the radiation of placental mammals was due to their good luck in outlasting the dinosaurs and then stepping into their evolutionary shoes, or whether they had natural advantages that spurred their radiation well before the dinosaurs left the stage.

Our understanding of the Mesozoic history of the mammalian lineage is constantly being updated, with new finds adding to an increasingly rich ecological picture. But, although there are eutherian mammals described from the Mesozoic, none of the fossils has been unambiguously assigned to a modern placental order.[3,4] Instead, fossils of nearly all modern mammalian orders appear within 20 million years of the end of the Cretaceous. This pattern of fossil occurrences has been interpreted as the signature of an explosive post-dinosaur radiation of placental mammals. Whatever the cause and nature of their extinction, the non-avian dinosaurs all disappear from the fossil record by the end of the Cretaceous. It is reasonable to assume that the vacation of ecological niches previously occupied by dinosaurs would provide the opportunity for end-Cretaceous survivors to evolve into diverse new forms. So it came as a great surprise when molecular dates for the timing of the radiation of placental mammals, derived by comparing proteins or DNA sequences from living mammal species, painted a completely different picture.

The advent of protein sequencing in the 1960s, and the parallel development of molecular evolutionary theory, paved the way for a new method of estimating a timescale for evolutionary events. If you compare the amino acid sequences of a particular protein, such as fibrinopeptide, between mammal species, you find that most of the sequence is the same in all species, but there will be some differences (**Figure 6.2**). The more distantly related the species, the more differences there will be between

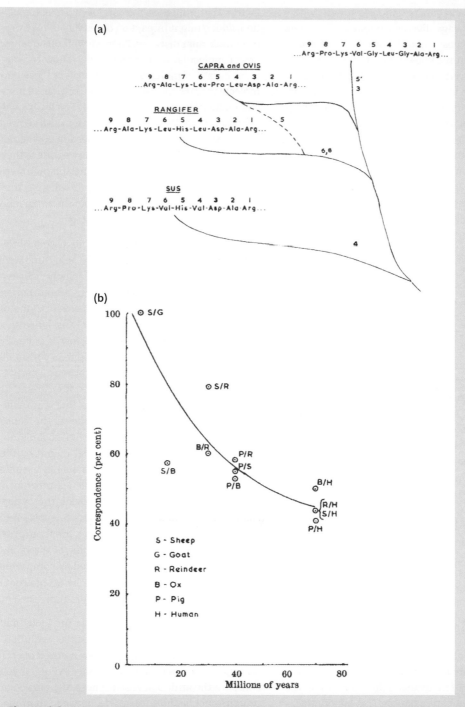

Figure 6.2 The molecular clock hypothesis was based on early observations of a relatively steady rate of change in proteins and genes. These figures are from an early study[5] which compared amino acid sequences of fibrinopeptides between mammal species (a) and calculated a rate of change based on assumed fossil dates of divergence between species (b).

Figures courtesy of Springer Nature.

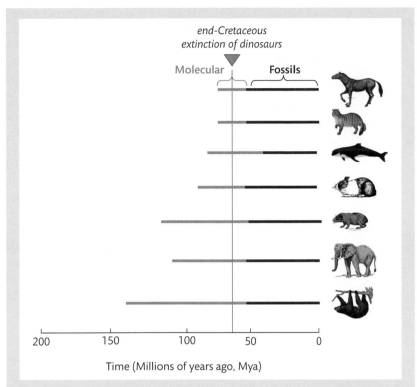

Figure 6.3 Age of mammal orders estimated from genetic distance assuming the same average rate of change in all lineages (grey lines) tend to be much older than the fossil evidence of those orders would suggest (black lines). These dates are from a 'strict clock' study that used a rate of change calculated by assuming that the bird and mammal lineages split 310 million years ago.[10]

their protein sequences.[5] If these changes accumulate constantly, and if the average rate of accumulation is more or less constant, then if you could estimate the rate of change, you could use the number of differences between proteins to estimate time. This use of protein or DNA sequences to infer the age of biological lineages became known as the 'molecular clock'.[6] The same logic could be applied to gene sequences, once it became possible to sequence DNA as well as proteins.

You can estimate the rate of change for proteins or DNA if you have some sequences for which you know how long it has been since they last shared a common ancestor. Fossil evidence of the earliest known member of a given lineage is usually used to 'calibrate' the molecular clock, allowing an absolute rate of change to be estimated (for example, average number of changes per million years). Once you have a calibration rate, the genetic distance between the genes of two different species will give an indication of how long it had been since they last shared a common ancestor.

Some early molecular dating studies challenged widely held beliefs. In the late 1960s, molecular dates suggested that humans and chimpanzees shared a common ancestor only four to five million years ago, less than half the generally accepted date at the time.[7] These initially shocking date estimates were later backed up not only by subsequent molecular studies but also by fossil discoveries.[8] As DNA sequencing became routine, there were more shocks in store. When amino acid sequences of proteins from different types of placental mammals were compared, and the differences between

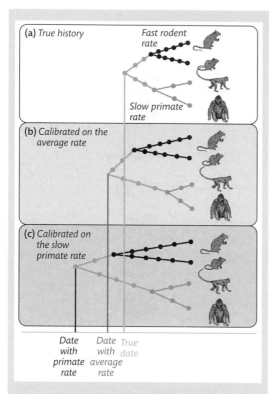

(a) *True history* Fast rodent rate

Slow primate rate

(b) *Calibrated on the average rate*

(c) *Calibrated on the slow primate rate*

Date with primate rate

Date with average rate

True date

Figure 6.4 Rate variation can influence date estimates made under a 'strict molecular clock' (assumption of uniform rates). Rats and mice tend to have faster rates of molecular evolution than apes so more substitutions (represented by solid circles) accumulate per unit time. Calculating the average substitution rate will produce date estimates that are too young for primates and too old for rodents. Calibrating the clock based on fossils from ape lineage results in overestimated rodent divergence dates.

dinosaurs was the primary cause of placental diversification.[11] Instead, these surprisingly old molecular date estimates were interpreted as implicating an earlier period of continental break-up as a major driver of global mammalian diversification.[10]

Why did the story told from molecules not match the story told from fossils? Molecular dates were rejected by some researchers on the grounds that the rate of molecular evolution was unlikely to be constant over time, making the 'molecular clock' an unreliable timepiece. Variation in the rate of molecular evolution need not invalidate molecular date estimates if the average rate is a fair reflection of change in most lineages. If the rate of change fluctuates randomly over time, molecular dates might still be accurate, even if not particularly precise. But if rates vary consistently and systematically between lineages, molecular dates based on average rates might be misleading. For example, it became apparent that genes in rats and mice evolved several times faster than the same genes evolved in apes.[12] This means that a rate calculated on rodent sequences would underestimate dates in the slower ape lineages, and vice versa, and an average calculated for both might not produce accurate date estimates in either (**Figure 6.4**). It also suggests that if early mammals were more like rodents, with fast rates of molecular evolution, they would have acquired more genetic changes in a shorter space of time than many modern mammals. This might mean that estimating rates of change from larger living mammals with slower rates of change might cause molecular dates to be overestimated. So the debate has raged: just how reliable are evolutionary dates estimated from proteins or DNA?

In this chapter, we are going to use the debate over the accuracy of molecular dates for the mammalian radiation as a case study for understanding how divergence dates are estimated from molecular data, and how those dates have influenced macroevolutionary hypotheses. We have to carefully consider the assumptions underlying such analyses in order to critically evaluate the accuracy and precision of the date estimates. Before we look at how we get a historical signal out of DNA sequences, we need to think about how the historical signal

them divided by the average rate of change per million years for those proteins (calculated by comparing bird and mammal sequences with their inferred fossil date of origin), there were far more differences than would be expected from 65 million years of post-dinosaur evolution. Instead, the genetic distance between them was commensurate with a much older origin of placental orders, deep in the Mesozoic (**Figure 6.3**).[9,10] These early molecular dates, and subsequent studies which produced similar results, challenged the widely accepted hypothesis that the disappearance of the non-avian got into the genome in the first place. To do that, we need to briefly review the way that genomes change over time, through the processes of mutation, substitution, and divergence. Because we are interested in molecular date estimates, we are going to focus specifically on aspects of molecular evolution that generate differences in the rate of DNA sequence change between lineages.

Key points

- Mammals evolved in the Jurassic, contemporary with dinosaurs.
- Mesozoic mammals occupied a diversity of niches, but most are not considered to be direct ancestors of modern mammals.
- Undisputed fossil evidence of modern placental mammal orders, such as rodents, primates, and carnivores, does not appear until after the extinction of the non-avian dinosaurs.
- The post-Cretaceous fossil diversity has been interpreted as evidence that mammalian diversification was triggered by the availability of vacant niche space.
- Molecular dates that place the origin of many placental lineages in the Mesozoic have been used to challenge this view by implying that placental diversification was not dependent on the extinction of dinosaurs.

Molecular evolution

Evolutionary change depends on heritable variation. Genetic variation comes in many forms, including whole genome duplications, chromosomal rearrangements, alteration of gene expression, and epigenetic modification. In this chapter, we will only consider single-nucleotide changes to DNA sequences, written in the four bases A, C, T, and G, since these are the changes that are most commonly analysed in modern molecular dating studies.

When we compare DNA sequences from the same gene in different species, we will usually be able to identify both similarities—sites where all species have the same nucleotide 'letter'—and differences—sites where species have different nucleotides (Figure 6.5). Each one of these nucleotide differences was originally generated by a mutation—an accidental change to the DNA sequence that occurred either as a mistake when the genome was copied, or as imperfectly repaired damage to DNA. Clearly, that mutation must not have been so severe that it resulted in the death of its carrier, nor did it stop that individual from successfully reproducing. We know this because that mutation must have been passed on to successful offspring, who then

also reproduced, giving rise to a line of descendants carrying that particular DNA sequence variation, all ultimately derived from the original mutation.

If the line of descent is broken, because none of the carriers of that mutation reproduce, all copies of that mutation disappear. But if the descendants increase in number, compared with those with other gene variants, this particular variant will form a greater percentage of the population in subsequent generations. If the frequency of the mutation reaches 100%, so that all offspring born in the population in that generation carry copies of that mutation, then we say that the mutation has gone to fixation. This is the process of substitution, where a particular variant in the DNA sequence replaces all other alternative sequences in that population. Now all members of the population carry a copy of the mutation that originally occurred in a single individual. It has become a defining feature of that population, a change in the genome that sets it apart from related populations. When we compare the DNA sequences from any member of that population with any other, we would expect to see that distinctive nucleotide sequence in their genome (until another mutation

Thylacine	A A A A A A G A G G A G A A A A G T C G T A A C A T G G T A A G T G T A C T G G A A G G T G
Tasmanian devil	C A A A A A G A G G A G A A A A G T C G T A A C A T G G T A A G T G T A C T G G A A A G T G
Numbat	T A C A A A G A G G A G A A A A G T C G T A A C A T G G T A A G T G T A C T G G A A A G T G
Kangaroo	A C A A A A G A G G A G A A A A G T C G T A A C A T G G T A A G T G T A C T G G A A G G T G
Dog	A C A C A A G A G G A G A C A A G T C G T A A C A A G G T A A G C A T A C C G G A A G G T G
Fox	A C A C G A G A G G A G A T A A G T C G T A A C A A G G T A A G C A T A C C G G A A G G T G
Giant panda	G C A T A A G A G G A G A C A A G T C G T A A C A A G G T A A G C A T A C T G G A A A G T G
Dormouse	T T G C A A G A G G A G A T A A G T C G T A A C A T G G T A A G C A T A C T G G A A A G T G

Figure 6.5 An alignment of a section of the 12s ribosomal RNA gene from eight different mammal species. In some places the species all have the same nucleotide sequence; in other places the sequence differs between species. The differences in DNA sequence provide the basis for dating divergence times (some dating studies use amino acid sequences of proteins instead).

at the same site goes to fixation, replacing all previous versions of that sequence).

Mutations occur every generation, and some of these mutations will go to fixation, so over time each population acquires a unique set of DNA changes. Eventually a population will have acquired such modifications to the genome that its members cannot successfully interbreed with individuals from other populations, because their genomes are too different to be able to function well together in a hybrid individual (see Chapter 7). They might also acquire genetic changes that prevent members of one population from mating with individuals from the neighbouring population, such as a change in flowering time or different mating calls. At that point, we are happy to say that the population is now so genetically distinct that we recognize it as a separate species. As long as populations do not exchange genetic variants with other populations, the inevitable processes of mutation and substitution will eventually lead to divergence of lineages into separate species.

When we compare the DNA sequences between species, we are observing one point in time in this continuous process of mutation, substitution, and divergence. Since changes accumulate continuously, we expect that the longer two lineages have been evolving separately, the more differences we should see between the DNA sequences of their genomes. So genetic distance—a measure of the amount of difference between two individuals for the same DNA sequences—provides an indication of how long it has been since they last shared a common ancestor. In other words, the genetic distance gives us information on evolutionary separation of lines of descent, and therefore the timing of diversification events.

But although the process of genomic divergence is continuous, the rate of change need not be the same in all lineages and in all time periods. For example, most genes evolve faster in rodents than in primates. How much variation in rate of molecular evolution should we expect? Can we use differences between DNA sequences to infer an evolutionary timescale if we cannot be sure whether the rate of change has been constant over time? Can we predict rate changes based on our knowledge of molecular evolution? We will first briefly consider the forces that shape differences in rate of molecular evolution, before considering how rate variation can be incorporated into molecular dating analyses of the timing of the mammalian radiation. The two key processes that can vary in rate between different species are mutation (generation of variation) and substitution (population genetic change).

Mutation

Mutation is the source of heritable changes in the DNA sequence, so we will first consider the way that mutation rate can vary between species. A mutation is any permanent change in the genome

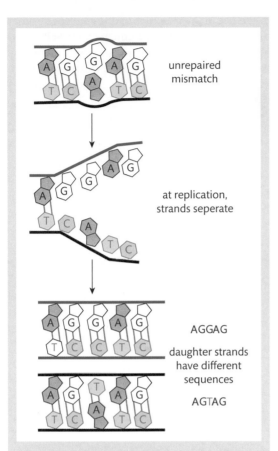

unrepaired mismatch

at replication, strands seperate

AGGAG

daughter strands have different sequences

AGTAG

Figure 6.6 Mutation occurs when either damage or copy error results in a change of sequence. Once the sequence is altered, to maintain base pairing, the original sequence can never be recovered and the new base sequence will be passed on to all copies made of that sequence.

that will be inherited by all copies of that genome. In the case of DNA sequences, mutations arise from two sources: damage and copy errors.

It is inevitable that a large and complex biomolecule like DNA will suffer occasional damage. The cell has a barrage of repair mechanisms, so most damage is fixed. But sometimes the repair itself may alter the precise sequence of nucleotides in the DNA molecule (**Figure 6.6**).

The other source of mutation is DNA replication. The whole genome must be copied every time a cell divides. Thanks to a sophisticated system for checking and correcting mistakes, DNA replication is amazingly accurate, with most cellular genomes having an error rate around one wrong nucleotide in every billion bases. But, given that the size of many genomes is measured in billions of nucleotides, this means that every round of replication is expected to introduce several changes to the DNA sequences due to copy errors. In multicellular organisms, like mammals, the genome must be copied many times per generation on the passage from zygote to adult to gamete.

Given these sources of changes to the nucleotide sequence of the genome, we can identify several factors that could drive differences in mutation rate between species. Since mutations can arise as DNA copy errors, differences in either the frequency of replication or the rate of replication error should influence the overall mutation rate. For any given error rate, the more copies are made, the more mistakes accumulate. Mammals with fast generation turnover tend to have faster rates of molecular evolution than their slower-reproducing relatives, which is consistent with an increased mutation rate due to more frequent DNA replication.[13]

Since mutations also arise from unrepaired damage, we might expect species that suffer more damage or have lower rates of repair to have higher mutation rates. Species living in mutagenic environments can suffer raised rates of DNA damage. For example, bacteria living in high-altitude lakes in the Andes must contend with a range of DNA-damaging mutagens, from high salt concentrations to arsenic to strong UV light.[14] They need enhanced DNA repair otherwise they could not survive and persist without undergoing mutational meltdown, with the population acquiring so many mutations that it drastically reduces reproductive output. And this brings us to an important point: the mutation rate is such an important part of the long-term survivability of a lineage that increased risk of mutation will usually need to be met with better damage protection or correction. So we find that species living in environments with more UV exposure tend to have more efficient UV damage repair mechanisms.[15,16] While mammal species may not differ so dramatically in their exposure to external mutagens, it has been suggested that they differ in the amount of mutagens generated internally by metabolism, with small-bodied species with fast metabolisms accumulating more DNA damage. The evidence for this hypothesis is equivocal, but it is consistent with a general pattern that small-bodied mammal species with rapid generation turnover, high fecundity, and short lifespans tend to have faster rates of DNA sequence evolution.[17]

Given that species are exposed to different mutation risks, selection could drive an optimum balance between the costs of mutation and the costs of repair. It takes more cell generations to make a large body, so an individual of a large-bodied species has a greater risk of acquiring a copy mutation in one of its cells during its lifetime. A large organism must also maintain more body cells over a longer life in order to reproduce successfully. Unrepaired damage in any one of these cells could lead to a disastrous failure and loss of evolutionary fitness. A large-bodied species might suffer both a greater risk of mutation and a greater cost associated with mutation than its smaller relatives. It is possible that large long-lived animals have low per-base mutation rates not just because they copy their DNA less often, but also because they have evolved to copy their genomes more carefully in order to avoid the amplified costs of mutation.[13]

What do you think?

If selection can fine-tune mutation rates, and given that most mutations are harmful, why does selection not reduce the copy-error rate to zero, and increase the damage repair efficiency to 100%?

There are many reasons to expect mutation rate to vary between species, and this means that species differ in the amount of heritable variation added to the population each generation. But mutation rate itself is not actually what we measure when we compare DNA sequences. Mutations happen in individuals, and if those mutations aren't ruinous they might be passed on to that individual's descendants. To be detected as a consistent difference in the genomes of different species that mutation has to rise in frequency until all members of the population carry a copy of the same genetic variant. This is the process of substitution.

Key points

- Rate of DNA sequence evolution depends on the rate of mutation (heritable change to the DNA sequence) and the rate of substitution (fixation of sequence variants in a population)

- The two causes of mutation—unrepaired damage and copy errors—are both modulated by DNA repair, which can vary in efficiency between species.

- Increasing the number of DNA copies per unit time can result in a higher mutation rate.

- A mutation becomes a substitution when it rises in frequency until all members of the population carry only copies of that mutation and all alternative sequence variants have been lost.

Substitution

Most new mutations are lost when their carriers fail to leave descendants. Some persist as polymorphisms, one of several alternative genetic variants in the population's gene pool. Few are passed on through enough generations to become a fixed part of the species' shared genetic information. What determines which mutations become substitutions, present in all members of a species? The most important determinant is the effect that the mutation has on its carrier. If a mutation harms its carrier's chance of survival or reproduction, then it is less likely to be passed on to future generations (this is called negative selection). If a mutation increases its carrier's chance of reproduction, then it is likely to increase in representation over the generations (positive selection).

What about mutations that have no effect on fitness? Mutations that occur in non-functional sequences will make little or no difference to the survival prospects of their organism (and non-functional sequences appear to make up a substantial proportion of mammalian genomes). Even mutations that change a functional sequence might not harm the organism if the resulting sequence works just as well as the original. Mutations that neither increase nor decrease their carrier's chance of survival and reproduction are said to be neutral with respect to fitness. 'Neutral' doesn't necessarily imply the mutations have no effect at all on the organism, simply that they don't affect their carrier's chance of passing the mutation on. It is simply a matter of chance whether these neutral mutations get passed on or not—they may be lucky enough to end up in an individual that survives to reproduce, or they might not.

While the frequency of advantageous mutations is expected to increase over generations until they replace all alternatives, and harmful mutations are expected to decrease in frequency and disappear, the frequency of neutral mutations will simply fluctuate from one generation to the next, sometimes increasing by chance, sometimes decreasing. By chance, these random fluctuations may eventually result in the frequency of the neutral mutation going all the way to 0% (in which case it disappears) or to 100% (in which case it is fixed as a substitution). This process of neutral mutations becoming fixed by chance is referred to as genetic drift. Drift is a random sampling effect; because not all individuals reproduce every generation, the sample of genetic variants passed to the next generation may not exactly match the proportions in the previous generations, leading to fluctuations in gene frequencies over time. Since all neutral mutations have the same chance of accidentally going to fixation, the rate of neutral substitutions is a function of their rate of production; that is, the neutral substitution rate is determined only by the mutation rate.

In fact, all genetic variants are subject to drift because no individual is immune to chance. The influence of random sampling is always present, as even the best adapted individual may fail to reproduce. The influence of random sampling on the frequencies of genetic variants in the population is modulated by the influence of the mutation on fitness and by population size. Imagine a mutation that has a small effect on fitness, such as an amino acid change that makes a particular enzyme 1% more efficient. Will it go to fixation? If the mutation occurs in an individual who lives in a large freely interbreeding population, and if it continues to be passed on to that individual's descendants, it has a fair chance of going to fixation. Because the advantage is small, we don't expect every individual carrying this mutation to do much better than every individual with alternative sequences. Each individual with the mutation is still subject to chance events, and may die or fail to reproduce for reasons entirely unconnected with this particular enzyme. But considered over a very large number of individuals, and over many generations, those with this small biochemical advantage might have a slightly more efficient metabolism and a slightly greater rate of successful reproduction, so even this small advantage should result in a greater number of carriers of the mutation in each passing generation. But in a small population, a slight advantage is less certain of going to fixation. When there are fewer individuals, the chance death of one or a few carriers has a greater proportional impact on the frequency of the mutation in the next generation. So selection is less efficient at promoting slightly advantageous mutations in a small population (see Chapter 3, p110).

The flipside is that slightly deleterious mutations are more likely to go to fixation in a small population than in a large population. If a mutation only slightly decreases its carrier's chances of survival and reproduction, it is possible for it to increase in frequency by chance. If a carrier of a mutation that makes an enzyme 1% less efficient just happens to find a large amount of food, or shelter from a storm, or happens to grow where there are fewer competing individuals, it might have more offspring than its neighbours who don't have the deleterious mutation, thus increasing the representation in the next generation despite its slightly negative effect. In a small population, it takes fewer chance increases in allele frequency to reach 100% fixation. So while we expect fewer slightly advantageous mutations to go to fixation in a small population, we also expect more slightly deleterious mutations to accidentally become fixed in a small population than in a large one (**Figure 6.7**).

Figure 6.7 Mutants on the loose. Most red-necked wallabies (*Macropus rufogriseus*) are brown, but a population on Bruny Island, off the coast of Tasmania, has a high frequency of individuals with white coats. Why? We could hypothesize that white fur is usually under negative selection, but the lack of predators on Bruny Island means that bright white fur is not a liability so is free to evolve by drift. However, red-necked wallabies in Tasmania are usually brown, even though many other Tasmanian mammals show changes associated with the lack of predation pressure, such as diurnal activity and reduced defensive structures. So absence of predation might not provide a full explanation of the rise of the white morph on Bruny Island, which might instead owe its high frequency to the relatively low power of selection to remove slightly deleterious mutations in small populations. How could you test which hypothesis provides a better explanation for the high frequency of white wallabies on Bruny Island?

Photo by Stefan Heinrich (CC2.0 license).

The upshot of this is that it's not just the effect of the mutation on the individual that influences whether it becomes a substitution. It is also the population it finds itself in. Which populations will have the greatest overall rate of change? There are far more ways for a random alteration to make a gene slightly worse at doing its job than there are to make it better, so we expect advantageous mutations to be less likely to occur than slightly deleterious mutations. Since slightly deleterious mutations are more numerous than advantageous ones, and slightly deleterious mutations have more chance of fixation in a small population where drift overwhelms selection, small populations will tend to have a faster overall substitution rate. Since population size is likely to vary between species, and also over time, we must also expect the rate of substitution to vary between species and over time. This is important when we want to use molecular divergence to infer evolutionary time.

Key points

- Substitution rate depends on the effect of the mutation on fitness: advantageous mutations are more likely to go to fixation and deleterious mutations are more likely to be lost.

- Mutations that have little or no influence on fitness fluctuate in frequency but will eventually be lost or fixed by chance.

- Population size influences rate of substitution: selection is more efficient in larger populations which will have a greater rate of fixation of advantageous mutations, but less efficient in small populations where random sampling effects can lead to slightly deleterious mutations going to fixation.
- Since there will be more slightly deleterious mutations than advantageous ones, the overall rate of change is expected to be faster in smaller populations.

At this point you may be wondering why a book on macroevolution and macroecology contains a chapter on the mechanics of mutation, population genetics, and genome change. There are two reasons why a familiarity with these processes should be part of every evolutionary biologist's mental furniture. One is that this is the engine room of evolutionary change: mutations create the variation on which selection acts, substitution drives the acquisition of genomic differences between populations, and those genomic differences fuel differential adaptation and genetic incompatibility which drive speciation. Of course, we can study the diversification of mammals, or any other group, without thinking specifically about genetic change, but genomic change is the ultimate source of evolutionary change. In fact, evidence is accumulating that in many groups the rate of molecular evolution is linked to the diversification rate, which might illustrate a fundamental link between change at the genomic level and the origins of biodiversity.[18]

The other reason that we need to cover genome evolution is a more practical application. Molecular dates are increasingly being used in macroevolutionary and macroecologial research. By learning to read the historical record in the genome, we are unlocking one of the most powerful tools we have ever had to explore the evolutionary past, patterns, and processes. In the case of mammalian radiation, DNA analysis allows us to reconstruct the history of all mammal species, including those that have relatively poor fossil records (such as bats), and those from areas or habitats where the fossil record is patchy (such as tropical forest). DNA analysis provides us with an alternative way of asking whether the absence of fossils—such as the lack of unambiguous placental fossils in the Cretaceous—is indicative of their absence from the biosphere, or whether it represents a failure of preservation or discovery.

There is a much broader reason why we have a chapter on molecular dating in a textbook on macroevolution and macroecology. All living organisms have DNA genomes, so we can potentially construct evolutionary histories for all species on earth, including those that have poor (or non-existent) fossil records. But learning to read the historical information in DNA intelligently requires us to understand how that record was written, and what we need to do to read it accurately. To do that, we need to become familiar with the richness and complexity of molecular evolution. Now that we have explored some of the factors that can change the rate of DNA sequence evolution, causing some species to have a faster rate of genomic change than others, we can consider how this variation in rate might impact on how we estimate evolutionary time from DNA sequences.

Molecular dating

The earliest molecular dating studies rested on the assumption that the accumulation of genetic differences could be adequately described using a constant average rate of change. Note that this doesn't imply that change occurs at exactly regular intervals, having a constant 'tick rate' like a clock.

Instead, it relies on assuming that, over time, the rate of change evens out to a predictable average. Clearly, this is not going to be the case for all parts of the genome. For example, genes for proteins that enable ultrasonic sound detection in the mammalian ear evolve faster in echolocating mammals (due to positive selection for improved function),[19] and genes that code for sweet taste receptors change rapidly in carnivores (due to lack of negative selection against changes to these dispensable traits).[20]

Early molecular dating studies focused on one or more 'housekeeping genes' that code for molecules with essential functions in all mammals, such as genes for metabolic proteins or for RNA molecules associated with gene expression. More recently, molecular dating studies of mammalian evolution have used phylogenomic datasets that include a large selection of sequences from across the genome, on the assumption that any patterns of selection specific to one gene in one lineage will tend to be overwhelmed by the overall historical signal.

Rates not only vary between genes, they also vary consistently between lineages. For example, genes from mice and rats accumulate changes several times faster than the same genes in primates (Figure 6.4). This is interesting from a biological point of view, because comparisons of the rate of change in different species can shed light on the drivers of genomic evolution.[13] But it also has great practical significance for molecular dating. If rates fluctuate randomly up and down over time, an average rate of change may still provide an accurate, though imprecise, indication of the amount of time over which genetic change has accumulated. But if some species have a consistently faster rate of change than others across their whole genome, using a single average rate of change to estimate time for all kinds of mammals might lead to errors in inferred dates (such as young dates in slow-changing lineages, and old dates in fast-changing lineages (Figure 6.4)).

To address this problem, molecular dating methods were developed that allowed different lineages to have different average substitution rates. Remember that to calculate an absolute rate of molecular change, we need a known date of divergence to 'calibrate the clock'. So if we want to estimate different rates in different lineages, we will need a calibration for each lineage. For example, the earliest unambiguous fossil proboscideans are from the Early Eocene, so this provides a minimum date of divergence of Proboscidea (elephants) and their sister clade Sirenia (dugongs and

manatees) that allows an average rate of change to be calculated for DNA sequences from present-day Proboscidea and Sirenia. Similarly, the first definitive fossil evidence of armadillos are scutes (amour plates) from the Early Eocene, providing a minimum divergence date for Cingulata (armadillos) and their sister clade Pilosa (sloths and anteaters), which allows an average rate for xenarthrans to be calculated from DNA sequences from armadillos, sloths, and anteaters. In addition to the fossil dates in each of these orders, we have some additional information we can use to estimate rates. We know that armadillos, sloths, elephants, and dugongs must have a common ancestor at some point in the past, so the rates must be solved to be compatible with both the fossil dates for each pair and the shared date between them (**Figure 6.8**).

This example shows how we can assign different rates to different parts of the phylogenetic tree. It's an example of a 'local clock' analysis, which allows a relatively small number of discrete rate categories, so that branches on the phylogeny can be considered to belong to a group of lineages that all share the same rate. These methods were developed to address criticisms of earlier studies that assumed all mammal lineages had the same average rate of change. However, local clock methods that allowed different orders of mammals to have distinct rates of molecular evolution did not solve the problem of the mismatch between molecular and fossil dates for the origin of placental orders. These local clock dates gave similar results to earlier dating studies, placing placental divergences deep in the Cretaceous.[21,22]

But there is a problem here. If we accept that mammalian lineages can have different rates of molecular evolution, then we must also accept that rates can evolve over time within lineages. Local clock models invoke a switch in average rate when two lineages split from each other—the elephants and dugongs are assigned a proboscidean rate throughout their evolutionary history and the sloths and armadillos have a xenarthran rate (Figure 6.8). In reality, we would expect two sister lineages to start with the same rate, inherited from their common ancestor, and then gradually diverge from each other in the substitution rate. This means that the earlier mammalian lineages could have

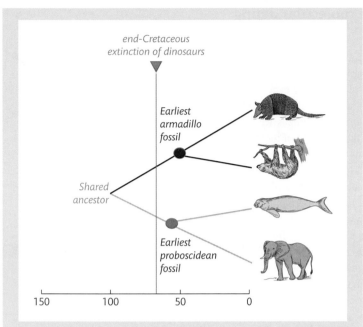

Figure 6.8 The age of the basal divergence of placental mammals has been estimated by using fossil evidence for the minimum age of divergence between pairs of mammal orders to estimate 'local clock' rates for different mammalian lineages. These dates are from a study that used quartets of taxa (each a pair of dated pairs) to estimate dates of the placental radiation.[21] One such quartet (Proboscidea/Sirenia vs Cingulate/Pilosa) is illustrated here. Note that in this study, calibrations are used as absolute dates of divergence, marking the probable age of the split between lineages.

had different average substitution rates than their more recent descendants. This might not be a problem if the average rate of change estimated by local clocks is a fair representation of their history. But it might cause dates to be misleading if there are systematic biases in rates, so that the average for each order does not provide a good estimate of their earlier history.

In particular, if rates of molecular evolution are associated with species traits, such as body size, and those species traits change over time, so will the rate of genomic change. We don't have fossil evidence of the shared ancestor of elephants, dugongs, sloths, and armadillos, so we can't be sure what it was like. But almost all placental orders increase in average body size from their first appearance in the fossil record to their later

descendants. So we can reasonably expect the shared ancestor of xenarthrans, proboscideans, and sirenians to have been smaller than any of the living species that DNA sequences have been sampled from. We know that smaller-bodied mammals tend to have faster rates of molecular evolution, so if these lineages increased in average size from the time of their shared ancestor to the present day, their average rate of molecular change might also have decreased.[13] The average rate calculated for each group might underestimate the rate in the early history, leading to the conclusion that more time was needed to achieve the observed level of genetic divergence, thereby making the molecular dates too old. If rates evolve along phylogenies like species traits do, we need a way of inferring dates of divergence that allows rate of molecular evolution to evolve along the phylogeny.

Rate-variable molecular dating methods

We have considered the way that some species traits, including population size and body size, can influence the rate of molecular evolution in mammal species. What we need is a molecular dating method that not only recognizes that mammals can differ in their average rate of molecular evolution, but also allows rates to evolve along lineages, as species evolve. Ideally, we want every branch in a phylogeny to be able to have a different average rate of change. But here we hit a practical problem. If rates can take a wide range of values, then for any observed amount of genetic difference between DNA sequences, we have to make a decision about whether it is best explained by an old divergence and a slow rate of change, or a recent divergence and a fast rate of change, or anything in between (**Figure 6.9**). There are three ways to make this decision: use independent estimates of dates to constrain the rate solutions (e.g. fossils provide information on the time it took to accrue the genetic differences); use knowledge of rates to constrain solutions (if we have an understanding of the likely substitution rates in these lineages); or use a model of evolution that allows us to predict how rates have changed throughout the evolutionary history of the group.

In most cases, we will be limited in our ability to apply the first two solutions for assigning different rates to all the branches in a molecular phylogeny. If we had fossil evidence that allowed us to accurately estimate the time period occupied by every lineage in the phylogeny, we could make a direct calculation of different rates on every branch. But, of course, it is precisely because we often lack such continuous fossil evidence that we need molecular dates in the first place. And for the same reason, we are unlikely to know about ancestral species traits for every point in the lineage's past, such as its body size, generation time, fecundity, population size, and so on. So even if we had a perfect understanding of the determinants of rate variation, we would still be unable to determine past patterns of rate variation with certainty.

Without complete knowledge of the evolutionary past of a group, or the exact determinants of rate of molecular evolution, there will always be a large

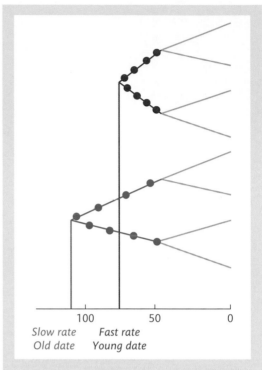

Figure 6.9 If rates can vary across the phylogeny, and we have no external information on the rates for a particular branch, any estimate of genetic divergence can be reconstructed as a wide range of a combination of rates and dates. Here the circles represent the inferred number of substitutions on each branch. If we asume a slow rate of change, we infer an old divergence date. If we assume a fast rate of change, we infer a young divergence date.

range of possible solutions for divergence dates estimated from molecular data. The only way that we can allow rates to vary over the phylogeny without having a detailed understanding of the causes of rate variation and complete information on all past states and dates is to use a set of assumptions to choose which of the possible solutions is most plausible. In other words, we need a set of rules that allows us to weigh the plausibility of different ways of explaining how the observed DNA sequences came to be.

There are a number of different techniques that allow the inference of evolutionary time scales from DNA sequences without assuming constant rates, and new methods are constantly being developed.

This is not the place for a review of molecular dating methods, but we will briefly consider one particular class of methods—Bayesian 'relaxed clocks'—which have been enthusiastically embraced by researchers in macroevolution and macroecology. These methods are popular because they allow the application of sophisticated models of molecular evolution, including variable rates of molecular evolution, and provide a convenient way to incorporate additional information from the fossil record. Bayesian methods also provide a framework for exploring and reporting uncertainty in molecular dates and for comparing the statistical support for alternative hypotheses. In particular, Bayesian molecular date have resulted in a shift in attitude towards mammalian molecular dates, by producing dates considered more compatible with the story inferred from fossil evidence. In Case Study 6, we go into more detail about how Bayesian molecular dating methods work, and examine one particular study where these methods are applied to uncovering the tempo and mode of the placental mammal radiation.

Here we will take a broad overview of the process of molecular dating using Bayesian methods, particularly because these techniques have led to date estimates that are considered to reconcile molecular and fossil-based date estimates for the mammalian radiation. Considering these methods also highlights some more general issues, such as the balance between using assumptions simple enough to allow tractable analyses, yet complicated enough to model the complexity of evolution. While molecular dating is typically treated as a computational, bioinformatics problem, it shares the same set of fundamental challenges that are common to all historical inference from present data: How is our understanding of past events shaped by our preconceptions of evolutionary processes? Can we ever reliably reconstruct the past, given that we cannot ever observe it directly? What do we do when different forms of evidence give conflicting histories for the same event?

Bayesian molecular dating

Bayesian methods represent a particular school of thought within statistics. They provide a way of comparing the plausibility of different hypotheses for explaining a particular observation, incorporating any prior information we may have about the probability of those hypotheses. We start with an assessment of the probability of each hypothesis being true based on what we already know (prior probability). We then consider how likely is it that this hypothesis could have produced the data that we have (likelihood), and then consider whether these data have changed our mind about the probability of the hypotheses (posterior probability). In a molecular dating analysis, the 'hypotheses' are different possible histories that could have given rise to the DNA sequences that we have observed (phylogeny), including the set of assumptions made about the way that evolution occurs (model).

Assigning prior probabilities to alternative hypotheses, independently of information derived from the data, is a key feature that sets Bayesian molecular dating analyses apart from other approaches. In most phylogenetic analyses, the number of possible solutions is vast. Even if you have a rather small dataset that contains only 20 species, there are more possible phylogenies than there are grains of sand in all the beaches on earth. And most datasets have far more than 20 DNA sequences. Yet in a Bayesian framework, we need to be able to give each of these possible solutions a prior probability score. Given the vast number of possible solutions, a Bayesian molecular dating analysis needs an efficient way of assigning a prior probability to every possible phylogeny and any combination of parameter values in the evolutionary model.

We can't separately evaluate the plausibility of every possible way of arranging the branches on the phylogeny. We could assign all possible trees the same probability (known as a flat prior). Or we can have a function that tells us what kind of trees we consider to be more reasonable descriptions of the data. A model of the branching process that describes the rate of speciation and extinction of lineages can be used to derive the kinds of 'shape' we expect our phylogenies to have. For example we could use a model of constant speciation and extinction rates to generate the range of tree shapes we consider most plausible (see Chapter 9). The speciation rate itself might also have a prior probability that describes the likely distribution of possible rates, with decreasing probabilities assigned to more extreme values. Of course, the real history of the species represented by the sequences might be quite different, with a burst of speciation in a given lineage or at a particular point in time. But the prior distribution gives an expectation that allows a probability to be assigned to each possible tree before we have considered what the data have to say.

Fossil evidence in Bayesian molecular dating

Some of the prior probabilities used in Bayesian molecular dating are derived from theoretical expectations, such as the expected shape of a tree if speciation and extinction have been constant throughout the lineage history. Other prior probabilities are described using statistical distributions for possible parameter values. For example, we might draw possible values for the substitution rate from a gamma distribution. This provides a convenient tool for modelling substitution rate variation, but is not based on any particular knowledge of the way rates evolve. But priors can also be based on knowledge we already have about the evolution of these sequences, independently of this analysis. Most importantly, information from the fossil record can be incorporated into the analysis to provide temporal information to calibrate the timescale inferred from molecular data. If we have fossil evidence that a particular lineage existed in a particular time period, then we can assume that the origin of that lineage must be at least as old as the fossil.

For example, fossil rodents have been described from the Late Palaeocene (58–55 Mya). As long as these fossils are correctly identified and accurately dated, we can be sure that the split between the rodent and primate lineages must have occurred before that time. But we wouldn't expect a lineage to instantly develop all the defining morphological characteristics of its later descendants as soon as it splits from its close relatives. When the lineage representing the most recent shared ancestor of primates and rodents first split into two separate lineages, individuals of both lineages would have resembled their common ancestor in almost all traits. Then, over time, as the lineages diverged and diversified, they each acquired distinct traits, such as gnawing teeth in rodents and forward-facing eyes in primates. Just as we don't expect rates of molecular evolution to instantly change as soon as a lineage divides in two, we also don't expect other characteristics to appear as soon as a lineage splits. This means that the first known fossil with all the key defining characteristics of a particular lineage is likely to be somewhat younger than the time the lineage diverged from its sister lineage (**Figure 6.10**).

But how much younger? Can we infer the length of time between the divergence of a lineage and the age of its oldest fossil? For mammals, as for most organisms, fossilization is rare. The first known fossil of a lineage is highly unlikely to represent the first member of that lineage that ever lived. In fact, we don't even have fossils of every mammal species that has ever existed. New mammal species are described from fossil evidence every year, so we know that there must still be gaps in our knowledge of extinct species. Fossil dates provide us with good evidence for minimum dates of origin, because the lineage must have originated by the time that identifiable fossils are present. But they don't tell us exactly when the split between that lineage from the common ancestor it shares with its sister lineages occurred. It is this splitting event that is marked by a bifurcation in the phylogeny, and a corresponding molecular date (Figure 6.10).

Although we can't know for sure when the split between two lineages occurred, we can use fossil evidence to put likely bounds on the age of the divergence. A fossil identified with a particular lineage marks a minimum age—the lineage split must have occurred before the first fossil member of that lineage is found. We might also be able to define a maximum age. For example, placentals are more closely related to marsupials than they are to monotremes. The split between the monotremes and the therians (placentals and marsupials) must have happened before the placentals split from the marsupials. But note that we are a lot less certain about maximum ages than about minimum ages. As it happens, the monotreme

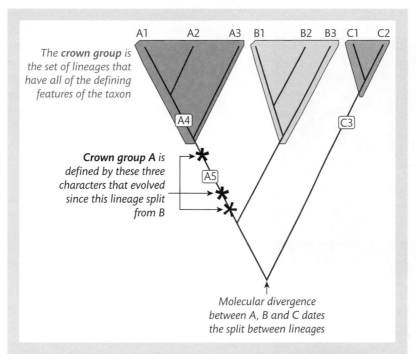

*The **crown group** is the set of lineages that have all of the defining features of the taxon*

Crown group A *is defined by these three characters that evolved since this lineage split from B*

Molecular divergence between A, B and C dates the split between lineages

Figure 6.10 Fossil evidence rarely marks the exact point of divergence between two lineages. Instead, the first fossil with recognizable features of a particular group is likely to be dated to some time after the genetic split between the lineages, but the length of time between the lineage split and the first fossil is generally unknown. Using fossil information to infer the possible dates of genetic divergence also requires an appreciation of crown groups and stem groups. Fossil A4 is a crown group taxon—it has all the defining features. Fossil A5 is a stem group taxon—it has some but not all of the defining features. The first fossil of a lineage may post-date the split with its relatives (e.g. C3). In some molecular dating analyses, fossil calibrations are treated as exact ages of nodes in the phylogeny. In other analyses, fossil information informs a probability distribution of possible divergence dates for a given node (see Case Study 6).

fossil record is relatively poor (**Figure 6.11**). Very few monotreme fossils have been described from the Mesozoic, with the most reliably identified fossils dated to around 120 million years ago.[23] Some of these fossils were described relatively recently. Before their discovery, the monotreme record started in the Palaeogene, long after the branching event that split the monotremes from their marsupial and placental cousins. Therefore we can see that the oldest evidence of an early-branching lineage might actually occur after evidence of later descendants (Figure 6.10). When does a lack of

fossils from a given period represent the absence of the species from the biota of the time (evidence of absence), rather than a failure to leave informative fossils (absence of evidence)? Maximum age bounds must be used with great caution, as we can rarely be sure that the oldest fossil represents the maximum age of a lineage.

We can sometimes define a minimum date for the age of a split, and in some cases we might feel we have sufficient reliable evidence to define a maximum. But we might not feel that every possible

Figure 6.11 Monotremata, which includes the platypus (a) and the echidna (b), is one of the three main lineages of mammals (along with Marsupialia and Placentalia). The monotreme fossil record is usually described as 'sparse'—until a few decades ago there were very few fossils at all, but the palaeontological information on monotreme history is growing. However, most monotreme fossils are fragmentary, such as isolated teeth or jaw bones, making inference on morphology and niche challenging. Monotreme fossils also demonstrate some of the biases of preservation: the semi-aquatic platypuses are better documented in the fossil record than the terrestrial echidnas, presumably due to the greater probability of preserving taxa in river or lake sediments. Monotremes also represent an interesting case of dissonance between palaeontological and molecular dates of divergence: the palaeontological record has been interpreted as indicating a Cretaceous split between the echidna and platypus lineages, but many molecular date estimates place this divergence in the Palaeogene.

(a) © Dave Watts/naturepl.com. (b) Kristian Bell/Shutterstock.com.

date between the maximum and the minimum are equally plausible. In some cases, we might think that it's more likely that the origin of the lineage was relatively close in age to the earliest fossil, and that older dates become less plausible the further back we go in time. In this case, we could describe our prior belief in the date of divergence using a probability distribution of possible dates, with the fossil evidence defining the minimum and maximum (Case Study 6).

Now we have some independent information that helps us place some evolutionary events on a timeline (measured in millions of years). We want to use this information to help us infer rates of molecular change for all the mammal lineages represented by our DNA sequences (typically expressed as number of substitutions per site per million years). If we can accurately estimate rates across the entire phylogeny, we can use the absolute timescale derived from fossil information to estimate the age of every divergence event in our phylogeny. However, if we want to allow rates to vary across the phylogeny, but we don't have sufficient fossil evidence to place the start and end of every branch on the timescale, then we face the problem of how to decide what the rate is on each branch. This will have a big effect on the date estimates—for the same amount of genetic divergence, assigning a branch a fast rate will imply a younger date of origin, and assigning it a slower rate will give that lineage an older date of origin (Figure 6.9). How do you decide which is the most believable solution?

Relaxed clocks

Molecular dating methods that allow rates of molecular evolution to change between lineages are commonly referred to as 'relaxed clocks' because they relax the assumption of rate constancy in the inference of evolutionary time.[24] These methods evaluate different possible patterns of assignment of rates to different branches on the phylogeny, using a probabilistic model to identify the most plausible explanations for the observed sequence data. For example, some models allow rates to step up or down from one branch to the next, but preference is given to solutions that minimize the number or size of these steps. Under this model, a solution that has fewer or smaller rate changes will be preferred over one that requires large and frequent rate changes. Other models for assigning rates to branches try to group lineages into shared rate categories, such that rates don't change at every node and related branches will often share the same rate.

All of these rate-variable models are statistical solutions to the problems of changing rates. For example, the aim may be to try to minimize rate changes, or smooth the changes in rates over the tree. These optimization strategies are not informed by any particular knowledge of the patterns of molecular evolution for a given group of species, or any particular understanding of the way the evolutionary history of this group might have shaped their rates of genetic change. These models are reasonable but arbitrary ways to make a decision between alternative solutions. In the absence of independent information about the rates on each branch, or the dates of each node, we can never know for sure if the optimal solution, given the model we have specified, represents the true historical pattern of rates.

Compare this with the relative simplicity of the earliest molecular clock studies. In some of the earliest studies, a single calibration point (for example, assuming that birds and mammals split exactly 310 million years ago) was used to calculate a rate of change for a particular gene, and then this rate was used to estimate the age of other nodes in the tree (Figure 6.3). When a single calibration is used to produce a uniform rate, there is a single solution for the age of every node in the phylogeny. But for molecular dating methods that allow every branch to have a different rate, there are uncountable possible solutions for the same dataset; any branch could have a slower rate and an older date, or a faster rate and a younger date (Figure 6.9). How can the vast array of alternative solutions be evaluated, and how do we decide which one to report as the most plausible solution? Most Bayesian dating analyses produce a set of possible solutions, each of which has almost the same probability, given the assumptions of the method. This 'credible set' represents the range of plausible solutions for one

particular dataset, combined with a given model of evolution. Running the analysis again with a different model, or different data, may produce different results.

It is important to remember that just because the assumptions made in Bayesian molecular dating are, in many cases, an unrealistic simplification of the evolutionary process or an imperfect description of the dataset, this does not necessarily compromise the accuracy of the molecular dates. In fact, we generally aim to have the simplest possible model that can adequately describe the pattern in the sequences, rather than a model that captures every complicated aspect of the real world. Considering the earliest molecular dating studies, few would expect the rate of change of protein sequences to be exactly clock-like, with an amino acid change occurring exactly every seven million years.[6] But this simple model would suffice if the average rate of change was a fair description of the history of the sequences, even if there are times when the rate of protein change accelerated, and other times when it slowed. Of course, the uniform rates model would be inadequate if changes in rate were not just random fluctuations, but showed systematic differences in different lineages. When this is the case we reject the uniform rates model as an inadequate description of the data. We aim to use the model that can most efficiently and reliably extract the information we need from the data we have, but to reject any model that is shown to be inadequate to derive the true story from the data.

Key points

- Bayesian molecular dating methods can incorporate a number of different models that allow rates of molecular evolution to vary over a phylogeny.

- These methods use a large number of assumptions about the evolutionary process and the nature of the dataset in order to assign a prior probability to any possible arrangement of the phylogeny, combination of branch lengths, or set of model parameters.

- The analysis allows us to evaluate whether the observed sequence data changes our opinion on the probability that date estimates are a true description of evolutionary history (posterior probability).

- Bayesian methods provide a convenient framework for evaluating the range of dates compatible with the sequence data given the assumptions. The influence of these assumptions on molecular dates must be investigated and evaluated for any dating study.

Capturing rate change

We have discussed how Bayesian molecular dating methods use a sophisticated set of assumptions about the evolutionary process in order to allow for variation in the rate of molecular evolution over the tree by combining models of changing rates with prior beliefs about divergence times from the fossil record. But we have also seen that, once you allow for changing rates, there are always many possible solutions to any molecular dating problem. At this stage, we don't know how well the 'relaxed clock' models capture real variation in rate of molecular evolution. For example, all relaxed clock models are based on stochastic change, where rates wander up and down without bias towards a particular direction. But given that variation in rate of molecular evolution is influenced by many different species characteristics, such as size, reproductive rate, and population size, we might expect rates to undergo consistent directional change in lineages where those species traits are also undergoing directional change. Will such directional change be adequately captured by current methods?

The mammals present a fascinating case study because several clear patterns in species-specific

rate variation have been identified. Recall that smaller-bodied mammal species tend to have faster substitution rates, probably due to higher mutation rates arising from short generations, possibly combined with less stringent selection on DNA repair or copy efficiency due to smaller bodies and shorter lives. Models of rate change employed in Bayesian molecular dating are stochastic, not giving any particular preference to increases or decreases in rate. But if the earliest placental mammals were smaller than their descendants, then rates might have slowed down in most lineages. This means that the rate of molecular evolution estimated using fossil calibrations from the Palaeogene history of mammals may be slower than the true rate in the earliest part of mammalian history, when most mammals were small, lived fast, and died young.

This potential for rate bias will be exaggerated if the selection of taxa in the molecular dating study is skewed toward larger-bodied mammals.[25] Large-bodied vertebrates tend to have a better fossil record than tiny ones, perhaps because the bones are better preserved, or maybe because they are preferentially collected. It has been suggested that the tendency to calibrate mammalian molecular dating analyses predominantly on large-bodied fossil taxa has led to date estimates of the placental radiation that are too old, and that calibrating on smaller-bodied lineages give date estimates more in line with the fossil evidence.[26] However, other studies have suggested that even when larger taxa are excluded, divergence dating still leads to deep Cretaceous date estimates for the diversification of placental orders.[27] So far, attempts to model the way that rates change with body size and mammal life history have led to somewhat peculiar results, for example predicting a placental ancestor the size of a rabbit (~1kg) but with the lifespan of a moose (~25 years).[28] However, it's early days for these methods, and much remains to be done to develop and test approaches to molecular dating that take the dynamic nature of mammalian molecular evolution into account.

What can DNA tell us about mammal radiations?

We started this chapter by asking whether the dinosaur extinction triggered the radiation of placental mammals. Let's recap what we have covered and remind ourselves why we have spent most of the chapter thinking about mutations, substitutions, and genetic divergence, and getting submerged in the details of Bayesian phylogenetic analyses.

A widely accepted picture of mammalian evolution is that mammals emerged at the same time as the dinosaurs, but that placental mammals did not diversify until the Palaeogene when they rapidly radiated to fill the niches left vacant by the extinction of the dinosaurs. But this hypothesis was challenged by molecular dates that placed the origin of many placental mammal orders deep in the Cretaceous, contemporary with a diverse dinosaur fauna. Those molecular dates have themselves been challenged on the grounds that variation in the rate of molecular evolution over time makes molecular dates based on the assumption of constant rates of molecular evolution unreliable. New methods, particularly those based on Bayesian analyses, allow for variation in the rate of molecular evolution. These methods involve a complicated set of assumptions about the evolutionary process, and the incorporation of prior beliefs about the timing of some divergence events. Modifying these assumptions can result in substantially different date estimates. Nonetheless, many recent molecular dating studies place the origin of many placental orders in the Late Cretaceous, lessening the gap between fossil evidence and molecular dates. Can we now conclude that the problem of the mismatch between molecular and palaeontological dates for the placental radiation has been solved by new improved molecular dating methods that overcome the limitations of older methods? Let's examine the newer date estimates a little more carefully to see if we can be confident that the right answer has now been reached.

The most obvious aspect of these new methods that might lead to a narrowing of the gap between molecular and palaeontological date estimates is the allowance for variation in rate of molecular evolution. But remember that these methods don't 'read' rate variation from the data (unless there are known calibration dates at the beginning and end of each branch in the phylogeny). These methods assess many possible solutions, with different rates on different branches, and then report the solutions that optimize the probability that the hypothesis is true, given the assumptions made. Therefore the date estimates must be interpreted as the best solution given a particular model (or models). Are these models a fair description of molecular evolution? We don't need models to resemble reality in all its complexity. We just need to know if they are adequate to do the job of allowing us to infer time from differences in DNA sequences.

If you compare recently published studies reporting molecular dates for placental mammal diversification, you will see that using different sequence datasets or alternative evolutionary models can give different date estimates.[29,30] But the greatest differences between date estimates for the placental radiation seem to be the result of differences in the way that the dating analyses are calibrated using fossil evidence.[31] Calibrations do two kinds of work in a Bayesian analysis. First, in common with other molecular dating methods, calibrations allow us to convert branch lengths from units of molecular sequence change to units of absolute time (e.g. millions of years). You can run a Bayesian molecular phylogenetic analysis without any calibrations and it will find an optimum distribution of rates across branches, given the model and assumptions, but the branch lengths will not have a time unit. But in a Bayesian molecular dating analysis, calibrations do more than provide a way of converting branch lengths to time. They narrow the range of acceptable solutions by weighting some solutions as more probable than others.

For example, if a molecular phylogeny of placental orders is calibrated by fixing the age of nodes to the first known fossil of each order,[32] in order to match the nodes in the phylogeny to the fossil evidence, the analysis must infer astonishingly high rates of molecular evolution during the early part of placental mammalian radiation, faster than rate estimates for any known mammal species today.[33]

But when calibrations are given as distributions of possible ages, the possible solutions to the dating analyses are 'relaxed' in that they allow a greater range of combinations of dates and rates. This relaxation of the possible ages of calibrated nodes results in older date estimates for the divergence of placental orders than when the calibrations are used as fixed ages[34] (see Case Study 6).

There is a quandary here. If you conduct a molecular dating analysis on mammals where the maximum dates are unconstrained, you get date estimates for the origin of placental orders far older than the fossil record would suggest. If you set the dates of divergence of placental orders to the earliest known fossil, you get unreasonably fast molecular rate estimates. If you use a Bayesian molecular dating analysis that uses prior information based on fossil evidence to constrain the solutions, you tend to end up with dates that fall within the acceptable range of dates specified in the assumptions you made before looking at the DNA sequence data. If this is the case, the molecular data have not really challenged the beliefs you already had based on fossil evidence alone. Rather than a way of reading a timescale from molecular data, perhaps we should consider these Bayesian molecular dating analyses as a way of encapsulating what we have to believe about the molecular data in order to fit it to a specified range of possible diversification dates for certain nodes in the tree.

Which is correct, fossils or molecules?

We are left with alternative explanations for the relationship between molecular and fossil evidence. We can consider three broad interpretations of what the relationship between molecular dates and fossil evidence tells us about the timing and mechanisms of the placental mammal radiation.

It could be that Cretaceous molecular dates for placental mammal divergences are wrong because rates of molecular evolution were much faster in the early part of the history, making date estimates based on recent mammalian lineages too old. This interpretation is commensurate with the pattern in the fossil record (mammal diversity explodes after dinosaur extinctions), and broadly compatible with our understanding of molecular evolution (smaller mammals have faster rates, and ancestral placentals are likely to have been small). But how do we model these changes in life history and molecular evolution into the past? Just how fast should we consider the ancestral rates to be? How can we be sure that 'relaxed clocks' provide an adequate description of these rate changes? One approach is to examine the accuracy of molecular date estimates throughout the mammalian phylogeny, not just for the divergence times of the placental orders. We can look at studies that produce date estimates that are compatible with a post-dinosaur radiation, and see if the inferred ages of other nodes are also compatible with fossil evidence. In some cases, the molecular dates for placental lineages that diverged long after the early Palaeogene are surprisingly young, in conflict with known fossil evidence. For example, a molecular dating study that reconciles the molecular dates of the placental radiation to the earliest fossil evidence also places the origin of the Perissodactyla in the late Oligocene in contrast with fossil evidence of early horses in the early Eocene[27] (Chapter 3).

An alternative explanation for the common mismatch between molecular dates and fossil dates is that the surprisingly old molecular date estimates reveal that the radiation of placentals began long before the dinosaur extinction, even though unambiguous fossil evidence of Cretaceous placentals is missing. We know that the fossil record is patchy in time and space; some areas and time periods are more likely to be represented in the fossil record than others. Could it be that the radiation of placental mammals took place somewhere that is currently poorly sampled for late Cretaceous terrestrial vertebrate fossils? Descriptions of eutherian fossils from Gondwana have led to suggestions that the early history of placentals may have occurred in the Gondwanan supercontinent, where the Cretaceous mammal fossil record is poorer than in Laurasia. This hypothesis could be confirmed by finding unambiguous Cretaceous fossils of modern placental orders, which so far have refused to be found.[4] But, even if Gondwanan placentals existed, it is possible that they might never be found if there are preservation biases that have effectively removed them from the known record. Is there any other corroborating evidence we can draw on? One approach is to ask whether other groups show a similar history of Cretaceous radiation. For example, lice tend to co-speciate with their hosts, so the dates of louse diversification are expected to reflect the diversification of mammals. Molecular dates for lice suggest they radiated in the Cretaceous.[35] This result has been interpreted as giving some independent support to the mammalian molecular dates: because we don't necessarily expect the same patterns of molecular rate variation in lice as we do in mammals, the lice dates may be free from some of the suspected biases in the mammal dates.

A third scenario is that both the molecular record and the fossil record can be read as literal and accurate records of the history of placental mammals, but that they mark different events in their evolution, with the molecular dates marking the separation of the lineages, and the fossil dates marking the adaptive radiation into different ecological niches. In this scenario, the ancestors of placental mammals diversified in the Cretaceous, generating many lineages, but they were constrained in their ecological disparity. Once the dinosaurs were out of the way, these lineages developed a much greater range of ecological roles and morphological diversity. While this last scenario is attractive and harmonious, it too is not without its problems. We still need to be able to explain why there are no confirmed fossils of the dozen or more placental lineages that molecular dates place in the time of the dinosaurs. To believe molecular dates that predict a late Cretaceous radiation of mammals, we need to have confidence in the assumptions made in the analysis concerning the way rates evolve and the acceptable range of date estimates based on the fossil record.

This may seem unsatisfying. What is the point of spending time and money sequencing DNA if we can't use it to make a definitive statement about timing, or to overcome possible deficiencies of the fossil record? To reject molecular dating due to its imperfections would be like rejecting palaeontological evidence because we know that the fossil record is incomplete. There is valuable information in both the fossil record and the molecular record. The value of molecular dating is the potential for putting a timescale on all evolutionary divergences, not just lineages for which we have reliable fossil evidence. The reason that people have spent so much time arguing about molecular dating techniques and invested so much time in developing and improving them is that a timescale for evolution is vitally important for understanding the patterns and processes of macroevolution and macroecology. Molecular dates allow us an unprecedented opportunity to gain a new view of the evolutionary past.

But the reality of scientific research in macroevolution and macroecology (shared with many other fields of science) is that there are few model-free, assumption-free statements we can make about events or processes that we cannot witness directly. Instead, we observe the outcomes of those events and processes, and we consider the range of possible processes that could have led to those outcomes. There will rarely be a single plausible explanation for a set of observations, so we need to be able to evaluate and compare alternative explanations in light of what we know—and what we don't know—about evolution. Where we are uncertain—where there is more than one possible explanation and we cannot say for certain which one is true—then we should report that uncertainty as part of our findings. Otherwise we risk basing our understanding of the natural world on false premises. Statements of uncertainty should never be interpreted as scientific failure but as summaries of our current understanding.

What do you think?

Can you think of evidence we could find that would resolve this debate about placental mammal radiations resolutely in favour of one hypothesis or the other? If we fail to find evidence in favour of a particular hypothesis, should we concede that the hypothesis is incorrect? Or is it inevitable that the evidence will always be open to interpretation and people will side with one hypothesis over the other along lines of prior belief?

Conclusion

Molecular date estimates have been used to challenge the long-held hypothesis that the adaptive radiation of placental mammals was triggered by the end-Cretaceous extinction of the dinosaurs, which released mammals to exploit a wide range of vacated ecological niches. Early molecular dating studies suggested that the divergence of the main lineages of placental mammals occurred deep in the Cretaceous, contemporary with a diverse dinosaur fauna. These molecular date estimates were challenged on the grounds that the rate of molecular evolution varies over time and between lineages, which may complicate attempts to estimate time from genetic distance. Although genetic differences must inevitably accumulate between lineages over time, molecular change need not accumulate at a uniform rate in all lineages. Species traits can influence both the rate of mutation and the rate of substitution, so we should expect rates to vary between species and over time. Date estimates from Bayesian phylogenetic analyses that allow for variation in the rate of molecular evolution between lineages have produced date estimates which, while younger than the earliest known fossils, do not suggest such a huge gap between the origin of placental lineages and their first known fossils in the Palaeogene. These newer molecular dating studies have led to the conviction that the earlier discrepancy between molecular date estimates and the fossil record has been solved by the application of models that allow variation in rates of molecular evolution.

But, as with all macroevolutionary analyses, we must be careful to identify where the assumptions of the analytical method have a significant impact on the answer we get, and, if they do, we need to carefully consider the reliability of the assumptions that we have made. In Bayesian molecular dating, we have to make a very large number of assumptions about the nature of the data and the processes of evolution, including the sampling of taxa, the state of knowledge of the evolutionary timescale given fossil evidence, the macroevolutionary model of speciation and extinction, and the way that rates of genomic change vary over the phylogeny.

For example, the models of rate variation are based on stochastic change in rate over the tree, so that a descendant branch is as likely to increase or decrease in rate from the parent branch. But we have seen that many factors influence the rate of molecular evolution. In particular, we expect small mammals to have faster rates of molecular evolution than their larger relatives. Given that most orders of placental mammals are likely to have started small and increased in average body size, we might expect rates of molecular evolution to have been faster in the earlier part of the mammalian radiation, making molecular date estimates based on more recent rates of genetic change too old. Unless we know that our models can capture non-random directional rate change such as this, we may be led astray by biased date estimates. We need a priori information on the timing of divergence events to 'ground truth' substitution rate estimates. But given the uncertainty in estimating timing of divergence events from both the palaeontological record and the molecular evolutionary record, there will always be a wide range of possible ways of fitting the two together, depending on what we are prepared to believe about the gaps in the fossil record or the variation in rates of molecular evolution.

There is a broader problem here. If we accept only those molecular date estimates that agree with pre-existing fossil dates, molecular dates will only provide new information in the case of divergence events for which we have no fossil evidence. But for nodes with no fossil evidence, we have no way of knowing whether the molecular dates are correct or not. You might be tempted to say that we can never trust molecular dates, but it would be a tragedy to lose such a potentially useful tool which can give us previously undreamed-of insight into the evolution of any biological group for which we can obtain DNA sequence data and some way of establishing rates of molecular evolution.

There is valuable temporal information in the genome, but there will always be uncertainty associated with attempts to infer time from comparisons between DNA sequences, just as there is uncertainty in fossil dates of divergence. For any molecular dating analysis, we need to explore the range of solutions that is compatible with reasonable assumptions. The molecular data are telling us something interesting about the mammalian radiation— either that molecular change can run faster in some parts of the history than in others, or that the history of mammals goes back deeper than previously thought, or possibly both. More work is needed to untangle these possibilities before we can be certain we know the real story of the diversification of placental mammals.

Further reading

If you want to learn more about the use of DNA sequence analysis in evolutionary biology, including molecular dating, there is a friendly introduction to the field in the companion volume to this book: *An Introduction to Molecular Evolution and Phylogenetics*, Lindell Bromham, 2016, Oxford University Press.

Q Points for discussion

1. What could cause the 'molecular clock' to speed up? How could you test for faster rates in the past? What information would you need?

2. Do you think that adaptive radiation and divergence into new niches will occur more rapidly in lineages that tend to have small populations or large populations? Can we predict the 'evolvability' of lineages if we know their mutation rate and population size?

3. Can we use molecular dates to estimate divergence times in groups for which there are no fossils?

✳ References

1. Kemp TS (2016) *The Origin of Higher Taxa: Palaeobiological, Developmental, and Ecological Perspectives*. Oxford University Press, Oxford.

2. Close RA, Friedman M, Lloyd GT, Benson RBJ (2015) Evidence for a mid-Jurassic adaptive radiation in mammals. *Current Biology* 25(16): 2137–42.

3. Archibald JD, Averianov AO, Ekdale EG (2001) Late Cretaceous relatives of rabbits, rodents, and other extant eutherian mammals. *Nature* 414(6859): 62.

4. Goswami A, Prasad GV, Upchurch P, Boyer DM, Seiffert ER, Verma O, Gheerbrant E, Flynn JJ (2011) A radiation of arboreal basal eutherian mammals beginning in the Late Cretaceous of India. *Proceedings of the National Academy of Sciences of the USA* 108(39): 16333–8.

5. Doolittle RF, Blombäck B (1964) Amino-acid sequence investigations of fibrinopeptides from various mammals: evolutionary implications. *Nature* 202: 147–52.

6. Zuckerkandl E, Pauling L (1965) Evolutionary divergence and convergence in proteins. In Bryson V, Vogel HJ (eds), *Evolving Genes and Proteins*. Academic Press, New York, pp.97–165.

7. Wilson AC, Sarich VM (1969) A molecular timescale for human evolution. *Proceedings of the National Academy of Sciences of the USA* 63(4): 1088–93.

8. Bradley BJ (2008) Reconstructing phylogenies and phenotypes: a molecular view of human evolution. *Journal of Anatomy* 212(4): 337–53.

9. Li W-H, Gouy M, Sharp PM, O'hUigin C, Yang YW (1990) Molecular phylogeny of Rodentia, Lagomorpha, Primates, Artiodactyla, and Carnivora and molecular clocks. *Proceedings of the National Academy of Sciences of the USA* 87: 6703–7.

10. Hedges SB, Parker PH, Sibley CG, Kumar S (1996) Continental breakup and the diversification of birds and mammals. *Nature* 381: 226–9.

11. Bromham L, Phillips MJ, Penny D (1999) Growing up with dinosaurs: molecular dates and the mammalian radiation. *Trends in Ecology & Evolution* 14(3): 113–18.

12. Ohta T (1993) An examination of the generation time effect on molecular evolution. *Proceedings of the National Academy of Sciences of the USA* 90(22): 10676–80.

13. Bromham L (2011) The genome as a life-history character: why rate of molecular evolution varies between mammal species. *Philosophi-*

cal Transactions of the Royal Society B. Biological Sciences 366(1577): 2503–13.

14. Albarracin V, Pathak G, Douki T, Cadet J, Borsarelli CD, Gärtner W, Farias ME (2012) Extremophilic *Acinetobacter* strains from high-altitude lakes in Argentinean Puna: remarkable UV-B resistance and efficient DNA damage repair. *Origins of Life and Evolution of Biospheres* 42(2-3): 201–21.

15. Miner BE, Kulling PM, Beer KD, Kerr B (2015) Divergence in DNA photorepair efficiency among genotypes from contrasting UV radiation environments in nature. *Molecular Ecology* 24(24): 6177–87.

16. Svetec N, Cridland JM, Zhao L, Begun DJ (2016) The adaptive significance of natural genetic variation in the DNA damage response of *Drosophila melanogaster. PLoS Genetics* 12(3): e1005869.

17. Welch JJ, Bininda-Emonds ORP, Bromham L (2008) Correlates of substitution rate variation in mammalian protein-coding sequences. *BMC Evolutionary Biology* 8: 53.

18. Hua X, Bromham L (2017) Darwinism for the genomic age: connecting mutation to diversification. *Frontiers in Genetics* 8: 12.

19. Liu Y, Rossiter SJ, Han X, Cotton JA, Zhang S (2010) Cetaceans on a molecular fast track to ultrasonic hearing. *Current Biology* 20(20): 1834–9.

20. Jiang P, Josue J, Li X, Glaser D, Li W, Brand JG, Margolskee RF, Reed DR, Beauchamp GK (2012) Major taste loss in carnivorous mammals. *Proceedings of the National Academy of Sciences of the USA* 109(13): 4956–61.

21. Murphy WJ, Eizirik E, Johnson WE (2001) Molecular phylogenetics and the origins of placental mammals. *Nature* 409: 614.

22. Bininda-Emonds ORP, Cardillo M, Jones KE, MacPhee RD, Beck RM, Grenyer R, Price SA, Vos RA, Gittleman JL, Purvis A (2007) The delayed rise of present-day mammals. *Nature* 446: 507–12.

23. Weisbecker V, Beck R (2015) Marsupial and monotreme evolution and biogeography. In Klieve A, Hogan L, Johnston S, Murray P (eds), *Marsupials and Monotremes*. Nova Science, Hauppauge, NY, pp.1-32.

24. Lepage T, Bryant D, Philippe H, Lartillot N (2007) A general comparison of relaxed molecular clock models. *Molecular Biology and Evolution* 24(12): 2669–80.

25. Bromham L (2003) Molecular clocks and explosive radiations. *Journal of Molecular Evolution* 57(1): S13–20.

26. Phillips MJ (2015) Geomolecular dating and the origin of placental mammals. *Systematic Biology* 65: 546–57.

27. Springer MS, Emerling CA, Meredith RW, Janečka JE, Eizirik E, Murphy WJ (2017) Waking the undead: implications of a soft explosive model for the timing of placental mammal diversification. *Molecular Phylogenetics and Evolution* 106: 86–102.

28. Romiguier J, Ranwez V, Douzery EJP, Galtier N (2013) Genomic evidence for large, long-lived ancestors to placental mammals. *Molecular Biology and Evolution* 30(1): 5–13.

29. dos Reis M, Inoue J, Hasegawa M, Asher RJ, Donoghue PC, Yang Z (2012) Phylogenomic datasets provide both precision and accuracy in estimating the timescale of placental mammal phylogeny. *Proceedings of the Royal Society of London B. Biological Sciences* 279(1742): 3491–500.

30. Ronquist F, Lartillot N, Phillips MJ (2016) Closing the gap between rocks and clocks using total-evidence dating. *Philosophical Transactions of the Royal Society of London B. Biological Sciences* 371(1699): 20150136.

31. Meredith RW, Janečka JE, Gatesy J, Ryder OA, Fisher CA, Teeling EC, Goodbla A, Eizirik E, Simão TL, Stadler T, Rabosky DL, Honeycutt RL, Flynn JJ, Ingram CM, Steiner C, Williams TL, Robinson TJ, Burk-Herrick A, Westerman M, Ayoub NA, Springer MS, Murphy WJ (2011) Impacts of the Cretaceous terrestrial revolution and KPg extinction on mammal diversification. *Science* 334(6055): 521–4.

32. O'Leary MA, Bloch JI, Flynn JJ, Gaudin TJ, Giallombardo A, Giannini NP, Goldberg SL, Kraatz BP, Luo ZX, Meng J, Ni X, Novacek MJ, Perini FA, Randall ZS, Rougier GW, Sargis EJ, Silcox MT, Simmons NB, Spaulding M, Velazco PM, Weksler M, Wible JR, Cirranello AL (2013) The placental mammal ancestor and the post-K-Pg radiation of placentals. *Science* 339(6120): 662–7.

33. Springer MS, Meredith RW, Teeling EC, Murphy WJ (2013) Technical comment on 'The placental mammal ancestor and the Post-K-Pg radiation of placentals'. *Science* 341(6146): 613.

34. dos Reis M, Donoghue PCJ, Yang Z (2014) Neither phylogenomic nor palaeontological data support a Palaeogene origin of placental mammals. *Biology Letters* 10(1): 20131003.

35. Smith VS, Ford T, Johnson KP, Johnson PC, Yoshizawa K, Light JE (2011) Multiple lineages of lice pass through the K-Pg boundary. *Biology Letters* 7(5): 782–5.

Case Study 6
Bayesian molecular dates: timing of the placental mammal radiation

There are a range of methods for inferring dates of divergence from molecular data. Here we are going to consider the general class of Bayesian molecular dating methods that are rapidly becoming the most commonly used. There is a rich cultural history of debate over whether or not Bayesian statistics is the best approach to assessing the empirical support for hypotheses. In the end, though, most biologists choose these methods largely for practical reasons, rather than due to a philosophical commitment to Bayesian principles. Bayesian methods allow the adoption of sophisticated models of evolution, in particular allowing the use of a range of models of variation in rate of molecular evolution. They also provide a flexible framework for incorporating information such as fossil evidence into date estimation, and a convenient way of comparing alternative solutions and reporting on uncertainty.

What sets Bayesian methods apart from other approaches is that every hypothesis must be given a prior probability, which is the evaluation of how plausible an explanation it is, before we have considered what we learn from looking at the data. In the case of molecular dating, each hypothesis consists of a tree (relationships between sequences) with branch lengths (time duration of each lineage, with the start and end points defined by nodes which split a lineage into two or more descendant lineages), plus parameter values for many aspects of the model of evolutionary change (such as the substitution matrix and distribution of substitution rates).[1] If we want to produce absolute dates of divergence (for example, the age of a given divergence event in millions of years), we also need to include calibrations, providing information on the likely age of some nodes, usually derived from fossil evidence.

In Bayesian molecular dating analysis, we combine all these prior beliefs about the tree shape, the age of particular calibrated nodes, and the way rates change over the tree with some other assumptions about the evolutionary process, such as the frequency of different kinds of base changes and the distribution of changes across the sites in the DNA sequence. We use them to generate a prior probability for any hypothetical history of these DNA sequences: any possible topology (branching relationship between species), any combinations of branch lengths, and all parameter values of the model (such as the speciation rate or nucleotide substitution rates) (Figure 6.12).

Now we can consider our sequence data and see if it changes our mind about how plausible each of these alternative histories is. We might find that patterns in the data speak against particular prior

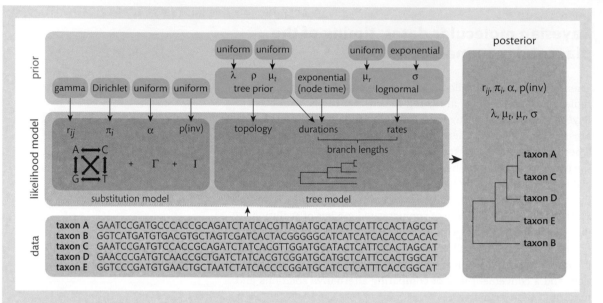

Figure 6.12 An example of the kinds of assumptions about the evolutionary process that might be included in a Bayesian dating analysis (although specific assumptions and parameters vary between analyses and methods).[1] The prior probabilities of different alternative histories for a set of sequences will be influenced by assumptions about the process of change in DNA sequences (substitution model), and may include a model of speciation and extinction of lineages in order to place prior probabilities on different phylogenies (tree model) including the distribution of branch lengths derived from assumptions about the distribution of substitution rates. These assumptions are used to evaluate the likelihood of the data, given the model of evolution. The posterior distribution of parameters, including the tree and branch lengths (dates), is derived from the combination of the prior and the data (likelihood).

Figure courtesy of John Wiley & Sons Inc.

assumptions. In this case, the probability assigned to those solutions after the analysis (posterior probability) will be different from the probability assigned to them before looking at the data (prior probability). If we have gained new information about evolutionary history from the sequence data, the posterior distribution of dates for any particular node should be different from the prior distribution. If it isn't, this tells us that the results of our analysis may be primarily reflecting the beliefs we expressed at the start of the analysis, in which case the DNA sequence data have not changed the answer we have derived using only those assumptions.

Calibrations

Many of the priors in a Bayesian molecular dating analysis are there for statistical convenience, to provide a tractable way of ranking the prior probability of the very large number of possible solutions. For example, the models of rate variation usually describe the set of possible rates with a relatively simple distribution that can easily be parameterized. This distribution is not based on any empirical knowledge of substitution rates developed from studies of molecular evolution. Instead, it provides a flexible but simple way to assign different rates to branches on the tree. But some priors do use empirical observations to 'ground truth' the search for plausible histories. In particular, the use of calibrations—'known dates' in the tree—is a critical part of molecular dating and provides a chance to use real-world information to inform molecular dates.

If we knew exactly when two lineages in the phylogeny last shared a common ancestor, we could use that date as a point estimate of the node age. But, usually, we don't have a precise age of the node. Instead, we have information that helps us to narrow down the range of dates for the node. For example, if we have a reliably dated and identified fossil of a particular lineage, we know that the lineage must have arisen before the date of that fossil. In a Bayesian molecular dating analysis, we can express this calibrating information as a probability distribution on the age of the node.[2] Calibrating information can be described by a 'flat' prior probability distribution that makes all dates between a maximum and minimum equally probable (uniform) or can be described using a distribution that makes some median date most probable and more extreme dates less so (such as a log-normal distribution). The bounds on the prior distributions can be 'hard' (dates outside those bounds are given zero probability) or 'soft' (older or younger dates are highly improbable but not impossible).

Calibrations are one of the most important parts of a molecular dating analysis. Analyses using the same data and method can give different divergence dates depending on the calibrations used and the way they are expressed.[3] **Figure 6.13** shows the results of a study that examined the influence of calibrations on mammalian molecular date estimates.[3] When fossil calibrations are used as point estimates (that is, the first fossil exactly marks the point of origin of the lineage), the nodes representing placental ordinal divergence are forced to occur in the early Palaeogene,[4] resulting in extremely high rates of molecular evolution on the early lineages (Figure 6.13(b)).[5] When the same fossil calibrations are used as minimum bounds, allowing older dates, the molecular dates for the placental orders are deeper into the Cretaceous, well before the final extinction of the dinosaurs (Figure 6.13(a)).

Searching tree space

One of the biggest problems we need to solve for any Bayesian molecular dating analysis is how to evaluate the phenomenally large number of possible solutions. It is simply impossible to consider and compare all the hypothetical histories that are consistent with a given set of sequences (including all possible combinations of parameter values). So we need a way of navigating through the multidimensional space

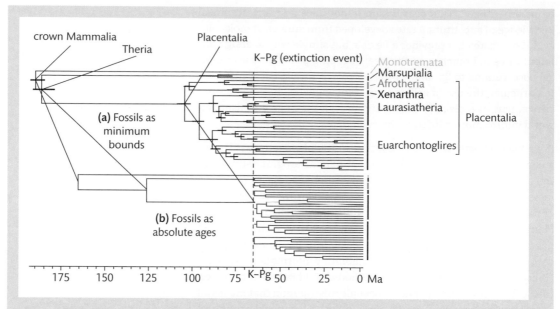

Figure 6.13 The effect of calibration priors on mammal molecular dates, for a phylogeny of 36 mammals (one monotreme, two marsupials, and 33 placentals[3]): (a) fossil dates are treated as soft-bounded minimum values; (b) the same calibrations are treated as absolute divergence times. The Cretaceous–Palaeogene (K–Pg) boundary at 66 Mya marks the final extinction of the dinosaurs (see Chapter 5).

Figure courtesy of the Royal Society.

of all possible solutions that doesn't require us to consider every possibility, yet has a good chance of allowing us to find the best solutions. The usual approach to this problem is to use a search strategy called a Monte Carlo Markov Chain (MCMC). 'Monte Carlo' refers to chance outcomes (like the famous casino of the same name), because the potential solutions are generated by random alterations in the phylogeny and parameter values of the evolutionary model. The 'Markov Chain' is a way of moving through solution space: after each random change, the search will either accept the altered solution and then propose another change, or stay in the previous state and make another change to it (**Figure 6.14**).

In a Bayesian molecular dating analysis, each new altered state in the MCMC is tested against the previous one by comparing their posterior probabilities, which represents the probability that the hypothesis (tree + model) is true, given the prior probabilities, moderated by what we have learned from the data (the likelihood). If the altered state has a higher posterior probability, the chain will move to that state. If it doesn't, then it may move to the new state, or stay in the previous state. The MCMC will keep proposing random changes and evaluating them, but will eventually converge on a set of phylogenies

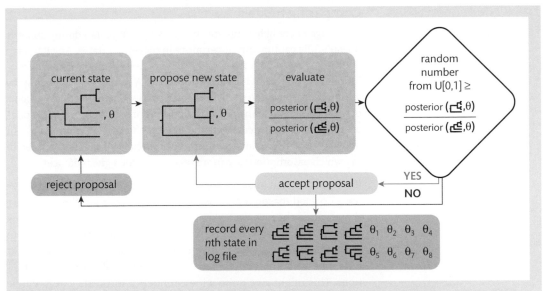

Figure 6.14 Bayesian molecular dating methods use an MCMC search of tree space to find solutions with high posterior probabilities.[3] Starting with a possible phylogeny (current state), a new hypothesis is proposed by making an alteration to the topology, branch lengths, or parameters of the evolutionary model (here represented by the Greek letter theta, θ). The likelihood of the data given the model is calculated, and this is used to calculate the ratio of posterior probabilities of the new hypothesis and the previous state. Unlike a hill-climbing procedure, which always moves to the higher likelihood solution, the MCMC chain has a chance element that means that it might move to the proposed state even if it has a slightly lower posterior probability.

Figure courtesy of John Wiley & Sons Inc.

that have the highest posterior probability, given the data and the assumptions. A Bayesian phylogenetic analysis explores tree space, but tends to move away from low probability solutions and spend more time on those solutions that have a higher posterior probability.

These high probability solutions may represent slightly different combinations of parameter values, topology or branch lengths that all give similar posterior probabilities. You could choose to report the 'best' phylogeny with the highest score, but given that alternative solutions may explain the data almost as well, it's better to report a range of solutions from the posterior distribution. You could, for example, give the range of dates for a particular node represented by the majority of solutions within the distribution of the best scoring phylogenies (conventionally, the range that encompasses the central 95% of the posterior distribution is reported). So Bayesian dating analyses give you a convenient way of not only finding the most plausible solutions, given the assumptions, but also reporting the uncertainty around those dates. It is important that the range of solutions suggested by a Bayesian analysis is inspected and evaluated with respect to known relationships and fossil ages to detect possible errors.[6]

Comparing models

The range of credible estimates from a single Bayesian dating analysis doesn't fully capture our uncertainty in molecular dates. Any Bayesian dating analysis relies on a large number of assumptions about the evolutionary process and the history of the DNA sequences. Changing the assumptions will often result in different dates. Given that many aspects of the models, and many prior distributions of parameters of the model, are chosen for statistical tractability rather than being informed by our knowledge of molecular evolution, we often have no clear way of deciding which assumptions are more reasonable for a given dataset.

If we get different answers for different models of evolution, and we have no good reason for choosing one model over another, there are three approaches we can take. The first is to recognize that our lack of decisive knowledge about the evolution of the sequences leads to unavoidable uncertainty in date estimates, and so we should report the range of estimates obtained under all reasonable sets of assumptions. This might give quite wide confidence intervals, but it will be an honest statement of what we know about the dates given what we know about the way the DNA sequences have evolved. Wide confidence intervals might not tell you exactly when splits between lineages occurred, but they can allow you to rule out dates outside those intervals as being incompatible with the data you have, given the particular assumptions you have made in your analysis. For example, if we try different combinations of genes and dates for divergences between placental mammal orders and get different results, but they are all earlier than the Cretaceous–Palaeogene boundary, then we might conclude that the molecular dates reject a post-dinosaur radiation, even if we don't know exactly when the placental radiation occurred. Given that we will rarely be able to estimate molecular dates with unerring accuracy, a hypothesis testing approach—asking whether the molecular data are compatible with a particular scenario given reasonable assumptions—will often be the most defensible use of molecular dating in macroevolution and macroecology.

The second approach is to formally test models and choose the one with the best statistical fit to the data. Statistical tests of goodness of fit assess the degree to which the observations deviate from what we would expect to see under that model. If the model was a fair description of the evolution of these lineages, would we expect to have got the data that we have? These tests have to allow for the fact that adding more free parameters to a model will usually result in a better fit to any dataset because they allow more variation in the data to be accounted for. Taking the number of parameters into account, we can compare models and reject those that are a significantly worse fit to the data than the best-fitting model. But although a model with a good fit provides a plausible explanation for the data, that doesn't mean that we know the model is true. After all, there might be other models that also fit the data just as well. So we might find that there are a number of alternative models that have equally good fit to the data, yet might give different date estimates. In this case, we could report the range of date estimates under a range of reasonable models.

Or we could take the third approach to modelling uncertainty, which is to use a formal statistical solution to average over all models within the Bayesian framework. The MCMC sampling procedure gives us a convenient way to analyse the data using a range of possible models. Remember how the MCMC chain makes one change at a time, which might be a change to the tree itself (changing the topology or branch lengths) or a change to one of the parameters in the model of evolution? We can use the same procedure to sample different models. Just as the phylogenetic solutions can be considered to occupy a multidimensional 'tree space', so variations on the models of evolution can be thought of as describing 'model space'. We can use the MCMC to explore model space as well as tree space. Now, when the MCMC proposes a change at each step of the analysis, it could change some part of the phylogeny, or modify a parameter of the model, or it could change an aspect of the model itself, for example the form of the distribution of rates. Just as before, the MCMC chain will move towards the solutions with higher posterior probability, so the phylogenies sampled at the end of the procedure represent the best possible solutions of topology, branch lengths, model, and model parameters. Of course, we have to bear in mind that this analysis can only give the range of solutions under the set of models we have explored, and we cannot guarantee that any of those models is a fair description of the actual history of change in those sequences.[1]

Influence of prior assumptions

The sophistication of the Bayesian molecular dating methods allows us to incorporate much more detail about the evolutionary process and a wider range of observations about evolutionary history than is possible with many other methods, but it also makes it more challenging to understand just what we have to believe in order to accept the molecular dates. In some ways, the inclusion of so many assumptions designed to capture the complexity of molecular change and evolutionary history might make Bayesian dates more believable, but the complexity of assumptions also means that it's not always easy to understand exactly how all that information is combined to give an answer.

For example, a researcher may specify a particular distribution of possible dates on a fossil-calibrated node. But during the analysis this information is combined with other assumptions of the analysis, such as the form of the 'tree model' that gives a prior distribution on branch lengths, as well as information from other calibrated nodes. This means that the calibrating information on one node in the tree can influence the inferred rates and dates on all other parts of the phylogeny, and can interact with other priors to produce a joint prior that ends up quite different from the information supplied by the

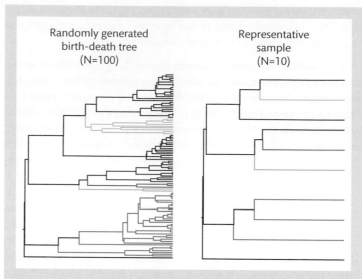

Figure 6.15 Representative sampling, where the phylogeny is constructed from a sample of sequences chosen to represent each taxonomic grouping, results in a skewed sampling of nodes in the tree.[7] On the left is a tree generated under a constant speciation and extinction rate, a process that leads to an increasing number of nodes toward the present. On the right is the same phylogeny with just a few samples per clade included; the nodes included in the phylogeny are now skewed toward the base of the tree.

Figure courtesy of Oxford University Press.

researcher. These interactions between different parts of the models and assumptions may not be obvious to the user of the methods, but might affect the date estimates obtained.

Take, for example, the assumption that the sequences being analysed represent a random sample of all living species descended from the common ancestor. Many people conducting a Bayesian molecular dating analysis will be unaware that random taxon sampling is an implicit assumption of some forms of the 'tree model' that allows a prior probability to be assigned to every possible phylogeny for a set of sequences. Few phylogenies contain 100% of the living species from that clade. If the number of species in the clade is known and species have been chosen at random, then the nodes in the tree should be a reflection of the speciation and extinction rates for the phylogeny. But if sampling is uneven—for example, if some genera are represented by 100% of living species but other genera are represented by only 20% of living species—the nodes in the reconstructed tree are a biased representation of the speciation and extinction rates across the tree, underestimating the number of divergence events in poorly sampled clades.[7] Few datasets are randomly sampled. Rare species may be less likely to be sampled than common ones. Often, researchers

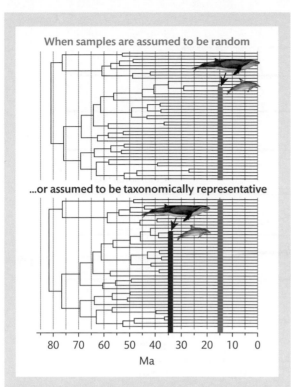

Figure 6.16 Taxon sampling strategy can affect Bayesian date estimates.[8] Here the same mammal phylogeny is analysed under the assumption that sequences have been sampled at random from all possible living descendants of the last common ancestor of placental mammals (top panel), and under the assumption that sampling has been chosen to be representative of major clades (bottom panel). Some nodes in the tree, such as the split between whales and dolphins, have very different dates under the two different sampling assumptions (the blue line marks the date inferred under a random sampling assumption, the red line under a representative sampling assumption).

Figure courtesy of the Royal Society.

represent each genus or family with the same number of sequences, inadvertently over-sampling species-poor clades and under-representing species-rich clades. The effect of 'representative sampling' is to preferentially sample deeper nodes in the tree (Figure 6.15), which violates the prior assumption that the nodes are a random sample of those in the phylogeny.[7] This can influence the inferred molecular dates. For example, dating the mammalian tree from a representative sample under the false assumption of random sampling results in doubling some of the age estimates (Figure 6.16).[8]

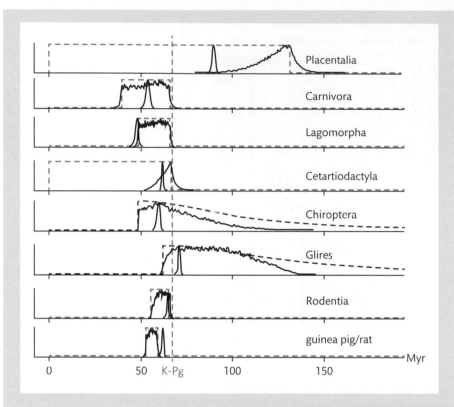

Figure 6.17 Prior and posterior distributions for placental mammal divergence dates from a Bayesian molecular dating study.[9] The dashed lines represent the specified probability densities for the calibrations: red when calibrations are specified as a minimum bound, green where they specify the maximum bound, and blue where both minimum and maximum bounds are specified. The purple lines are the joint priors (prior probability on node times, given all assumptions of the model including the tree model and other calibrating information), and the black lines are the posterior distributions (the inferred divergence dates).

From Reis M, Inoue J, Hasegawa M, et al. (2012), Phylogenomic datasets provide both precision and accuracy in estimating the timescale of placental mammal phylogeny, *Proceedings of the Royal Society of London B. Biological Sciences* 279(1742):3491–500. Republished with permission of the Royal Society.

Comparing the prior probability distribution with the posterior probability distribution for node heights allows you to judge whether the analysis of DNA sequence data has informed the date estimates, or whether the molecular dates primarily reflect the assumptions made on the basis of other data before the analysis was conducted. Figure 6.17 compares the calibrating information (range of dates specified by the researchers) with the joint prior (the distribution of prior probabilities based on the combination of all assumptions of the analysis, also referred to as the marginal prior) with the posterior probability distribution (dates derived by considering the DNA sequence data in light of the priors).[9] This allows us to ask how much our understanding of mammalian radiation has been changed by considering the molecular data. In some cases, such as Carnivora and Cetartiodactyla (the order containing whales, dolphins, and artiodactyls), the date estimates are a much narrower range of dates within the prior. In other cases (such as Lagomorpha) the date estimates strongly match the minimum bound, while in others (Rodentia) they match the maximum bound. But in almost all cases the molecular date estimates for the placental orders fall within the range of possible dates assumed by the researchers before starting the analysis (only one date estimate in that study, the within-Rodentia split between rat and guinea pig, gave a date estimate outside the soft bounds of the priors). If the analysis were run without any molecular data, sampling only from the prior (reflecting the assumptions), most inferred dates would not be very different from the analysis with molecular data. This emphasizes how careful researchers need to be when specifying calibrations as they can have a strongly determinant effect on the possible dates produced. Where dates at the maximum or minimum bound are specified, the veracity of the estimates depends on whether those calibrations are a true reflection of evolutionary history.

More generally, these results highlight a quandary for Bayesian molecular date analyses. There are two ways to interpret an analysis where the posterior distribution of dates is not very different from the prior distribution. On the one hand, we could say that the prior assumptions are confirmed, because the molecular data agree with the fossil dates of divergence. On the other hand, we might say that the lack of difference between prior and posterior means that the signal in the molecular data was too weak to overcome the specified calibrations. Either way, if the posterior has not changed much from the prior, then analysing the molecular data has not had much impact on our understanding of the placental mammal radiation. If we were to run the analysis without the molecular data, just sampling from the prior distribution, we would get roughly the same date estimates for many of the nodes as we do when considering the sequence data. It is important to compare the prior and posterior, and to investigate whether changing the assumptions of the analysis changes the dates obtained. If it does, and you have no independent way of assessing which assumptions are most reasonable, then you should report the results obtained over all reasonable assumptions. In the context of the radiation of placental mammals, we can see that the Bayesian molecular dates are shaped not just by the molecular data or by our understanding of molecular evolution but also by what we believe the fossil evidence is telling us.

Questions to ponder

1. Should molecular dating analyses start with the simplest set of assumptions—such as constant rates of change—adding more sophisticated assumptions when the simpler models are shown to be inadequate? Or should we always start with the most sophisticated analysis, allowing for as many possible complexities of evolutionary processes and history?

2. If the Cretaceous dates for the radiation of placental mammals are correct, should we expect to find fossil evidence to support this hypothesis? If we don't, does that disprove the molecular dates?

3. If the posterior distribution of dates for a molecular dating study is not different from the prior, then should we even call them 'molecular dates', given that the molecular data have not changed the dates from those we started with given our calibrations and other assumptions?

Further investigation

Just as for the mammalian radiation, many molecular dates for the radiation of modern birds are older than consideration of fossil evidence alone would suggest. Do the Cretaceous molecular dates for both birds and mammals suggest a shared mechanism for the discrepancy between molecular and fossil date estimates? Or shared drivers of adaptive radiation? Or is the resemblance between the cases due to chance?

References

1. Bromham L, Duchêne S, Hua X, Ritchie AM, Duchêne DA, Ho SYW (2018) Bayesian molecular dating: opening up the black box. *Biological Reviews* 93(2): 1165–91.

2. Ho SYW, Phillips MJ (2009) Accounting for calibration uncertainty in phylogenetic estimation of evolutionary divergence times. *Systematic Biology* 58(3): 367–80.

3. dos Reis M, Donoghue PCJ, Yang Z (2014) Neither phylogenomic nor palaeontological data support a Palaeogene origin of placental mammals. *Biology Letters* 10(1): 20131003.

4. O'Leary MA, Bloch JI, Flynn JJ, Gaudin TJ, Giallombardo A, Giannini NP, Goldberg SL, Kraatz BP, Luo ZX, Meng J, Ni X, Novacek MJ, Perini FA, Randall ZS, Rougier GW, Sargis EJ, Silcox MT, Simmons NB, Spaulding M, Velazco PM, Weksler M, Wible JR, Cirranello AL (2013) The placental mammal ancestor and the post-K–Pg radiation of placentals. *Science* 339(6120): 662–7.

5. Springer MS, Meredith RW, Teeling EC, Murphy WJ (2013) Technical Comment on 'The placental mammal ancestor and the post-K–Pg radiation of placentals'. *Science* 341(6146): 613.

6. Springer MS, Murphy WJ, Roca AL (2018) Appropriate fossil calibrations and tree constraints uphold the Mesozoic divergence of solenodons from other extant mammals. *Molecular Phylogenetics and Evolution* 121: 158–65

7. Beaulieu JM, O'Meara BC, Crane P, Donoghue MJ (2015) Heterogeneous rates of molecular evolution and diversification could explain the Triassic Age estimate for angiosperms. *Systematic Biology* 64(5): 869–78.

8. Ronquist F, Lartillot N, Phillips MJ (2016) Closing the gap between rocks and clocks using total-evidence dating. *Philosophical Transactions of the Royal Society of London B: Biological Sciences* 371(1699): 20150136.

9. dos Reis M, Inoue J, Hasegawa M, Asher RJ, Donoghue PC, Yang Z (2012) Phylogenomic datasets provide both precision and accuracy in estimating the timescale of placental mammal phylogeny. *Proceedings of the Royal Society of London B: Biological Sciences* 279(1742): 3491–500.

7

Is sex good for survival?

Roadmap

Why study the evolution of sex?

Sex results in offspring with DNA from more than one parent. Sexual reproduction is so common and familiar that it may come as a surprise to find that it has been an enduring puzzle in evolutionary biology. Given that many species can reproduce uniparentally, without needing to combine genes with other individuals, why would sex evolve? And given that asexual lineages frequently evolve from sexual species, why do so many different species retain sexual reproduction? Many different solutions have been proposed for both the origin and maintenance of sexual reproduction, ranging from genome repair to long-term evolvability. This broad range of proposed mechanisms requires us to think about how evolution works at different levels of biological organization, from 'selfish genes' to evolutionary lineages, and at different timescales, from individual lifetimes to millions of years.

What are the main points?

- When reproducing entities (such as individuals) are made up of multiple units which also have the capacity for copying themselves (such as genes), we have to think about the way selection acts at different levels of biological organization.
- Fitness benefits at one level of biological organization may come at the cost of the reproductive capacity of other levels, and different players in a system may benefit from different strategies.
- We might seek separate explanations for the origin and maintenance of complex traits. Uncovering the selective pressures that generated the capacity for genetic exchange may tell us little about why most living eukaryotic species retain sexual reproduction.

Photograph: Wasps pollinating a sexually-deceptive orchid. © Rod Peakall.

What techniques are covered?

- **Evolutionarily stable strategies:** the fittest variant needs to be considered in terms of the other possible variants in the population.
- **Genetic conflict:** genes within a genome may have different ideal strategies for reproduction, and this may be reflected in evolution of genomic architecture, individual behaviour, and long-term lineage evolvability.

What case studies will be included?

- Phylogenetic models of macroevolution: 'tippy' distribution of asexuality on the tree of life.

"The impossibility of sex being an immediate reproductive adaptation in higher organisms would seem to be as firmly established a conclusion as can be found in current evolutionary thought. Yet this conclusion must surely be wrong. All around us are plant and animal populations with both asexual and sexual reproduction. . . So that which must surely be false, by the method of deductive analysis, must as surely be true by comparative evidence, and vice versa."

George C. Williams (1975) *Sex and Evolution.*
Princeton University Press.

Why bother with sex?

Reproduction is essential to evolution. Any organism that has the capacity and resources it needs to survive and successfully reproduce must be well adapted to its environment. Its offspring are likely to have inherited the same features that made their parent so successful. As a bacterium divides in two, or a sponge makes a new individual by budding, or a plant sends out horizontal shoots from which new individuals can grow, each will have the same genome as its successful parent (barring occasional mutation: see Chapter 2).

Why then would a well-adjusted individual take the risk of reproducing sexually, combining its DNA with that of another individual? There are so many potential costs associated with sexual reproduction that it is an evolutionary puzzle why any species would have sex. A sexually reproducing individual must invest its resources in offspring that only carry half its DNA. Not only that, the sexual individual may have paid the cost of investing in structures and behaviours designed to attract a mate, such as flowers, plumage, or bowers, and competition for mates may have resulted in it producing orders of magnitude more gametes than could possibly turn into children. A sexual reproducer might potentially pay the ultimate cost of failing to reproduce if a suitable mate is not found, or a reduction in offspring quality or quantity if the other parent's DNA has less favourable, or incompatible, genetic variants. Given all these costs of sex, and more, why do species not reproduce asexually by simply copying their own successful genome?

Many species do reproduce asexually, either exclusively (such as bdelloid rotifers: Figure 7.1) or predominantly (as in some aphids). But the majority of eukaryotic species have the capacity for sexual reproduction, and many cannot reproduce in any other way. If sexual reproduction is widespread, despite the obvious costs, it seems likely that it has benefits—or at least that the costs of not being sexual are even greater than the costs of having sex.

The puzzling popularity of sexual reproduction provides an excellent case study in macroevolutionary thinking, because it requires us to keep many challenging concepts in mind as we generate and test hypotheses. We need to be able to consider why and how sexual reproduction could have initially evolved, and to do this we will need to think at the level of genes, genomes, and single-celled organisms. But we will also have to consider that hypotheses for the origin of sexual reproduction may provide less convincing arguments for why so many species alive today rely on sex; in other words, we may have to seek a different explanation for the maintenance of sex than we do for its origin. And to do this, we may have to

think not only at the level of the costs and benefits to individuals, but also to populations and lineages. As we consider these different levels, we will confront both cooperation and conflict, and these topics are central to many macroevolutionary puzzles. All the genes in an organism cooperate to produce complex morphology and behaviour—but what happens when one gene, or one cell, replicates at the expense of the others? Individuals in a population might also have to cooperate to produce offspring, bringing male and female gametes together, or acting together to aid the survival of shared offspring—but what if those individuals differ in their optimum investment in those offspring? Conflicts may not always be apparent when we first look at communities, or organisms, or genomes. But the competing demands of different replicating units, or selection at different levels of biological organisation, might explain some of the peculiar features manifest in biological systems, from genetic imprinting to sexual displays to the relative numbers of males and females in populations.

Finally, considering the evolution and maintenance of sex requires us to consider both short- and long-term costs and benefits. Is sexual reproduction maintained by the need to keep genes or cells from going rogue and killing the individual or their offspring? Or is it a way of generating genetically diverse offspring to maximize the chances of leaving descendants in an uncertain world? Or are these gene-level and individual-level benefits overwhelmed by long-term evolvability, because sexual lineages persist for longer than their asexual relatives? These are the levels of selection we must keep in mind as we consider the fundamental question: why have sex? Perhaps the best place to start is to consider the alternative to sex. So let us begin by thinking about the costs and benefits of reproducing by simple copying.

Mutation accumulation

Copying is at the heart of evolution. A cell divides, a genome is copied, and now there are two instead of one. Those copies that have features that promote copying will make more copies of themselves, and be present in increasing proportions in the population (Chapter 2). But no copying process is perfect. Eventually a mistake will be made, the wrong base inserted, or the DNA molecule damaged so that the sequence is unreadable. Of course, the cell has a barrage of equipment for dealing with such eventualities. If the mistake is detected when first made, the newly copied DNA strand can be removed so that a fair copy of the opposite DNA strand can be made. But if the mismatch between the strands is not detected immediately, it won't be clear which base is correct and which is the mistake. When a mismatch is repaired, by changing the base on one strand to pair with the opposite strand, it may be returned to the original sequence, or possibly changed to the new altered sequence. This change to the base sequence overwrites the original, leaving no trace of the previous sequence. It is no longer possible to recover the original sequence; the mutation has become a permanent part of genome, and will be copied every time the genome is copied.

In each generation, as the genome is copied, there is the chance to acquire more of these accidental changes to the sequence. The result is that, over time, the genome must steadily accumulate mutations. Occasionally one of these mutations will enhance the copyability of the genome, increasing the chances of survival and reproduction. But most mutations will, at best, have no influence on fitness or, at worst, make survival and reproduction less likely. This is because a random change to organized genetic information has very little chance of making it better than it was before (see Chapter 2). The occasional win (beneficial mutation) comes at the cost of accumulation of many less favourable mutations.

Because any permanent change to the genome is passed on to the next generation, the number of mutations can only go up and up as the genome is copied again and again (a phenomenon sometimes referred to as Muller's ratchet). But what if undamaged DNA could be patched in from another source? Cells have repair proteins that allow broken DNA to be mended by rejoining the double strands of the helix. Double-strand break repair is error prone because there is no template to guide the construction of the new nucleotide sequence. But the repair machinery could be adapted to replace the damaged DNA with a spare copy of that sequence.[1] Where would the cell get a spare copy of the sequence for repair? The cell could carry an extra copy of its genome for such eventualities.

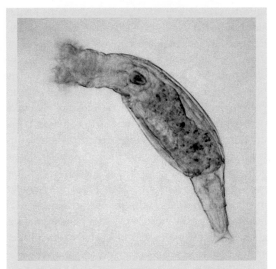

Figure 7.1 Queens of the asexual world. Bdelloid rotifers are microscopic animals, less than a millimetre long, that predominantly live in a wide range of freshwater habitats. The species shown (*Habrotrocha rosa*) is found in soil, moss, and the water that collects in pitcher plants. Males have never been observed in any bdelloid species.

Photo by Rkitko (CC3.0 license).

For example, bdelloid rotifers have evolved to survive desiccation, which shears DNA (Figure 7.1). Bdelloid rotifers survive genome-smashing desiccation by having extremely efficient systems for repairing double-strand breaks. But they also have multiple copies of the whole genome,[2] which may act as the templates for repairing their DNA. Interestingly, the same strategy seems to have evolved independently in *Deinococcus radiodurans*, otherwise known as 'Conan the bacterium' for its remarkable ability to withstand radiation, desiccation, extremes of temperature, and even acid. Like bdelloid rotifers, *Deinococcus* keeps four copies of its genome, and has fabulously efficient systems for double-strand break repair.

Alternatively, the spare copy used for repair could come from another organism. All DNA is structurally the same, regardless of the species of origin. If an organism has the equipment needed to cut and rejoin double-stranded DNA in order to repair a damaged section, then it also has the basis for incorporating DNA from another source. There is a lot of DNA about, left in the environment after the unfortunate demise of other cells. Genomic analysis shows that nearly a tenth of the genes in bdelloid rotifer genomes are derived from sources as diverse as bacteria, fungi, and plants.[3] The uptake of cosmopolitan DNA might be facilitated by the phenomenal ability of bdelloids to rejoin broken DNA helices. But many other species can also take up DNA from their environment. In some bacteria, uptake of DNA can be so rampant that it can be difficult to define exactly what the composition of the genome of a particular bacterial species is, because individuals can vary in which genes they carry.[4] In some cases, DNA exchange between bacteria is facilitated by specific structures for gene exchange, such as the formation of bridges between cells, or vesicles that carry DNA from one cell to another.

Meiosis

One possible explanation for the origin of sex is that it regulates a process of DNA exchange between individuals as a way of getting around the grinding inevitability of mutation accumulation. If you could keep a spare copy of your genome, you might be able to use it to patch the genome back together and restore sequences by gene conversion (using one copy of the gene as a template to replace the other). If you have the biochemical equipment to cut and paste DNA strands, then you might also be able to take up spare DNA from other organisms, whether incidentally (random uptake from environment) or deliberately (facilitated DNA transfer). If, by chance, these horizontally acquired sequences work better than the version already in the genome, or provide a new function, they may give that individual a fitness advantage.

DNA exchange can be more formally coordinated by a regulated system of combining whole copies of the genome from different individuals. Carrying two identical copies of the same genome risks doubling the copies of harmful mutations. But if an individual has two entire genome copies from different sources, there is the potential to mask the deleterious effects of some mutations (if one good copy is sufficient to maintain function). There is also the potential to use recombination between DNA strands to bring together mutations derived from more than one source (a point we will return to later in the chapter). But you can't just keep adding more genome copies to a cell, or the amount of DNA will become unmanageable. If you are going to combine whole genome copies from more than one individual, you need a way of managing copy number, otherwise the amount of DNA in the genome would double each generation.

To maintain the same amount of DNA in sexual reproduction, it is necessary to reduce the amount of DNA received from each of the parents. One way to do this would be to take both parental copies and then throw half of the combined DNA out of the developing embryo. But, typically, genome reduction occurs in the parents before fertilization through a special kind of cell division called meiosis. If you can't remember what meiosis is (reduction–division of one diploid cell to give four haploid cells with different combinations of the parent cell's alleles), and how it differs from the usual cell division of mitosis (in which the genome is copied in its entirety and inherited by two identical daughter cells), now would be a good time to pick up a general biology text and refresh your memory.

The upshot of meiosis is that offspring are not identical to their parents, for two reasons. Firstly, each offspring receives only half of the available alleles from each parent. They might be unlucky and fail to inherit the advantageous traits; they might be lucky and fail to inherit a disadvantageous trait. Secondly, sexual offspring have DNA from two different sources, so they will have some alleles that their mother didn't have, and some that their father didn't have. By taking a risk by combining your successful genome with another individual, you might end up with offspring with lower fitness than yourself if they inherit unfavourable alleles from the other parent. But there is also some chance that they will end up fitter than you, either because they are fortunate enough to get better alleles from their other parent, or because the combination of both parents' alleles they inherit is particularly propitious.

Meiotic drive

For sexually reproducing organisms, meiotic cell division is needed to reduce the parental contribution of DNA to each offspring, so that the child of two parents ends up with the same amount of DNA as each parent, because only half of each parent's genetic variants are contributed to each offspring. This introduces an interesting problem, which forces us to think about the way that different players in a system can have different ideal reproduction strategies, a theme we will return to again and again as we think about the evolutionary biology of sex. Meoisis is a shuffling of alleles, with a random deal of each parent's genetic variants to each offspring. But what would happen if one allele had a way of stacking the deck, making sure it ended up in more than half the offspring?

For example, production of female gametes often involves asymmetric cell division, where only one daughter cell forms a viable gamete and the other daughter cell becomes the polar body, a small non-reproductive sister cell. Normally, each paired chromosome has a 50:50 chance of ending up in the gamete or the polar body. But any allele that could reduce its chances of ending up in the polar body would be present in more than 50% of gametes, and so would have an increased chance of ending up in an offspring. Such alleles do exist, and they are often referred to as meiotic drivers or segregation distorters (**Figure 7.2**).[5] For example, many species contain additional 'B chromosomes' which don't carry essential genes, yet are able to be preferentially transmitted to offspring. The biased transmission of the B chromosome increases its own copy number, regardless of whether it has any benefit to the organism (and occasionally despite reducing fitness). Some meiotic drivers succeed in biasing transmission even though they are harmful to the organism. For example, there are meiotic driver alleles that act to kill all sperm that carry the alternative alleles, resulting in biased transmission of the driver allele (Figure 7.2). This can occur even if the driver allele is linked to alleles that cause male sterility or even death.

Given the selective advantages to an allele of subverting meiosis to gain an increase in frequency, what keeps meiosis fair? Why does meiosis fairly divide the chromosomes into gametes in most organisms, most of the time? There are several evolutionary forces that might kick back against meiotic distorters. The increase in frequency of meiotic driver alleles comes at the expense of the frequency of other alleles, so if other alleles acquire mutations that restore the balance they will increase in frequency as they regain their share of representation. Selection will favour alleles that suppress meiosis distorters. This gives rise to a 'parliament of genes', where actions by one allele to increase its own representation may be counteracted by others to create an uneasy democratic balance. This may explain why meiotic distorters can skew their representation to 98% of seeds in hybrid monkeyflowers (offspring generated from gametes from two closely related species), but have a minimal effect in pure crosses (from gametes of the same species) (Figure 7.2).[6] It seems likely that the presence of a meiotic distorter in a population will result in strong selection for alleles at other loci that suppress it, but these suppressor alleles might not be present in unaffected populations, leaving hybrids between the two populations with the driver but not the suppressor.

Key points

- Reproduction is the core process of evolution. Why would an organism reproduce sexually (blend its DNA with that of another individual) rather than simply copy its own genome to give to its offspring?

- The genome accumulates copy errors, so one possible benefit of sex is as a way of restoring sequences using DNA templates from another individual.

- Combining DNA from two individuals requires a method of avoiding doubling DNA every generation.

- Meiosis creates non-identical gametes which each carry half the parents' genetic variants.

- There is a selective advantage to any allele that can increase its representation in gametes, by either preferentially ending up in the reproductive cell, or killing gametes with any alternative allele. But there is also a selective advantage to other alleles to restore their share of representation in the gametes by suppressing the action of meiosis distorters.

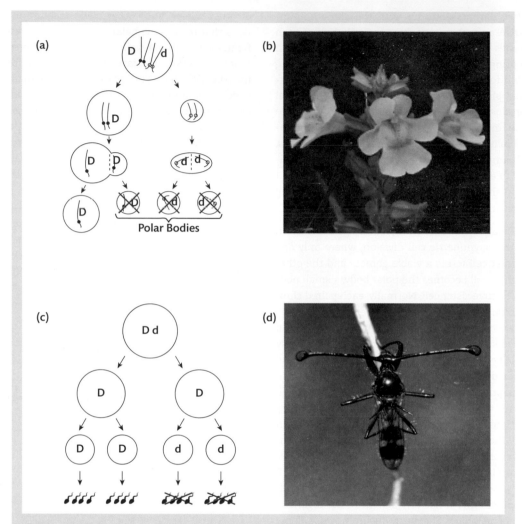

Figure 7.2 Examples of meiotic drive.[5] (a) If female gametes are formed by asymmetric cell divisions, so not all daughter cells become gametes, then any gene that can make sure it ends up in the gamete will be present in more offspring. This is how a driving allele in hybrid monkeyflowers (*Mimulus guttatus × nasutus*) (b) ends up in 98% of progeny. (c) Male killer alleles ('D') cause sperm that carry the rival chromosome ('d') to die. Several species of stalk-eyed flies (d) have X-linked meiotic drivers which alter the sex ratio of offspring. Wide eyespan may provide an indicator for females to preferentially mate with males that lack the meiotic drive gene.

From Lindholm AK, Dyer KA, Firman RC, et al. (2016), The ecology and evolutionary dynamics of meiotic drive, *Trends in Ecology & Evolution* 31(4): 315–26. © 2016, with permission from Elsevier. Photos by H. Zell (b) and G.S. Wilkinson (d).

This conflict between alleles to increase their chance of being included in successful offspring extends to any genes, not just those in the nuclear chromosomes. Consider a parasite that lives in the cellular cytoplasm. Now imagine that it can be passed on to the offspring of its host only if it happens to reside in a female, because sperm cytoplasm is discarded at fertilization. What does a parasite do if it has infected a male? One strategy would be to turn the male into a female—then the parasite

Figure 7.3 The egg fly butterfly is distributed across Australasia. In most populations, males infected with *Wolbachia* die. In a Samoan population of these butterflies, the effect of the parasite was so severe that males almost disappeared in the wild population. But then resistant male butterflies emerged, which presumably carried a suppression allele in their genomes, and so the proportion of males began to increase, rapidly regaining parity with females.[28]

Photo by Vtbijoy (CC3.0 license).

can be passed on. Indeed there are many parasites that can 'feminize' their hosts to ensure their own reproduction (e.g. Chapter 3, page 105). An alternative strategy is less obviously helpful; parasites that find themselves in males, and therefore at an evolutionary dead end, could kill their hosts. What is the selective advantage of killing your host, and therefore yourself? If the parasite has siblings who have infected female hosts in the same population, self-destruction may remove some of the competing host individuals and increase the chance of survival and reproduction of those same parasite genes in female hosts. Counter-intuitively, a type of parasite that kills its male host might increase in frequency if it results in greater success of the same kind of parasite via the enhanced rate of reproduction in female hosts.

For example, many insect species will produce female-dominated broods when infected with the bacterium *Wolbachia*, because male embryos infected by *Wolbachia* die before hatching. Of course, this places strong selective pressure on the insect host; any trait

in the host that allows a male to survive infection can be passed on to his offspring and rise in frequency in an infected population (**Figure 7.3**). So if the host genome evolves a 'suppression' allele that counteracts the effect of the parasite, it will rise in frequency. Now there is selection pressure on *Wolbachia* to be able to be transmitted even when the host has a suppression allele. As it happens, *Wolbachia* has another trick up its evolutionary sleeve. If the host evolves so that infected males survive infection, *Wolbachia* can causes incompatibility between the infected male sperm and uninfected female eggs.[7] This means that the uninfected eggs cannot give rise to a functioning embryo, increasing the reproductive success of infected males and females at the expense of uninfected females—and of course increasing the transmission rate of *Wolbachia*. These host-manipulating shenanigans have resulted in *Wolbachia* being developed as a biological control agent for disease-carrying mosquitoes. If large numbers of males carrying *Wolbachia* are released into the mosquito population, they will mate with the wild females and stop them from reproducing.

What do you think?

Does parasite-induced manipulation of sex ratio in disease vectors represent an 'unbeatable' control strategy? Or will vectors evolve 'resistance' to such manipulations? Is it possible to come up with an evolution-proof mosquito control strategy that allows no room for selection for resistance?

Levels of selection

Selective advantage looks different depending on the perspective. From a single allele's point of view, increasing copies at the expense of other alleles by distorting meiosis is a selective advantage. But from the perspective of the genome taken as a whole, as a collection of genes that all contribute to a joint reproductive effort, a meiotic driver allele is a 'cheater' that increases its own fitness at the expense of others. So there is selective advantage

Figure 7.4 Stalk-eyed flies from the family Diopsidae have become a focus of much research into sexual reproduction because they demonstrate a number of key features, including meiotic drive alleles that bias the sex ratio towards females (and suppression alleles that restore balance towards males), strong genetically determined sexual selection (females with long-eyed fathers prefer long-eyed mates), male–male competition (some species have competitions where males aggressively compare eye stalk length), and the handicap principle of sexual selection (longer eye stalks may indicate males in better condition and act as a marker for 'good genes'). Stalk-eyed flies should be enough to convince anyone that, as the great evolutionary biologist J. B. S. Haldane said, 'the Universe is not only queerer than we suppose, but queerer than we *can* suppose'.[29]

Photo by Bernard Dupont (CC2.0 license).

Figure 7.5 Escaping the constraints of levels of selection: cancer cells increase their representation in an individual by unrestrained cell duplication, but usually have limited evolutionary potential as they die with their host. But some cancer lines have escaped from this bind and outlived their original hosts. This Tasmanian Devil (*Sarcophilus harrisii*) is lucky to have avoided infection by Devil facial tumour diseas (DFTD), a cancerous cell line transmitted between individuals, probably by biting, or by eating infected carcasses. The cell line may have originated in a single individual, but has now caused a reduction of 80–90% in the Tasmanian Devil population in just over a decade, so that the Devil is suddenly facing extinction. Other transmissible cancers include a venereal disease in dogs and several forms of transmissible leukaemia in marine bivalves. One of the most successful cancer lines to spread beyond its initial host is the HeLa cell line, originally derived from cervical cancer in Henrietta Lacks in 1951 and maintained as an immortal laboratory cell line for use in research. Not only have HeLa cells been distributed to research laboratories all over the world, but they have infected many other laboratory cell lines and now rival many free-living species in biomass produced per year (for example, it is a fair guess that there are more tons of HeLa cells right now than there are of the blue whale *Balaenoptera musculus*).

Photo by Chen Wu (CC2.0 license).

to other alleles that can suppress segregation distorters and restore a fair meiosis. From the point of view of the persistence of the population, a distorter that skews the sex ratio might reduce population growth, with the extreme case that a distorter allele that results in the production of only one sex can lead to the extinction of a sexually reproducing population, unless suppression alleles evolve to counter the sex ratio distortion. This applies whether the sex ratio distorter comes from within the host genome (like male-killer alleles in stalk-eyed flies, **Figure 7.4**), is additional to the host genome (like supernumerary B chromosomes), or comes from a parasite gene (like *Wolbachia*).

Whenever a complex entity reproduces as a whole but it is composed of smaller units that are also capable of making copies of themselves, then you have to think about the different levels at which selection acts. Why do the smaller units work towards cooperative reproduction of the larger unit? Why don't they cheat, and copy themselves at the expense of others? Maximizing individual copy representation in the short term might not maximize long-term representation at the collective level if the cooperatives can ultimately out-reproduce the selfish renegades (see Chapter 2). Genomes that contain meiosis distorters might not be replicated as efficiently as those with fair meiosis. A cancer cell copies at the expense of the multicellular organism in which it arises; it is successful in the short term, as the numbers of copies of the cancer cell and its genome increase, but if it kills its host and dies with it, it won't have long-term evolutionary persistence. Actually, there are a small number of cancers that have escaped this bind by being transmissible to other individuals, so can keep copying after the death of their host (**Figure 7.5**).

Segregation distorters and cancer cells represent cases of conflict between players in a system. Conflict occurs when not all players benefit from the same strategy. The short-term fitness gains for the renegades come at the cost of the other alleles in the genome, and other cells in the body. In other cases, the existence of conflict between players is less obvious. Sex creates offspring that are genetically different from their parents. This opens the door for parents and offspring to benefit from different strategies. While the parent may wish to spread their resources evenly between all their offspring, current and future, each offspring wants to maximize its own share of parental investment. This may be particularly pronounced for internal fertilization and gestation, where the interest of the fetus may be to maximize the transfer of nutrients from mother to offspring, even at the expense of maternal health. For example, in humans, the fetus regulates the transfer of calcium across the placenta, even though the calcium will be drawn from the mother's own skeleton if there is insufficient dietary intake[8] (hence the old adage on the costs of pregnancy: one baby, one tooth). Of course, if action by the fetus causes the death of its mother before it can live independently, it will reduce its fitness to zero.

Evolutionarily stable strategies

66 *Roughly, an ESS is a strategy such that, if most of the members of the population adopt it, there is no 'mutant' strategy that would give higher fitness.* **99**
John Maynard Smith, George Price (1973) The logic of animal conflict. *Nature* 246(5427): 15–18.

Considering the evolution of sex, we keep coming to situations where we are confronted by systems that have a superficial appearance of harmony, but on closer inspection we see that not all players in a system benefit from the same strategy. Meiosis is a democratic model of gene replication, but any selfish gene that can replicate itself at the expense of others can gain an advantage. Mothers, fathers, and babies may all wish the best possible future for the child, but they may vary in their ideal investment in that future. Conflict arises when the best possible strategy may not be the same for all players in the system. In complicated reproductive systems, made of components that have some potential to copy at the expense of others, can we predict how the conflicting strategies will play out?

John Maynard Smith and George Price realized that this problem could be framed in terms of game theory.[9] Game theory had previously been formulated in terms of the intelligent action of rational players and therefore was presumed to apply only to humans. As it turns out, data suggest that people rarely play entirely rationally. But, as Maynard Smith suspected, organisms are ideal game players because, when the benefits are considered in terms of fitness, natural selection acts as the referee. In such games, 'strategy' is typically understood to be a particular inherited phenotype or behavioural trait, rather than a conscious decision to act in a particular way.[10]

Maynard Smith and Price demonstrated the game theoretical approach in evolution by considering the question: why don't animal conflicts escalate? In

Figure 7.6 Male oryx (*Oryx gazella*) locked in combat. Maynard Smith and Price[9] pointed out that horns, while they look fabulous, are inefficient as weaponry. 'And in the Arabian oryx (*Oryx leucoryx*) the extremely long, backward pointing horns are so inefficient for combat that in order for two males to fight they are forced to kneel down with the heads between their knees to direct their horns forward . . . How can one explain such oddities as . . . antelope that kneel down to fight?'

Photo by Jean and Natalie (CC2.0 license).

many species, males undergo combat for access to territories and mates, but often this combat appears like a ritual, with a period of highly stylized competition followed by one male backing down and ceding to the victor (Figure 7.6). Instead of males locking horns and making a bit of a show, why doesn't the stronger one just kill the weaker? The advantage to the victor would be a permanent end to that particular loser's challenge, preventing any further contribution of those alternative alleles to the gene pool.

Showy, non-lethal male conflict had been explained by some scientists in terms of group selection: the species as a whole benefits if conflicts can be resolved without blood spilt, because it provides regulation of reproduction to prevent the population growing too big and over-exploiting available resources. But Maynard Smith saw that this explanation could not beat the logic of natural selection. A mutant male that lacked the self-restraint allele

and slaughtered his opponents would leave more offspring than his gentlemanly cohort, so in due time the population would become dominated by selfish fighters. So why do real populations not get taken over by no-holds-barred fighters?

Maynard Smith and Price ran a computer model where individuals met in pairs to contest a resource that will increase their fitness (say, a prime territory). There are three possible 'moves' in this game: attack (fight with the intention of doing damage), display (contest in a way that does not involve damage), and retreat (run away and cede resource to opponent). Each individual adopts one strategy that is a set of rules about how to respond to an opponent, for example, 'Mouse' (never attack, retreat if attacked, display otherwise) or 'Hawk' (always attack, never retreat, fight until seriously injured or opponent runs away) (Figure 7.7). Each contest had a pay-off in terms of the fitness benefits accrued by

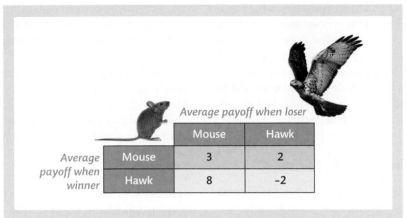

	Average payoff when loser	
	Mouse	Hawk
Mouse	3	2
Hawk	8	–2

Average payoff when winner

Figure 7.7 In some games, the best strategy depends on what others are doing. In a population of 'Hawks,' it pays to be a 'Mouse,' but in a population of individuals playing Mouse it pays to be a Hawk. This is one of the games from Maynard Smith and Price's classic paper 'The logic of animal conflict,' which launched the concept of the Evolutionarily Stable Strategy (ESS)[10] (later papers used the terms Hawk and Dove). In this example the numbers represent the average fitness payoff over a series of contests if the winner of a contest gets 6, a player that retreats gets 0, and a player that fights and is injured gets -10. The fighting abilities of all players are assumed to be the same, they just differ in their balance of attack, display and retreat tactics. Note that this is not a 'zero-sum game'—when a player with the 'Mouse' strategy cedes victory to one playing 'Hawk,' the fitness of the mouse is lower than it would have been if it had won, but it need not be zero. For example, Maynard Smith and Price[10] imagined the players competing for territories: the winner gets the best territory which can attract high quality mates and support many healthy offspring, but the loser may still get a marginal territory and breed at a lower reproductive output. The ESS will change depending on the costs and benefits assigned to each tactic. For example, if a single 'attack' has a high chance of inflicting mortal injury, then Hawk becomes an ESS due to the advantage of taking 'pre-emptive strike' action.

Photos by U.S. Fish and Wildlife Service Southeast Region (CC22.0) (hawk); Eric Isselee/Shutterstock.com (mouse).

gaining the resource, so the average benefit of each strategy could be calculated over a series of contests between members of the population.

Needless to say when a Mouse meets a Hawk, the chances are that the Mouse either cedes immediately or pays the cost of getting walloped. So will the Hawk strategy completely replace Mouse? Surprisingly, no. As the Mouse strategy decreases in frequency due to repeated losses, a Hawk will increasingly meet with other Hawks. And when a Hawk fights a Hawk, things go badly for at least one Hawk. Hawk is not always an evolutionarily stable strategy (ESS) due to the costs of fighting

other Hawks, leaving a Hawk population open to invasion by the humble Mouse strategy that wins less often but also doesn't pay the cost of fighting. But neither is Mouse an ESS, because a population of individuals playing Mouse will be vulnerable to being invaded by the aggressive Hawk strategy. So what strategy is stable in this game? A 'limited war' strategy where the individual typically fights without harming unless attacked, but will fight hard if attacked and run away if the conflict lasts too long.

'Games' do not have to involve direct combat. The same kind of thinking can be applied in every situation in which the fitness benefits of a particular strategy depend on what everyone else in the population is doing. One of the clearest examples of this is the ratio of males to females. From the point of view of reproductive opportunity, in a population of males it pays to be a female, and in a population of females it pays to be a male. But before we look at the way ESS plays out for sex ratios, we first need to ask: why have males and females in the first place?

Why have males and females?

Sex requires the DNA of different parents to come together in one offspring. How is this going to happen? Prospective parents could cast their chances to the wind or water, with all individuals spawning gametes into the environment so that, if they meet another, they can fuse together. The chances of forming an embryo that will go on to grow into a successful adult will increase with the advent of strategies that help the gametes come together, or by strategies that give the embryo a head start in life by provisioning for early growth. But these two strategies are at odds: the first favours mobility of gametes, and the second favours increased gamete size. Maximizing gamete mobility will come at the cost of provisioning, and maximizing gamete provisioning will come at the cost of mobility. One solution to this problem is to split these two strategies into different kinds of gametes, with some travel-and-seek gametes, which can evolve to be mobile and small, and some stay-and-be-found gametes, which can be large and well-provisioned.

There are several possible advantages to this system. The stay-and-be-found gametes do not have to be adapted for travel, so they can have a large nutrient-rich store to feed the embryo growth if fertilized. Unlike broadcast spawning, which is inherently risky for the gametes, stay-at-home gametes can be retained within the parent's body, providing protection and a safe environment for early growth. Home-bound gametes can have a relatively low rate of metabolic turnover which might protect biomolecules, including the genome, from damage due to metabolic by-products (such as free oxygen radicals). On the other hand, the travel-and-seek gametes can evolve for activity

and movement, streamlined to have only what is needed to deliver the DNA to the larger, nutrient-rich, stay-at-home gamete. Many more small active cells can be produced for the cost of one large well-provisioned cell. They can have lots of mitochondria to power movement, yet those mitochondrial genomes could be ditched when they achieve fertilization, avoiding the potential for passing on organelle genomes that have acquired metabolic damage.[11]

We can consider the costs and benefits of investing in two alternative reproductive strategies: male reproduction (many small motile gametes) and female reproduction (fewer resource-rich gametes). In some species, individuals adopt only one strategy, either male or female. In other species, individuals produce both kinds of gametes—most plants produce both pollen and ovules, and many animals are hermaphrodites that produce both eggs and sperm. Hermaphroditism brings the potential for self-fertilization, which has both benefits and costs. On the one hand, it's handy to be able to reproduce uniparentally, without relying on finding another parent. On the other hand, self-fertilization won't bring the benefits of combining alleles from different parents, and risks accruing the genetic costs of inbreeding (loss of genetic variation and high risk of being homozygous for deleterious recessives). But the costs of self-fertilization can be avoided through self-incompatibility, or by serial hermaphroditism (produce only one kind of gamete at a time). For example, if larger females lay more eggs, individuals might switch to being female when they attain large size. Or if larger males can breed with more females, then individuals might switch to male when they get big enough to be competitive. Sex might also be adjusted to opportunity; coral gobies form exclusive pair bonds, but if one partner dies, the survivor will seek a new partner (**Figure 7.8**). If it meets a partner of the same sex, it can change sex in order to form a new sexual partnership.[12]

Whether an individual has one fixed sex, or whether both male and female gametes are produced in the same body at the same time or sequentially, the production of male gametes and female gametes represents different strategies

Figure 7.8 Gender fluid. Yellow coral gobies (*Gobiodon okinawae*) are bidirectional protogynous hermaphrodites, meaning that they are sequential hermaphrodites that begin life as females but can change sex from female to male, or back again. If two yellow coral gobies pair up, and they are both the same sex, one of them changes sex to form a breeding pair. How do they decide? The larger one becomes a male (or the smaller one of two males becomes a female).

Photo by Sean McGrath on Flickr via Wikimedia Commons (CC2.0 license).

for investment in reproduction: many cheap versus few expensive gametes. When we see that there are different strategies with variable fitness pay-offs, we can ask: what is the optimum investment that maximizes fitness benefits? Considering these strategies gives us a convenient approach to thinking about the way that selection operates at different levels of biological organization.

Sex ratio as an evolutionarily stable strategy

We have seen that the ratio of males to females in a population is open to manipulation. Parasites can distort the sex ratio to produce only the sex that is optimal for their own transmission. The same strategy can play out for meiotic driver alleles or selfish chromosomes, which raises the question: why shouldn't the same logic apply to the genes in the host's own genome? Alleles will rise in frequency in the population if they end up in more offspring than alternative alleles, so they should manipulate the sex ratio in order to maximize their own chances of reproduction. We can consider the balance between reproductive strategies—males with many small mobile gametes and females with fewer well-provisioned gametes—as a classic example of an ESS.

In a sexually reproducing species, each offspring must have one father and one mother. By definition, male gametes are cheaper to produce than female gametes. Therefore a single male can produce enough gametes to fertilize many female gametes, so it is not necessary to have equal numbers of male and female individuals. This is, of course, why farmers tend to keep more cows than bulls, and more hens than roosters. In a species where the females provision the young, production of offspring is limited by the number of females because they provide the energy for offspring growth. A population that contains mostly females will produce more offspring per head of population than one with equal numbers of males and females. If female-skewed populations can grow at a faster rate, why don't they take over the world and replace those with equal sex ratios?

To think about whether a particular strategy is an ESS, you need to ask whether it could be 'invaded' by a mutant. If you introduce a heritable variant that has a different strategy, will it rise in frequency at the expense of the existing strategies? We saw earlier that even if Hawks have the highest pay-off in combat with Mice, their fitness advantage diminishes as the proportion of Hawks in the population increases. In a population of Hawks, it pays to be a Mouse. Now consider the sex allocation strategy of a reproductive individual in a female-biased population: should she have a male or female child? When males are rare, any given male has the potential to get a relatively high share of fertilization opportunities, so in a female-biased population, a mother having a son is likely to have more grandchildren. But if all the mothers in the female-biased population optimize their number of grandchildren by having sons, there won't be a lack of males any more. The advantage to being male will diminish as the proportion of males in the population rises until eventually the fitness advantage of male offspring approaches that of females as their frequencies equalize. In this situation, a sex ratio of 1:1 males to females is the only ESS, because any other strategy can be invaded by a mutant that gains a reproductive advantage by producing the rarer sex. As long as each offspring has one father and one mother, in a population of females it pays to be a male, and vice versa.

> ### Key points
>
> - Game theory is a useful framework for thinking about the likely evolutionary outcome when you have 'players' in a system that benefit from different 'strategies'—that is, when there are multiple heritable alternatives in a population that have different ways of maximizing their own fitness, potentially at the expense of others in the population.
>
> - An evolutionarily stable strategy (ESS) is a heritable type, or mix of types, that maximizes fitness outcomes, so cannot be displaced by an alternative type.
>
> - Equal numbers of males and females is a classic case of an ESS; since it always pays to produce the rarer sex, the only stable equilibrium is an even investment in both.

Balancing costs and benefits

ESS is also a useful way of thinking about the persistence of sexual species. Can a population of sexually reproducing individuals be invaded by an asexual mutant? Such a mutant is possible in many species. For example, many sexual plant species have given rise to asexual lineages that reproduce by self-fertilization or clonally (see Figure 4.10). Asexual species also arise within many animal lineages. For example, there are about 50 parthenogenetic (all-female) lizard species. Many other species contain a mix of reproductive strategies, with the potential for both sexual and asexual reproduction. Pond fleas (*Daphnia*: Figure 3.15) typically reproduce asexually, hatching diploid unfertilized eggs that usually develop into females. But at certain times, some eggs hatch into males that produce haploid sperm and females lay haploid eggs, and so diploid sexual offspring can be formed. Some species of *Daphnia* seem to have given up sex entirely, and never produce males: why haven't the rest?

In a species that does not have any paternal care, so that the father contributes only his DNA and no other resources to his offspring, an asexual female and a sexual female may use the available resources to produce the same number of offspring. But the asexual female's offspring carry only their mother's genes, as will her grandchildren, whereas the sexual female's children carry only half her DNA, and her grandchildren only a quarter (assuming no inbreeding). All other things being equal, we expect the asexual female to double the representation of her genes in each generation (including alleles for asexuality). She does not have to gamble her genes by blending them with another individual, and avoids the cost of investing in another parent's DNA in sexual offspring. Of course, a sexual female that produces sons may still have as many grandchildren, in which case other females may invest in raising her son's offspring. But we still expect a population of asexual females to have reproductive advantages over their sexual sisters in terms of genetic representation in future generations. In addition to the advantages in terms of genetic legacy, most asexual females do not have to pay the cost of finding, attracting, and interacting with a mate.

Given that we know that asexual mutants can occur in many lineages, and that they appear to have so many benefits over sexual individuals, then why don't they take over? Why isn't asexuality always an ESS? There must be costs to asexuality, or benefits to sexuality, that prevent asexual mutants from rising in frequency until they replace all sexual females. But we might have to look for costs and benefits at different levels of organization.

We saw how the selective advantage of sex ratio varies across different levels of genetic organization. There are parasites, meiotic drivers, and selfish chromosomes that can increase their own copy rates if they skew the sex ratio, but other alleles in the genome regain relative advantage if they can suppress meiotic drivers. Individuals maximize reproductive success in the long-term by investing equally in males and female, even if population growth as a whole benefits from having more females than males. All-female lineages may have a short-term advantage by avoiding the costs of sexual reproduction, but there must be long-term costs that limit their evolutionary potential. We know this because asexual species are rare and most asexual lineages are relatively young.

What do you think?

Are the selective pressures that maintain a 'fair meiosis' the same as those that maintain a 50:50 sex ratio? And are they the same as cancer suppression mechanisms? Are these different processes that have a superficial resemblance, or exactly the same process playing out at different levels of selection?

Evolvability

66 *Asexual progenies are mitotically standardised, and sexual ones meiotically diversified.* 99
George C. Williams (1975) *Sex and Evolution*. Princeton University Press, Princeton, NJ.

The most convincing evidence that asexuality has long-term evolutionary costs is the phylogenetic distribution of asexual species. Asexual species tend to occur on the 'twigs' of the tree of life, rather than grouping on deep-branching lineages (see Case Study 7). In other words, asexuality arises often, but these species rarely become the ancestors of a long line of asexual descendants, and asexual lineages generally do not show the branching diversification that gives rise to families of related species. Here we see an important macroevolutionary phenomenon at play: selection at the lineage level.

Selection occurs when entities have heritable variations that influence their relative success at making copies of themselves. Lineages make copies by the process of diversification (one lineage splits into two or more descendant lineages) and they have heritability (descendant lineages tend to resemble their parent lineages more closely than other unrelated lineages). If the descendants of sexual lineages also tend to be sexual, and the descendants of asexual lineages tend to be asexual, then we say that the reproductive mode is a heritable feature of a lineage. Of course, the reproductive mode of a lineage can change; a sexual lineage can give rise to an asexual descendant lineage, just as a sexual individual can give rise to a mutant asexual offspring. So we can see that there is heritable variation in reproductive mode at the level of lineages.

Now that we have established that there is is heritable variation between lineages for reproductive mode, we need to ask: does that variation have a fitness difference at the lineage level by influencing the relative proportion of lineages that are produced by diversification over time? Sexual lineages seem to have a selective advantage over asexual lineages; they make more copies of themselves, because they have more descendant lineages (higher net diversification rate). If asexuality had equal fitness to sexual reproduction, at the lineage level, then we ought to see that sexual and asexual lineages produce the same average number of descendants. But we don't see that. We see that asexual lineages rarely give rise to many descendant lineages, and asexual species rarely replace related sexual lineages. In fact, asexual species often coexist alongside their sexual sister species. If asexuality evolves often and has short-term benefits in terms of reproductive advantage to females that can reproduce uniparentally, then its limited distribution and failure to replace similar sexual species tells us that there must be macroevolutionary costs to being asexual, or relative benefits of being sexual.

First let us consider the possible benefits of sexual reproduction to an individual parent. An asexual parent produces offspring that are genetically similar to itself. Not only will sexual offspring be different from their parent, they will also be different from each other. Since offspring will differ slightly from each other, there is some chance that, instead of all competing for exactly the same resources as their parents, such diverse offspring might spread out and find slightly different ways of living. This might reduce competition between siblings, which is good for the parent if it results in more grandchildren. If this was the case, then we might expect asexuality to be more common in lineages that are specialized to specific niches, with little variation in resource availability or conditions, and sexual species to be more common where individuals encounter a wide range of conditions and resources. The evidence suggests that this is not always the case. For example, asexuality frequently evolves in hymenopteran lineages (**Figure 7.9**). But parthenogenetic (all-female) hymenoptera species tend to have broader host ranges if they are parasitoids, more host plants if they are phytophagous, and larger geographic distributions than sexual hymenoptera.[13] This is not what we would expect to see if a major benefit of sexual reproduction over asexuality was the exploitation of varied resources.

In addition to providing genetic diversity to exploit spatial variation in resources, sexual reproduction might provide a buffer against temporal variation in conditions. If a parent is right at the adaptive peak, with the best possible genotype given the conditions, then it pays for it to produce offspring just like itself (see Figure 4.9). What if the conditions change and the adaptive peak shifts? Now a sexual parent might have some advantages; its offspring are not all like itself and they are not all like each other. Some will be worse off, but there is a chance that, by virtue of their novel genetic combinations, some will be better adapted to the shifting conditions. Reproducing sexually is a gamble, but it's a gamble that might pay off in uncertain times. If this was the main advantage of sex, then we might expect to see

that sexual reproduction is more common in unpredictable environments, or where conditions change rapidly over time. Again, there is little support for this prediction when we consider the geographic distribution of asexual species, which often occupy marginal habitats or disturbed environments. Asexuals are generally more common at high latitudes, high altitudes, and on the edges of the geographic range of their parent species.[14] This does not seem to be the pattern we would expect if sexual reproduction had particular advantages in response to environmental challenges. However, we can't rule out that this geographic pattern is really a reflection of the possible advantages of asexual reproduction in dispersal and establishing populations at the edges of a species range, where densities might be low and uniparental reproduction useful.

If we just think of 'environment' as the physical conditions an organism faces—temperature, water availability, and so on—we might miss an important component of changing conditions. An important aspect of the environment of all organisms has the potential to change every generation. Each species interacts with many other species—predators, prey, parasites, pathogens, competitors—and every other species is also evolving. Therefore each species can be seen as being in a constant race to keep up with the changes in other species in order to outwit or outcompete the others. This coevolutionary race is sometimes referred to as the Red Queen model, after the character in *Alice Through the Looking Glass* who said, 'It takes all the running you can do, to keep in the same place'.[15] The Red Queen hypothesis suggests that lineages must constantly change to maintain the same level of fitness and avoid extinction. Changing conditions may place a premium on heritable variation, allowing rapid response through selection.

By producing offspring with a greater genetic diversity, sex may help keep a species one step ahead. This may be particularly valuable when a large long-lived species is locked in a coevolutionary race with a smaller short-lived species. The short-lived species has the advantage of speed of change; every generation can throw up new variants, challenging the static genotype of the long-lived

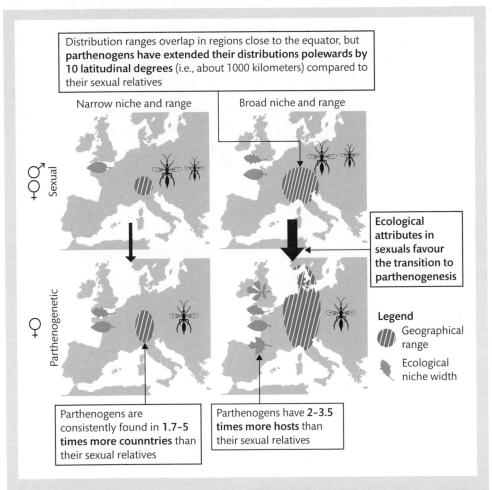

Distribution ranges overlap in regions close to the equator, but **parthenogens have extended their distributions polewards by 10 latitudinal degrees** (i.e., about 1000 kilometers) compared to their sexual relatives

Narrow niche and range

Broad niche and range

Sexual

Parthenogenetic

Ecological attributes in sexuals favour the transition to parthenogenesis

Legend
Geographical range

Ecological niche width

Parthenogens are consistently found in **1.7–5 times more counntries** than their sexual relatives

Parthenogens have **2–3.5 times more hosts** than their sexual relatives

Figure 7.9 All-female asexual lineages have arisen many times in arthropods, especially in haplodiploid lineages, such as many Hymenoptera (ants, wasps, and bees). Perhaps this is because the haplodiploidy, where females develop from fertilized diploid eggs and males from unfertilized haploid eggs, makes transitions to parthenogenesis less of a leap. A study of 765 cases of parthenogenesis in haplodiploid arthropods found that parthenogenetic species tend to have wider niches (for example, using more host plants) and distributions that extend further towards the poles than sexual species.[30] However, this might be partly because asexual lineages are more likely to arise in sexual species with relatively wide niches and distribution ranges. Parthenogens are found in more countries than sexuals (distribution marked with diagonal stripes), and have a wider niche (represented by green leaves) and polewards extended distribution. The width of the purple arrows represent the relative rate of evolutionary transitions from sexual to asexual species: parthenogenesis is more likely to evolve in sexuals with relatively wide niches and distribution ranges.

Casper J. van der Kooi, Cyril Matthey-Doret, and Tanja Schwander. Evolution and comparative ecology of parthenogenesis in haplodiploid arthropods. *Evolution Letters*, 1: 304-316, 2018. Wiley Online Library (CC BY 4.0).

species. Such asymmetric contests are common—so common that they may be one of the major evolutionary pressures on many species. Every species is plagued by parasites, and parasites can exert a major fitness cost on their host. If you doubt that something as humble as parasites could drive the evolution of a feature as complex, costly, and widespread as sexual reproduction, consider that every vertebrate species is likely to carry around half a dozen species of internal parasites (such as cestodes and trematodes) plus several species of ectoparasites (such as mites),[16,17] let alone the disease burden from viral, bacterial, protistan, and fungal pathogens. And the larger the host species, the longer its generation turnover time and the more parasite species it tends to support.

Parasites, with their short generation times and high fecundity, have a clear advantage in adapting to the defences of their larger-bodied hosts who are slower to reproduce and have fewer offspring. Most vertebrates have an adaptive immune system that undergoes a process akin to selection, replicating those antigens that are useful in defence against pathogens, which provides one way of increasing the speed of response to pathogens. But perhaps sexual reproduction, by generating genetic diversity every generation, is also a way for long-lived hosts to keep up with the fast-evolving parasites. Asexual individuals may have a fitness advantage against parasite attack when rare, because they can reliably transmit resistance alleles and out-reproduce their sexual sisters. But once a successful asexual clone rises in frequency, it becomes a target for parasites.[18] We expect alleles for parasite resistance to have frequency-dependent fitness effects—beneficial when rare, less so when common. Sexual reproduction, by producing genetically varied individuals, presents a moving target generating new variants every generation, each of which has some chance of avoiding parasite attack. For example, a study of fish living in small rock pools, where sexual populations coexisted with their hybrid clonal offshoots, showed that the most common asexual clone accumulated parasites at a greater rate than its sexual sisters. Inbred sexual populations, with reduced genetic variation, likewise suffered higher parasite numbers.[19]

If sex allows rapid evolution in the face of changing environment, arms races with other species, or parasitic challenge, do sexual lineages have a faster overall rate of change than their asexual relatives? And if they do, could this explain why sexual lineages seem to have an advantage over asexual lineages in terms of diversification and persistence?

Key points

- Whenever entities have differential copying abilities by virtue of particular heritable variations, we expect to see a rise in representation of those heritable variants.

- Sexual reproducing lineages typically have more descendant species that asexual lineages, so we can conclude that sexual reproduction has a macroevolutionary advantage.

- Proposed macroevolutionary benefits of sexual reproduction include production of genetically variable offspring reducing competition between offspring, allowing exploitation of a wider range of resources and more rapid adaptation when selection acts on multiple genes.

Sex and the pace of adaptation

If selection acts on only one gene at a time—for example, selection on an allele that confers resistance to a common parasite—asexual reproduction will be effective in increasing the prevalence of that allele. Individuals with the fittest allele will have the highest average reproduction rate, and each successful parent will pass on the winning allele to its lucky offspring. Now imagine that another gene is also under selection—for example, a mutation that increases metabolic efficiency or confers resistance to a different parasite. In an asexual population, to obtain the fitness benefits of both advantageous alleles, this second mutation has to occur in an individual already carrying

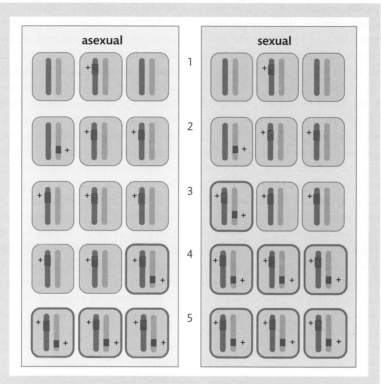

Figure 7.10 Faster response to multi-locus selection in sexual lineages. If selection acts on more than one locus (i.e. on variants occurring in several different genes at once), the asexual population must wait for both mutations to occur in the same genome. In the asexual population, the waiting time to the fittest combination (marked with a bold outline) will depend on the population size and the mutation rate (plus the relative selective advantage of the alleles, which will determine their rate of fixation in the population). In this example, the fittest combination of alleles is fixed in the asexual population at time 5. But in a sexual population, mutations occurring in different individuals can be combined together in a single offspring, so the ideal combination can go to fixation earlier (time 4).

the other beneficial allele. In practice, this means that fixation of positively selected mutations must occur serially (**Figure 7.10**). The waiting time to fix a multi-gene combination of alleles will be longer in smaller populations; fewer individual genomes means less chance that, at any given time point, a second mutation will occur in an individual who already has the first. Similarly, the lower the mutation rate (less chance of mutation occurring per copy) and the longer the generation time (longer waiting time between mutation

events), the less chance there is that these two optimal variants will occur together in one fortune-favoured individual. In addition, the lower the average fitness advantage of new mutations, the weaker the effect of selection, and the slower the pace of adaptation at multiple genes will be in a clonal population.

But in a sexually reproducing population, mutations that occur in different individuals can end up in the same individual, who has inherited them from two different parents (Figure 7.10). Sexual reproduction enables the reassortment of alleles, allowing different combinations to be trialled, with the most successful combinations resulting in individuals who are more likely to reproduce and contribute those alleles to the next generation. Sex can also separate the good mutations from their association with other less favourable alleles by reassortment of alleles at meisosis, and recombination of variants that co-occur on the same chromosome. Of course, the same process also produces individuals who have unfortunate combinations of less favourable alleles. This is one of the costs of sexual reproduction. Once a fortuitous combination of alleles has occurred in a single individual, an asexual parent can make many faithful copies that contain both high fitness alleles. But the sexual parent with two high fitness alleles may produce some offspring with both, some with only one, or even some with none. The cost of selection in terms of producing non-optimal gene combinations is referred to as 'load'.

In clonal reproduction, inheritance is all or nothing; the good comes attached to the bad. As selection promotes the favourable alleles, it will drag other less profitable variants with it. But in a sexual population, there will be some offspring who inherit the beneficial allele without the baggage of the less fit alleles, and these lucky variants will prosper at the expense of others, pushing the advantageous allele to fixation. If load slows evolution, then sexual reproduction may take the brake off. However, if optimum fitness depends on particular combinations of alleles (epistatic fitness effects), sexual reproduction may actually slow down adaptation because it will keep disrupting their inheritance together (**Figure 7.11**).

The net result of sexual reproduction shuffling alleles and promoting favourable combinations is that sexually reproducing populations will build up collections of co-adapted alleles that work well together in the current environment. This process contributes to genetic isolation between species. To see why, consider the fate of alleles

in neighbouring populations of the same species. If the two populations are isolated, so that their members never meet and mate, their particular alleles in each population will rarely if ever find themselves in the same individual. This means that alleles in one population are never tested against alleles in another population for negative epistatic fitness effects. Eventually, each population will accumulate genetic variants that work well with each other, but may be incompatible with those accumulated in the neighbouring population. Now, if individuals from the two populations do meet and mate, their hybrid offspring may be of low fitness due to negative interactions between their untested allele combinations. This, in turn, may prompt the evolution of traits that will stop members of the two populations wasting reproductive effort on hybrids, for example by promoting songs or behaviours or markings that distinguish members of one population from another. So one way that sexual reproduction might speed the rate of evolution is by increasing the speciation rate, generating selective pressure on traits that limit interbreeding between populations, or increasing the rate of formation of separate non-interbreeding species. In fact, many definitions of 'species' and 'speciation' are specific to sexually reproducing lineages. Of course, we still recognize 'species' in asexual populations as sets of genetically similar individuals with particular sets of features. Even in the robustly asexual bdelloid rotifers, species can be recognized as being genetically distinct populations with particular morphological traits and ecological niches.[20]

There are good theoretical reasons to believe that sexual reproduction will speed the rate of adaptation. Yet there is surprisingly little empirical evidence that can be used to test this prediction. Some experiments comparing the rate of adaptation in asexual and sexual populations support the prediction that sexual organisms are better at avoiding parasites and adapting to rapid change. For example, populations of facultatively sexual rotifers reared in environments with changes in multiple dimensions (salinity, temperature, and heavy metals) produced relatively more sexual

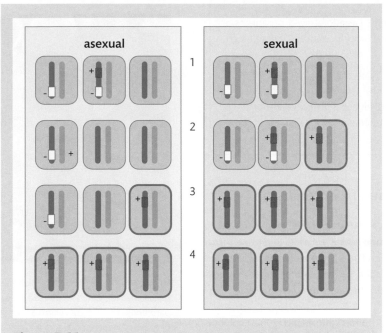

Figure 7.11 Sex can help to separate advantageous alleles from less helpful ones. In this example, there is an allele (–) in the population that is not harmful on its own, but reduces the fitness benefit of the advantageous allele (+). This is an example of negative epistasis—alleles with selective costs when in combination. In the asexual population, the advantageous allele is selected against when it occurs in the same genome as the interacting allele, so the waiting time for the fittest combination (marked with a bold outline) depends on the time taken for the mutation to occur in a genome without the interacting allele. But in the sexual population, reassortment of alleles each generation allows the advantageous allele (+) to be separated from the interacting allele (–). Note that the opposite would be true if both alleles were favourable; asexual parents could reliably pass on both alleles, whereas some sexual offspring will contain only one or two of them. Breaking up favourable allele combinations is one of the potential costs of sex (sometimes referred to as recombination load).

eggs than those raised in environments changing in only one dimension.[21] Whereas sexually reproducing flatworms could persist over many generations when raised in the presence of coevolving bacterial pathogens, the asexual populations all went extinct.[22] But, just as it is not always easy to extend the predictions of evolutionary models to real biological systems, so it is not always certain that experimental results reflect common occurrences in the natural world. Much remains to be done before the puzzle of sex can be resolved.

Key points

- Sex can bring together the best alleles for different genes without having to wait for all the mutations to occur in the same genome.

- Sex can help to separate favourable alleles from unfavourable ones, making selection more efficient and less costly.

- Sex may give large long-lived organisms a way of producing genetically variable offspring that have a greater chance of reducing parasite stress.

- There is currently insufficient empirical evidence to evaluate the relative importance of these mechanisms in maintaining and promoting sexual reproduction.

Origins versus maintenance

66 *... Transfer of DNA creates the potential for sex to evolve simply because genetic elements that cause themselves to be copied and transferred to other individuals can spread in a population, as long as they infect new cells faster than they kill their host or otherwise reduce host fitness.* 99

Sarah P. Otto and Thomas Lenormand (2002) Evolution of sex: resolving the paradox of sex and recombination. *Nature Reviews Genetics* 3(4): 252–61.

The evolution of sexual reproduction highlights an important distinction in macroevolutionary explanations: the evolutionary forces that lead to the origin of a particular trait are not necessarily the same as those that keep the trait in high frequency. The hypothesis for the origin of sexual reproduction as a means of acquiring DNA copies to use as templates for repair does not explain why sexual reproduction is maintained in species that routinely carry multiple genome copies (such as asexual diploids or clonal polyploids). Ancestral advantages of diploidy will not apply to organelle genes in contemporary organisms, nor to genes active in only one copy, such as sex-linked or differentially imprinted genes. Avoiding the cost of producing males does not explain the evolution of all-male asexual clam species, where diploid sperm, carrying a full complement of the father's genome, fertilize an egg cell from which the maternal genome is then ejected so that the embryo grows with only the paternal nuclear DNA (**Figure 7.12**).[23]

Furthermore, we should not conflate sex (combining DNA from different parents into a single

Figure 7.12 Asexual species often evolve from sexual lineages, but most derived asexuals are female parthenogens. *Corbicula* clams are an exception, because all-male asexual species have arisen many times in this genus. These asexual clams reproduce by androgenesis, where offspring carry only the paternal DNA. The androgenetic males produce diploid sperm, which fertilize and effectively parasitize eggs, using their resources but not the egg's DNA. Androgenesis has clear short-term genetic benefits for males, but it can lead to population extinction when the reproduction of asexual males leads to a lack of female gametes for reproduction. In the case of *Corbicula*, the maintenance of male asexuality may be a result of high rates of selfing (which favours maintenance of female gametes) or sexual parasitism of females of closely related species.[23]

Photo by Friedrich Boehringer via Wikimedia Commons (CC3.0 license).

offspring) with recombination (chromosome break and repair that shuffles genetic variants). While these two normally go together, they can be separated—some asexual lineages generate haploid gametes that fuse to create genetically non-identical diploid offspring (such as the parthenogenetic fruit fly *Drosophila mangaberei*), and (more rarely) some sexual organisms don't have recombination (e.g. the quill beaksedge *Rhynchospora tenuis*), or have very low rates of recombination (e.g. many conifers). We should consider the possibility that recombination originated as a mechanical adaptation for ensuring proper chromosome pairing at meiosis,[24] and that its influence on genetic variation is a side effect of that role. We have also seen that the presence of a particular trait needs to be examined in light of selective benefit to different entities. For example, any genetic element that was able to stimulate its transfer to other individuals could have a selective advantage as it increased in copy number, whether it was a selfish genetic element that spread by causing bacterial conjugation, a rogue gene that caused sexual reproduction in its host, or a parasite that benefited from the infection possibilities of mating.[25]

The origin and maintenance of a complex trait may have different explanations. Sex is ancient, and is present in most branches of the eukaryote family tree. But sex may be retained along some branches of the tree of life, not because of the benefits such as DNA repair or genetic variability, but because of more mundane benefits of sex-associated traits. For example, some species produce propagules (relatively large genetically identical copies that grow near the parent) by asexual reproduction but seeds (relatively small genetically variable offspring that disperse widely) by sexual reproduction. If there is an advantage to dispersal, then sexual reproduction will be favoured, either temporarily at appropriate times or permanently to avoid competition with parent and siblings.

Similarly, some species may predominantly reproduce asexually, but when times are hard they can produce hardy spores, which can lie dormant until better times, by sexual reproduction. In this case, can we tell if the advantage of sexual reproduction is the production of a spore to get through hard times, or the production of genetically diverse offspring which may by chance be better adapted to the changing conditions, or both, or neither? The pond flea *Daphnia* typically reproduces asexually, but occasionally undergoes sexual reproduction to produce dormant 'winter eggs' which can survive the pond drying, then hatch when it refills (Figure 3.15). But some wholly asexual *Daphnia* species have retained the 'winter egg' feature, without first requiring sexual reproduction.[24] Does this tell

us that in this case the adaptive advantages are not due to the genetic effects of sexual reproduction, but to the physical advantages of a sexual spore stage?

Sexual reproduction is a complex trait, which might not always be gained or lost as a single package, and species can show a range of traits that betray their evolutionary past. For example, some all-female asexual species still have to mate with males of a related sexual species, as sperm is needed to stimulate egg development, embryo formation, or the formation of the seed endosperm, even if the father's DNA is not incorporated into the offspring. This illustrates how gain or loss of a trait may be complicated by the interactions between features of the organism.

Genomic imprinting and genetic conflict

When we want to explain why many different lineages share a complex trait, we have to remember that species may share a trait because they inherited it from a common ancestor, and that trait has been retained. We also need to keep in mind that the explanation of the presence of the trait in those species might be quite different from hypotheses about the trait's origin in their ancestor. In the case of mammals, it may be that sex is a shared ancestral trait that is impossible to lose. While there are all-female species of fish, frogs, and reptiles, there are no confirmed cases of parthenogenesis in mammals. Mammals' inability to reproduce asexually might be a result of sex-specific methylation of the genome, otherwise known as genetic imprinting. We are going to briefly consider what imprinting is, and what the consequences are for mammalian reproduction, before taking a broader macroevolutionary perspective on why mammals have this complicated system of genome management and expression.

When mammalian gametes are formed, biochemical markers are added to some genes that silence their expression. Different genes are silenced in male and female gametes, so when an embryo is formed by the fusion of egg and sperm it is effectively 'hemizygous' for these genes (one working copy from the father and one silenced copy from the mother, or vice versa). Sex-specific imprinting has several important consequences. One is that functional hemizygosity makes individuals more vulnerable to inherited disease at those genes, because expression cannot be masked through heterozygosity (when a good copy from one parent might compensate for a faulty copy from the other parent). Sex-linked imprinting can result in different disease phenotypes depending on whether the male-derived or female-derived version of the gene is defective. For example, a section of human chromosome 15 is differentially imprinted in males and females. The absence of that region from the paternally derived chromosome leads to Prader–Willi syndrome (a disorder characterized by insatiable appetite, amongst other traits), but the absence of the maternally derived genes leads to Angelman syndrome (characterized by a happy disposition and love of water, amongst other traits). Imprinting also has interesting implications for heritability; unlike a mutation that permanently alters the sequence of a gene, imprinting can be reset every generation in the production of gametes.

Why would this complicated system of sex-specific gene imprinting evolve? One of the consequences of sex-specific genomic imprinting might provide a clue. Because of the sex-specific inactivation of essential genes, a mammalian embryo can't form from same-sex gametes. A diploid germ cell could form by duplication of a haploid gamete's chromosomes, or fusion of two haploid gametes from the same parent. But in mammals, a diploid germ cell with two copies of one parent's genome won't be capable of forming a working embryo because it needs both a male-derived and a female-derived genome copy in order to have a full complement of active genes. An all-female or all-male diploid germ cell may start dividing, resulting in a ball of cells that is recognized as a kind of cancer. These unregulated cellular growths can have the disturbing feature of recognizable tissue differentiation, such as hair, teeth, or eye tissue, but they cannot form a functional offspring. Some ovarian cysts or testicular cancers arise from these renegade embryos (often referred to as teratomas).

What does this tale of germ-cell cancers tell us about the macroevolution and conflict across levels of selection? It is possible that sex-specific imprinting evolved as a way of limiting conflict between the different genomes that must cooperate to make a new mammalian individual. Recall how the embryo's best strategy—gain as many resources from your mother as you can—is not necessarily the same as the mother's best strategy—spread your resources between current and future offspring.

We can see that the maternal and paternal genes within the embryo will likewise benefit from different strategies; while the mother's genes may wish to invest equally in all offspring, current and future, a male's genes will want his current offspring to derive more resources from their mother than her future offspring with other males do. If he is not guaranteed to be the father of her other offspring, the father's genes in the embryo may act to derive more than their share of resources, even at the mother's expense. The father's genome might gain from upregulating genes for early embryo growth, but the mother's genome may gain from silencing the same genes. Similarly, the paternal genome might profit from silencing any genes that inhibit early growth, although the maternal gene benefits from expressing them.[26] These differences in fitness strategy of maternal and paternal genes would not occur in a mammalian species that had lifelong strictly monogamous male–female pairing, because then both mother and father would share the same optimal investment in offspring. But such species are rare, if they exist at all.

Sex-specific imprinting might also have arisen through selection on the maternal genome to inhibit uncontrolled growth of a teratoma arising from the germ cells. We have already seen how the growth of a cancer can be interpreted as a selfish replication strategy of some cells within an individual (Figure 7.5). The genes in a teratoma gain a short-term reproductive advantage at potential costs to the mother's long-term fitness, so the mother's genes would benefit from inactivating genes for early embryo growth.[27] Thus a diploid cell that forms from the mother's cells alone will be limited in its capacity for growth. As with many other facets of sexual reproduction, we can see that the arguments for the origin of genetic imprinting are not necessarily the forces that maintain it in current mammalian species. But whatever the evolutionary origin of sex-specific imprinting, the long-term evolutionary result is that even if there were adaptive benefits to asexual reproduction in mammals, it seems unlikely that it could easily arise by a mutation.

So we can see that a feature of genomic architecture like imprinting, which we might consider the province of study of geneticists, can be interpreted in the light of macroevolution as a way of resolving conflict between different players in a system. Imprinting might have evolved as a consequence of conflict between genes within a genome, as the father's genes and mother's genes might differ in their ideal strategy of investment in the growth and development of the offspring. Or imprinting might have evolved in response to conflict between generations, to protect individuals from selfish replication of their germ cells. This is an illustration of how a macroevolutionary perspective may prove useful in unexpected places, such as understanding the origin of genetic disorders such as Prader–Willi and Angelman syndromes, and why it currently seems unlikely that we can form viable mammalian embryos from two parental genomes of the same sex. Perhaps this macroevolutionary perspective will lead to unexpected solutions to these apparently intractable problems arising from genomic conflict.

Conclusion

Sex is an enduring evolutionary puzzle. The costs of sex are substantial. An asexual female can, all things being equal, have twice the genetic representation in future generations as a sexual female, and can ensure that all her offspring inherit her well-adapted genotype. An asexual female may avoid the cost of seeking and interacting with mates, and she does not have to gamble her genetic inheritance by combining it with DNA from another individual. It seems obvious that, when a mutation that causes parthenogenesis occurs, it should give rise to a new lineage that rapidly out-reproduces and replaces the related sexual forms. But this does not often happen. In many organisms, asexual lineages often arise, and they coexist with sexual forms. Yet asexual lineages seem to have a short evolutionary lifespan; they rarely give rise to long-lived lineages that produce many descendant species. In most lineages, the predominance of sexual reproduction cannot be due to the impossibility of asexual reproduction, so it must be because sexual lineages have some kind of macroevolutionary advantage over clonal reproduction.

Consideration of the evolution of sex requires us to think about selection operating at many different levels of organization. Selfish genes copy themselves at the expense of others, but the parliament of genes evolves strategies that enforce a fair meiosis. Diploidy has advantages of keeping spare copies of genes and masking deleterious mutations, but introduces the challenges of balancing the potentially conflicting strategy of genes from different parents. Sexual individuals must pay the costs and reap the benefits of having offspring different from their parents and siblings, but the balance between costs and benefits is not always clear. Populations containing much genetic variation might speed the rate of adaptive evolution, yet also introduce the problem of conflict between players with different heritable strategies. The phylogenetic pattern of sex supports the idea that it has long-term advantages over asexual lineages, yet it is not clear why the sexual lineages are not replaced by asexuals in the short term. How do sexual species out-diversify their clonal sisters in the first place in order to gain long-term advantage? Asexual offshoots of sexual species could be seen as something akin to a benign cancer; short-term reproductive advantage through rapid rates of reproduction, but no long-term evolutionary future.

The costs and benefits of sex will vary with environment, and with population size and composition. Many of the theoretical models of the evolution of sex have based their conclusions on infinite population sizes, but their conclusions may not hold when populations are small and dispersed. The costs of asexuality might also be greater in a small population due to increased waiting times to adaptation. We can make predictions about which kinds of species should benefit the most from sex—those with small population sizes, long generation time, low mutation rate, and generally low benefits to each adaptive allele. So why are there not more asexual species, especially amongst organisms with large population sizes, short generation times, high mutation rates, and strong selective advantage to alleles? The prevalence of sexual reproduction also requires us to consider different possible explanations for origin and maintenance. Even if it has fitness benefits, asexuality cannot successfully 'invade' a sexual population unless it can arise by mutation. Why do most species have sex? A definitive answer to this question is yet to be found.

○ Points for discussion

1. Why only two sexes (male and female)? Why not three? Or more?

2. It has been suggested that genomic imprinting evolved as a mechanism for enhanced evolvability because it allows the effects of some alleles to be masked against natural selection in some individuals, therefore permitting genetic variation to be maintained in the population. How could you test this hypothesis?

3. Will selection at a lower level always trump selection at a higher level? For example, will group selection always give way to selection on individuals? And will selection on genes always win over selection on individuals?

❋ References

1. Marcon E, Moens PB (2005) The evolution of meiosis: recruitment and modification of somatic DNA repair proteins. *Bioessays* 27(8): 795–808.

2. Hespeels B, Knapen M, Hanot-Mambres D, Heuskin AC, Pineux F, Lucas S, Koszul R, Van Doninck K (2014) Gateway to genetic exchange? DNA double-strand breaks in the bdelloid rotifer *Adineta vaga* submitted to desiccation. *Journal of Evolutionary Biology* 27(7): 1334–45.

3. Nowell RW, Almeida P, Wilson CG, Smith TP, Fontaneto D, Crisp A, Micklem G, Tunnacliffe A, Boschetti C, Barraclough TG (2018) Comparative genomics of bdelloid rotifers: evaluating the effects of asexuality and desiccation tolerance on genome evolution. *PLoS Biology* 16(4): e2004830.

4. Ochman H, Lawrence JG, Groisman EA (2000) Lateral gene transfer and the nature of bacterial innovation. *Nature* 405(6784): 299–304.

5. Lindholm AK, Dyer KA, Firman RC, Fishman L, Forstmeier W, Holman L, Johannesson H, Knief U5, Kokko H, Larracuente AM, Manser A, Montchamp-Moreau C, Petrosyan VG, Pomiankowski A, Presgraves DC, Safronova LD, Sutter A, Unckless RL, Verspoor RL, Wedell N, Wilkinson GS, Price TAR (2016) The ecology and evolutionary dynamics of meiotic drive. *Trends in Ecology & Evolution* 31(4): 315–26.

6. Fishman L, Saunders A (2008) Centromere-associated female meiotic drive entails male fitness costs in monkeyflowers. *Science* 322(5907): 1559–62.

7. Hornett EA, Duplouy AM, Davies N, Roderick GK, Wedell N, Hurst GD, Charlat S (2008) You can't keep a good parasite down: Evolution of a male-killer suppressor uncovers cytoplasmic incompatibility. *Evolution* 62(5): 1258–63.

8. Haig D (2004) Evolutionary conflicts in pregnancy and calcium metabolism—a review. *Placenta* 25: S10–15.

9. Maynard Smith J, Price GR (1973) The logic of animal conflict. *Nature* 246(5427): 15–18.

10. Maynard Smith J (1982) *Evolution and the Theory of Games.* Cambridge University Press, Cambridge.

11. Allen J (1995) Separate sexes and the mitochondrial theory of ageing. *J Theor Biol* 180(2): 135–40.

12. Munday PL, Buston PM, Warner RR (2006) Diversity and flexibility of sex-change strategies in animals. *Trends in Ecology & Evolution* 21(2): 89–95.

13. van der Kooi CJ, Matthey-Doret C, Schwander T (2017) Evolution and comparative ecology of parthenogenesis in haplodiploid arthropods. *Evolution Letters* 1(6): 304–16.

14. Tilquin A, Kokko H (2016) What does the geography of parthenogenesis teach us about sex? *Philosophical Transactions of the Royal Society B: Biological Sciences* 371(1706).

15. Carroll L (1871) *Through the Looking Glass: And What Alice Found There.* Macmillan, London.

16. Poulin R, Morand S (2000) The diversity of parasites. *Quarterly Review of Biology* 75(3): 277–93.

17. Dobson A, Lafferty KD, Kuris AM, Hechinger RF, Jetz W (2008) Homage to Linnaeus. How many parasites? How many hosts? *Proceedings of the National Academy of Sciences of the USA* 105(Supplement 1): 11482–9.

18. Jokela J, Dybdahl MF, Lively CM (2009) The maintenance of sex, clonal dynamics, and host–parasite coevolution in a mixed population of sexual and asexual snails. *American Naturalist* 174(S1): S43–53.

19. Lively CM, Craddock C, Vrijenhoek RC (1990) Red Queen hypothesis supported by parasitism in sexual and clonal fish. *Nature* 344(6269): 864.

20. Fontaneto D, Herniou EA, Boschetti C, Caprioli M, Melone G, Ricci C, Barraclough TG (2007) Independently evolving species in asexual bdelloid rotifers. *PLOS Biology* 5(4): e87.

21. Luijckx P, Ho EKH, Gasim M, Chen S, Stanic A, Yanchus C, Kim YS, Agrawal AF (2017) Higher rates of sex evolve during adaptation to more complex environments. *Proceedings of the National Academy of Sciences of the USA* 114(3): 534–9.

22. Morran LT, Schmidt OG, Gelarden IA, Parrish RC 2nd, Lively CM (2011) Running with the Red Queen: host-parasite coevolution selects for biparental sex. *Science (New York, N.y.)* 333(6039):216–8.

23. Hedtke SM, Stanger-Hall K, Baker RJ, Hillis DM (2008) All-male asexuality: origin and maintenance of androgenesis in the Asian clam *Corbicula. Evolution* 62(5): 1119–36.

24. Maynard Smith J (1978) *The Evolution of Sex.* Cambridge University Press, Cambridge.

25. Otto SP, Lenormand T (2002) Evolution of sex: resolving the paradox of sex and recombination. *Nature Reviews Genetics* 3(4): 252.

26. Wilkins JF, Haig D (2003) What good is genomic imprinting: the function of parent-specific gene expression. *Nature Reviews Genetics* 4(5): 359.

27. Morison IM, Ramsay JP, Spencer HG (2005) A census of mammalian imprinting. *Trends in Genetics* 21(8): 457–65.

28. Charlat S, Hornett EA, Fullard JH, Davies N, Roderick GK, Wedell N, Hurst GD (2007) Extraordinary flux in sex ratio. *Science* 317(5835): 214.

29. Haldane JBS (1927) *Possible Worlds and Other Essays.* Chatto & Windus, London.

30. Neiman M, Meirmans PG, Schwander T, Meirmans S (2018) Sex in the wild: How and why field-based studies contribute to solving the problem of sex. *Evolution* 72: 1194–1203.

Case Study 7
Phylogenetic models of macroevolution: 'tippy' distribution of asexuality on the tree of life

One of the most compelling observations supporting the idea that asexuality has long-term evolutionary costs is that asexual lineages tend to occur on the 'twigs' of the tree of life. If you map asexual lineages onto a phylogeny, you typically see that the asexual species are scattered across the 'tips' of the tree (the branches on the edge of the phylogeny, not the internal connecting branches), and each one has few asexual relatives (asexual lineages don't usually form large clades of related asexual species) (Figure 7.13).

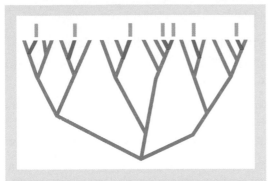

Figure 7.13 Hypothetical phylogeny with asexual lineages, marked in blue, on the 'tips'.

This has been interpreted as a sign that asexual lineages arise often (which is why they are scattered across the phylogeny) but they also have a rapid rate of extinction (which is why they do not form large clades). Of course, in biology, there are always exceptions to the rule, such as the large and diverse class of bdelloid rotifers (Figure 7.1). But the distribution of asexual species across the tips of phylogenies is sufficiently common that it has been considered one of the strongest arguments for the long-term evolutionary costs of asexuality.[1,2] But frequent gain plus raised extinction risk is not the only process that will cause a trait to be distributed on the tips of a phylogeny.[3] For example, a recent environmental change can cause parallel adaptation in many different lineages at once.[4] A tippy distribution might also arise simply by chance. So how can we be sure that the distribution of asexuality on the tips of phylogenies is telling us about the macroevolutionary costs of asexual reproduction? Can we formally test the strength of the phylogenetic evidence?

To use phylogenetic distribution to test the hypothesis that asexuality is subject to negative lineage selection, we need to be able to do three things. Firstly, we need a way of measuring the degree of 'tippiness'. There is a limit to our ability to compare phylogenetic distributions or test alternative explanations if we simply look at a phylogeny and say 'Well that looks quite tippy, doesn't it?' If we don't have an objective way of quantifying the tippiness of the distribution, we will be limited in our ability to determine when a phylogenetic distribution is too tippy to be explained by chance. And that is the second thing we need to be able to do: work out just how tippy we expect traits to be under normal circumstances, when the processes we are interested in are not occurring (often referred to as a null distribution). If evolution often produces tippy patterns, even when there is no effect of that trait on the evolutionary success of a lineage, then we may be erroneously interpreting a chance tippy pattern as a sign of an interesting macroevolutionary process. The third thing we need to be able to do is assure ourselves that the processes we are interested in really would produce a tippy distribution on phylogenies.

Quantifying phylogenetic distributions

The hypothesis that asexual lineages arise often but are relatively short-lived makes a number of predictions:

(1) asexual species should have a younger average age than sexual lineages;

(2) asexual species should have few close asexual relatives.

We can't know how many asexual lineages have evolved in the past and then gone extinct without leaving descendants, but we can test if the pattern of evolution of current asexual species conforms to these predictions.

Can we compare the average age of asexual and sexual species?

The length of the branch connecting a species to its nearest relative is often taken as a measure of species age, on the assumption that the branching point marks the speciation event that created the species. We don't know whether asexuality arose at the point of speciation or some time after it split from its sister species, and we also can't tell if there were other more closely related species that have since gone extinct. But as long as the transition to asexuality reduces the longevity of lineages, we might expect that asexual lineages in the phylogeny will tend to be younger than sexual species.[5]

Imagine cutting off all the tips of the tree in Figure 7.13 at the point where each extant species meets another lineage and then lining them up, shortest to longest (Figure 7.14). If asexuality has no effect on tip length, then the asexual species should be randomly scattered through that distribution. But if asexual species tend to be younger, there will be more asexual tips in the short end of the tip distribution than expected by chance. We can test this using a rank sum statistical test.

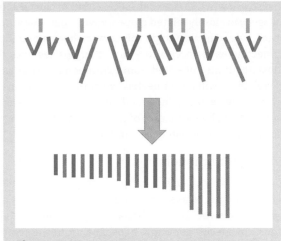

Figure 7.14 The tips of the phylogeny in Figure 7.13, cut at the first node and then reordered by length.

Can we compare how many close relatives asexual and sexual species have?

When we see closely related species that all share a complex trait, the most parsimonious explanation is that they all inherited that trait from their common ancestor. If asexuality increases the extinction rate, we would expect most asexual lineages to give rise to only one or a few species before succumbing to extinction. We can ask how many descendant species each inferred origin of asexuality gives rise to. For simplicity, we could choose to express this as a ratio, as the average number of tips (species with the trait) per origin (inferred ancestral gain of the trait).[4] We can compare two values of tips per origin in **Figure 7.15**.

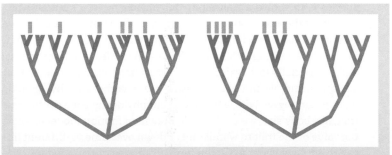

Figure 7.15 The tree on the right has the same number of asexual species as the tree on the left (Fig 7.13), but they arise from fewer evolutionary origins (3.5 species per origin, compared with the more 'tippy' tree with 1.2 species per origin).

We can also ask whether asexual species tend to cluster together on the phylogeny in closely related groups, or whether they are scattered distant relatives. One way to do this is trace lines along the phylogeny connecting each species to every other species with the same trait. Clustered traits, inherited from a common ancestor and shared by close relatives, will be connected by shorter paths than scattered traits which have evolved independently in distant lineages. There are several established measures of phylogenetic clustering we could use, such as the sum of sister clade differences (SSCD).[6]

Generating null models

For tip length, we can use a simple non-parametric test to ask whether the length of asexual tips in the phylogeny is randomly distributed or skewed toward shorter lengths (Figure 7.14). But for the number of species per origin of asexuality, or the scatter across the phylogeny, we need a baseline for comparison. We need to know what kind of values could be generated if asexuality has no effect on phylogenetic distribution. We can generate a null distribution by asking what pattern we would expect to see if asexuality evolves randomly along the phylogeny, having no effect on speciation or extinction rates. One way to do this is to evolve alternative phylogenies, starting with an ancestral lineage that, at any point in time, has a chance of splitting to form two descendant lineages, and also has a chance of changing from sexuality to asexuality or vice versa (we could also add a risk of extinction to the model). This would give us a range of evolutionary outcomes under a model where sex or asex makes no difference to diversification. We then face the challenge of comparing the outcome on our particular tree—with a certain number of asexual and sexual species and a particular tree topology (pattern of branching)—with this set of random trees with a range of outcomes.

An alternative way of generating a null distribution of traits with the same tree topology and the same relative number of each type of species is to evolve a trait along our phylogeny in a random walk. Starting at the ancestral root of the tree, moving towards the tips, we can use a model where the trait will have some probability of arising in a given lineage, then inherited by that lineage's descendants, and occasionally lost again. This is a Brownian motion model of trait evolution, and we can use it to generate hypothetical phylogenetic distributions by evolving a trait along our phylogeny. For a continuous trait, like body size, we can use the Brownian motion model to allow the value of the trait to wander up or down over time in different lineages. But how about a discrete trait, like mode of reproduction? We could model a binary trait that switches between sex and asex with some defined probability, or we could evolve a continuous trait and then set a threshold value above which we call the lineage 'asexual'.

The advantage of the threshold model is that we can set the threshold value to give us the same number of extant asexual species as in our real data.[7] Under this random model, we expect some species to occur in related groups that all inherited the trait from their ancestor.

If we run this random model of evolution again and again, we get a range of possible outcomes given a random model of evolution (this approach of generating different hypothetical outcomes is referred to as a simulation study). From this we can generate a null distribution for the distribution of number of species per origin, and the measures of phylogenetic scatter. Then we can ask if the values we measure from our real data fit within this randomly generated data. If we do, we have no convincing reason to think that there is anything special going on. But if our observed values fall outside the null distribution, we can conclude that this random model doesn't provide a convincing description of the evolution of asexuality in this case.

Testing outcomes of lineage selection

We have come up with some potential measures of tippiness (tip length, number of species per origin, phylogenetic scatter) and null models that tell us the range of values we would expect under a random model of evolution. If the values we get from our real data lie outside the values in the null distribution, we can reject this random model. Does this prove our hypothesis true? No, it just tells us that it doesn't look as if asexuality evolved by a random walk of this kind along this phylogeny.

If we reject the random model, we can ask if the macroevolutionary model of high gain rate plus high extinction rate provides a better explanation of our data. We can fit a range of models to our phylogeny, varying in their speciation and extinction rate, and the rate of gain and loss of asexuality.[4,8] Then we can choose the model that provides the best statistical description of our data. The 'best-fit' model is the one that gives the best predictive power for our observations, so that the difference between our data and the values predicted under the model is minimized. To chose the best-fit model, we rank all our candidate models (in this case, consisting of different values of speciation and extinction rates and rates of gain and loss of the trait) by how much of the variation in the data they explain.

It's important to remember that a model-fitting approach ranks the available models by how well they explain the data. We can reject any model that provides a much poorer fit to the data than alternative models. But while the best-fit model (or models) may be better than the other models, it still might not be an adequate description of the data. In other words, is the best-fit model good enough to explain our data? Would we really have got the patterns we observe if this model was a true description of the evolutionary process? Or is it simply the least worst choice from among a set of unrealistic evolutionary models, none of which is a good description of our data. So, to avoid falling into the trap of accepting a best-fit model that does not fit very well, we need to do some additional testing of model adequacy. We need to ask: could we reasonably expect the chosen model to have produced the data we see? We need to evolve phylogenies under different macroevolutionary models—varying the speciation, extinction, trait gain and loss rates—and see which models could produce data like ours. And we need to make sure that they have the same number of asexual and sexual species as our real dataset, to make sure that it is a fair comparison.[7] If our observed values fall outside the range of possible values generated under a particular model, we reject that model as an inadequate description of our data.

Selfing in Solanaceae

Let us apply this phylogenetic approach to a real group of organisms. Asexuality evolves frequently in plants, often through the loss of genetic self-incompatibility so that an individual's own pollen and ovules can combine to make a functioning zygote (referred to as 'selfing'). It has been suggested that selfing evolves frequently in the Solanaceae (the nightshade family, which includes tomatoes and potatoes: **Figure 7.16**), but leads to a raised extinction rate.[8] Does the phylogenetic pattern support this hypothesis? The asexual tips on the phylogeny of Solanaceae are shorter than would be expected under a model where the trait has no effect on speciation, extinction, and trait loss rates, there are fewer descendant asexual species per origin than expected, and asexuality is more scattered on the phylogeny than expected under a random model.[4] The best-fit macroevolutionary models have a high trait gain and/or a high extinction rate.[4,8] So this looks like good evidence that asexual lineages in Solanaceae evolve often but are short-lived. But when you test for model adequacy, asking if a high extinction rate and/or high loss rate could produce the observed pattern, all models are rejected except for one with high trait transition rates. In other words, the best model for the distribution of asexuals in the Solanaceae is one of frequent gain and loss of selfing. It has been assumed that loss of self-incompatibility is a one-way street, so that once a lineage becomes a selfer, it can't revert to biparental reproduction.[8] So either that conclusion needs to be revised,[9] or a wider range of evolutionary models needs to be tested against the data in order to come up with a more convincing story for patterns of evolution of asexuality in this group.

Questions to ponder

1. Asexuality is considered to represent an 'evolutionary dead-end' because it has short-term advantages but long-term costs. Specialization has also been described as an evolutionary dead-end. Is this another case of the same evolutionary mechanisms?

Figure 7.16 The reproductive mode has been evolutionarily labile in the Solanaceae, so even closely related species can differ in their mode of reproduction. For example, tomatoes (*Solanum lycopersicum*) (a) are capable of selfing, whereas the potato vine (*Solanum jasminoides*) (b) is self-incompatible, and so needs two parents to produce an offspring. However, the potato vine can also reproduce clonally, whereas the tomato cannot.[10]

(a) YuriyK/Shutterstock.com. (b) Photo by C T Johansson via Wikimedia Commons (CC3.0 license).

2. Should we expect to see 'tippy' phylogenetic patterns of inbreeding (such as species with frequent self-fertilization or sibling mating)? Should the same patterns also apply to sexual species with consistently small populations?

3. Why are transitions from sexual reproduction to asexuality common, yet transitions from asexuality to sexual reproduction are considered rare, or even wholly absent in some groups?

Further investigation

In this Case Study we have applied several tests of 'tippiness' to one particular biological group, the Solanaceae, in order to test the hypothesis that asexual lineages arise often but have a short evolutionary lifespan. There are many other groups which contain both asexual and sexual species. See if you can find another dataset and perform the same analyses. You can find the R scripts you need to do these tests at https://cran.r-project.org/web/packages/phylometrics/index.html

References

1. Williams GC (1975) *Sex and Evolution.* Princeton University Press, Princeton, NJ.

2. Maynard Smith J (1978) *The Evolution of Sex.* Cambridge University Press, Cambridge.

3. Schwander T, Crespi BJ (2008) Twigs on the tree of life? Neutral and selective models for integrating macroevolutionary patterns with microevolutionary processes in the analysis of asexuality. *Molecular Ecology* 18(1): 28–42.

4. Bromham L, Hua X, Cardillo M (2016) Detecting macroevolutionary self-destruction from phylogenies. *Systematic Biology* 65(1): 109–27.

5. Neiman M, Meirmans S, Meirmans PG (2009) What can asexual lineage age tell us about the maintenance of sex? *Annals of the New York Academy of Sciences* 1168(1): 185–200.

6. Day E, Hua X, Bromham L (2016) Is specialization an evolutionary dead end? Testing for differences in speciation, extinction and trait transition rates across diverse phylogenies of specialists and generalists. *Journal of Evolutionary Biology* 29(6): 1257–67.

7. Hua X, Bromham L (2015) Phylometrics: an R package for detecting macroevolutionary patterns, estimating errors on phylogenetic metrics, and backward tree simulation. *Methods in Ecology and Evolution* 7(7): 806–10.

8. Goldberg EE, Kohn JR, Lande R, Robertson KA, Smith SA, Igić B (2010) Species selection maintains self-incompatibility. *Science* 330(6003): 493–5.

9. Hanschen ER, Herron MD, Wiens JJ, Nozaki H, Michod RE (2018) Repeated evolution and reversibility of self-fertilization in the volvocine green algae. *Evolution* 72(2): 386–98.

10. Vallejo-Marín M, O'Brien HE (2007) Correlated evolution of self-incompatibility and clonal reproduction in *Solanum* (Solanaceae). *New Phytologist* 173(2): 415–21.

Why are most species small?

8

Roadmap

Why study body size and diversity?

Body size is the most obvious and one of the most easily measured features of an organism's biology. It is also one of the most important; body size is strongly connected with the strategies that an organism can employ to obtain resources from the environment and use them for growth and reproduction. What is fascinating is the idea that a biological trait that figures so prominently in the day-to-day lives of individual organisms could have a profound influence on large-scale patterns of biodiversity. One of the clearest macroecological patterns is the preponderance of small compared with large species. Yet we still do not fully understand why there are so many species of small size and so few of large size. Explanations range from macroevolutionary processes that play out over large scales of space and time, to the microevolutionary optimization of body size within populations by natural selection.

What are the main points?

- An organism's body size places severe constraints on its fundamental architecture and physiology, and therefore the way it obtains energy and resources from the environment, assimilates energy into body tissue, and uses it for growth and reproduction.

- The frequency distribution (histogram) of body sizes for large groups of species is often right-skewed on a logarithmic scale. This indicates a strong bias in species numbers towards species of smaller size. But it also requires us to explain the low species richness at the very smallest sizes.

- There are many hypotheses for this uneven distribution of body size, some focusing on the relative lack of small species, and others on the low numbers of both very small and very large species.

Photograph: weaver ants. frank60/Shutterstock.com.

What techniques are covered?

- Using life history optimization models to determine optimum body sizes for different species.
- Using phylogenetic methods to test for associations between body size and rates of diversification.

What case studies will be included?

- Testing for a link between body size and diversification rates in insects.

"An insect going for a drink is in as great danger as a man leaning out over a precipice in search of food. If it once falls into the grip of the surface tension of the water—that is to say, gets wet—it is likely to remain so until it drowns. A few insects, such as water-beetles, contrive to be unwettable; the majority keep well away from their drink by means of a long proboscis."

J.B.S. Haldane (1926) On being the right size. *Harper's Magazine.*

There is no more elegant and entertaining synopsis of the mechanical, physiological, and energetic implications of body size than the 1926 essay 'On being the right size', by British biologist J.B.S. Haldane.[1] Reading Haldane's essay, you are immediately struck by the profound differences in the way that an ant and a buffalo interact with their physical environments. Where surface tension presents danger for an insect, gravity does the same for a 500kg mammal. The insect has no need for a closed circulatory system, but a large mammal requires a complex branching network of vessels to convey blood under pressure to every corner of its body. The same is true for plants: a tree requires a tension-driven hydraulic system to lift water through vessels from its roots to its uppermost branches, but a liverwort does not. The consequences of body size for an organism's architecture are clear. The organism's architecture and physiology, in turn, dictate the manner and efficiency with which it obtains energy from the environment and uses this energy to grow and reproduce. Less well understood are the evolutionary mechanisms that connect body size with patterns of biodiversity. The simple question that forms the title of this chapter has proved surprisingly resistant to an easy answer. In attempting to explain why there are so many more species of small body size than of large body size, we must consider some of the processes that lie at the heart of macroevolution and macroecology, such as speciation, extinction, selection, adaptation, and niche theory. In this chapter we will explore how researchers have used creative ways of proposing good questions about the link between body size and species richness, and powerful and elegant ways of attempting to answer them.

The physiology and ecology of body size

Haldane's 1926 essay is full of pithy insights into the implications of body size. He noted, for example, that a tenfold expansion in the linear dimensions of an organism in every direction (length, width, and height) leads to a thousandfold increase in weight. To power its muscles with the same efficiency as a smaller animal, a tenfold larger animal would need a thousand times more food and oxygen, and excrete a thousand times more waste. To compensate, larger animals have developed complex gut structures that vastly increase the size of the surface through which food is absorbed into the bloodstream, and complex respiratory structures, such as lungs or gills (or leaves in plants), with large surfaces for gas exchange. So, while a larger organism requires more energy to function than a smaller one, it assimilates that energy into body tissue more efficiently. Not only that, but larger organisms are also better at energy conservation. The energy an organism loses to its environment as heat—a by-product of metabolism—is proportional to its surface area. Because a larger organism has less surface area in proportion to its weight than a smaller one (5000 mice weigh as much as one person, but have the surface area of 17 people), the larger organism loses comparatively less heat than the smaller one. Overall, this means that mass-specific energy use (i.e. energy use per gram of body mass) decreases as the total body mass of an organism increases.

If efficiency of energy use increases with size, why are there still any small creatures left at all? Why is there not a strong selective pressure on all populations of organisms to evolve towards larger sizes? One answer is that there are also costs involved in becoming larger. Increased size demands an increase in the proportion of an organism's energy budget that needs to be allocated to specialized structures for support and energy transport. Unless a lineage is capable of evolving appropriate solutions to these problems, this imposes constraints on the maximum sizes possible for different kinds of organisms. As Haldane put it, 'If the insects had hit on a plan for driving air through their tissues instead of letting it soak in, they might well have become as large as lobsters, though other considerations would have prevented them from becoming as large as man'. Trade-offs between an organism's structural strength and energy use mean that 'for every type of animal there is an optimum size'. Later in this chapter we will return to the idea of body size optimization, and we will examine whether this is, or isn't, of importance when it comes to explaining species richness.

What do we mean by 'body size'?

It will be apparent from this brief overview of Haldane's essay that when we speak of an organism's 'body size' there are a few different things that this could mean. We may be talking about its mass (weight) in grams or kilograms, or about its linear dimensions in millimetres or centimetres. The size of a plant is usually described by its maximum height above the ground. Less often, we refer to an organism's volume, for example in litres or cubic centimetres, or for some two-dimensional creatures, its area (square millimetres or square centimetres). All of these measures of an organism's size are proportional to one another, but (as we have seen) they are proportional in non-linear ways that can be interesting and informative. Throughout this chapter we will refer simply to 'body size' when the concept we are talking about is general, applying to all measures of body size. We will refer to a particular measure of body size when the concept we are discussing requires a more specific definition.

From viruses to redwoods: 24 orders of magnitude

Virus particles are the smallest organisms we know of. The mass of a single HIV-1 virus particle is about one femtogram, or 10^{-18}kg. The largest individual organisms are giant redwood trees, which reach in excess of 100m in height and over 1000 tons in mass. In between these extremes is the whole array of life on earth, spanning 24 orders of magnitude in body mass and representing at least 3.5 billion years of evolutionary history. A microscopic particle and a 1000-ton tree interact with the world on scales that are inconceivably different. The life cycle of a virus plays out within a few minutes, largely within the cells of host organisms. A giant redwood can live for thousands of years, a period long enough for the climate into which it emerged as a seedling to change around it. The difference in the net energy flux between each kind of organism and its respective environment is vast.

Even within far more restricted domains of life, the variation in sizes of individual organisms can be massive. The reptiles are a good example. Adult individuals of the smallest reptile species (dwarf geckoes and dwarf chameleons) can fit comfortably on the tip of a person's finger, but an entire person may fit comfortably inside the belly of the largest reptile, the saltwater crocodile. One particularly size-diverse group of reptiles, the family Varanidae (monitors), includes species ranging across 3.5 orders of magnitude in body mass, from the Dampier Peninsula goanna, at 16g and 115mm long, up to the Komodo dragon, at 90kg and 3m long (**Figure 8.1**). In the relatively recent past (up to 50,000 years ago), the size range of varanids was even more impressive, with the now extinct Megalania probably reaching a mass of at least 500kg and a length of 5m or more.

Figure 8.1 The smallest known monitor lizard, the Dampier Peninsula goanna (*Varanus sparnus*), shown roughly to scale with an Indonesian 50 rupiah coin. The coin depicts the largest living monitor, the Komodo dragon (*Varanus komodoensis*), which can reach over 3m in length.

Photo of lizard: Ryan Ellis. Coin: Fluorite, My Own Photo Works.

The simple fact of such great variation in the sizes of fairly closely related species tells us how effectively body size encapsulates the fundamental processes of energy acquisition and reproduction that determine the fitness of organisms. Whether body size itself is the target of natural selection, or body size covaries with other physiological features that are subject to selection, it is clear that for every way of making a living, there is a suitable size. For example, most geckos have adhesive toe pads that allow them to forage for insects in otherwise inaccessible places (such as vertical or overhanging parts of vegetation or rock surfaces), providing access to energy sources not available to other lizards. However, this strategy becomes impossible once the force of gravity exceeds the adhesive power of the toe pads, so geckos with adhesive toe pads tend to weigh less than 100g. Their small dimensions also give them access to feeding and hiding places such as small rock crevices or beneath loose tree bark. Larger lizards use other strategies to acquire energy. For example, predators of vertebrate animals, such as the majority of the monitor species, must be quite large in order to overpower their prey, but their body mass is proportional to that of their main prey species. Smaller monitors eat invertebrates and small mammals such as mice, while the largest monitors (such as the perentie) may take larger mammals such as rabbits or wallabies. The match between body size and ecological niche forms the basis for many of the hypotheses explaining the great diversity of small-sized organisms.

> ### Key points
>
> - An organism's body size dictates its fundamental architecture and physiology. Larger organisms need to invest energy in support structures such as skeletons and transport networks such as blood vessels, but are more efficient at energy assimilation and conservation.
>
> - The organism's physiology determines the way it obtains energy and resources from the environment, assimilates energy into body tissue and uses it for growth and reproduction.
>
> - Body size also determines more directly the strategies an organism can use to obtain energy as well as the ecological niche it can occupy. Different body sizes best suit different ecological strategies.
>
> - Living creatures span a massive range of body sizes. Even some groups of closely related species have evolved into a range of body sizes spanning several orders of magnitude.

All creatures great and small: body size and diversity

Body size and species numbers in Metazoan families

There are more kinds of organisms of small size than there are of larger size. This general observation applies to the diversity of living things in its entirety, but also to many smaller taxonomic subsets of living things. Indeed, higher diversity of smaller creatures is one of the major generalizations of macroecology—a fundamental pattern of biodiversity. But the relationship between body size and species numbers is not always simple, linear, or consistent. In some cases, exceptions to the general pattern make the job of explaining it

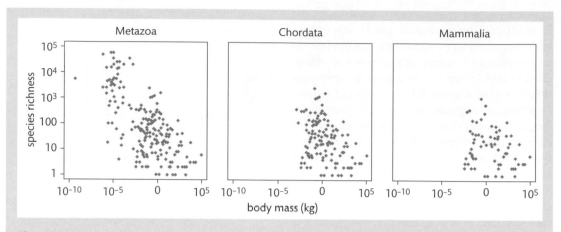

Figure 8.2 The mean body mass of species within 198 families of animals, including families of insects, mammals, birds, fish, molluscs, and other classes, plotted against their species richness. When families of all phyla and classes included in this dataset are plotted (left panel), a clear negative relationship between body size and species richness is visible. This relationship becomes less clear when we plot subsets of the data for the phylum Chordata (middle panel) and the class Mammalia (right panel).

more difficult, but in other cases the exceptions and inconsistencies provide useful additional information that can prove helpful. In this section we will take a broad overview of the pattern, to try to develop a sense of how general the relationship between body size and species richness is, and in what ways the relationship varies.

The left-hand panel in **Figure 8.2** shows the mean body mass of 198 families from various phyla of Metazoa (animals) plotted against their estimated species richness. These are not all the metazoan families, just a selection of ones for which we have reasonably confident estimates of species number.[2] Across this broad sweep of the world's biodiversity, the negative association between body size and species richness is quite obvious, notwithstanding the large variation in species richness among families of any given body size.

The centre and right-hand panels in Figure 8.2 are subsets of the left-hand; they show the same pattern for one phylum (Chordata) and one class (Mammalia). Among families within the Chordata and the Mammalia, the negative relationship between mean body size and species richness is less

clear. This is because, within each of these groups, there is a large spread of species richness values among families with small mean body size, while families with large mean body size are all species poor. Based on the plots for chordate families and mammal families, it would be difficult to draw the general conclusion that the mean body size of species in a family closely predicts the family's species richness. But we can also see from all three plots that there are simply fewer families towards the upper end of the body-size scale—and, presumably, fewer species in total, across all families, at the upper end.

Frequency distributions of body size

An alternative way of visualizing the pattern of species richness variation with body size is to look at the frequency distribution (histogram) of body size, where the frequencies are given as the total number of species within each size class, ignoring their division into higher taxonomic groups. To do this, we need data on the body sizes of all species in the groups we are interested in, not just the mean body sizes within families or other subtaxa. These kinds of datasets have been compiled and

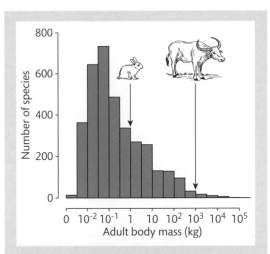

Figure 8.3 Frequency distribution (histogram) of body mass for the world's mammal species. In this graph, the body mass axis is presented on a logarithmic scale for easier visualization. This is typical practice when dealing with continuously varying biological traits, but it does make the comparison of different body-size classes a bit less intuitive. For reference, the mean adult body masses of a rabbit (*Oryctolagus cuniculus*) and a water buffalo (*Bubalus bubalis*) are indicated.

are easily accessible only for a few large taxonomic groups, mostly vertebrate animals. One such freely available dataset is PanTHERIA,[3] which provides summary values of body mass and other traits for the great majority of mammal species. We can use data from PanTHERIA to construct a body-size frequency distribution for the world's mammal species (**Figure 8.3**).

Now that we have dispensed with a particular taxonomic rank (family) as an arbitrary unit of observation, it becomes more clear that there are many more mammal species of small size (say, below 1kg, or about the size of a rabbit) than there are of larger size. In fact, the modal body-size class (about 32–100g) is quite close to the lower end of the size axis, and the number of species declines reasonably steadily in each of the successively larger size classes. At the upper end of the body-size scale, there is not much diversity at all: there are only 173 mammal species (3.2%) with a mean adult mass above 100kg.

There is something else very important to note about the body size distribution of mammals shown in Figure 8.3. Although the modal size class is close to the lower end of the body-size scale, the very smallest mammals are not the most diverse; there is a steep decline in species richness from the mode (32–100g) down to the smallest size class (<3.5g). This kind of unimodal log-right-skewed frequency distribution of body sizes is not a peculiarity of mammals. In fact, a very similar kind of distribution is seen for many other taxonomic groups, including birds, reptiles, amphibians, and some insects. Body-size distributions of this shape are also seen in many geographically defined assemblages of species, such as Australian birds or North American terrestrial mammals.

The body-size distribution for some other large taxonomic groups is less strongly skewed, or not skewed at all, on a logarithmic scale. For example, **Figure 8.4** (left panel) shows the distribution of log-transformed maximum body lengths for 2188 species of fish, obtained from another freely available database, FishBase.[4]

Unlike the distribution of mammal log(body mass) values, the distribution of fish log(body length) values is only very slightly right-skewed; indeed, it is almost symmetrical (this is not simply because of the different measures of body size—the plot looks the same if we use the cube of body length to approximate body mass). But if we plot the histogram on an untransformed linear body-length axis (Figure 8.4, right panel), we see that there are still far more fish species of small length than there are of greater length. The log-transformed distribution needs to be interpreted carefully, because at first sight it can look as if the smaller size classes are no more species-rich than the larger ones. So why do we log-transform the body size data at all, and present the patterns in a way that could mislead? We do this partly because log-transformation is often what we do before analysing the data; highly skewed variables such as body size can violate

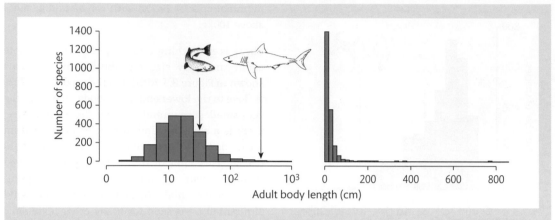

Figure 8.4 Frequency distribution of maximum adult body lengths of 2188 fish species. When the body-length axis is logarithmically transformed (left panel), the distribution is more symmetrical than that of mammals. However, when the body-length axis is shown on a linear scale (right panel), it is clear that there are many more species of small size than of large size. Species indicated for reference are a rainbow trout (*Oncorhynchus mykiss*) and a great white shark (*Carcharodon carcharias*).

the assumptions of many parametric statistical tests (such as normally distributed errors and homogeneous variances). But log-transformation also makes it easier to visualize patterns on plots such as Figures 8.2–8.4, because it spreads the values more evenly along the x-axis. It also makes biological sense to present body-size data in this way: a 100g difference in body mass has profound ecological significance if we are comparing two animals that weigh less than 200g, but is trivial if we compare two animals that weigh more than 100kg. It could, in fact, be argued that a log-normal distribution such as the one for fish lengths indicates the *absence* of any body-size bias in species richness, because the proportional change in body size from one class to the next remains constant along the body-size axis.

Key points

- In many groups of organisms, the frequency distribution (histogram) of body sizes among species shows a characteristic right-skewed shape, even when body size is plotted on a logarithmic scale. This indicates a great preponderance of small species over large ones.

- The log-right-skewed body-size distribution gives us two phenomena to explain: the decline in species numbers as body size increases towards the largest size class, and the rapid decline in species numbers towards the smallest size class.

Explaining the diversity of small creatures

We saw in the previous section that, in general, there are more species with a small body size than with a large body size. So one of the problems that

we wish to solve as biodiversity scientists is the question of why, for example, there are around 1700 mammal species weighing less than 0.5kg, but only about 170 species that weigh more than 100kg. An additional problem is evident in the frequency distributions of body size shown in Figures 8.3 and 8.4: why does the number of species also drop off at the very lowest end of the body-size scale? These two patterns—the small number of large species and the small number of very small species—could be viewed as separate questions, each with their own independent answer. But the relationship between body size and species richness could also be viewed as a single phenomenon, perhaps with a single answer: one process that results in an accumulation of species at the modal size. In the remainder of this chapter we will explore some of the explanations that have been offered for the connections between body size and species richness. We will see that some hypotheses focus on the general scarcity of large-sized species and some on the scarcity of the very smallest species, while others have presented a single model to account for both these patterns.

Habitat specificity, habitat diversity, and body size

❝ *It would seem intuitively that the environment does not provide adequate room for a very large number of species of large animals while there is much more room for an abundance of smaller species. Moreover it is quite obvious in many cases that the large species roam about over a number of biotopes specific for smaller species. A large ungulate may require a water hole, a grazing area and some degree of cover; the wet marginal area of the water hole, the open grazing area and the cover might provide specific biotopes for three species of rodents.* ❞
G.E. Hutchinson and R.H. MacArthur (1959) *American Naturalist* 93: 117–25.

One of the earliest attempts to provide a theoretical framework for the link between body size and species richness in animals was in a 1959 paper by the prominent ecologists G.E. Hutchinson and Robert MacArthur.[5] Their simple and intuitive hypothesis is summed up succinctly in the above quotation from the introduction to their paper: larger animals need more room, but also utilize a greater variety of habitat types (biotopes) than

do smaller animals. This verbal summary of the hypothesis seems to offer an explanation for the decline in species numbers with increasing size, but not for the paucity of very small species.

However, Hutchinson and MacArthur developed this idea into a simple mathematical theory that explains not only the geometric decline in species numbers in successive body-size classes above the modal size, but also the low number of species in the smallest body size classes. They assumed that the environment of a given region was composed of a set of randomly arranged 'mosaic elements' corresponding to spatial chunks of distinct habitat; for example, the waterhole, grazing area, and area of cover they mention in their introduction. Suppose that we select a set of n contiguous mosaic elements. Hutchinson and MacArthur derived the probability p that, for each value of n, the selection will contain any given number (up to n) of different types of mosaic elements, and the probability p_c that the selection will contain a given number of mosaic elements that form a qualitatively unique set. Working up from $n = 1$, these probabilities start off small, rise rapidly, and then fall off steadily towards progressively higher values of n. In fact, the shape of the plot of these probabilities against n looks rather similar to the log-right-skewed frequency distribution of body sizes. From this point, it is a fairly small conceptual jump to assume that (1) n is proportional to the body sizes of species occupying the region, and (2) a unique set of contiguous mosaic elements represents a unique ecological niche, able to be occupied by one distinct species. With an adjustment to relate n to the body lengths of individual animals in a more realistic fashion, the probability p_c represents the number of such unique niches available to species of each body size, and hence the number of species that can occur (**Figure 8.5**).

When the Hutchinson–MacArthur theory was published in the late 1950s, ecology was developing beyond its natural history origins and maturing into a quantitative model-based science, thanks largely to the work of Hutchinson and MacArthur themselves. Their body size–diversity theory fell squarely into the emerging paradigm of this time, which emphasized ecological determinism in

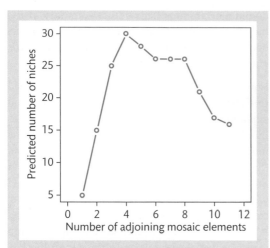

Figure 8.5 The number of unique niches in a landscape as a function of the number of adjoining 'mosaic elements', predicted by the Hutchinson–MacArthur model. The number of adjoining mosaic elements is assumed to be proportional to the body sizes of different species, and the number of niches proportional to the number of species. This right-skewed curve is remarkably similar in shape to the frequency distribution of body sizes in many taxa; compare with the body size distribution of mammals in Figure 8.3.

patterns of biodiversity, i.e. the patterns we see in a region are the inevitable outcome of predictable ecological processes playing out until equilibrium diversity is reached, given the particular environment of that region. Although this remains a powerful way of thinking about diversity, supported by a large body of theory, it is no longer the prevailing ecological 'paradigm'. The range of processes entertained by ecologists nowadays has expanded to include non-equilibrium dynamics, and historical, macroevolutionary, and biogeographic phenomena. Later in this chapter we will explore more historical hypotheses for the species richness of small creatures, but before we do we will look at two important extensions of Hutchinson and MacArthur's pioneering theory: the fractal dimension of habitat surfaces, and the evolutionary optimization of body sizes.

The fractal dimension of vegetation: more space for smaller creatures

The Hutchinson–MacArthur theory explained the relationship between body size and species richness in terms of the number of different kinds of habitat available for different kinds of organisms to occupy. As well as habitat diversity, the amount of physical space (area) each habitat element represents is also likely to be important for determining the number of species. A given area can support more individuals of smaller species than larger species, and more individuals of a given size can occupy a larger area than a smaller one. So we need to think about the relationships between habitat area, number of individual organisms, and species richness. What we need to know is whether there are general rules that govern (1) the way habitat area is apportioned among individual organisms, and (2) the way individuals are apportioned among species.

Animals forage or hunt in the space provided by habitat surfaces; plants and other sessile organisms use habitat surfaces as substrates. Some habitat features, such as trees, are three-dimensional, but the area of the foraging, hunting, or substrate surface they offer still reduces to the sum of the two-dimensional surfaces of their branches, twigs, and leaves. In other words, as a habitat for other organisms, what matters is not the volume of a tree (or a rock, coral reef, or soil particle), but its surface area. It seems intuitively obvious that there should be a simple linear relationship between habitat surface area and number of individual organisms that habitat can support, and that this relationship should be invariant with respect to body size.

However, interesting complexities arise from the interaction of body size with the measurement of habitat surface area itself. This is because the amount of habitat surface varies depending on the spatial grain (or resolution) of measurement. The degree to which measurement resolution changes the measured value of a dimension, such as the area of a surface or the length of an edge, is known as the *fractal dimension*. A good demonstration of the fractal dimension is the length of a coastline on a map (usually of Britain, following

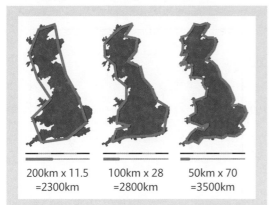

Figure 8.6 The coastline of Britain is the classic example used to illustrate the concept of fractals. The shorter the stick used to measure the length of the coast, the longer the coast becomes. The relevance for biodiversity is that habitats also have a fractal dimension, so that smaller species have a greater area of habitat surface available for foraging.

200km x 11.5 =2300km   100km x 28 =2800km   50km x 70 =3500km

the original example given by mathematician Benoit Mandelbrot). If we use a 200km-long measuring stick to trace around the British coast and thereby measure its length, the total length we get is 2300km; a 100km-long stick follows more of the indentations in the coastline and the total length increases to 2800km; a 50km-long stick gives us 3500km (**Figure 8.6**).

To put this in an ecological context, earlier in this chapter we saw how smaller organisms are able to access habitat surfaces that are not available to larger ones. A large adult Komodo dragon can't climb trees, and is limited to the foraging space offered by the ground surface. A large goanna (somewhat smaller than a Komodo dragon) can climb, extending its foraging surface to the trunk, limbs, and larger branches of trees. A small skink or gecko can additionally access the smaller twigs, and can forage underneath loose pieces of bark as well as on top of them. A small beetle, mite, or collembolan can occupy both upper and lower surfaces of the tree's leaves, as well as the smaller cracks and fissures in the bark that even a skink can't get into. And so on. In effect, smaller creatures have a much greater surface area of habitat available to them than larger ones (**Figure 8.7**). The fractal dimension of a habitat can be estimated and accounted for to modify the expected relationship between the size of organisms and the number of individuals a given area can support.

This is what David Morse and colleagues did for the first time in the 1980s.[6] They estimated the fractal dimension of vegetation using photographs of plants taken from different angles. They placed the photos on a grid, counted the number of grid cells intersected by the outline of the plant, and repeated this for grids of different sizes. Across a range of

(a)   (b)   (c)

Figure 8.7 The surface area available for foraging depends on an organism's body size. The Komodo dragon (a) can't climb trees, and is limited to the area provided by the ground surface. The smaller goanna (b) can climb trees, but can only access the trunk and larger branches. The tiny gecko (c) can forage on smaller twigs and leaves. The fractal dimension of habitat surfaces describes the way the size of surface increases as the scale of measurement declines.

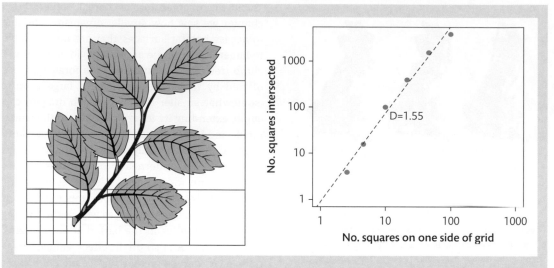

Figure 8.8 The method used in 1985 by David Morse and colleagues to estimate the fractal dimension of vegetation sampled from the University of York. By overlying photos of branches on grids of varying resolutions (left), they counted the number of grid cells intersected by the outline of the branch. Plotted against the grid resolution on a double-logarithmic scale, this yields a straight line with slope *D*, the fractal dimension (right). Nowadays the same thing could no doubt be done far more quickly, easily, and accurately using computers, although it might not be as much fun[6].

From Morse, D.R., Lawton, J.H., Dodson, M.M., & Williamson, M.H., Fractal dimension of vegetation and the distribution of arthropod body lengths, *Nature* 314, 731–3 (1985). © 1985. Reprinted by permission from Springer Nature.

plant species sampled from the grounds of the University of York, they estimated an average fractal dimension *D* of around 1.5—this is the slope of the relationship between grid resolution and the number of grid squares intersected, on a log-log scale (**Figure 8.8**). Substituting animal body length (*L*) for grid resolution, Morse and colleagues estimated that for *D* = 1.5, the habitat surface area accessible to an animal 3mm long is around an order of magnitude larger than that for an animal 30mm long. Next, they calculated the expected increase in number of individuals per unit area (population density) with decreasing body length, without taking fractal dimension of habitat into account. Based on some generalizations about the way body weight (~*L*³) scales with individual resource use and population density, this gives us a 178-fold increase in density of individuals with every 10-fold decrease in body length. Finally, Morse and colleagues combined this with the increased habitat surface area for smaller organisms. They arrived at an estimated increase of a factor of 560–1780 in the density of individual animals with an order of magnitude decrease in body

length. In other words, animals 3mm long ought to be hundreds, or even thousands, of times more abundant than animals 30mm long.

Morse and colleagues stopped short of extending their analysis to numbers of species, but their discovery had obvious, and intriguing, implications for biodiversity. If each species requires a certain minimum population size to remain viable in the long term, a greater number of individuals can, potentially, be divided among a greater number of species. If there are regular, predictable rules that govern the relationship between numbers of individuals and numbers of species, this could help to explain the relationship between body size and species number. The findings by Morse and colleagues led to the exciting suggestion that in complex fractal habitats, the number of small species may have been enormously underestimated. For empirical evidence, attention naturally turned to tropical forests, where new sampling methods such as canopy fogging were already beginning to reveal previously undreamt-of levels of arthropod diversity.

In a follow-up study, Morse and colleagues analysed data on the sizes, abundances, and species numbers of beetles sampled by canopy fogging of tropical rainforest trees in Borneo.[7] In these samples, the increase in numbers of individuals with decreasing body size was even more dramatic than predicted by their earlier study. An order of magnitude decrease in body length was associated with a 10,000-fold increase in the number of individual beetles. It was also apparent from the Bornean beetle data that numbers of species and numbers of individuals both reach a maximum at a similar body length—around 2–4mm. This suggests that there is a correspondence between total numbers of individual animals and the number of species. But how general is this association, and could a general species–individuals relationship be used to predict species numbers from the area of habitat available for organisms of different sizes?

Some of the first clear evidence for a general species–individuals relationship came from a very different kind of ecosystem: the grasslands of Minnesota, USA. In a study published in 1996, Evan Siemann and colleagues showed that both species richness and the number of individuals of grassland insects reached a maximum at a coincident intermediate body size.[8] These data supported a power-law relationship between species number S and number of individuals I across body-size classes: $S = I^{0.5}$. This relation leads to the prediction that a size class with 10 times more individuals will have 3.2 times more species. This relation appeared to be independent of body size itself, and to hold within different insect orders spanning a wide range of body sizes. This led the authors to suggest that $S = I^{0.5}$ could be a general rule across assemblages of species. Since then, similar findings of coincident maxima in numbers of species and numbers of individuals at intermediate body sizes have been made for a range of other taxa, including fish and marine gastropods. Most of these studies support a body-size-invariant species–individuals relationship, with exponents that are sometimes, but not always, similar to 0.5. However, the body sizes at which numbers of species and numbers of individuals reach their maxima tend to differ for each taxon, even among orders of insects.

We shouldn't really be surprised by this; it simply suggests that the trade-offs between the various ecological, structural, and energetic constraints that influence body size are different for groups of organisms with different evolutionary histories.

> **Key points**
>
> - One explanation for the greater species richness of small organisms is that they specialize on finer subdivisions of habitat. In this way a given area can support a larger number of unique combinations of habitat, providing more niches for smaller species.
>
> - Habitat surfaces provide the foraging or substrate area for species. The fractal nature of habitat surfaces means that the habitat area available to small species is far greater than that for large species. A 10-fold size reduction can mean an increase in the number of individual organisms by a factor of several hundred.
>
> - The relationship between numbers of individuals and numbers of species seems to be very general across many different taxa. In principle, this means that the number of species a habitat can support should be predictable from the average body sizes.

Where do all the small species come from?

We have now seen how the population density (and, by implication, the total number of individual organisms) increases as the body size of organisms decreases, and does so more sharply than we might expect because of the fractal nature of habitat surfaces. We have also seen how the number of individuals predicts the number of species in what looks like a reasonably consistent fashion across different taxonomic groups. If the habitat surface available to organisms around 1mm long allows them to be over a thousand times more abundant than organisms 10mm long, the relation $S = I^{0.5}$ predicts that the 1mm organisms should be at least 32 times more species-rich than the 10mm organisms. Now we have a theory of body size and species richness: there are more species of smaller creatures than larger ones because, within any given region, there is more habitat space available to smaller creatures, allowing them to become more abundant.

But this theory is incomplete, because it leaves two important questions unattended to. First, it doesn't explain the low diversity of the smallest species, and thus the unimodal log-right-skewed shape of body size frequency distributions. If habitats are fractal, then habitat space should keep increasing right down to the smallest scales of measurement, and so population densities should keep increasing right down to the smallest species. If this is the case, then why are the smallest body-size classes not the most species-rich? Second, it doesn't explain where all these small species *come from*. For a small-scale community of species (e.g. beetles in the canopy of a rainforest tree) we can imagine that the numerous niches available to small species are filled by species that already inhabit the broader surrounding region, which move in and colonize the tree. But where do *those* species come from? We can keep scaling up the source pool for colonization, but at the continental and even global scales, there are still more small species than large ones. At this point, we need to start thinking past colonization as a source of species to fill our niches, and consider evolutionary processes—both microevolution and macroevolution.

What do you think?

Up to this point in this chapter, the hypotheses for the greater diversity of small compared with large species could all be considered *equilibrium* explanations. The basic assumption is that whatever the mechanism by which small species become so diverse, this mechanism has already played out to its inevitable conclusion wherever we happen to look.

How reasonable is this way of thinking about biodiversity? Can you imagine any scenarios in which the biodiversity we observe is best explained by *non-equilibrium* dynamics (i.e. all possible niches have not yet been filled, and there are fewer species in an area than the number that could potentially be supported)?

Functional trade-offs and body-size optimization

Are there interspecific optimum body sizes?

At the beginning of this chapter, we looked briefly at J.B.S. Haldane's intimation that there is an optimum size for every kind of organism. Within the constraint of an organism's basic architecture—the legacy of its evolutionary history—there will be a particular body size at which the right balance is struck between energy intake, assimilation, reproduction, and other considerations so as to maximize the organism's evolutionary fitness. We would expect natural selection to cause the average body size of individuals in a population to evolve towards the optimum body size for that population. If there is a body size optimum which applies across multiple species, it is quite easy to explain the modal shape of the interspecific body size distribution. We would expect to find more species with average body sizes close to the optimum value because a species with average body size removed from the optimum is likely to be under stronger selective pressure to evolve towards the optimum—or otherwise go extinct. It follows that if the optimum body size for a whole assemblage of species lies partway along the body-size axis, then this is where the modal species richness should be, with fewer species of smaller and larger size.

So, to assess the idea that interspecific body size distributions are the result of optimization for maximum fitness, perhaps the first key question we need to answer is at what phylogenetic level do body size optima exist—within species, or across species, or perhaps within clades of related species or multispecies regional assemblages? One way of testing if there are multispecies body size optima is to ask what kind of patterns we would expect to see if this were the case, and then test for those patterns in comparative analyses. One prediction is that the components of life history that determine fitness should tend to scale positively with body size for species below the proposed optimum body size, but negatively for species above the optimum (or the other way round for some life history variables). Scaling of variables with body size is called allometry. In 1997, Kate Jones and Andy Purvis used data for hundreds of species of bats to test the interspecific allometry of a range of life history variables, such as litter size, weaning age, and maximum longevity.[9] Using phylogenetically independent contrasts to account for the non-independence of the data due to shared evolutionary history (see later in this chapter, and Chapter 9), Jones and Purvis found that all variables scaled monotonically with body size (either a continuous increase or a continuous decrease). There was no evidence for a change in the direction of any of the allometric relationships at any particular body size.

We should probably not be surprised that Jones and Purvis's analysis for bats did not detect the signature of an interspecific body-size optimum. The idea of a single optimum body size for a wide range of species seems counterintuitive and unrealistic. Even within many closely related groups of species, the range of environments occupied and ecological strategies employed by different species can be large. We have already seen in this chapter how ecologically varied the different species of the monitor lizard family Varanidae are (in fact, all living members of this family are in one genus, *Varanus*). The massive bulk of the Komodo dragon (*Varanus komodoensis*) has no doubt evolved by natural selection because larger individuals are more capable predators of small and medium-sized vertebrates. The diminutive Dampier Peninsula goanna (*Varanus sparnus*) makes a living consuming invertebrates and burrowing under rocks. Each species has a body size that appears well-suited to its particular way of life, and it seems hard to accept that the average body size of each of these species is wildly suboptimal. It is probably more likely that body size is independently tuned by

natural selection within the far more ecologically uniform groups of individuals represented by a species population. In other words, it seems more reasonable that every species should have its own independent optimum body size.

Optimum body sizes within species

If each species has its own optimum size, how can this tell us anything about the interspecific patterns of body size and species richness? To answer this question, we will turn to a model devised by Jan Kozłowski and January Weiner.[10] This model predicted optimum body sizes *within* species based on the relative amounts of energy allocated to the growth of organisms and to reproduction. In all species of animals, individuals grow from the juvenile stage to maturity. In some species, individuals keep growing after they become mature (indeterminate growth), while in others, growth stops when maturity is reached and they are able to start reproducing (determinate growth).

For animals with determinate growth, when is the right point for individuals to stop growing and start reproducing? In the Kozłowski–Weiner model this depends on two key processes. The first is the production rate, the difference between the rates of assimilation (incorporation of nutrients from food sources into the body's cells) and respiration (the cellular processes that release energy from those nutrients). The second is the mortality rate—the expected length of survival after maturity is reached. The rates of assimilation, respiration, and mortality are dependent on body size, and their variation with body size can be described by simple power functions (which are straight lines on a log-log scale). This means that the timing of the switch in the allocation of energy from growth to reproduction will determine the organism's adult body size. Natural selection optimizes this timing, and thereby the organism's adult body size, to maximize the organism's lifetime production of offspring. The allocation of, for example, one calorie of energy to growth represents a sound investment in lifetime fitness if it results in an increase in the organism's future reproductive output by more than one calorie's worth (taking into account mortality rate—how long it is likely to live). If not,

Figure 8.9 Small mammal species such as the meadow vole (*Microtus pennsylvanicus*) tend to live fast and die young. The meadow vole reaches sexual maturity at only 31 days of age and has a maximum lifespan of less than a year. Its strategy to maximize lifetime fitness by an early shift of energy allocation from growth to reproduction means that it attains an adult body mass of only 40g. In contrast, the dugong (*Dugong dugon*) takes over 10 years to reach sexual maturity, can live for over 50 years, and reaches an adult body mass of 300kg.

Vole photo by Daderot (CCO 1.0 licens). Dugong photo by Christian Haugen via Flickr (CC 2.0 license).

lifetime fitness will be maximized by starting to reproduce sooner (**Figure 8.9**).

The Kozłowski–Weiner model assumes that the allometric relationships for assimilation, respiration, and mortality are different for every species, for reasons that we discussed above: different species occupy different environments and have different strategies for obtaining energy from the environment. This means that the optimum body size will be different for every species. The unexpected consequence of this intraspecific optimization is the way that these species-level patterns combine to produce an emerging interspecific pattern. Even though the parameters of the model (the slopes and intercepts of the intraspecific life history allometries) are normally distributed, the distribution of optimal body sizes among species that emerges is log-right-skewed. According to the Kozłowski–Weiner model, then, the modal body size in the interspecific distribution, and the shape of the distribution, have no functional significance themselves—they are simply the mathematical consequence of the independent optimization of body sizes for different species with different ecologies.

Do smaller creatures diversify faster than larger ones?

We began our exploration of the great diversity of small creatures by looking at why there is so much more habitat space available for organisms of smaller size. But explaining why more species of small size are able to occupy a given area does not answer the question of where the small species come from in the first place. Colonization from surrounding regions is one answer, but this can only get us so far before we need to start thinking about evolutionary mechanisms. In the previous section, we examined one such mechanism—the evolutionary optimization of body size for each species as a result of natural selection balancing the allocation of energy to growth and reproduction. This is not a mechanism that generates new biodiversity; it changes or maintains the average body sizes of existing species. Biodiversity is generated only by speciation, and is decreased by extinction. It is shown in other chapters that rates of speciation and extinction are sometimes closely connected to patterns of species richness, either across clades (Chapter 9) or across geographic regions (Chapter 10). Can the high diversity of small species be explained by rates of speciation or extinction (or the rate of diversification—the difference between the two) which vary depending on body size?

How should we summarize body size?

Before going further, it is worth giving some thought to the ways in which this question can best be answered. If we wish to test for an association between body size and rates of diversification, one of the first things to think about is what the scope and units of analysis should be. Should we expect to find an association between diversification rate and body size when we compare taxonomic genera within a family? Between families within an order? Or even across the phyla of Metazoa? Do the units of analysis even need to be taxonomic groups: could they be clades or lineages? Actually, there is no obvious answer to these questions, and whether an association between body size and diversification rate exists at one level of analysis and not at another is likely to be a purely empirical matter. In practice, researchers have sought such associations at most of the levels just mentioned, and the choice is often made for practical reasons—the availability of data on body size and phylogeny, or the convenience of using taxonomy to define the units of analysis.

Another key issue is how body size should be measured and summarized for each of the units of analysis. Researchers have used a variety of different measures of body size—mostly mass or length, because these are the most widely available data for large numbers of species. Some researchers have used the product of length, width, and height as an estimate of 'biovolume', which may be useful when data on mass are scarce. In the end, the measure of body size probably doesn't matter very much for a broad-scale comparative analysis, since most of the measures will be highly intercorrelated and will most likely lead to very similar conclusions. But then we need to decide how to summarize the body-size measure across the species within each of our analysis units. For groups in which the body-size frequency distributions are strongly modal (e.g. mammals), summarizing body size using the median or mode should do an acceptable job of capturing the differences in body size among different groups. For groups without a unimodal body-size distribution, it might be wiser to use measures that better capture the range of sizes of species within the group; this could require, for example, separate analyses for the maximum and minimum species sizes within the group.

How should we infer and compare diversification rates?

Methods for inferring and comparing diversification rates are covered in Chapters 9 and 10, but here we will briefly summarize the key issues. Perhaps the most important concept to remember is that diversification rate is not synonymous with species richness. A taxon or clade with more species than another has not necessarily diversified more rapidly, because the two clades might be different ages. If the rate of diversification has been the same for the two clades, the older one is expected to have more species (all else being equal) because it has had more time in which to accumulate diversity. Not only that, but we expect the number of species to accumulate exponentially with time. This is because a very simple process of diversification, in which diversification is stochastically constant through time and across lineages, yields exponential growth in the number of lineages through time. What this means is that if we logarithmically transform the number of species (N), the exponential growth curve turns into a straight line, so when we divide by clade age ($\log N/t$), we can obtain an estimate of the clade's average diversification rate from its origin to the present day. This requires us to assume that diversification in the clade has happened in a way that can be described by a constant-rate process—and there is debate over whether violation of this assumption invalidates methods of inferring diversification rate from clade age and current species numbers. These issues are described in more detail in Chapter 10, but for now we will put this debate to one side and assume that $\log N/t$ provides an acceptable summary of a clade's diversification rate.

We can illustrate the difference between the patterns of species richness and the patterns of diversification rate using the data for Metazoan

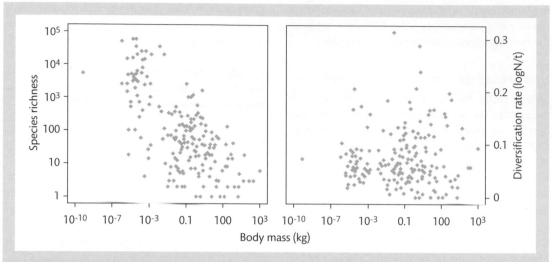

Figure 8.10 Species richness does not necessarily coincide with rate of diversification. The plot on the left shows the numbers of species in families of Metazoa (animal kingdom), while the plot on the right shows the average diversification rates within families, estimated as the logarithm of current species richness (N) divided by the age of families (t). While small-body families tend to have the most species, they have not, on the whole, diversified more rapidly than families of larger body size.

families that we looked at earlier in this chapter. In **Figure 8.10**, the panel on the left shows the relationship between the number of extant species in Metazoan families and their mean body mass, with both axes log-transformed. This is the broad-scale pattern of decreasing species richness with body size that we began the chapter with: it simply presents the basic pattern we have been trying to explain.

Recall that although this plot shows a clear negative relationship between mean body sizes of families and their species richness, it doesn't mean that there are more species overall at the smallest sizes, because there are fewer families at the small end of the size axis. Nonetheless, the mean species richness of small-bodied families is clearly higher than for large-bodied families. Now these units of observation are all of the same taxonomic rank (family). If all families tend to be about the same age, then we might reasonably infer from this plot that diversification rates have been higher in families of small

body size; they have all accumulated more species in roughly the same amount of time as families of large body size. Can we assume that across the Metazoa, families are all of roughly similar age? No—this becomes clear when look at the plot on the right, showing mean body mass of families against diversification rate (logN/t). In this plot, a negative relationship between body mass and diversification rate across families appears non-existent. In fact, the regression of log(body mass) against logN/t is non-significant, with slope 0.001, $p = 0.35$. What this tells us is that small-bodied families are more diverse, but they also tend to be older. The data confirm this: the average age of families with mean body mass below the median (80g) is 102 million years, whereas for families with mean body mass above 80g it is only 55 million years. The reason for this is not clear, but perhaps the morphological differences that systematists use to distinguish families of arthropods and other small-body taxa tend to have deeper evolutionary origins compared with those used to distinguish families of mammals or birds.

The problem of phylogenetic non-independence

At this point, there is still one important issue that we haven't dealt with: phylogenetic non-independence. In Chapters 9 and 10 we shall see how closely related species or clades are expected to be more similar in their morphological or ecological attributes than more distant relatives, because they are less distantly removed from their common ancestor. Under a simple non-directional, random-drift model of phenotypic evolution (such as Brownian motion), the difference between species in body size, or any other continuously varying trait, reflects the time elapsed since they diverged. The degree to which interspecies trait differences match divergence times is known as phylogenetic signal. The presence of phylogenetic signal can be a problem, because statistical tests assume independence of observations. If our observations are species, families, or other taxonomic units, then different degrees of shared ancestry means that knowing the value of one observation allows us to predict the approximate value of another—they are non-independent. Body mass usually shows strong phylogenetic signal. If we know that the body mass of a short-tailed shrew tenrec (*Microgale brevicaudata*) is about 10g and that of a crested genet (*Genetta cristata*) is around 1.8kg, then we can be fairly sure that all species in the genus *Microgale* are likely to weigh a lot less than species of *Genetta*.

Life history traits such as body size typically show phylogenetic signal, but should we expect diversification rate–not a biological 'trait' but a clade-level property–also to show phylogenetic signal? In fact, there are good biological reasons why we should expect closely related clades to be more similar with respect to diversification rate than more distantly related clades. Theory suggests that speciation and extinction rates should be influenced by heritable biological traits (e.g. sexual ornamentation or ecological specialization) and the environment (e.g. climate or topography). Closely related species share similar biological traits, and often inhabit similar environments because they arose in the same part of the world.

So it is certainly plausible that diversification rates of clades show phylogenetic signal, and we may need to account for this when using statistical tests to make comparisons across clades.

Dealing with phylogenetic non-independence: phylogenetic generalized least squares

Statistical methods to account for phylogenetic non-independence when testing for an association between two variables across species or clades are covered in Chapters 9 and 10, but here we will briefly summarize two of the most common approaches. The most widely used approach in recent years has been phylogenetic generalized least squares (PGLS). In PGLS a statistical model is fitted to describe the association between response and predictor variables, with model coefficients estimated by optimizing the association so as to minimize variance in the model's residuals (in the same way as a standard regression). However, in PGLS the coefficients of the model are estimated in a way that incorporates a covariance structure describing the varying degrees of relatedness among the units of observation (**Figure 8.11**).

We can use PGLS to test for an association between body size and diversification rate ($logN/t$) among the Metazoan families. Recall that the standard regression, assuming independence of observations (families), was non-significant. However, when we use PGLS to account for the phylogenetic covariance structure in the data, we now obtain a significant negative relationship between body size and diversification rate, with slope -0.004, $p = 0.026$. This suggests that when we looked at the raw data in Figure 8.10 and fitted a regression to these data, the negative association was obscured by the patterns of relatedness among families.

Dealing with phylogenetic non-independence: contrasts between sister clades

Another approach to dealing with phylogenetic non-independence is the method of sister-clade contrasts (also known as comparisons). Under this method, the units of analysis are not the species- or clade-level values of each variable, but differences (or ratios) in values between pairs of sister

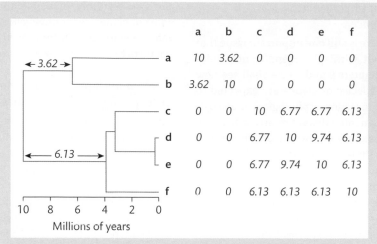

Figure 8.11 Hypothetical example of a phylogenetic variance–covariance matrix as used in PGLS to modify the parameter estimates of a regression model. The values in the matrix elements represent the amount of branch length (from the root of the tree to the tips) that is shared by each pair of species. For example, species a and b share 3.62 million years, and species c and f share 6.13 million years. Under a Brownian motion model of trait evolution, we expect the value of a trait such as body size to be more similar between species d and e, than between species d and a.

clades. The differences are phylogenetically independent because each difference represents evolutionary change that happened in only one part of the phylogeny; therefore the change is independent of change in any other part of the phylogeny. If one of the variables we are interested in is diversification rate, we can calculate the difference or ratio in log(species richness) between each pair of sister clades. Because sister clades are the same age, this gives us an estimate of the *relative* diversification rate between each pair of clades, and not the absolute rate for each clade that we get when we use logN/t. But we can still use relative diversification rate to test for an association with body size, and in some ways this is a more conservative measure to use compared with calculating absolute rate. For example, to estimate absolute rates, we need estimates of the ages of each clade, and these are rarely estimated without error (and in many cases, not known at all). Sister-clade contrasts avoid much of this uncertainty. However, there are other

drawbacks of the sister-clade method, which we examine in Chapter 10.

The pairs of sister clades used in tests of diversification rate are often chosen so that no sister-clade pair overlaps with any other pair. In many cases, taxonomic groupings provide a convenient way of designating the clades, so we might choose, for example, pairs of sister genera or sister families. However, methods are also available to calculate contrasts of species richness at every node in a phylogeny (nested contrasts).[11] When we fit a regression model to a set of nested contrasts for the Metazoa, we find a significant negative association between contrasts in log(species richness) and contrasts in mean log(body mass), with slope –0.3, $p = 0.004$. This is consistent with the result we obtained using PGLS and absolute diversification rate. However, it is not a very tight relationship, as we can see by plotting the contrasts in log(body mass) against the contrasts in log(species richness) (**Figure 8.12**).

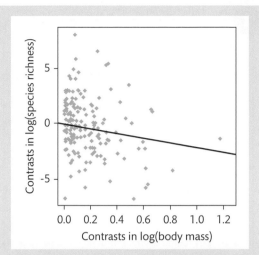

Figure 8.12 Testing the association between body size and diversification rate between Metazoan families. The points in this plot are not the raw values of mean body mass and diversification rate for each family, but contrasts between pairs of sister clades. Using sister-clade contrasts is one way of ensuring that the observations used in a statistical test are phylogenetically independent, and thus more likely to be statistically independent. The plot is messy, as plots of contrasts often are, but shows a slight negative trend, hinting at a negative association between mean body mass and diversification rate.

A non-linear relationship between body size and diversification?

Based on these tests, we might conclude, tentatively, that small body size does tend to be associated with a higher rate of diversification among families of Metazoa. But if variation in diversification rate is the primary mechanism that underlies differences in species richness, we would expect clades at the smallest end of the body-size scale to be the most species-rich. This means that a negative correlation between body size and diversification rate still doesn't fully account for the typical distribution of body sizes that we see in many groups of organisms, in which the smallest species are not the most species-rich. What if this correlation was not monotonic negative, but hump-shaped? Diversification rate might be highest in

families at the modal body size, and scale negatively with body size for families above the mode, but positively for families below the mode.

There is in fact a plausible theoretical reason to postulate a hump-shaped body size–diversification rate relationship. First, a general decrease in diversification rate in groups with increasing body size can be imagined for a number of reasons. For example, smaller organisms tend to have faster life cycles and shorter generations, which could speed the processes involved in genomic and phenotypic divergence of populations, elevating speciation rates. But it has also been suggested that for organisms at the very lowest end of the body-size scale (<1mm long) rates of speciation are depressed. This is because very small organisms are usually both abundant and easily dispersed over long distances, giving them widespread, even global, distributions. This limits the opportunities for disruption of gene flow between geographically separate populations, and hence the formation of species endemic to different regions.[12] If we superimpose the reduced diversification of the smallest creatures on to the overall negative body size–diversification rate trend, we may see a pattern that approximates, and perhaps explains, the log-right-skewed body size–species richness pattern.

One simple test of this idea can be done by splitting our dataset of Metazoan families into those with average body size below the overall modal value (about 18g), and those with average body size above the modal value. We would predict a positive correlation between size and diversification rate among the small-body families, and a negative correlation among the large-body families. But this is not what we find. Using the method of nested contrasts between sister clades, we find negative body size–diversification rate correlations among both small-body and large-body families. Using the PGLS method, we find no significant association in either subset of the families. There is no suggestion of a positive relationship between size and diversification rate among the small-body Metazoan families. However, we must keep in mind that the smallest of the families in this dataset are insects (beetles, flies, ants, wasps, and bees), very few of which are less than a millimetre or so in length. Perhaps these groups are not small enough for the hypothesis of widespread dispersal/low speciation to apply.

What do you think?

One hypothesis for the low diversity of the smallest organisms, and hence the modal shape of many interspecific body size frequency distributions, is that very small organisms are simply under-sampled, and their true diversity is much higher than the number of species that have been collected and described to date. How plausible is this hypothesis for groups like mammals and birds, which are far more likely to have been comprehensively sampled? Are there ways that you could test the influence of under-sampling on the shape of body size distributions either for one taxon, such as birds, or by comparing different taxa?

Key points

- Hypotheses for the size–diversity relationship based on habitat and niche availability can explain how it is possible for so many small species to occupy a landscape or a region, but cannot fully explain the origins of the diversity of small species.

- Optimization of body size by natural selection changes the mean body sizes of species, but does not give rise to more diversity. Only speciation can generate new diversity.

- It is possible to test the hypothesis that rates of diversification (the difference between speciation and extinction) have varied with body size using phylogenetic and body size data for species or higher phylogenetic units. We must always be aware of the limitations on what we can infer from comparative tests of this kind.

Conclusion

In terms of the diversity of species, the abundance of individual organisms, and the variety of morphological types and adaptations, there can be no doubt that the world is dominated by small creatures. But explaining exactly why this is so has remained a persistent challenge for biodiversity scientists. Even more challenging, perhaps, is finding a solution to the puzzle of the log-right-skewed frequency distribution of body sizes across species, which seems to be a very common feature of biodiversity. Some hypothesized mechanisms for increasing diversity with decreasing size seem plausible and powerful, but break down at the lower end of the body-size scale where species numbers fall off rapidly. Perhaps the answer lies in single mechanisms, such as Hutchinson and MacArthur's mosaic elements, or Kozłowski and Weiner's intraspecific optimization model, which simultaneously explain the decline in species numbers both above and below the modal body size. Or perhaps there are multiple mechanisms at work, superimposed onto one another: a general decrease in diversification rate with size, for example, but a depressed speciation rate in the smallest organisms as a result of superabundance and ubiquitous dispersal.

In this chapter, we have considered a selection of the most prominent hypotheses for the body size–diversity relationship. We have looked at so-called equilibrium hypotheses, such as habitat area and the species–individuals relationship, and the non-equilibrium hypothesis of diversification rates. We looked in particular at phylogeny-based methods for testing the association between body size and diversification rate, but we must keep in mind the limitations of phylogenetic approaches. For example, estimates of divergence times, and even sister-clade relationships, are often fraught with uncertainty. Further, it can be easy to forget that a phylogeny reconstructed from molecular data can only ever be an estimate of the history of lineages ancestral to present-day species. A great deal of diversity may have come and gone, but remains invisible to our phylogenies. In other chapters, we explore ways of using the fossil record to estimate patterns of diversification through time and the macroevolution of body size.

Points for discussion

1. This chapter has explored the many ways that body size could influence the processes that generate biodiversity. But what about body shape? Elongated creatures like worms or snakes have a greater surface area than animals of similar mass that are not elongated. In what ways does shape influence an animal's or plant's physiology and interactions with its environment? Could this have an effect on the processes that control species richness?

2. In the Cretaceous period there was a great diversity of very large terrestrial vertebrates, many over 100kg in mass and several metres in height. Why has such a diversity of large terrestrial animals not evolved again since the end of the Cretaceous?

3. It is hypothesized that small-bodied organisms diversify faster than large-bodied organisms, for reasons including faster life cycles and shorter generations. According to Bergmann's rule, the average body size of animals in colder environments, such as high latitudes, tends to be larger than in warmer environments, such as the tropics. What are the possible directions of causation in the associations between body size, species richness, and latitude?

References

1. Haldane JBS (1926) On being the right size. *Harper's Magazine*.

2. Etienne RS, de Visser SN, Janzen T, Olsen J, Olff H, Rosindell J (2012) Can clade age alone explain the relationship between body size and diversity? *Interface Focus* 2(2): 170–9.

3. Jones KE, Bielby, Cardillo M, Fritz SA, O'Dell J, Orme CDL, Safi K, Sechrest W, Boakes EH, Carbone C, Connolly C, Cutts MJ, Foster JK, Grenyer R, Habib M, Plaster CA, Price SA, Rigby EA, Rist J, Teacher A, Bininda-Emonds ORP, Gittleman JL, Mace GM, Purvis A (2009) PanTHERIA: a species-level database of life history, ecology, and geography of extant and recently extinct mammals. *Ecology* 90(9): 2648.

4. Froese R, Pauly D (eds) (2000) *FishBase 2000: Concepts, Design and Data Sources*. ICLARM, Los Baños, Laguna, Philippines.

5. Hutchinson GE, MacArthur RH (1959) A theoretical ecological model of size distribution among species of animals. *American Naturalist* 93: 117–25.

6. Morse DR, Lawton JH, Dodson MM, Williamson MH (1985) Fractal dimension of vegetation and the distribution of arthropod body lengths. *Nature* 314: 731–3.

7. Morse DR, Stork NE, Lawton JH (1988) Species number, species abundance and body length relationships of arboreal beetles in Bornean lowland rain forest trees. *Ecological Entomology* 13: 25–37.

8. Siemann E, Tilman, D, Haarstad J (1999) Abundance, diversity and body size: patterns from a grassland arthropod community. *Journal of Animal Ecology* 68: 824–35.

9. Jones KE, Purvis A (1997) An optimum body size for mammals? Comparative evidence from bats. *Functional Ecology* 11: 751–6.

10. Kozłowski J, Weiner J (1997) Interspecific allometries are by-products of body size optimization. *American Naturalist* 149: 352–80.

11. Isaac N, Agapow P-M, Harvey PH, Purvis A (2003). Phylogenetically nested comparisons for testing correlates of species richness: a simulation study of continuous variables. *Evolution* 57: 18–26.

12. Fenchel T, Finlay BJ (2004) The ubiquity of small species: patterns of local and global diversity. *BioScience* 54: 777–84.

Case Study 8

Phylogenetic tests of diversification: body size and diversity in insects

Most of the examples of empirical research that we have looked at in this chapter have been based on studies of vertebrate animals for the simple reason that these groups are generally well known and well studied, and are represented by the most comprehensive biological, geographic, and phylogenetic datasets. However, by focusing on vertebrates we run the risk of developing an understanding of body size and diversity that may not be very representative of the majority of the world's biodiversity. For this reason, detailed large-scale studies of non-vertebrate groups are very important. Insects form the bulk of the world's animal diversity, but large-scale macroevolutionary and macroecological analyses of insects have always been limited by a scarcity of data. This is changing, and in recent years new phylogenetic and biological datasets have made it possible to apply large-scale analyses to test some of the hypotheses for the link between size and diversity that we have considered in this chapter. Nonetheless, the sheer magnitude of insect diversity means that large datasets are still very incomplete, and short cuts need to be taken to summarize the available data for analysis.

James Rainford and colleagues presented one of the first large-scale phylogeny-based analyses of the link between body size and diversification rate in insects.[1] The massive diversity of insects, and the fact they are far less well sampled or well studied than larger animals, means that there are still very few large insect phylogenies that are resolved to the level of species. The phylogeny that Rainford and colleagues worked with is resolved to the level of families or superfamilies, and was constructed using maximum likelihood analysis of nuclear and mitochondrial DNA from representative species within each family. This means that their phylogeny represents relationships between insect families, but carries no information on relationships among genera or species within families. However, such is the diversity of insects that even a family-level phylogeny still includes over 800 tips, giving the authors good statistical power to test large-scale patterns of diversification. At the family level, their phylogeny is well sampled, representing around 80% of the 1100 recognized insect families.

The authors' first task was to summarize the distribution of body sizes of species within each family to a single family-representative value. Data on body mass are rare for insects, so the authors used length, which is commonly recorded or else easily measured from preserved specimens (**Figure 8.13**). Because many insect families contain hundreds or thousands of species, and the body sizes of many of them

Figure 8.13 A disadvantage of macroevolutionary research on insects is that their sheer diversity usually means working with datasets that are very incomplete at lower taxonomic levels. One advantage is that body length measurements can often be obtained easily from mounted museum specimens, even ones such as these that are more than a century old.

Photo by Andrew Moore via Wikimedia Commons (CC2.0 license).

are likely to be unrecorded, the authors did not have information on the full frequency distributions of species body lengths for each family. Instead, they inferred the distribution of body lengths from the limited data available. As a point estimate of body size for each family, they calculated the mean of the log-transformed minimum and maximum species body length values. This required them to assume that body lengths within families are distributed log-normally, not log-right-skewed as is the case in many other animal taxa. Was this assumption reasonable? This is difficult to judge, simply because of the limited amount of data on full body size distributions in large taxonomic groups of insects. Rainford et al. justified their assumption by citing empirical support for log-normal body size distributions in insect taxa in previous studies. In the absence of much empirical evidence, perhaps assuming a log-normal distribution is no less arbitrary than assuming a particular degree of skew. These 'pseudo-distributions' allowed the authors to calculate a point estimate summary value of body length for each family, and also provided a way of incorporating uncertainty into the estimates by randomly sampling body size values from the distribution and constructing a confidence interval. This in turn allowed the authors to place confidence bounds on the parameter estimates for the body size–diversification rate relationships.

Rainford et al. first drew attention to the inter-family frequency distribution of minimum, maximum, and mean body sizes. Rather than being log-right-skewed, indicating a preponderance of small taxa, the distributions are symmetrical on a logarithmic body size scale— even slightly left-skewed when the mean values for families are weighted by the species richness of families. This is consistent with the assumption of log-normal distributions within families, although it is difficult to say whether or not it supports this assumption.

To model the relationship between mean body size and diversification rate among insect families, Rainford et al. used the approach based on nested sister-clade contrasts that we have looked at in this chapter (see also Chapter 10). Relative diversification rates were estimated using two methods of calculating contrasts in species richness between sister clades: relative rate difference (RRD) is the log-transformed ratio of species richness values between two sister clades ($\log(N_1/N_2)$); proportional dominance index (PDI) is calculated as $N_1/(N_1+N_2) - 0.5$. In both metrics, N_1 is the sister clade with the larger body size. To test the association, the authors fitted regression models of the contrasts in mean body size against RRD and PDI. They fitted these models separately for each major clade of insects.

Plots of the body size contrasts against diversification rate for the major insect clades examined by Rainford et al. are shown in **Figure 8.14**, with the slopes from the fitted regression models, and upper and lower confidence bounds on the slopes. For some taxa (Hexapoda and Holometabola), the slopes look quite flat and do not seem to support an association between body size and diversification rate. But for other taxa, slopes are a bit more steeply positive (negative in the case of Palaeoptera). Do these support significant associations between body size and diversification rates? In a standard regression model, the significance of the association between two variables is tested by comparing the elevation of the slope to a null model slope of zero. But a zero slope is not always the appropriate null model to use. In some datasets, the structure of the data (e.g. the phylogenetic or spatial structure) means that a non-zero slope is seen even when two variables are not causally connected. This could mislead us into falsely rejecting a null hypothesis: in other words, inferring an association between two variables, when the data do not support one.

One way to deal with this is to use a custom null model to generate the null slope, rather than assume a null slope of zero. This is what Rainford et al. did. They generated a custom null model for each of their datasets, by randomly shuffling the mean body size values across the tips of the phylogeny (i.e. among the families), and then recalculating the contrasts in body size and performing another regression against diversification rate to estimate a 'null' slope. By doing this 1000 times, they generated a distribution of null slopes—the range of slopes for the body size–diversification rate association that we would expect to see if there was no causal connection between the two. Where the observed slopes (solid lines in Figure 8.14) fell within

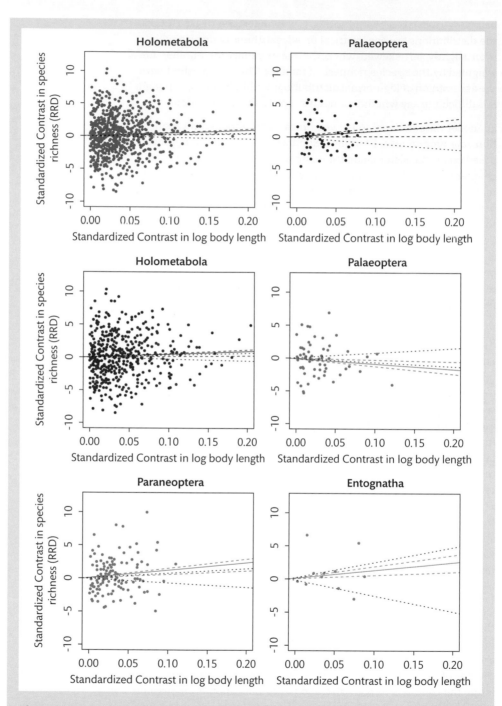

Figure 8.14 There is little evidence for relationships between body length and rates of diversification within major taxonomic divisions of insects. The plots show phylogenetically independent contrasts in family-level mean body lengths on the x axis, and contrasts in species richness of clades on the y axis. The species richness contrasts are a measure of the relative diversification rate of two sister clades[1].

From James L. Rainford, Michael Hofreiter and Peter J. Mayhew, Phylogenetic analyses suggest that diversification and body size evolution are independent in insects, *BMC Evolutionary Biology* 2016 16:8 (CC BY 4.0).

the upper and lower bounds of 95% of the null slopes (dotted lines in Figure 8.14), the relationship was considered non-significant. This is because it is not necessary to infer a causal association between body size and diversification rate—the slopes are within the range that we expect from chance alone. And this was what Rainford et al. found in each major insect taxon, as can be seen in Figure 8.14.

Based on these results, Rainford et al. conclude that, in insects, the evolution of body size is independent of variation in rates of diversification. They urge caution against the inference of general macro-evolutionary patterns and processes from analyses of well-studied groups such as vertebrates, which may be atypical of the majority of biodiversity. Their results raise the obvious question of why body size should be independent of species richness in insects. They offer several suggestions, including the possibility that the increase in habitat surface area afforded by the fractal nature of habitats becomes inapplicable for organisms below a certain size. One reason that this could be the case is that a large proportion of insect species are parasitic or live on the surface of larger organisms, so to a large extent the area of available habitat is homogenized with respect to body size.

Questions to ponder

1. The authors speculate that the fractal geometry of habitat surfaces has little influence on parasitic insects. What kind of tests could be devised to explore this?

2. Within insects we do not see the strong bias in species richness towards smaller body sizes that we see in larger organisms. How could you test the hypothesis that this is not the true pattern, but the result of under-sampling of the smaller-bodied insect taxa?

3. Could the rate at which insects speciate to form geographically endemic species be lower than expected because their small size permits wide dispersal? How could this hypothesis be tested?

Further investigation

A key hypothesis for insect diversification is the coevolution of plant-feeding beetles with angiosperms (see Chapter 9). This could be a major driver of insect diversity, overriding any influence of body size. If we compared tests of body size and diversification rate in phytophagous and non-phytophagous insect groups, should we expect to find any differences?

Reference

1. Rainford JL, Hofreiter M, Nicholson DB, Mayhew PJ (2014) Phylogenetic distribution of extant richness suggests metamorphosis is a key innovation driving diversification in insects. *PLoS One* 9: e109085.

9 Why are there so many kinds of beetles?

Roadmap

Why study beetle diversity?

Beetles are the largest order of animals on earth; their diversity eclipses other insect orders, even their nearest relatives. Why are beetles, of all groups, so diverse? Is there something about their biology or ecology that has promoted diversification, do they have a longer evolutionary history than other groups, or is the great diversity of beetles simply the outcome of random diversification processes? By investigating some of the proposed explanations for the great diversity of beetles, we can get to grips with the basic logic of setting up comparative analyses to discover why diversity is so unevenly distributed among clades. Exploring beetle diversity is a good way to become familiar with the ways that evidence from phylogenetics, biogeography, and palaeontology can be used to describe patterns and test hypotheses for variation in diversity.

What are the main points?

- Random processes of diversification, in which all lineages have equal probabilities of speciating and going extinct, can produce substantial variation in diversity across a phylogeny.

- Beetles have more species than expected under null models, giving us a reason to postulate and test deterministic hypotheses such as coevolution between beetles and angiosperms.

- Phylogenetic comparative methods can be used to test for associations between high diversity and the evolution of traits such as phytophagy (plant-feeding).

- Reconstructing the timing of key events in the history of clades can also contribute evidence that can be used to evaluate hypotheses of diversity.

What techniques are covered?

- Constructing null models to generate expected distributions of diversity and to identify unusually diverse clades.
- Using phylogenetic methods to pinpoint shifts in diversification rates on the phylogeny.
- Sister-clade comparisons to test associations between diversity and evolution of traits.

What case studies will be included?

- Reconstructing and comparing the timing of the beetle and angiosperm radiations.

"The only rules of scientific method are honest observations and accurate logic. To be great science it must also be guided by a judgement, almost an instinct, for what is worth studying. No one should feel that honesty and accuracy guided by imagination have any power to take away nature's beauty."

Robert H. MacArthur (1973) *Geographical Ecology: Patterns in the Distribution of Species*. Princeton University Press, Princeton, NJ.

One of the striking features of biodiversity is how much it varies. Diversity does not stay constant through time (see Chapter 3), and it varies enormously from place to place (see Chapter 10). Diversity also varies enormously between different groups of organisms, which is the focus of this chapter. Some groups of organisms have many more species than others. How can we explain this variation? Can great differences in species richness between groups arise simply by chance, as a result of the stochastic nature of diversification? Or do these differences arise deterministically, perhaps driven by the evolution of key adaptations, as has been proposed for dinosaurs (see Chapter 5) and mammals (see Chapter 6)? In this chapter, we examine ways of distinguishing stochastic patterns of variation in diversity from patterns driven by known biological processes. We will examine some hypotheses that have been proposed to explain variation in diversity between taxa, and some approaches that can be taken to testing these hypotheses. As a case study to illustrate these concepts, we will focus on one of the world's most stunning and extraordinary evolutionary radiations: the beetles.

314 | *9 Why are there so many kinds of beetles?*

The spectacular radiation of the beetles

Naturalists have been aware of the great diversity of beetles for centuries. Alfred Russel Wallace partly attributed his and Charles Darwin's co-discovery of evolution by natural selection to their early passion for collecting and studying beetles:

> ❝ *First (and most important, as I believe), in early life both Darwin and myself became ardent beetle-hunters. Now there is certainly no group of organisms that so impresses the collector by the almost infinite number of its specific forms, the endless modifications of structure, shape, colour, and surface-markings that distinguish them from each other, and their innumerable adaptations to diverse environments.* ❞
> **Alfred Russel Wallace** (1909) The origin of the theory of natural selection. *Popular Science Monthly* 1909: 396–400.

All beetles have hardened forewings (elytra) and a life cycle of complete metamorphosis, characteristic features that distinguish them as members of the order Coleoptera. As Wallace discovered, beetles have evolved an enormous variety of morphological forms and sizes, they occupy a huge range of environments, and they fill a wide range of ecological niches within those environments. The smallest beetles are probably the featherwings (family Ptiliidae), some of which reach less than 1mm in length as adults. Among the largest beetles are the scarabs (Scarabaeidae) such as Goliath beetles and Hercules beetles, which can reach up to 18cm in length and 100g in weight, and the longicorns (Cerambycidae) such as titan beetles, which have been reported up to 20cm in length (**Figure 9.1**).

Many beetles are herbivorous, but beetles also occupy most other consumer trophic levels. Beetles include fungivores, scavengers, detritivores, carnivores, and hyper-carnivores (carnivores that feed on other carnivores). Dung-feeding is also common, especially among the scarabs. Beetles occupy nearly all major terrestrial and freshwater habitat types. They are particularly diverse in tropical rainforests, from the ground litter to the forest canopy, but there are also many species with unique adaptations to dry environments. Perhaps the best known of these

Figure 9.1 Titan beetles (*Titanus giganteus*) are among the largest insects known. They are found in the tropical rainforests of South America.
Photo by Bernard Dupont (CC2.0 license).

are the darkling beetles (Tenebrionidae) of the Namib Desert in southwest Africa. Some of these beetles collect atmospheric moisture from fogs that roll in from the Atlantic Ocean, using microscale grooves and ridges on their elytra. The beetles orient their bodies so that the condensed moisture is directed towards their mouths, a behaviour termed 'fog-basking'. Other darkling beetles have evolved behavioural adaptations, such as constructing sand trenches, to capture and condense fog.

In fact, the Coleoptera are the most diverse order of organisms in the world. With over 350,000 described species of beetles, this one order appropriates around a third of all insect diversity and, indeed, over a quarter of known animal diversity. We can compare beetle diversity with that of the remaining 29 or so orders of insects, by plotting a bar chart (**Figure 9.2**).

There are two key things to note about this plot. First, it is clear that one group (beetles) has a disproportionate share of insect diversity. Secondly, there is a rapid drop-off in diversity as we move through the remaining orders from left to right. The majority of orders have very few species, giving the bar chart a very long tail. This kind of pattern, known as a 'hollow-curve' distribution, is not confined to the insects; indeed, it seems to be very general, appearing almost consistently in a wide variety of different groups of organisms. So, not only is there a common pattern of great variation in diversity among taxa, but this variation is typically of a characteristic form. As we shall see later in this chapter, this is useful to know because it may help us uncover the underlying processes of diversification.

We can contrast the diversity of beetles with that of one of the insect orders in the tail of the hollow-curve distribution. One of the least diverse insect orders is the scorpionflies (Mecoptera), with a mere 550 described species. The public profile of scorpionflies is so low that most non-entomologists probably haven't even heard of them. Even though they are of the same taxonomic rank (order), the beetles have about 600 times more species than the scorpionflies. Why are there so many kinds of beetles, and why is there such an enormous difference in diversity between beetles and scorpionflies? Answering these questions might help us understand a great deal about beetles, scorpionflies, and the evolution of insects. It might also bring us a step closer to understanding the more general theoretical issue of why diversity shows such great variation.

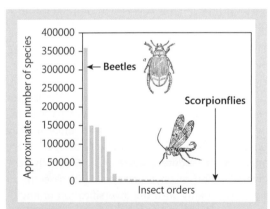

Figure 9.2 The distribution of species among insect orders shows a typical 'hollow-curve' pattern. One order, Coleoptera (beetles), accounts for a disproportionate share of insect species, and the number of species drops off rapidly through the remaining orders. Can such unevenness in the distribution of diversity have arisen by chance, or must there be something special about beetles that has caused them to be so diverse?

Beetle image: *Webster's New International Dictionary of the English Language*, 1911. Internet Archive. Scorpionfly image: © barbulat/VectorStock.

Why are there so many kinds of beetles?

Not surprisingly, many different explanations for beetle diversity have been suggested. One idea is that beetles are highly adaptable and readily able to evolve into different forms. Beetles exhibit a generalized basic body form that (it is suggested) can easily be shaped by natural selection to suit many different habitats, and their small size means that they can occupy many niches not available to larger animals (see Chapter 8). Their versatile mouthparts suit beetles to many different modes of feeding, including plants, fungi, carrion,

Figure 9.3 The leaf beetles (family Chrysomelidae) are one of the largest of the phytophagous (plant-feeding) beetles clades, and include (a) rainforest species such as the one shown from Papua, and (b) agricultural pests such as the Colorado potato beetle (*Leptinotarsa decemlineata*). The enormous diversity of phytophagous beetle groups has led to suggestions of an evolutionary connection between the radiations of angiosperms and beetles.

(a) Photo by gbohne (CC2.0 license). (b) Photo by Scott Bauer, US Department of Agriculture.

decaying vegetation, and dung. It is unusual to find beetle species with radically modified mouthparts, suggesting that dietary specialization in beetles doesn't necessarily rely on large evolutionary changes in the shape of the mouthparts.

Another prominent hypothesis is that the massive diversification of beetles was precipitated by the evolution of plant-feeding (phytophagy), and in particular the shift from feeding on gymnosperms (a group of plants that includes conifers and cycads) to feeding on angiosperms (flowering plants).[1] In fact, these two hypotheses (adaptability and angiosperm-feeding) go hand in hand, because the key idea is that the rise of angiosperms opened up a vast range of new habitats and feeding niches into which beetles were well placed to diversify. But more than this, many researchers have suggested that beetles and angiosperms diversified together in a two-way process known as coevolution. This is a process of reciprocal diversification resulting from an evolutionary arms race between plants and their herbivorous attackers.[2] Plants develop secondary chemical compounds as mechanisms of defence against attack by herbivores, allowing plant species to diversify. Herbivores then evolve mechanisms to cope with chemical

defences, opening up new ecological opportunities and allowing them to diversify in turn. Plants then develop new chemical defences, and so on. This process is also known as 'escape and radiation'.

The idea that beetle diversification was driven by phytophagy is certainly compelling, for several reasons. First, about half of all beetle species are plant-feeders. Within some of the largest beetle radiations, such as the weevils (superfamily Curculionoidea) and the leaf beetles (family Chrysomelidae), the great majority of species are angiosperm-feeders (Figure 9.3). Secondly, the diversification of phytophagous beetles seems to coincide broadly with the diversification of angiosperms during the Mid-Cretaceous Period.[1,3] Thirdly, the theory of coevolution provides a plausible evolutionary mechanism for a link between angiosperm and beetle diversification. Perhaps, then, we can explain the great diversity of beetles as the result of their evolution of angiosperm-feeding in the Early Cretaceous, which allowed them to tap into the rapidly developing angiosperm radiation through a process of coevolutionary escape and radiation. If so, can we then also explain the paucity of scorpionfly species by their lack of a largely phytophagous feeding mode? Although some scorpionfly species do feed on plant material, such as fruit or pollen, the majority are predators or carrion-feeders. For this reason, scorpionflies may not have been presented with the same opportunities for diversification into phytophagous niches as the beetles.

Does this mean that we have now answered the question of beetle diversity and come to the end of our investigation? On the contrary, we have now arrived at the starting point. What we have so far is an observed pattern that we want to explain (the massive diversity of phytophagous beetles), and a plausible evolutionary mechanism that might provide the explanation (coevolution with angiosperms). By itself, this is enough to develop a *hypothesis* that beetle diversity was driven by angiosperm-feeding, but it should not be enough to draw the *conclusion* that beetle diversity was driven by angiosperm-feeding. To reach that point, there are a few extra steps we need to take, and questions we need to ask. The first question is: do we even need to postulate a biological mechanism in order to explain beetle diversity?

What do you think?

Coevolution with angiosperms has been suggested as a possible mechanism underlying the enormous diversity of beetles. But why should the evolution of tolerance to plant defence compounds in a beetle species lead to diversification into numerous species and to great morphological diversity? Why would we not simply expect that the species that evolves tolerance becomes more abundant and numerically dominant, without necessarily diversifying?

Can a 600-fold difference in diversity arise by chance?

When we want to understand the processes underlying the diversity of a group like beetles, or the difference in diversity between two groups, we naturally turn first to potential biological explanations such as angiosperm-feeding and coevolution. But before we start invoking biological explanations like these, we need to step back and ask if such biological processes are even needed to explain the diversity difference. Is it possible that the differences in species numbers between beetles and scorpionflies could have arisen simply by chance, as a result of the stochastic (random) nature of diversification?

To answer this question, we need to think about the process of diversification. Diversification is really pretty simple—it consists only of two processes, speciation and extinction (although both these processes are themselves far from simple, and our understanding of them is still incomplete). Speciation occurs when a single genetic lineage splits into two or more lineages, and each then proceeds along its own independent evolutionary path. This is represented in a phylogenetic tree by the bifurcation of a branch. Extinction occurs when the last remaining individual of a lineage dies, so the evolutionary path of this lineage terminates, and it leaves no descendants. It follows that the rate at which a clade grows (the net rate of diversification) is the difference between the rate at which lineages are added to the clade by speciation, and the rate at which they are removed by extinction. If the rate of speciation exceeds the extinction rate, the number of species in the clade will increase through time. If the extinction rate exceeds the speciation rate, the clade will shrink and ultimately the entire clade will go extinct.

For a given lineage, there will be a probability of speciating, and a probability of going extinct, at every point in time (time steps can be defined arbitrarily, e.g. millions of years). A random model of clade diversification is one in which speciation and extinction probabilities of all lineages are drawn from the same distribution, and are not linked to the biological attributes of particular clades, or to the environments in which members of the clade live. This is usually known as an equal-rates Markov (ERM) model. If rates of speciation and extinction remain constant through time as well as across lineages, it is often called a constant-rates birth–death model (if the extinction rate is assumed to be zero, it is a pure-birth or Yule model). In a random model, therefore, diversification rates do not vary *predictably* between clades. Nonetheless, the number of speciation and extinction events will still vary among lineages because speciation and extinction are stochastic events. So, a phylogenetic tree that has grown under a random diversification process will show variation in the density of lineages across different clades within the tree. The key question is: can a randomly generated tree produce as much variation in the number of lineages (and thereby the richness of species or other taxonomic units) as we see in real-world phylogenies?

If a random model could produce differences in species richness as great as those between the beetles and scorpionflies, then we should acknowledge that we might not need to invoke deterministic biological processes, such as angiosperm-feeding and coevolution, to explain beetle diversity. The same differences in diversity might have arisen even if these processes were not operating. The random model would be a more parsimonious explanation than a deterministic model, because it would require us to make fewer novel assumptions. A widely used guiding principle for choosing between alternative explanations in science is 'Occam's razor'. Under this principle, we should accept the simplest model that accounts for the observed data, only adding more complexity (more processes, parameters, or assumptions) if this provides better explanatory power. For example, to accept the null model for beetle diversity we must assume that diversification rates vary stochastically. For the coevolution model, we must additionally assume that diversification in angiosperms is linked to diversification in beetles, and that the radiations of the two groups were coincident in time. The principle of Occam's razor provides a check against adding more and more complexity to an explanatory model when a simpler model would do an equally good job of accounting for the patterns.

Key points

- Beetles are one of the largest and most ecologically diverse evolutionary radiations, making them a good case study for investigating the processes that generate diversity.
- Because of the stochastic (random) nature of the diversification process, there will always be a certain amount of variation in diversity among taxa or clades that arises purely by chance.
- To understand diversity variation we need to ask whether random processes are sufficient to account for all the variation in diversity that we observe.
- If the answer is no, this may justify the investigation of non-random (deterministic) processes that may have elevated or depressed speciation or extinction rates in some clades, but not others.

Null models: distinguishing random from non-random patterns

What are null models, and why are they useful?

We know that a high proportion of beetle species are phytophagous (plant-feeders). Scorpionflies, on the other hand, are fungivores, scavengers, and predators. Beetles are around 600 times more species-rich than scorpionflies. By considering these patterns, can we conclude that phytophagy, and in particular angiosperm-feeding, is responsible for the massive diversity of beetles? Actually, simply looking at the descriptive patterns of diversity and feeding modes would, by itself, be a weak basis for inference. To draw a conclusion that beetle diversity was driven by phytophagy, by simply observing the diversity pattern, does not give equal weight to the possibility that the difference in diversity of beetles and scorpionflies might actually be no greater than we would expect under a process of random diversification. There is no a priori reason to assume that the deterministic explanation (the phytophagy hypothesis) is more likely to be true than the non-deterministic explanation (random diversification). A model that explicitly excludes the deterministic processes we are interested in is called a null model. A null model is similar to an experimental control, because it provides us with an appropriate baseline against which to compare the patterns we see when the process of interest is in operation (see Chapter 2).

Null models are the analytical workhorses of research in macroecology and macroevolution. A null model provides us with a description of the kinds of patterns we would expect to see in the absence of the deterministic biological processes we want to test. If the pattern we see in the real world (the observed pattern) is indistinguishable from a set of patterns generated at random under a null model, we have little basis for inferring that the pattern was the result of the deterministic processes we are testing. Alternatively, if the real-world pattern is significantly different from that generated under the null model, we can reject the possibility that the processes included in the null model are sufficient to have generated the pattern. We can then begin to look for alternative explanations.

Singular versus general associations

There are further reasons to be cautious about drawing conclusions directly from the comparison of beetle and scorpionfly diversity. The conclusion would be based on a single comparison only, so it lacks replication. Without replicated observations, we can't perform a statistical test of a link between angiosperm-feeding and high diversity. This doesn't mean that it is never worth asking questions about a single comparison. To reconstruct the unique series of events in the history of a single group of organisms is often a worthwhile aim—indeed, much research in palaeobiology and historical biogeography has precisely this kind of aim. But we need to remember that reconstructing history and testing general cause–effect associations represent two different kinds of questions.

A historical question might be: 'Was the massive radiation of beetles closely preceded by an evolutionary shift to angiosperm-feeding?' If the answer is yes, this might be indicative of a cause–effect relationship between angiosperm-feeding and elevated diversification rate, but it could also be an incidental association. If we wish to test the general question 'Do increases in diversification rate result from shifts to angiosperm-feeding?', we need to identify multiple independent instances of the evolution of angiosperm-feeding. We will see later in this chapter that both historical reconstruction and general hypothesis testing can form part of the body of evidence used to assess the angiosperm-feeding and coevolution hypothesis.

Additionally, tests of association between two variables may be confounded or obscured by other unmeasured variables. With only one comparison, we cannot statistically distinguish the possible effects of angiosperm-feeding on diversity from those of any number of other physiological, ecological, or behavioural features possessed by beetles but not by scorpionflies. The more independent comparisons we have at our disposal, the more statistical power we have to rule out such confounding factors. So, again, replication is the key.

Key points

- Inferring a deterministic cause (a known non-random process) for an observed diversity pattern without comparing it with a random pattern generated by a null model has several pitfalls. There is no basis for assuming, a priori, that a pattern is more likely to have had a deterministic cause than to be the result of random processes.

- We need to analyse replicated patterns in order to infer a general cause–effect association between a pattern and a particular deterministic process. A single comparison makes it hard to rule out an incidental association or possible confounding factors.

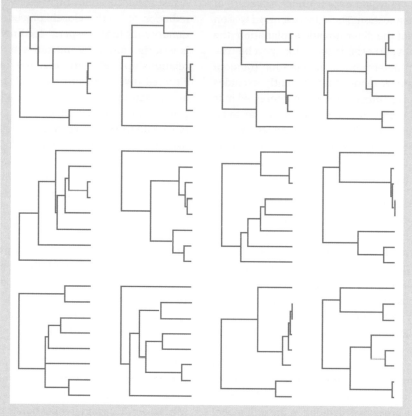

Figure 9.4 These 12 phylogenies were all produced under the same process of random diversification, all with the same branching probability. Because the branching process is stochastic, a wide range of outcomes are produced.

How do we compare real observations with a null model?

How do we actually compare a real pattern with one generated under a null model? The first thing we need is something to compare—a test statistic, or some quantifiable pattern, that we can measure in both the real dataset and the null dataset. For example, the test statistic might be the difference in species richness between two clades, or perhaps the rate parameter of the geometric distribution describing the distribution of species among higher taxa. What we want to know is whether the value of the test statistic measured from our real-world observed data is significantly different from the value obtained under a null model.

This is where the probabilistic nature of null models is important. In a null model, whether an event (such as one lineage branching, or splitting into two) occurs or does not occur at any given time is governed by chance. In other words, the occurrence of an event depends on the outcome of a random draw from a set of numbers that follow a specified distribution (e.g. normal, uniform, geometric). This generates variation in the final outcome (e.g. a randomly generated phylogeny), and no two runs of a null model will be the same. As an example, look at the 12 phylogenies in **Figure 9.4**. These were all generated under the same random diversification process, with the same specified diversification rate (a 20% chance of any lineage branching at any time step), until the same number of tips (eight) is generated. As you can see, the branching patterns, branch lengths, and overall shape of these trees vary quite a lot. For this reason, we can't compare our real-world pattern with the pattern generated by a single run of a null model. Instead, we must repeat the null model numerous times (usually 1000 or more) to generate a distribution of random outcomes. We then see if our test statistic measured from real data lies within the distribution of test statistics measured from the randomly generated data.

As an example, suppose we count the number of phytophagous and fungivorous species in a clade of beetles, and find that there are 15 times more phytophagous than fungivorous species: this is our test statistic. We want to know if a diversity ratio of this size (15:1) could plausibly have been generated by chance. One way of generating a null model for this statistic is to simulate the evolution of feeding modes along our phylogeny. We might start by assuming a phytophagous ancestral state at the root of the phylogeny, and then move in increments along the branches of the phylogeny, towards the tips. At each time step we assume a particular probability (for example 0.1) that the feeding mode shifts from phytophagy to fungivory. We assume that a shift from phytophagy to fungivory has the same probability as a shift from fungivory back to phytophagy. We run the simulation until we reach the tips of the tree, then count the numbers of phytophagous and fungivorous species we end up with. These numbers will be different each time we run the simulation, so the test statistic will also be different every time. If we repeat this procedure 1000 times, we get a null distribution of test statistics. In the example shown in **Figure 9.5**, the values range from zero to 24, but most values are between 8 and 16 (left panel). The arrow shows our observed value of 15, and we see that it lies well within the null distribution: 24% of the null values are greater than 15, giving us a *p*-value of 0.24. This leads us to conclude that a 15:1 ratio in species numbers between the two feeding modes is not very unusual, and would be generated quite frequently by the processes incorporated in our null model.

Now suppose we modify the null model in some way; for example, we might assume a fungivorous ancestral state, or that the probability of a reversal (character shift from fungivory back to phytophagy) is lower than the probability of a shift from phytophagy to fungivory. In the right panel of **Figure 9.5**, the observed value (15) hasn't changed but the whole null distribution has shifted to the left, so the observed value is now unusual, with a probability of only 0.04 that it could have arisen under the null model. We can see that the way the null model is constructed can change the baseline values that we use for comparison, and thereby influence the conclusion we draw about biological processes.

If our test statistic is inconsistent with a null model, it suggests the pattern we summarized with the statistic was unlikely to have been generated under the processes incorporated in that null model. This gives us a rationale for further investigation of the link between the pattern and a biological process that was not part of the null model, such as an elevated diversification rate associated with a particular feeding mode. Rejecting a null model does not necessarily mean that a particular biological process was responsible for producing the pattern; it simply gives us grounds for further testing.

What do you think?

Suppose that you wish to determine whether an observed diversity pattern could plausibly have been produced by chance. You generate random patterns under two null models with different assumptions, and find that your test statistic is rejected by one null model but consistent with the other. How will you decide whether your diversity pattern could have been produced by chance? Should you aim to construct a single most plausible null model by including some extra biologically informed assumptions? Or is it OK to accept a pluralistic outcome, where the answer depends on the way you ask the question?

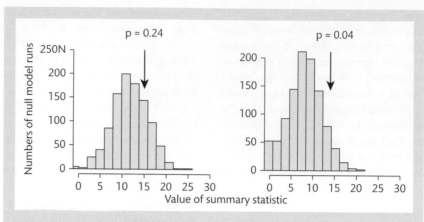

Figure 9.5 A test statistic with a value of 15 (indicated by the arrow) is consistent with the null distribution in the left panel ($p = 0.24$) but rejects the null distribution in the right panel ($p = 0.04$). Although the test statistic hasn't changed, the two null models lead to different conclusions about whether a value of 15 for this test statistic could plausibly have been generated by chance. In the example in the left panel, we would conclude that the diversity pattern summarized by the test statistic is nothing unusual; in the example in the right panel, we might conclude that the pattern is unusual and worth further investigation.

Testing hypotheses of diversity

To test whether or not the massive diversity of beetles is the result of the evolution of angiosperm-feeding, we should, ideally, do four things.

1. We should test if the overall distribution of species among insect orders is more uneven than would be expected under a null model in which all lineages speciate or go extinct with the same probabilities.

2. We should test if beetles, specifically, have an excess of species compared with the expectation under a null model (or several alternative null models).

3. We should estimate the timing of the beetle and angiosperm radiations, and the timing of the shift to angiosperm-feeding in beetles, and determine whether these historical events are closely coincident.

4. We should test for a general statistical association between the evolution of angiosperm-feeding and elevated species richness or elevated diversification rate, using replicate comparisons.

Is diversity distributed more unevenly than expected by chance?

We have now seen that the starting point for developing explanations about differences in diversity between clades should usually be a comparison of real patterns against a null model. By applying a null model we are asking if we can reproduce the pattern we wish to explain under a model that excludes the process we want to test as the possible cause of the pattern. Applying a null model is

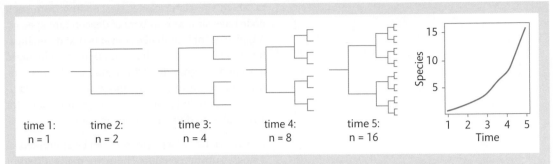

Figure 9.6 Bifurcation of lineages in a phylogeny at a constant rate generates a geometric increase in the number of species (*n*) through time. This is a feature of any hierarchical branching process, not just phylogenetic diversification.

not always straightforward. In fact, it can be quite challenging to design a null model in a clever and creative way, which effectively does the job we need it to do. As we just saw, the same question can be asked using different null models that make different assumptions about processes, or that allow different parameters to vary randomly while keeping other parameters fixed. It may well be that there is no particular null model design that is clearly the correct one for the question you want to ask. If this is the case, there is nothing wrong with applying a few alternative null models and comparing the results—if the results for all null models consistently lead to the same conclusions, so much the better. If the results differ depending on the null model you apply, drawing conclusions becomes trickier. In such cases, it is always important to be aware of the assumptions being made under each null model, and the way that the assumptions could influence the outcome. We will now examine two basic kinds of null models that have been used to test the degree of variation in diversity among clades.

Comparing the distribution of diversity with a theoretical frequency distribution

Imagine a single ancestral lineage that splits into two daughter lineages, which each then split into two daughters, and so on (**Figure 9.6**). If every lineage is guaranteed to split at every time interval, the growth in diversity under this process can be represented by the doubling series 1, 2, 4, 8, 16, . . .

When you plot this series on a graph (on the y-axis) against a linear time series (1, 2, 3, 4, 5, . . .) on the x-axis, you will see that while the rate of increase (doubling) remains constant, the numbers themselves increase in an accelerating manner through time. In fact, the increase in diversity through time under this process is geometric (or exponential if measured on a continuous timescale). The chance of splitting doesn't have to be 100% at every time interval for geometric growth to ensue—it could be 50%, 1%, or 0.1%. The important thing is that the splitting rate remains constant through time, and equal across lineages.

A stochastically constant branching rate through time leads to the expectation that the sizes of present-day clades will also follow a geometric distribution. So if we compare the distribution of clade or taxon sizes in our phylogeny with a geometric distribution, we can test whether or not the phylogeny has diversified at random (assuming that the geometric distribution is an appropriate way to generate the random distribution). The bar chart in **Figure 9.7** shows the frequency distribution of species numbers among insect orders. The dashed line represents the geometric curve that best fits these data. The fit between the insect data and the geometric curve is not very close; from the plot, it looks as if there are many more very species-poor orders than expected under the geometric distribution. It also looks as if the geometric distribution would be extremely unlikely to produce any order

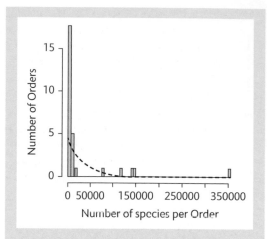

Figure 9.7 The number of insect orders that contain different numbers of species. The tallest bar on the left shows that 18 orders are in the smallest category (0–5000 species). Beetles are way out on their own on the far right, the only order in the category 355,000–360,000 species. The dashed line shows the expected shape of the distribution if species were assigned to orders according to the geometric distribution.

with as many species as the beetles (the lonely bar on the far right of the plot). So, using the geometric distribution as the null model, the distribution of species among insect orders appears to be more uneven than expected by chance.

Phylogenetic imbalance: an explicitly phylogenetic null model for diversity

If you think of a phylogenetic tree as a mobile, suspended from the ceiling by a string attached to the root node with the tips hanging down, it should be easy to grasp the concept of phylogenetic imbalance. If one of the two lineages descending from the basal split has many more species than the other, it will be 'heavier' and will weigh the mobile down on that side. If the two lineages are equally species-rich, the mobile will be balanced. Within each of the two basal clades, there may also be balance or imbalance among subclades, and so on down through the tree to the tips. The imbalance of a phylogeny is defined as the overall degree

to which the sister clades descending from each node have similar numbers of descendant species. A high level of imbalance suggests that diversification has been non-random, and that some lineages have had a higher diversification rate than others. A high level of imbalance has often been considered a rationale for seeking biological reasons for differences in diversity between clades.

The imbalance of a phylogeny can be expressed as a single summary value. There are many different metrics for calculating phylogenetic imbalance. One of the most widely used is Colless' I: this is simply the set of absolute differences in species richness between each pair of sister clades, summed across all nodes in the phylogeny. Therefore the greater the imbalance, the higher the value of I will be. Once we have used a metric such as I to summarize imbalance in a real phylogeny, how do we know whether this value of I is unusual or not? We need to compare it with a distribution of imbalance scores generated under a null model. Intuitively, we might expect that the appropriate comparison should be with a phylogeny that shows perfect balance; that is, at every node in the phylogeny, the two sister clades descending from the node have the same number of species. However, this is not the case. A phylogeny generated under a random diversification process—one in which no lineage has any more chance of splitting than any other—is very unlikely to be perfectly balanced. In fact, a high degree of balance (or very low imbalance score) would represent an extreme outcome under a random diversification process. When you think about it, this is not surprising; the chance of a stochastic diversification process resulting in a tree in which every node has two daughter clades with identical numbers of descendants must be very slight. Not only that, but a diversity difference between sister clades that arises by chance early in the history of a phylogeny will tend to be maintained and propagated through the phylogeny. This effect can bias larger phylogenies towards higher imbalance scores, so it is inadvisable to compare imbalance scores for phylogenies of different sizes. For these reasons, it is only meaningful to interpret an imbalance score by comparing it with a

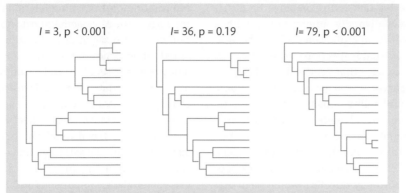

Figure 9.8 Three 16-tip phylogenies with different degrees of imbalance. Above each tree is the value of the imbalance metric Colless' *I* and the probability that this degree of imbalance could have been generated by chance. The tree on the left is almost (but not quite) maximally balanced, the tree on the right is almost maximally imbalanced, and the tree in the middle is more typical of a randomly generated tree. Note that the branch lengths play no part in the measurement of imbalance; it is purely a measure of the relative numbers of tips descending from the two daughter lineages of each node.

distribution of scores from phylogenies of the same size generated under a random unbiased diversification process. In **Figure 9.8**, you can see that the almost perfectly balanced tree on the left rejects the null model with a very low value of $I = 3$, while the highly imbalanced tree on the right rejects the null model with a very high value of $I = 79$.

Comparing the distribution of species among higher taxa with a theoretical distribution, such as the geometric, or the imbalance of a phylogeny with a null model such as ERM, tells us whether the shape of the phylogeny as a whole is consistent with a process in which all lineages have had the same chance of splitting. But if our observations reject these tests, they cannot tell us *where* in the phylogeny the non-randomness is found: which lineages have undergone a significantly higher rate of diversification, or which clades are more diverse than expected. If we want to test more focused hypotheses, such as whether the beetles are more diverse than we would expect under a null model, we need to move a step beyond imbalance metrics and use tools for pinpointing the location of diversification rate shifts on the phylogeny.

Key points

- A useful first step in analysing diversity patterns is to test whether the overall degree of unevenness in the distribution of species among clades is more extreme than we would expect if all lineages have had equal probabilities of splitting.

- Even if the diversification rate has been equal among lineages, we don't expect species to be apportioned equally to clades: models such as the geometric distribution or the equal-rates Markov (ERM) model can be used to generate the amount of diversity variation expected under an equal-rates model.

Do beetles have more species than expected by chance?

Theoretical distributions again

As before, probably the simplest way to test whether a particular clade such as the beetles is unusually species-rich is to compare the distribution of insect diversity among orders with a suitable theoretical distribution such as the geometric. From a visual inspection of the bar plot in Figure 9.7, we can see that beetles lie well outside the expected curve under a geometric distribution. By using the probability density of the geometric distribution, we can calculate that if 932,045 species (one estimate of the number of described insect species) are randomly assigned to 29 orders, the probability of any order containing 360,000 species or more is virtually zero. So it would seem that if the geometric distribution is an appropriate null model, the diversity of beetles is highly unlikely to have arisen by chance.

But once again, although the geometric distribution has some basis in the diversification process, it is not an explicitly phylogenetic model; we could apply the same test to any kind of data produced by successive, or simultaneous, splitting of groups into smaller groups. Another limitation is that while we can identify clades with unexpectedly high diversity, we cannot infer anything about *diversification rates* from this kind of test. This is because this test ignores any information we have about the timing of diversification events. Why is this important? Well, beetles may be unexpectedly diverse, but what if they are also much older, with deeper evolutionary origins, than other insect orders? If this were the case, beetles may not have diversified any more rapidly than other insect orders—they may be so diverse simply through having had a vast amount of time to accumulate species. It is important to distinguish high diversity that has resulted from a long evolutionary history from elevated diversification rates, because this will have implications for many hypotheses about the mechanisms underlying diversity of taxa such as beetles.

Phylogenetic tests of diversification rates using topology

One way of testing whether a particular clade is unusually diverse under a phylogenetic model is to use what have been called 'topological' tests for diversification rate variation.[4] Such tests use information on branching relationships only (the topology of the phylogeny), and do not use information on the timing of diversification events or the length of branches in the phylogeny. Nonetheless, a phylogenetic topology still permits tests for variation in diversification rates if we apply the simple principle that a difference in the species richness of two sister clades must reflect a difference in their average rate of diversification. This is because sister clades are the same age, and so have had the same amount of time in which to accumulate species (sister-clade comparisons are also discussed in Chapters 8 and 10).

Topological tests of diversification rate build upon tests of phylogenetic imbalance by calculating an explicit probability for the degree of imbalance at each node. Under an ERM process of diversification, all possible ways of dividing the N species descending from a node among its two daughter clades are equally probable. For example, if a node has six descendant species, there will be the same chance of the two daughter clades containing five and one, four and two, or three and three species. Therefore we can calculate the probability of a particular degree of imbalance using the uniform probability density function. For example, we can calculate the imbalance probability for beetles and their sister clade, the super-order Neuropterida (which consists of the orders Neuroptera, Megaloptera, and Rhaphidioptera). With the 6000 or so species of Neuropterida against the beetles' 360,000 species, we get a nodal probability of 0.032. This suggests there is only a 3.2% chance that a difference in diversity of this size between sister clades could have arisen if all lineages had equal probabilities of diversifying. This low probability implies that there was a significant increase in diversification rate somewhere in the beetle lineage after they diverged from Neuropterida. Actually, some of the inter-ordinal relationships in insects are uncertain, and recent molecular evidence suggests that the sister clade to beetles may in fact be the Strepsiptera, with only around 600 species. This would give us an even lower probability (0.0032 or 0.32%) that the difference in diversity between beetles and their sister clade could have arisen by

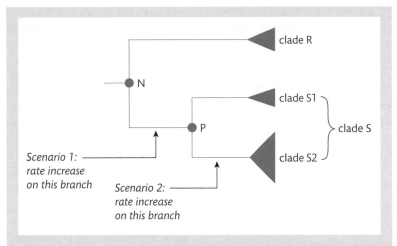

Figure 9.9 The 'trickle-down' effect can complicate the detection of diversification rate shifts on a phylogeny. Clade S is more diverse than its sister clade R (indicated by the width of the triangles), producing a high degree of imbalance at node N. This may lead us to infer a diversification rate increase on the branch between N and P (scenario 1), but imbalance at node N might also result from a rate increase below node P, on the branch leading to clade S2 (scenario 2).

chance, if diversification rates were equal. All the individual nodal probabilities for a given phylogeny can be combined (for example, by summing or multiplying) to obtain a whole-tree test for diversification rate heterogeneity, similar to an imbalance score.[4]

Nodal probabilities derived from the uniform probability density function give us a rough way to test the imbalance at a node against ERM, but don't have much power to detect shifts in diversification rate. A more powerful approach is to calculate the probability of a node N with n descendant species containing two clades R and S, with r and s species, respectively (so that $r + s = n$), given an ERM model of diversification and a particular value of the branching rate (often symbolized as λ). The value of λ is chosen so as to maximize the likelihood of the data (the product of the imbalance probabilities of all nodes). The likelihood of the data under a model with a single λ value can then be compared with that under a second model in which each of the two subclades is permitted to have a different

λ value. If the likelihood of the two-rate model is significantly higher than that of the one-rate model, we infer that a diversification rate shift has occurred below the node N.[4]

One issue that complicates the interpretation of nodal probabilities and likelihoods is the 'trickle-down' effect, which arises from the fact that many of the nodes in a phylogeny are nested, and therefore are not independent. In **Figure 9.9**, a high degree of imbalance at node N might be the result of a diversification rate shift along one of the two branches immediately descending from that node (scenario 1). However, it might also be driven by a shift at a shallower level (within clade S), so that the imbalance at node N is influenced by the imbalance at its descendant node P (scenario 2). This makes it difficult to determine exactly where the shift has occurred. We can deal with this issue by testing for rate shifts at both node N and the crown nodes (the node at which the earliest branching event occurred) of R and S. If a significant shift is detected at N and also at either R or

S, it suggests that the shift at N is likely to result from the trickle-down effect. If a shift is detected at N but not at R or S, then the shift is more likely to have occurred immediately below N.

Phylogenetic tests of diversification rates using branch lengths

Topological tests for shifts in diversification rate on phylogenies make use of information on branching relationships among clades, but not divergence times or branch lengths. A range of other methods utilize branch lengths as well as topology for detecting shifts in diversification rates.

If you divide the logarithm of a clade's species richness (N) by its age (t), you get an estimate of the clade's average net diversification rate throughout the period of its history. The clade's age could be either the age of its crown node, indicating the point when the descendants of present-day species began diversifying, or the stem age, indicating the point when the clade's ancestral lineage diverged from that of its sister clade. By itself, though, a set of diversification rate estimates for the different clades or taxa in a phylogeny doesn't allow us to determine which clades have undergone unusually rapid diversification. To do this, we need to estimate the probability that the single ancestral lineage of each clade could have produced the observed number of species in the time available under a random diversification process.

Recall that the net diversification rate is simply the difference between the rates of speciation and extinction. If we adopt the typical notation for speciation rate λ and extinction rate μ, the probability of a single ancestral lineage diversifying into a given number of descendant species after a certain amount of time has elapsed depends on the value of $\lambda - \mu$. It also depends on the ratio of speciation to extinction μ/λ, also known as the extinction fraction. A very high extinction fraction (say 0.99) will result in a low net diversification rate, even if the absolute values of λ and μ are quite high, because many lineages will go extinct. The equations to estimate the probability of observing the present-day diversity of a clade can be quite complex, but

typically they combine information on the clade age and species richness with a guess of the extinction fraction. The extinction fraction is a guess because a phylogeny reconstructed from present-day species does not include extinct lineages, and we usually have no way of knowing how many lineages went extinct without leaving living descendants. The value for the extinction fraction is usually arbitrary, with little or no information to guide us on what a reasonable value might be. Some studies deal with this by comparing results using a few different values for the extinction fraction; for example, a very low value (e.g. zero) and a very high vaue (e.g. 0.99). At the very least, this can tell us how sensitive the results and conclusions are to different assumptions about the extinction rate.

Multiplying the probabilities for all clades gives us the likelihood of the data (the observed set of clade diversities) for a whole phylogeny, assuming a constant-rates diversification model, the set of clade ages, and our arbitrary fixed value for the extinction fraction. We can then iteratively adjust the value of $\lambda - \mu$ until we get the highest possible likelihood score. This gives us a maximum likelihood estimate of the diversification rate (we don't have to hold the extinction fraction constant—we could also estimate it by maximum likelihood). But this is a single estimate of diversification rate for the whole tree—it still doesn't allow us to pinpoint rate shifts. To do this, we need to increase the complexity of the model by adding some extra parameters. A good way to start would be to have a model with two diversification rates r_1 and r_2, and another parameter B that denotes a particular branch in the phylogeny. Beginning with one branch B, we separately optimize the values of r_1 and r_2 for clades above and below B to find their maximum likelihood estimates. We then compare the overall likelihoods of the single-rate and two-rate models, using a likelihood ratio test or the Akaike information criterion (AIC). If the two-rate model is a significantly better fit to the data, we can conclude that a rate shift occurred along the branch B. If the single-rate and two-rate models fit equally well, the principle of Occam's razor tells us that we should favour the simpler single-rate model. We can repeat this process for all branches in the phylogeny to identify any branches along which a rate shift has occurred.[5]

> 🔒 **Key points**
>
> - Testing whether a particular clade has greater diversity than expected by chance can be done in several ways:
> - Comparing the clade's diversity with the expectation under a geometric distribution.
> - Topological tests that use information on phylogenetic branching relationships, but not branch lengths or divergence times.
> - Tests that use information on branch lengths as well as branching relationships.

Is the timing right? Dating the key historical events

So far in this chapter, we have explored ways of testing whether:

1. species are distributed among insect orders less evenly than expected by chance;

2. beetles in general, and angiosperm-feeding subclades of beetles in particular, have significantly more species than would be expected by chance;

3. the beetle radiation (or the radiation of angiosperm-feeding beetles) was unusually rapid.

These are important steps in establishing whether beetle diversity is, as we initially wondered, remarkable in comparison with other insect groups, and hence whether we are justified in seeking a deterministic mechanism to explain it. We can now begin to examine more directly the mechanism we pondered earlier in the chapter: that the advent of phytophagy, and more particularly the shift to angiosperm-feeding, allowed some beetle lineages to radiate in concert with the massive radiation of angiosperms in the Mid-Cretaceous Period.

For the coevolution hypothesis to be supported, the timing of the key historical events—the shift to angiosperm-feeding in beetle lineages, the radiation of angiosperms, and the radiation of beetles—has to fit. If the timing of the three events was more or less coincident, that would be consistent with the coevolution hypothesis. What if angiosperm-feeding arose long after the angiosperm radiation began, and the beetle radiation took off some time after that? This would fit less neatly with the coevolution hypothesis, because it would suggest that beetles and angiosperms did not diversify in lockstep with each other. However, this scenario would not be inconsistent with the idea that beetle diversification was driven largely by phytophagy. It may simply mean that beetle lineages did not take up the opportunity to switch to angiosperm-feeding until after angiosperms were already diverse, but when they did make the switch it had the effect of elevating beetle diversification. On the other hand, if beetles began their massive radiation long before the shift to angiosperm-feeding, or long before the origin and early diversification of angiosperms, the coevolution hypothesis would be on fairly shaky ground. So, how do we estimate the timing of these events? Perhaps more importantly, how precisely can we estimate them, and what degree of confidence can we have in our estimates?

We will briefly recap some of the kinds of data and approaches that we can use to estimate the age of origin and the timing of divergences in a clade of organisms (see also Chapter 6). To begin with, we could base our estimates on fossil data alone, assuming that the age of the oldest known fossil attributable to our clade represents a minimum age of origin of the clade. But if the fossil record of our clade is sparse, geographically or temporally biased, or just poorly or unevenly sampled, it is possible that the true age of origin is much higher than the oldest fossil. Estimating node ages from molecular phylogenies provides another way of dating divergences. A common approach is to use fossil dates (or in some cases, dates of geological events such as continental break-up) to calibrate the times estimated from molecular sequences to an absolute timescale. Typically, if a fossil exists that can be attributed with confidence to a particular clade, the age of this fossil can be used to constrain the minimum age of origin of that clade. If the rate of molecular evolution varies across the tree, including multiple calibrations should help to provide more robust molecular date estimates. Of course, fossil-calibrated molecular date estimates are not independent of the ages of the oldest known fossils; if older fossils are discovered, the calibrations change and so do the molecular date estimates. Yet molecular and fossil dates do sometimes disagree substantially. Before accepting the molecular date estimates at face value, it is important to understand how, and under what assumptions, these estimates were derived from the molecular data. In the worked example in Case Study 9, we examine the approaches that some studies have taken to estimating the timing of the beetle and angiosperm radiations, and of the origins of angiosperm-feeding in beetles.

Angiosperm-feeding and rapid diversification: is there a correlation?

There is a lot to consider before drawing firm conclusions about the relative timing of the key historical events in this story. For the moment, let us accept that the radiation of one major angiosperm-feeding clade of beetles, the weevils (superfamily Curculionoidea) occurred contemporaneously with that of angiosperms. This is of course consistent with the coevolution hypothesis. If this was the *only* shift to angiosperm-feeding in beetles (we know it wasn't—this is just a thought experiment), would this be enough to conclude that there is a cause–effect link between angiosperm-feeding and elevated diversification in beetles generally?

Now is probably a good time to recall the difference between singular and general questions that we touched on earlier in this chapter. The question we addressed in the previous section is a singular historical one—we were interested in reconstructing unique events in the history of beetles and angiosperms to establish whether the timing was coincident. As we have seen, this is certainly an important part of the evidence we need to

evaluate the phytophagy/coevolution hypothesis for beetle diversity. If the timing was coincident, it may indicate that there is a cause–effect association between angiosperm-feeding and beetle diversity—but alternatively, it could be an incidental association. We can't rely on a single historical event to provide good evidence for a more general cause–effect relationship between phytophagy or angiosperm-feeding and elevated diversification—we need to show that the connection between the two phenomena has happened more widely. To search for this kind of evidence we need replication, in the form of multiple independent origins of angiosperm-feeding in different lineages of beetles, or insects more generally.

Contrasts between sister clades

Figure 9.10 shows a hypothetical phylogeny. Assume that the triangles at the tips are beetle families, six of which are angiosperm-feeders (grey) and ten of which are fungivorous (black). The width of the triangles indicates the species richness of each family. Suppose that the arrangement of families with different feeding modes looked like the phylogeny on the left. If we do a simple statistical test for a difference in the mean species richness of families of each feeding mode, we will find that angiosperm-feeding families are significantly more species-rich, on average, than fungivorous families. This may then lead us to the conclusion that angiosperm-feeding leads to higher species richness.

But this test would be misleading, because it assumes that every family represents an independent piece of evidence with which to evaluate the association between species richness and feeding mode. In effect, it assumes that the feeding mode of each family has evolved separately and independently on 16 occasions, with an influence on diversification rate each time. In fact, it is far more parsimonious to assume that there was only a single shift between feeding modes, in either of the lineages descending from the basal node of the tree. So we do not have 16 independent observations for a statistical test. We have one shift between feeding modes, and hence one

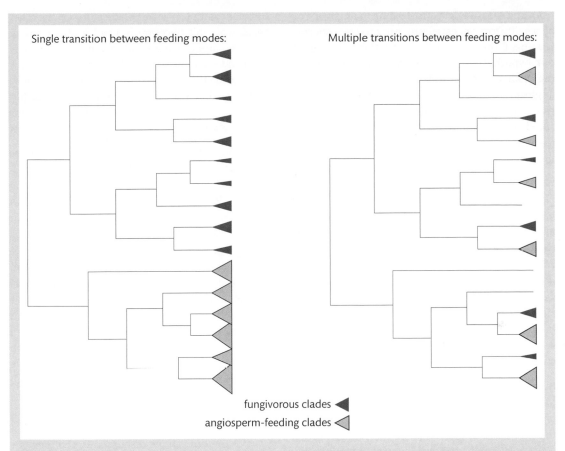

Single transition between feeding modes:

Multiple transitions between feeding modes:

fungivorous clades ◀

angiosperm-feeding clades ◁

Figure 9.10 The problem of phylogenetic non-independence in a hypothetical beetle phylogeny. Species richness of fungivorous (black) and angiosperm-feeding (grey) beetle families is indicated by the width of the triangles. In the tree on the left, there was only one evolutionary transition between feeding modes, so we have only one phylogenetically independent contrast and hence only one statistically valid observation. In the tree on the right, multiple evolutionary shifts between feeding modes provide us with six phylogenetically and statistically independent observations. In each case, the shift is associated with a difference in species richness.

observation—and we can't do a statistical test with only one observation. This doesn't necessarily mean that feeding mode doesn't influence diversity, just that it is difficult to rule out the possibility that the association is incidental because any number of other changes, which could influence diversity, might also have occurred following the basal divergence. With respect to the character we are interested in, the families in this tree are *phylogenetically non-independent*, and therefore statistically non-independent.

Now consider the phylogeny on the right in Figure 9.10. Here there are six pairs of sister families, in which each pair consists of one angiosperm-feeding family together with a fungivorous sister family. In this case, there cannot have been a single shift between feeding modes, because this would not be consistent with the data. We can assume that the difference in feeding mode arose independently six times; this is the simplest explanation for the data. It doesn't matter whether the difference in feeding mode in each pair arose by a shift from ancestral angiosperm-feeding to fungivory, or by a shift from ancestral fungivory to angiosperm-feeding. Either route is valid for the purposes of our hypothesis of an association between feeding mode and present-day species richness. If the angiosperm-feeding clade has a higher diversity than the fungivorous clade in all, or nearly all, the six sister-clade pairs, it seems unlikely that the association between feeding mode and diversity is incidental.

When we are testing a hypothesis using sister-clade pairs, the observations in our test are no longer individual clades—they are pairs of clades. So for each pair of sister clades, we first need to calculate a contrast in species richness. This could be a simple difference in species richness between the two sister clades. As we have already seen, the difference in species richness between sister clades can be interpreted as a difference in the average rate of diversification since the two clades diverged, because sister clades are the same age.

In one family-level beetle phylogeny there are eight pairs of sister clades in which one clade in each pair is phytophagous and the other is non-phytophagous.[1] In six of the pairs, the phytophagous clade is more diverse than its non-phytophagous sister clade. This is not sufficient to obtain a significant association between feeding mode and species richness under a simple statistical test (Wilcoxon test). When pairs of sister clades are selected in which one clade is angiosperm-feeding and the other is not, the association is closer to being significant, but this still does not represent compelling evidence for a link between angiosperm-feeding and diversification rate.

This example perhaps serves to highlight one of the limitations of using sister-clade contrasts to test diversification rate hypotheses. Because this approach relies on identifying phylogenetically independent origins of the character we are testing, the number of possible contrasts (hence the power of the statistical tests) for a particular group of organisms is often limited. This means that huge phylogenies are often needed in order to obtain a

sufficient sample of sister-clade pairs. On the other hand, we don't necessarily need a phylogeny to identify pairs of clades that are phylogenetically independent of other pairs. A taxonomic hierarchy can provide this information, as long as we can be confident that we can identify pairs of taxa (for example, families or genera) that are each other's closest relatives.

There are other assumptions we need to make when using sister-clade contrasts. If the predictor variable is a discrete character (such as feeding mode), it is not always straightforward to assign a particular character state to a particular clade or taxon. For example, if we are calculating contrasts between genera, and a genus consists of 90 angiosperm-feeding species and 10 non-angiosperm feeders, do we call this an angiosperm-feeding genus? In addition to this, we might have clades in which there have been reversals—shifts from gymnosperm-feeding to angiosperm-feeding, then back to gymnosperm-feeding. We can't deal with this using the sister-clade approach because we are only using information about the current state of taxa at the tips of the phylogeny.

Trait-dependent speciation and extinction

Despite the limitations discussed in the previous section, the sister-clade approach does have some clear advantages. We can obtain contrasts from phylogenies constructed at the level of higher taxa such as families, provided that we have an estimate of their species richness. And we don't need estimates of divergence times between sister clades, or even, necessarily, a phylogeny. But as large dated phylogenies with a high level of completeness have become more common, other methods have been developed that make fuller use of the information provided by phylogenies to estimate speciation and extinction rates. One widely used approach is BiSSE (Binary State Speciation and Extinction) and related methods.[6] The key feature of these methods is that instead of calculating diversification rate independently of the phenotypic character we are interested in, and then testing for an association between diversification rate and origin of the character, they model the two together. That is, they reconstruct the evolutionary history of the character, and allow speciation and extinction rates to change with each shift in the character state. By calculating the maximum likelihood of the data under alternative models that allow different parameters to vary and then comparing the different models, we can make inferences about the influence of a character on speciation and extinction rates.[6]

For example, we might construct two models to explain the effect of a binary character (e.g. phytophagous/non-phytophagous) on diversification rates. In model 1, we constrain both speciation and extinction rates to be the same for lineages with each character state. In model 2, we constrain the extinction rate to be the same, but we allow speciation rates to vary between lineages with each state. We can then build likelihood functions for the two models, use these to compute the maximum likelihood of the data under each model, and compare the fit of the models using likelihood ratio tests. If model 1 is favoured, it suggests that neither speciation nor extinction rate varies predictably between the two character states. If model 2 is favoured, it suggests that speciation rate varies between the two character states, because a model that estimates two separate values for speciation rate provides a better fit to the data than one with a single value for speciation rate. A more complete investigation would fit and compare a larger number of models that estimate either one or two parameter values for speciation rate, extinction rate, and the rate of transition from one character state to the other.

BiSSE and related methods have become widely used in studies of associations between traits and diversification rates. However, although these methods are in some ways more powerful than methods based on comparing sister clades (because they use more of the information provided by a phylogeny), they also have their own shortcomings. For example, these methods are more sensitive to inaccuracy or imprecision in branch length estimates, and they are more reliant on the completeness of phylogenetic sampling.[7]

Key points

- Testing for an association between a character and diversification rate can be done in several ways. Comparisons of multiple pairs of sister clades are based on the principle that a difference in species numbers between sister clades must reflect a difference in diversification rate.

- BiSSE and related methods allow the rate of diversification to change as a character evolves along the branches of a phylogeny.

Conclusion

We have now arrived at the end of our exploration of beetle diversity and some approaches that can be taken to explaining it. At this point, we should recap the steps that we have taken in this chapter. We began with an interesting observation which deserves explanation: the order Coleoptera is massively species-rich, far more so than any other insect order. Then we brought in a few leads that might provide some clues to the great diversity of beetles: most beetles are phytophagous (plant-feeders), and the largest beetle radiations, such as the weevils, are almost entirely angiosperm-feeders. Then we considered a plausible theoretical explanation for the diversity of angio-sperm-feeding beetles: they diversified in concert with angiosperms through a process of coevolutionary 'escape and radiation'. Based on this theory, we would expect that angiosperm-feeding beetle clades diversified more rapidly than other clades, and that the timing of their diversification was coincident with the timing of the angiosperm radiation.

Before we set out to test the expectations of the coevolution hypothesis, we first needed to establish that beetles are actually more diverse than would be expected by chance. We did this by comparing the observed patterns of insect diversity with various null models: ways of distributing species among higher taxonomic groups at random, or at least in ways that exclude any role for coevolution in the diversification of insects. We then considered the evidence that the timing of the radiations of angiosperms and angiosperm-feeding beetles was coincident. Finally, we explored ways of testing for a statistical association between diversity or diversification rate and feeding mode in beetles.

What can we say now about the mechanisms underlying the massive diversity of beetles that we see today? Do we have sufficient evidence to support the idea that shifts to feeding on angiosperms in the Mid-Cretaceous Period allowed beetles to diversify into an immense variety of morphological and ecological types? Or are there still more questions that we need to answer, and complexities that we need to explore more fully? For example, if there was a causal connection between angiosperm and beetle diversification, was it a case of the beetle radiation 'hitch-hiking' on the building angiosperm radiation, diversifying in response to the new opportunities for exploiting plant habitat and food resources? Or was it truly reciprocal coevolution, with beetle diversification and evolution of tolerance to plant defence compounds feeding back into the diversification of angiosperms? What further evidence would we need in order to distinguish these two scenarios?

● Points for discussion

1. In this chapter we examined an approach to testing hypotheses of diversity that consists of asking four questions: (1) Is diversity distributed among taxa non-randomly? (2) Is the diversity of a particular clade higher than expected? (3) Does the timing of key historical events fit the hypothesis? (4) Is there a statistical association between diversity or diversification rates and the suggested explanatory factor? But should we always need to answer all four questions? Can we arrive at a robust conclusion if we only have evidence bearing upon two or three of these? Do any of these four questions carry more weight than the others?

2. If we are interested in understanding the underlying causes of beetle diversity, can analyses of diversity patterns of broader taxonomic scope contribute to this understanding—for example, testing for associations between feeding mode and diversity across clades within Holometabola, within insects, or within arthropods? Or does the uniqueness of the beetle radiation mean that we should confine our analyses to beetles alone?

3. This chapter has focused on testing whether beetle diversification is linked to the evolution of angiosperm-feeding, as expected under the coevolution hypothesis. But what might have been the underlying mechanisms that triggered the apparently sudden diversification of angiosperms in the first place? The evolution of flowers as a key innovation may have kick-started the radiation, and it has been suggested that polyploidy and hybridization provided the fuel for continued diversification. Could such mechanisms have also contributed to the massive diversification of beetles?

✳ References

1. Farrell BD (1998) 'Inordinate fondness' explained: why are there so many beetles? *Science* 281: 555–9.

2. Ehrlich PR, Raven PH (1964) Butterflies and plants: a study in coevolution. *Evolution* 18: 586–608.

3. Misof B, and 100 other authors (2014) Phylogenomics resolves the timing and pattern of insect evolution. *Science* 346: 763–7.

4. Moore BR, Chan KMA, Donoghue MF (2007) Detecting diversification rate variation in super trees. In Bininda-Emonds ORP (ed.), *Phylogenetic Supertrees: Combining Information to Reveal the Tree of Life*. Kluwer Academic, Dordrecht: pp.487–533.

5. Alfaro ME, Santini F, Brock C, Alamillo H, Dornburg A, Rabosky DL, Carnevale G, Harmon LJ (2009) Nine exceptional radiations plus high turnover explain species diversity in jawed vertebrates. *Proceedings of the National Academy of Sciences of the USA* 106: 13,410–14.

6. Maddison WP, Midford PE, Otto SP (2007) Estimating a binary character's effect on speciation and extinction. *Systematic Biology* 56: 701–10.

7. Davis M, Midford P, Maddison W (2013) Exploring power and parameter estimation of the BiSSE method for analyzing species diversification. *BMC Evolutionary Biology* 13: 38.

Case Study 9
Timing of diversification: the beetle and angiosperm radiations

Investigating the causes of unusually high diversity in some groups of organisms often involves reconstructing historical events as well as testing for general associations between diversity and a suspected causal process. A prominent hypothesis for the massive diversity of beetles is coevolution with angiosperms. The proposal is that from the Early Cretaceous period, the diversification of angiosperms and beetles proceeded synchronously, each driving the other. Alternatively (or additionally), the increasing diversity of angiosperms simply created many new ecological opportunities which stimulated the diversification of beetles. Either way, part of the evidence that we can use to evaluate these hypotheses is the relative timing of the key events: the diversification of angiosperms, the shift to angiosperm-feeding in beetle lineages, and the diversification of angiosperm-feeding beetles. Various lines of evidence can be used to infer the timing of historical events such as these, including the fossil record, molecular phylogenetic data, and model-based reconstruction of the evolution of phenotypic or ecological characters.

Molecular estimates of divergence times are usually calibrated against an absolute timescale using estimated ages of fossils that can be assigned confidently to a particular present-day clade or taxon. However, the age of a fossil is rarely known with enough precision to be described by a single point estimate. Typically, upper and lower bounds of the geological stratum in which the fossil was found are provided. One of the most thorough molecular dating studies of the major insect groups, by Bernhard Misof and colleagues,[1] used a calibration for the origin of Coleoptera based on a fossil with estimated age bounds of 235–222Mya. Following phylogenetic analysis and molecular dating, this produced an estimate for the divergence of Coleoptera from their sister clade Strepsiptera of 245–290Mya (median 270Mya). However, the earliest reported beetle fossil, *Adiphlebia lacoana* from the Pennsylvanian Epoch of the Carboniferous Period, is quite a lot older than this (323–298Mya),[2] although some authors have disputed whether *Adiphlebia* is actually a beetle.[3] Nonetheless, it is commonly accepted that the earliest beetle fossils are from the Early Permian Period (298–272Mya). How would Misof et al.'s estimate change if Early Permian beetles, or even *Adiphlebia lacoana*, had been considered suitable to use as the calibration for the Coleoptera–Strepsiptera divergence? Presumably, the inferred age of the split would be pushed back, as would the inferred timing of the radiation of the major angiosperm-feeding lineages. This might be enough to alter our judgement of the relative timing of beetle and angiosperm diversification.

When did angiosperms undergo their massive radiation? For a long time, the sudden appearance and diversification of angiosperms in the fossil record of the Mid-Cretaceous Period, in a form already substantially differentiated from their nearest relatives, was considered a great mystery. Indeed, in a letter to Joseph Hooker in 1879, Charles Darwin famously referred to the origins of angiosperms as an 'abominable mystery'. It is not surprising that with the development of molecular dating and new ways of analysing fossil data, botanists were particularly keen to resolve the origin and early diversification of angiosperms. This job is by no means complete, but a great deal of progress has been made since the 1970s.

One important difference between angiosperms and beetles is that angiosperms produce pollen, which tends to fossilize well because it is quite resistant to degradation. Not only that, but pollen grains are so small, abundant, and easily dispersed, that if angiosperms were present in reasonable numbers at a particular time and place, we really should expect to find fossil pollen in the appropriate sediments. The earliest fossil angiosperm pollen is from the Valanginian Age of the Early Cretaceous Period (140–136Mya). This doesn't necessarily mean that angiosperms didn't exist before the Cretaceous, but if they did, they were probably not very common or perhaps restricted to a limited geographic area or habitat type. The 'dark and disturbed' hypothesis suggests that pre-Cretaceous angiosperms were restricted to the dark understoreys of tropical and subtropical forests, environments that do not usually lend themselves well to fossilization. Molecular dating studies very often place the origin of angiosperms in the Jurassic (201-145Mya), Triassic (252-201Mya), or even the Early Permian (298–272Mya). The discrepancy between the molecular dates and the age of the oldest known fossils has led some authors to conclude that there must be older angiosperm fossils out there, still waiting to be discovered.[4]

From the body of evidence we have, then, it seems that beetles originated well before the angiosperms became globally diverse and widespread. But the questions of more relevance to the beetle–angiosperm coevolution hypothesis are: (1) When did beetles shift to angiosperm-feeding? (2) When did the massive radiations of angiosperm-feeding beetles occur? Toby Hunt et al.[5] reconstructed the phylogeny of most of the world's beetle families, and then estimated divergence times using the 'rate-smoothing' procedure penalized likelihood (PL). Essentially, this means taking the branch lengths that were derived solely from the amount of molecular change, and adjusting them, based on some assumptions about sequence evolution, so that the tips of the phylogeny all line up at time zero (the present day). This is a rather different approach to divergence dating than that taken by Misof et al.[1] Rather than estimating the age of the root of the beetle phylogeny from the data with a minimum age constraint, as done by Misof et al., Hunt et al. assumed an origin of beetles at 285Mya, *a priori*, and fixed the root of their phylogeny at this age. With a fixed age of 285Mya at the root and a fixed age of zero at the tips, the PL method then transformed all branch lengths and node heights into units of absolute time.

Divergence times estimated in this way will depend heavily on the assumptions that the PL method makes about the rate of sequence evolution, and the variability in this rate across the phylogeny. For this reason, Hunt et al. cross-validated the PL parameters using seven fossil calibration points. The dated phylogeny that they obtained in this way allowed them to reconstruct the number of modern beetle lineages that were present at different points in time. We use the term 'modern' beetle lineages here because we are talking about a phylogeny reconstructed from present-day taxa. This means that the lineages represented by the phylogeny are the ancestors of present-day species, and the phylogeny carries no information about lineages that may have existed in the past, but are now extinct and left no present-day descendants. The plot of Hunt et al.'s phylogeny shows that 145 modern beetle lineages were already present at 140Mya, their accepted age for the origin of angiosperms (see **Figure 9.11**).

From the evidence of Hunt et al.'s analysis, it would seem that the overall beetle radiation was well and truly under way by the time the angiosperms originated. The age of origin of crown-group angiosperms may well have been earlier than 140Mya, as we have just seen, but from the fossil pollen evidence it seems unlikely that angiosperms were widespread, or had begun their massive diversification, prior to the Cretaceous. What if Hunt et al. had used an older date than 285Mya for the origin of beetles, perhaps an estimate based on Early Permian fossil calibrations? Then it would be likely that an even greater number of beetle lineages would be inferred to have been present at the time the angiosperms originated. On the other hand, we should also keep in mind that most of the tips of Hunt et al.'s phylogeny are beetle families. Even if most of the beetle families did arise by the end of the Jurassic, the switch to phytophagy, and the rapid diversification of clades that are mostly angiosperm-feeders today, may still not have occurred until the Mid-Cretaceous. So we need to examine the evidence for the origin of phytophagous, and particularly angiosperm-feeding, beetle clades more closely.

The largest radiations of phytophagous beetles are the superfamilies Curculionoidea (weevils) and Chrysomeloidea (leaf and longhorn beetles), which together account for at least 110,000 species—about a third of known beetle species. In recent phylogenetic reconstructions, the sister clades of both superfamilies are non-phytophagous (mostly fungivorous). Within the Curculionoidea, one family (Curculionidae) are angiosperm-feeders, and these comprise more than 80% of the species in the superfamily. The remaining five families (all much less species-rich) are gymnosperm-feeders. This suggests two things that are important: (1) the common ancestor of present-day weevils was

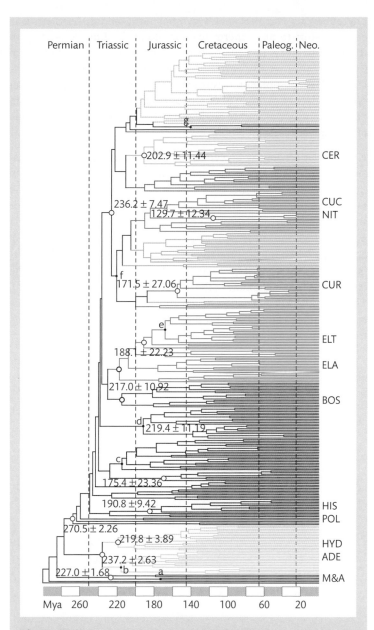

Figure 9.11 A phylogeny of 340 higher taxa (mostly families) of Coleoptera. The ages of nodes were estimated by assuming an a priori age of 285Mya for the origin of Coleoptera, and then using the rate-smoothing method penalized likelihood, to stretch the branches so that the tips all line up at time zero. The numbers show the age estimates for some of the nodes with 95% confidence intervals, and the different colours of the branches indicate different taxonomic subdivisions of Coleoptera.[5]

From Toby Hunt, Johannes Bergsten, Zuzana Levkanicova, et al, (2007) A comprehensive phylogeny of beetles reveals the evolutionary origins of a superradiation, *Science* 318: 1913–1916. Reprinted with permission from AAAS.

phytophagous and probably a gymnosperm-feeder; (2) there was a single shift to angiosperm-feeding along the lineage leading to the Curculionidae. So if we have an estimate of the age of origin of Curculionoidea, we also have an estimate of the timing of one particular shift from non-phytophagy to gymnosperm-feeding. And if we can estimate the age of origin of Curculionidae, we also have an estimate of the timing of the shift to angiosperm-feeding which preceded one of the largest beetle radiations.

A molecular phylogeny of weevil families, calibrated with eight fossil dates, by Duane McKenna et al.[6] estimated the age of origin of the Curculionoidea at around 175–164Mya, and the age of origin of the family Curculionidae at around 140–150Mya. These estimates are for *stem ages*, the age of divergence of a clade from its sister clade. The ages at which the ancestral lineage of Curculionidae first bifurcated, and hence when the descendants of modern species began diversifying (the *crown age*), is somewhat later. McKenna et al. estimated the crown age of Curculionidae in the late Early Cretaceous (125–112Mya). This puts the diversification, if not the origin, of this major angiosperm-feeding clade at around the same time that the angiosperms themselves were diversifying.

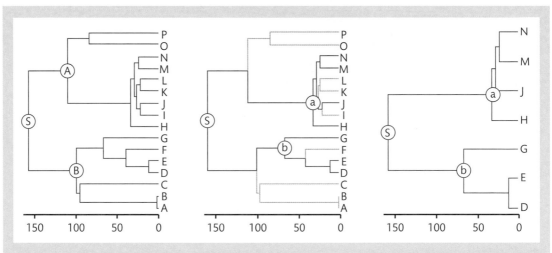

Figure 9.12 Under-sampling of lineages can affect crown age estimates. The tree on the left shows the relationships among 16 taxa (A–P) of two clades. The crown ages of the two clades are 110Mya (node A) and 99Mya (node B), and the stem age is 158Mya (node S). This is the tree we would recover if we had samples of all 16 taxa in our database. But if we only had samples for a small selection of taxa (D, E, G, H, J, M, and N), the only branches we can reconstruct are the ones shown in black in the middle tree. Therefore we would underestimate the ages of the crown nodes of the two clades: 32Mya (node a) and 66Mya (node b), although the stem age estimate will still be the same. More correctly, we may accurately estimate the ages of the nodes a and b, but wrongly infer that these are the crown nodes of the two clades. When we straighten out the branches and get rid of the grey lines, the tree we will reconstruct from this limited set of samples will look like the one on the right; much of the true history of the group will be hidden.

McKenna et al.'s findings certainly seem consistent with the beetle–angiosperm coevolution hypothesis. But there are a couple of points that we should keep in mind. First, it is worth reiterating that their phylogeny, and any phylogeny reconstructed from extant species, only depicts the history of the lineages that survived to leave present-day descendants. What if there was a pre-Cretaceous radiation of Curculionid species that are now extinct? These will have left no signal in the macroevolutionary patterns recovered by the phylogeny. Second, McKenna et al.'s phylogeny only includes a fraction of the 4600 genera, and an even smaller fraction of the 51,000 species, of the family Curculionidae. This means that much of the diversification history of even the descendants of present-day species remains hidden. Most importantly, under-sampling can produce under-estimates of the crown age, unless we are sure that the sampled lineages are connected to one another through the root of the Curculionidae (**Figure 9.12**). Also, there may have been periods of rapid diversification more recently than the Early Cretaceous which simply cannot be recovered from a phylogeny that represents only a sample of the major lineages. Of course, this is just one of the drawbacks of macroevolutionary analyses on such a hugely diverse group of organisms. It will be a long time before we can reconstruct the evolutionary history of more than a small sample of known beetle species; in the meantime, we must be aware of the limitations on what we can reasonably infer from incomplete data.

Questions to ponder

1. How confident in the identity and age of a fossil should we be before we use it to calibrate divergence times in a molecular phylogeny? Is a fossil still useful if we can only place it in a higher taxon, and estimate its age within very broad limits?

2. Suppose we had a good fossil record for a group, with lots of species confidently placed within genera and tightly bound age estimates. Apart from their use in calibrating the timescale of divergences, could fossils be integrated with molecular phylogenies in other ways?

3. Are there other ways (apart from fossils) that we might be able to calibrate the timing of the beetle and angiosperm radiations?

Further investigation

Does the sudden appearance of a diversified angiosperm flora in the Mid-Cretaceous fossil record suggest that there must be a hidden true history of angiosperms stretching back into the Jurassic or beyond? Or is it plausible that angiosperms really did arise and diversify so rapidly that their appearance in the fossil record seems almost instantaneous? How might the lessons we learn from the Cambrian explosion (see Chapter 4) help us approach the investigation of Darwin's 'abominable mystery?'

References

1. Misof B, and 100 other authors (2014) Phylogenomics resolves the timing and pattern of insect evolution. *Science* 346: 763–7

2. Béthoux O (2009) The earliest beetle identified. *Journal of Paleontology* 83: 931–7.

3. Kukalová-Peck J, Beutel R (2012) Is the Carboniferous *Adiphlebia lacoana* really the 'oldest beetle'? Critical reassessment and description of a new Permian beetle family. *European Journal of Entomology* 109, 633–45.

4. Soltis DE, Bell CD, Kim S, Soltis PS (2008) Origin and early evolution of angiosperms. *Annals of the New York Academy of Sciences* 1133, 3–25.

5. Hunt T, Bergsten J, Levkanicova Z, Papadopoulou A, John OS, Wild R, Hammond PM, Ahrens D, Balke M, Caterino MS, Gómez-Zurita J, Ribera I, Barraclough TG, Bocakova M, Bocak L, Vogler AP (2007) A comprehensive phylogeny of beetles reveals the evolutionary origins of a superradiation. *Science* 318: 1913–16.

6. McKenna DD, Sequeira AS, Marvaldi AE, Farrell BD (2009) Temporal lags and overlap in the diversification of weevils and flowering plants. *Proceedings of the National Academy of Sciences of the USA* 106: 7083–8.

Why are there so many species in the tropics?

10

Roadmap

Why study the latitudinal diversity gradient?

Biodiversity varies enormously from region to region. One of the challenges for macroecology and macroevolution is to discover general, predictable patterns within this variation that might shed light on the processes responsible for generating biodiversity and explaining why it varies among regions. Probably the most impressive geographical pattern of biodiversity is the decline in species richness from the tropics towards high latitudes. However, even the most fundamental questions about why this pattern exists are not fully answered, so we are still not sure whether the latitudinal diversity gradient has a historical explanation (e.g. tropical ecosystems are older) or an ecological one (e.g. there are more available niches in tropical ecosystems). Organizing, testing, and comparing the multitude of possible explanations for global diversity patterns is an exercise in the careful application of scientific logic. It requires us to consider carefully how we can extract information from fossil, phylogenetic, and geographical data.

What are the main points?

- The latitudinal diversity gradient (LDG) is both very general (most major groups of organisms show the pattern, and it is found in the marine and terrestrial realms) and very ancient.
- Proposed explanations for the LDG can be classified into historical (based on the accumulation of diversity through time) and ecological (based on current environmental conditions).
- An underlying assumption of historical explanations is non-equilibrium diversification: diversity is still increasing through evolutionary time.

Photograph: Ecuador rainforest. © Dr Morley Read/Shutterstock.com.

Ecological explanations assume equilibrium dynamics: diversity has levelled off to a 'carrying capacity'.

- Fossil and phylogenetic data provide complementary approaches to testing hypotheses about large-scale geographical patterns in diversity, each with its own strengths and weaknesses.

What techniques are covered?

- Using fossils and phylogenies to test whether diversity is at equilibrium, or still increasing through time.
- Using fossils and phylogenies to test for differences in rates of diversification across latitudes.

What case studies will be included?

- Models that explain the LDG through the global dynamics of speciation, extinction, and geographical range expansion.

". . . there is an inexhaustible variety of almost all animals. There are few places in England where during one summer more than 30 different kinds of butterflies can be collected; but here, in about 2 months we obtained more than 400 distinct species . . ."

Alfred Russel Wallace (1853) *A Narrative of Travels on the Amazon and Rio Negro.* Reeve & Co., London.

The tropics are home to the world's great repositories of biological diversity. In the terrestrial realm, the luxuriant and chaotic growth of tropical rainforests is emblematic of the rich diversity of nature. In the oceans, the same is true of coral reefs. But why do tropical ecosystems harbour such great diversity? When we think about what distinguishes the tropics from higher latitudes, probably the first thing that comes to mind is climate. The tropics are typically warm, wet, and humid, with weather conditions that are fairly consistent and predictable from year to year. But when we start to think about the actual mechanisms that might link such environments with the existence of large numbers of species, it soon becomes apparent that the explanation for tropical diversity is not easily revealed. In fact, the tendency for biodiversity to decline from the equator towards high latitudes -the latitudinal diversity gradient (LDG)- remains one of the great mysteries of macroevolution and macroecology. In this chapter, we will explore this pattern and try to make sense of the numerous ideas that have been put forward to explain it. We will then focus in depth on one particular hypothesis, that rates of evolutionary diversification are higher in the tropics, as a way of examining some of the approaches, data, and techniques that we can use to try and explain tropical diversity.

Explaining tropical diversity

Inexhaustible variety: the biological richness of the tropics

One way to appreciate the richness of tropical ecosystems is to compare them with non-tropical ones. Białowieża Forest in Poland, 52° north of the equator, is one of the largest remaining areas of the primeval forest that once covered most of Europe. Białowieża's 1400km² are home to around 25 tree species. Across the whole of Europe, there are around 124 tree species. In contrast, a single hectare of tropical forest in the upper Amazon Basin or in Southeast Asia can support more than twice the number of tree species as in the whole of Europe. Perhaps the most species-rich tree community in the world is the tropical forest at Yasuní

Figure 10.1 This white-lined leaf frog (*Phyllomedusa vaillantii*) is one of at least 150 amphibian species recorded in Ecuador's Yasuní National Park. Yasuní's equatorial forest habitat is home to some of the richest assemblages of animals and plants on earth. Some of this park's habitat and biodiversity may now be under threat from oil drilling.

Photo by Geoff Gallice (CC2.0 license).

National Park, on the equator in eastern Ecuador. Here, nearly 300 tree species have been recorded on a one-hectare plot, and over 1000 species have been identified from a network of 15 plots across the park.[1] As is typical of tropical forests, the number of tree species within each small plot is not much less than the number of individual trees. If you stand at a particular point in the forest and look around you, it might well be the case that of the trees that you can see, no two are of the same species. Because of the complex structure of the rainforest, most trees support vines, creepers, and epiphytes, further boosting the count of plant species; up to 4000 plant species are estimated to occur in the park.[1]

Yasuní's immense diversity also extends to animals. There are 150 species of amphibians, 121 reptiles, 596 birds, and an estimated 204 mammals and 499 fish—all either the highest recorded, or close to the highest, for similar-sized areas worldwide.[1] In fact, Yasuní National Park lies within a region of South America in which the centres of maximum global diversity of amphibians, birds, mammals, and vascular plants overlap (Figure 10.1). And then there are the invertebrates—although sampling has been very limited, one estimate puts the probable number of insect species in the park at 100,000 or more, as many as in the whole of North America.[1] We have, of course, singled out Yasuní National Park as one of the most exceptionally biodiverse places on the planet. But many other equatorial forests in America, Asia, Africa, and New Guinea are not all that far behind Yasuní in diversity of particular taxa, if not in the congruence of world-record diversity of so many different taxa.

Scientific awareness of the tropical–temperate disparity in biodiversity is comparatively recent. Although ancient Greek philosophers were aware of environmental and biological differences between tropical and temperate zones, few Western naturalists had direct experience of the tropics until the age of exploration and colonization from the fifteenth to the nineteenth centuries. One of the earliest was Walter Raleigh, the English explorer and courtier, who travelled in tropical South America in the 1590s. Although his mind was more focused on mineral than biological riches, Raleigh was impressed by the diversity of plant and bird life he encountered in the forests of Guiana. The first explicit statement of a continuous gradient in biological diversity across latitudes is usually attributed to Alexander von Humboldt, who travelled in Central and South America at the end of the eighteenth century.

> 66 *The nearer we approach the tropics, the greater the increase in the variety of structure, grace of form, and mixture of colors, as also in perpetual youth and vigor of organic life.* 99
> **Alexander von Humboldt** (1807) *Views of Nature: or Contemplations on the sublime phenomena of creation* (English translation, 1850).

By the late nineteenth century, thanks to popular travelogues and vivid descriptions of tropical environments by Humboldt and other naturalists such as Alfred Russel Wallace, Henry Walter Bates, and Charles Darwin, the idea of a global gradient in biological diversity had become widely recognized.

Mapping global diversity patterns

Since the end of the nineteenth century, the opportunistic collection of specimens from a handful of localities has gradually given way to more systematic, thorough, and widespread sampling, and a fuller understanding of the global distributions of many species. Researchers have now compiled data on the distributions of nearly all species of mammals, birds, reptiles, and amphibians, and made this information freely available in large online databases. By overlaying the distributions of species we can produce maps of species richness, allowing us to visualize global patterns of diversity in a way that was not previously possible. Our best estimates of the geographical distribution of many less well-known species are still highly uncertain, and might change substantially as more data are collected from field surveys. But at large geographical scales, the gross patterns of species richness of major groups of vertebrate animals, such as birds or mammals, are unlikely to change much, even with new information on the distributions of many species. For well-known taxa like these, our current picture of global diversity is probably quite accurate. Unsurprisingly, much of the research that aims to understand global diversity patterns has been based on large vertebrate taxa or flowering plants, for which we also have a fairly comprehensive picture of diversity patterns at a global scale.

This move from estimates of species richness at particular localities to large-scale species richness maps changes the way we picture global diversity patterns. For example, **Figure 10.2** displays the species richness of terrestrial mammals found within latitudinal bands of 5° width in the Americas.[2]

The prevailing pattern is a reasonably steady decline in species richness from the equator towards higher latitudes. There is some deviation from the line of best fit here and there, but the sense we get from this figure is that latitude is the dominant predictor of species numbers in mammal assemblages. When we look at the two-dimensional map-based pattern of species richness of mammals in the Americas (**Figure 10.3**), we see more complexity in the way species richness is distributed. It is certainly still clear that species numbers are highest in the tropics, but within the tropics there are some areas (the arc of the Andes Mountains) that have far higher diversity than others (the adjacent Pacific coastal region). Species numbers also vary considerably across the same

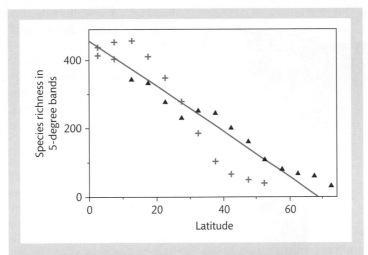

Figure 10.2 Number of terrestrial mammal species in latitudinal bands declines from the equator (0°) towards high latitudes (>60°): ▲ latitudinal bands of 5° width in North America; + latitudinal bands in South America.[2]

From Dawn M. Kaufman and Michael R. Willig, (2003) Latitudinal patterns of mammalian species richness in the New World: the effects of sampling method and faunal group, *Journal of Biogeography*, 25: 795–805. © 2003, John Wiley & Sons.

latitudes within the temperate zone, as we can see by looking at the patterns from east to west within North America.

These two ways of displaying large-scale diversity patterns are basically one-dimensional and two-dimensional views, respectively, of diversity gradients. This is an important difference. The pattern displayed in Figure 10.2 might well lead us to the view that mammal species richness at large scales is driven almost exclusively by something—a climatic or environmental feature—that varies smoothly and continuously with latitude. From Figure 10.3, on the other hand, it seems clearer that there are a variety of factors at work: topography is clearly important, and differences between major biomes (e.g. deserts of the Pacific coast of South America, equatorial forest of the Amazon Basin, savannas of southeastern Brazil) are also apparent. The map-based view conveys a sense of diversity determined by a complexity of interacting factors, rather than a simple monotonic association with one or two environmental features.

What do you think?

The two-dimensional view of biodiversity in Figure 10.3 presents a complex pattern where species richness seems to follow the boundaries of major biomes as much as latitude. So are we justified in pursuing a research programme that aims to explain the 'latitudinal diversity gradient' as a pattern of variation in one spatial dimension? Could this simplification of the geographical patterns constrain our thinking about the underlying processes by hiding much of the interesting and informative variation?

Why are there so many species in the tropics?

The length of the list of explanations proposed to explain the LDG is legendary. Moreover, the number of hypotheses seems to continue to

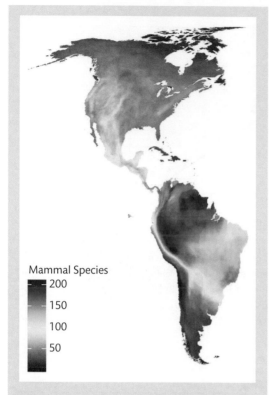

Figure 10.3 The two-dimensional species richness pattern: number of terrestrial mammal species within squares of 100 × 100 km in the Americas.

dominant. At high latitudes, in contrast, the struggle is not against other species but against a harsh climate, and the few species able to thrive are those that have evolved specialized adaptations to the extreme environment. To Wallace, this state of affairs was not simply the ecological consequence of climatic differences between lower and higher latitudes, but the historical outcome of climatic and environmental change:

> **❝** *The causes of these essentially tropical features are not to be found in the comparatively simple influence of solar light and heat, but rather in the uniformity and permanence with which these and all other terrestrial conditions have acted; neither varying prejudicially throughout the year, nor having undergone any important change for countless past ages. While successive glacial periods have devastated the temperate zones, and destroyed most of the larger and more specialized forms which during more favourable epochs had been developed, the equatorial lands must always have remained thronged with life; and have been unintermittingly subject to those complex influences of organism upon organism, which seem the main agents in developing the greatest variety of forms and filling up every vacant place in nature.* **❞**
> **Alfred Russel Wallace** (1878) *Tropical Nature, and Other Essays*. Macmillan, London.

Dobzhanzky: ecology and evolution of innovation

In 1950, the geneticist and evolutionary biologist Theodosius Dobzhansky expanded on Wallace's distinction between the kinds of 'struggle' species face at different latitudes.[4] According to Dobzhansky, the harsh physical environment of high latitudes—low temperatures, short growing seasons, lack of available moisture—excludes species of all kinds, apart from those few with adaptations to withstand such extreme physical conditions. Hence, the primary selective pressure on organisms at high latitudes is for increased rates of development and reproduction within short growing or breeding seasons, and major evolutionary innovations are rare. In the more benign conditions of the tropics, evolutionary mechanisms to cope with large numbers of competing, predatory, or parasitic species lead

increase—surely a healthy sign that this is an active and popular topic of research. Before moving on to explore ways of testing hypotheses, it is worth briefly examining some of the main ideas and historical currents in the thinking about tropical biodiversity.

Wallace: time and stability

One of the first detailed scholarly treatments of tropical biodiversity was by Alfred Russel Wallace in his 1878 book *Tropical Nature, and Other Essays*.[3] Wallace painted a richly descriptive picture of tropical forests teeming with life, in which all possible ecological roles are fulfilled. The benign physical conditions of the tropics allow all species to prosper, argued Wallace, putting them in a perpetual competitive struggle against each other and preventing any particular species from becoming numerically to a more diverse range of solutions, and 'natural selection becomes a creative process which may lead to emergence of new modes of life and of more advanced types of organization'.[4]

Evidence from populations of fruit flies (*Drosophila*) seemed to suggest that the seasonally variable climates of higher latitudes promote adaptive polymorphisms as a way of permitting a species to cope with a wide variety of environmental conditions. In the biotically diverse, strongly competitive environments of the tropics, Dobzhansky argued, ecological specialism is usually a more favourable strategy. In this way, temperate and high latitude regions have become occupied by a few wide-ranging ecological generalist species, while the tropics are occupied by numerous narrowly distributed specialists. In Dobzhansky's view, tropical diversity was driven by ecological and microevolutionary processes, but it also included a historical dimension; the reappearance of high latitude habitats following the retreat of the glaciers has favoured species able to rapidly evolve adaptations to extreme cold and short growing seasons, but often at the expense of genetic variability and future evolutionary potential.

Fischer: diversification trajectories

Another important treatment of the tropical diversity issue—certainly one that is still widely cited today—was by Alfred Fischer in 1960.[5] Fischer called attention to the logical circularity of ecological explanations of the kind advanced by Dobzhansky. If the diversity of species in the tropics applies strong competitive pressure to specialize, this doesn't explain why diversity of competitors is so high in the first place. Similarly, structurally complex tropical habitats may provide many more niches for plant, animal, and other species, but this doesn't explain the origination of the high plant diversity that led to complex habitat structures. Fischer took up Wallace's historical, rather than ecological, ideas about tropical diversity, and developed these into an explicitly macroevolutionary model of diversity.

Fischer's model focused on the large-scale processes that determine the diversity of a region:

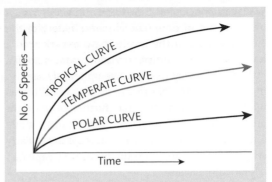

Figure 10.4 Alfred Fischer's diversity through time curves show his suggested trajectory of diversification for biotas in tropical, temperate, and polar regions.[5]

From Alfred G. Fischer (1960) Latitudinal variations in organic diversity, *Evolution*, 14: pp. 64–81. © 2017, John Wiley & Sons.

addition of species by speciation or immigration from other regions, and removal of species by extinction. The tropics have more species for one or both of two reasons: either the *rate* of diversification (the difference between speciation and extinction rates) is higher in the tropics, or tropical biotas have been evolving for a longer period of time. One of the figures in Fischer's paper presents the diversification of a biota in a way that is still repeated in papers today (**Figure 10.4**). As represented in this figure, diversity increases rapidly following the origin of a biota, as groups of organisms radiate into newly available niches. In the later stages of the trajectory, the rate of diversification slows as niches become filled, and as species become extinct the extinction rate approaches the rate at which new species originate. This is what is now referred to as 'diversification slowdown' or 'density-dependent diversification'.

Fischer saw no reason to suppose that the rate of speciation should be any higher in tropical regions. If speciation rate depends primarily on the amount or rate of generation of genetic variability, and on the time between successive generations, Fischer doubted that there was evidence to suggest that either of these varied between tropical and temperate organisms. In this way, Fischer supported

Wallace's conclusion that tropical diversity must be the result of time. The climatic history of the earth means that the tropics have experienced a long period of relative environmental stability, little disturbed by the massive upheavals seen in higher latitudes. This means that tropical biotas tend to be more mature (i.e. further to the right on the trajectories shown in Figure 10.4), while temperate biotas, in contrast, are kept in the early stages of radiation, continually reset by repeated climatic upheavals.

MacArthur: ecological niches and carrying capacities

Robert MacArthur, one of the most influential ecologists of the twentieth century, presented a more ecologically deterministic view of tropical diversity in his 1972 book *Geographical Ecology*.[6] Building on recent advances in community ecology theory, MacArthur stated that the number of species a region is able to support must be determined by a combination of four things.

1. The dimensionality of the environment, i.e. the number of different kinds of resources available for species to exploit.

2. The length of the spectrum of each resource dimension; for example, a wider range of insect sizes provides food for a wider range of bird species.

3. The average amount of resources utilized per species; if the average is lower, a given quantity of resource can be divided more finely among a greater number of species.

4. The degree of overlap in the resources used by different species; if all species use a mutually exclusive distribution of resource types, fewer species can be 'packed in' than if species overlap in their resource use.

Using examples from the bird faunas of different regions, MacArthur suggested that all four of these factors contribute to the higher diversity of tropical compared with temperate ecosystems. Not only are there more resources of more kinds, but species are more tightly packed into resource space because resource use per species is lower and

species overlap more in resource use. MacArthur's view was clearly one that saw tropical environments as having a higher 'carrying capacity' for species than temperate environments. Contrasting this view of tropical diversity with the more historical hypothesis promoted by Fischer, MacArthur introduced the critical question of whether or not ecosystems are 'full' of species. If the number of species in a region is at (or close to) that region's carrying capacity, differences in diversity among regions must be the result of the ecological factors that determine carrying capacity. On the other hand, if species numbers are below carrying capacity and still increasing through time, the historical explanations advanced by Wallace and Fischer are likely to be important.

Rohde: speciation rates

In 1992, the parasitologist Klaus Rohde published what was then probably the most thorough review of explanations for the LDG.[7] Building on ideas originally discussed by Bernhard Rensch in the 1950s, Rohde came down firmly on the side of historical explanations for tropical diversity. First, Rohde reiterated and expanded Fischer's claim that many ecological explanations suffer from circularity. For example, reduced niche width cannot be the *cause* of denser species packing (and thus of higher species richness in a given area), but must be the *result* of a greater diversity of competitors within the same area. In addition, much of Rohde's argument was based on pointing out exceptions to supposed general patterns, undermining their generality. Many of the examples of latitudinal patterns in features of ecological niches do not apply, for example, to parasites or in marine ecosystems. Yet we still see latitudinal gradients in species richness in these systems.

Rohde then addressed the Wallace and Fischer idea that tropical regions have remained relatively undisturbed by major climatic fluctuations and periods of glaciation, so that tropical biotas have had the opportunity to become 'mature' and accumulate large numbers of species. Rohde did not dispute that more species accumulate in older habitats, but he argued against this mechanism as a general driver of the latitudinal diversity gradient. If species were able to shift their distributions towards the equator as the climate cooled, then they ought to have been able to rapidly recolonize higher latitudes when the climate warmed again and the glaciers retreated. In some regions such as Europe, east–west mountain ranges blocked the latitudinal shift of species, leading to extinctions, but in other regions such as North America, as well as in the oceans, there should have been little impediment to the latitudinal shift of species.

In the decade preceding Rohde's article, a number of papers had been published demonstrating large-scale positive correlations between species numbers and various measures of environmental energy availability, such as temperature, potential evapotranspiration, and solar radiation. One explanation for these patterns is that species richness is ecologically limited by energy supply. But this leaves unexplained the question of why the available energy is not simply appropriated by a few species that are superior competitors, or particularly well adapted to exploiting resources. In Rohde's view, the only plausible general explanation for the LDG was greater evolutionary *speed* in the tropics; tropical biotas have accumulated more species because the origination of new species happens faster. He outlined a series of mechanisms by which higher environmental energy levels in the tropics could lead to faster generation of species.

First, Rohde suggested that higher levels of solar radiation lead to higher mutation rates. Secondly, faster physiological processes at higher temperatures lead to faster growth and development, and shorter generation times. According to Rohde, both these mechanisms (higher mutation rate plus shorter generations) lead to an elevated speed of phenotypic change, and hence more rapid speciation, in environments with higher levels of environmental energy. Rohde restated MacArthur's point that the validity of this kind of historical hypothesis depends critically on the assumption of non-equilibrium diversity dynamics—that the world is not yet full of species. Rohde argued that a large body of evidence supports non-equilibrium diversity in a range of major taxonomic groups. He

made the further point that there might not even be any such thing as an upper limit to diversity, given that new species often represent new niches (e.g. for parasites), creating new opportunities for further speciation.

> ### Key points
>
> - Proposed explanations for the LDG have included historical explanations and ecological explanations.
> - Ecological explanations have sometimes been criticized as circular because it is suggested that some mechanisms by which high diversity arises rely on the existence of high diversity in the first place (e.g. stronger competition in the tropics).
> - Ecological explanations rest on the assumption that the world is 'full' of species, and diversity differences among regions are explicable by the factors that influence the upper limits to diversity.
> - Historical explanations assume that the world is not full of species, and diversity is still increasing through evolutionary time. In principle, this means that the equilibrium versus non-equilibrium question is of critical importance in explaining the LDG.

Explaining tropical diversity: a synthesis of ideas

The previous section presented only a few of the many ideas for the causes of the LDG, but gives us some sense of how explanations have alternated between the historical and the ecological. Further, we can see that each of these two kinds of explanation has been presented in terms of several alternative mechanisms. Thus we have the LDG as a product of history, either because tropical ecosystems, uninterrupted by glacial cycles, have had plenty of time to develop to the fullest complexity and diversity (as described by Wallace and Fischer), or because the pace of evolutionary diversification

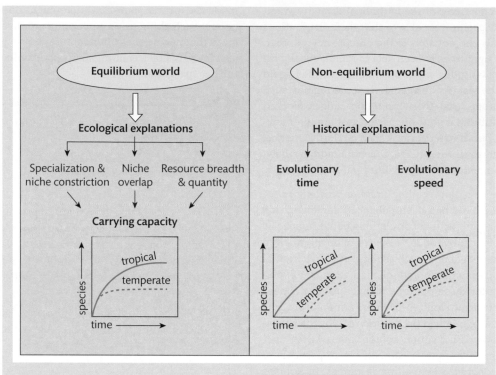

Figure 10.5 Different routes to high tropical diversity. The basic assumption of equilibrium or non-equilibrium diversity determines the most plausible kinds of explanations (ecological or historical) for the latitudinal diversity gradient. In principle, each kind of explanation should lead to a different trajectory of diversity through time for tropical and temperate biotas.

is higher in higher-energy environments (as described by Rohde). Alternatively, we have the LDG as the ecologically-inevitable outcome of climatic differences between tropical and temperate zones, either because of fundamentally different selective pressures for evolutionary innovation and specialization (as described by Dobzhansky), or because of differences in the way ecological resources are utilized and partitioned to allow different regions to support different numbers of species (as described by MacArthur). In fact, Dobzhansky's explanation collapses into MacArthur's resource-partitioning framework because greater specialization (niche constriction) allows resources to be divided more finely among species and a given ecosystem to support more species. MacArthur considered specialization as a fixed attribute of species, the units of

ecological niche theory, but Dobzhansky provided a mechanism by which specialization may have evolved in the first place, under tropical conditions.

MacArthur and Rohde both pointed out that the kind of explanation (historical or ecological) that is most plausible depends critically on the question of whether or not ecosystems or regions are 'full' of species. In other words, do we live in an equilibrium world, where species numbers are at carrying capacity and maintained by a balance between speciation and extinction, or in a non-equilibrium world, where speciation rates exceed extinction rates and species numbers are still on an upward trajectory? We could present a caricature of these concepts in a kind of hierarchical framework, as in **Figure 10.5**.

In the equilibrium world, to explain the LDG we need to explain why the tropics have a higher carrying capacity for species. In the non-equilibrium world, we need to explain why the tropics had a head start in accumulating species, or why the rate of species accumulation is faster in the tropics. Fischer's diversity through time curves provide an intuitive way of picturing these different scenarios. In Figure 10.5, the ecological mechanisms (niche constriction, niche overlap, and resource breadth and quantity) lead to the expectation that the diversity curve flattens out at different levels in tropical and temperate environments. Under both historical mechanisms (evolutionary time and evolutionary speed), the curves have not yet flattened out, and the difference in present-day diversity results from either an earlier starting point in the tropics, or a more rapid growth rate.

As already mentioned, there is a famously long list of hypotheses to explain the LDG; Klaus Rohde listed 28 in his 1992 review.[7] But do these all describe independent mechanisms? Not really—nearly all of the hypotheses proposed to explain the gradient are variations on the few basic ideas that we have just examined. For example, one hypothesis is that the tropics have more species simply because the 'tropical' bioclimatic zone occupies a larger geographical area than those at higher latitudes, and we know that species richness nearly always increases with area.[8] This is really just a variation on the higher carrying capacity theme: greater area means more resources to be partitioned among species.

Likewise, consider the hypothesis that the tropics have more species because the average size of species' geographical ranges is smaller.[9] This is also a carrying capacity hypothesis, because reduced geographical range size is a form of niche constriction. The idea that extinction rates are lower in the tropics (because, for example, higher environmental energy allows species to achieve higher and more stable population sizes) is a form of the evolutionary speed mechanism. This doesn't mean that any of these specific hypotheses are necessarily unoriginal or logically invalid. It means that the scheme presented in Figure 10.5 provides an organizing framework for the abundance of particular mechanisms that have been proposed, over the years, as explanations for the LDG.

In principle, then, the scheme presented in Figure 10.5 should provide a guiding framework for research on the causes of the LDG. First, research should aim to establish whether species numbers are still increasing over time, or are at a steady state. If the former, then the next step would be to test whether tropical environments have existed undisturbed for longer periods and tropical biotas are more 'mature', or alternatively if the speed of diversification is higher in the tropics. If species numbers are at steady state, research should focus on distinguishing the ecological mechanisms by which tropical ecosystems can support more species. But (as you have probably already guessed) the story isn't really as simple as this. In fact, explaining the LDG is a fiendishly complex endeavour, for two very general reasons. First, many of the hypothesized mechanisms for the LDG are not mutually exclusive, and there is no logical reason why the gradient could not have multiple causes, including both ecological and historical ones. Secondly, species diversity is determined by multiple mechanisms, including ones that do not vary predictably with latitude, as we can discern from Figure 10.3.

There is no real reason why all three of the basic mechanisms we have examined (carrying capacity, evolutionary time, and evolutionary speed) should not all be applicable. Tropical biotas may have had both an earlier start and experienced a higher rate of diversification. Although in principle the ecological and historical mechanisms should be distinguishable by the equilibrium/non-equilibrium criterion, tropical ecosystems may still have the capacity to support more species even if species numbers are still increasing. What's more, the evolutionary time hypothesis implies that tropical ecosystems could be at carrying capacity, while those at high latitudes have been kept below carrying capacity by repeated glacial cycles. This means that the ecological and historical explanations might be confounded. And finally, as Rohde pointed out, the concept of a carrying capacity for species numbers is itself somewhat ambiguous if more species provide more opportunities for parasites, predators, and other novel ways of making a living.

Rohde's article was entitled 'Latitudinal gradients in species diversity: the search for the primary cause'. The idea of a primary cause of the LDG is appealing because it is a virtually ubiquitous pattern; we see an increase in species numbers from the poles to the equator in vertebrate animals, marine molluscs, trees, fish, and termites. It is tempting to assume that such a general pattern must have an overarching general explanation. But because the tropics differ from high latitudes in so many ways, both ecological and historical, we need to consider the possibility that there is no single primary cause of the LDG. The upshot is that research evidence that supports a particular mechanism cannot really establish this mechanism as the primary cause, to the exclusion of alternative mechanisms. If we wish to tackle a big problem like the latitudinal diversity gradient, we need to consider carefully not only the kinds of evidence required to support a particular hypothesis, but also the kinds of evidence required to rule out alternative hypotheses.

What do you think?

The LDG is a very widespread general pattern, occurring in such divergent groups as trees and marine molluscs. Does it make sense to suppose that, because of this generality, the pattern is likely to have a single primary cause? Is it plausible that such a general pattern has multiple independent explanations? Are there any examples of other general patterns in biology that are driven by multiple mechanisms?

Is the world full of species?

The question of whether the world is full of species and diversity is at a steady state (or equilibrium level) could have a number of different answers. Possible answers to this question might include 'yes', 'no', 'sometimes', 'in some places', or 'for some groups of organisms'. Moreover, the kinds of data we have, and the methods we have of analysing them, might not permit us to answer the question with confidence at all (particularly given suggestions that there might not even be upper limits to species numbers). Nonetheless, the question is of critical importance for understanding the LDG because, as we have seen, it determines the very way that we think about the problem: is it primarily an ecological phenomenon, or primarily a historical one? So, it is certainly a question that deserves some attention, and in this section we will take a look at some of the kinds of data and methodological approaches that might be used to explore this issue.

To begin with, we need to decide what it is that we are imagining becomes filled up with species.

Is it a geographical region, a particular ecosystem type, or a particular biome? Are we talking about complete ecosystems, including plants, animals, invertebrates, and microbes? Or is it a particular subset of species, either a taxonomically defined group (e.g. primates) or an ecological/functional group (e.g. trees)? In the context of the LDG, many authors have been somewhat vague on this, but have mostly taken a holistic view of ecological niche space becoming occupied by species of all kinds. Fischer, for example, described an adaptive radiation process in which early diversification is rapid as species evolve into 'empty' ecological niches. He then imagined a kind of positive feedback as new species provide new niches for additional species; for example, plants evolve into the available physical niches, and then provide habitat structure for a new wave of animal evolution. As the radiation matures, the extinction rate of old species increases and so the rate of diversity increase slows down. So Fischer (and the other authors we have discussed) seem to be thinking of the diversity through time curve as the diversification trajectory of a biota in its entirety, not just a particular taxonomic or functional guild.

Fossil-based approaches

But how feasible is it to quantify the growth in diversity of an entire biota through evolutionary time? To answer this question, let us look at the kinds of data we can use if we wish to ask whether species diversity is steady or still increasing. Perhaps the most intuitive approach is to turn to the fossil record; by counting the numbers of fossil taxa dated to different time periods, it ought to be possible to plot a diversity though time curve. Indeed, there is a long history in palaeobiology of doing just that. Fossil diversity through time curves are explored in more detail in Chapter 3, but here we will take a brief look at what they could tell us about the equilibrium versus non-equilibrium issue.

The most comprehensive and widely used compilations of fossil diversity through time were constructed by palaeontologist Jack Sepkoski in the 1970s. Sepkoski initially compiled fossil data at a high taxonomic level (marine Metazoan orders), on a global scale, and across vast periods of time (from around 600 million years ago to the present).[10] On this grand scale, there seems to be a clear pattern of initial rapid increase in the number of orders, followed by a levelling off and a steady state of diversity for at least 400 million years (**Figure 10.6**, upper panel). This pattern seems to support the equilibrium model. The picture becomes a bit more complex when we examine the curves for diversity at lower taxonomic levels. The number of marine animal families and genera (Figure 10.6, middle and lower panels) also show the same long period of stasis up to the end-Permian, but they are then followed by an apparently exponential increase in diversity up to very recent times.

The family level curve for marine animals is probably the most famous and widely reproduced of Sepkoski's diversity curves. This curve has been interpreted as evidence for global equilibrium diversity dynamics because it looks like a series of successive logistic growth curves. Under this interpretation, a mass extinction at the end of one

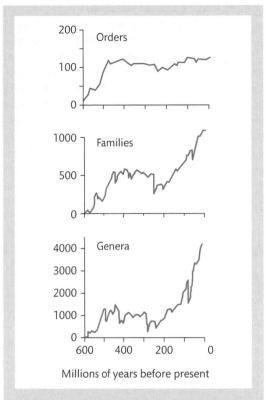

Figure 10.6 The number of marine Metazoan orders, families, and genera known from fossils dated to periods of geological time from 600 million years ago (the left of the *x*-axis) to the present (the right of the *x*-axis).[21]

From Benton MJ (1997), Models for the diversification of life, *Trends in Ecology and Evolution*, 12, 490–5. © 1997, with permission from Elsevier.

phase of logistic growth wipes clean much of the ecological space and paves the way for a subsequent radiation that elevates diversity to a new, higher equilibrium level.

But we have to be careful in our interpretation of these curves. We must keep in mind that the fossil record cannot be a perfect reflection of the true patterns of diversity through time. Rocks don't last for ever; they weather into sediment, or are destroyed by tectonic or volcanic activity, and recycled. The deeper in time a rock stratum was formed, the smaller the chance that it has survived (with its fossils) to be uncovered by palaeontologists. So a decrease in the diversity of known fossil taxa as we move further back in time is exactly what we expect to see. This kind of preservation bias in the fossil record will become more apparent at lower taxonomic levels, because the distribution and abundance of an average species or genus is far lower than for the average order or class. Palaeontologists have developed statistical methods to account for this and other kinds of preservation biases in the fossil record before examining patterns of diversity through time, and strong arguments have been advanced for the accuracy of the fossil record once these biases are accounted for (see Chapter 3). But it is tricky to completely rule out a bias when it predicts patterns that coincide closely with the ones we observe, and this is a debate that will no doubt continue.

What do you think?

Why do the fossil diversity through time curves become successively more exponential as the taxonomic level decreases? Could it be because all the basic ecological strategies that distinguish order-level groups were set in place early in the radiation of animals, and the more recent and ongoing evolutionary innovation is at the level of species, genera, and families? Or is this a circular argument because species and higher-level taxonomic groups are formed by exactly the same processes, just on different timescales?

Phylogeny-based approaches

An alternative way of understanding the temporal dynamics of diversity is by using phylogenies. Compared with a fossil-based approach, a phylogeny-based approach permits us to examine patterns with a much more restricted spatial, temporal, and ecological focus. This might be more suitable for questions about saturation of ecological niche space than low resolution global patterns. On the down side, most phylogenies are reconstructed from the DNA or phenotypic characters of living species, and therefore carry little information about species that went extinct—they are reconstructions of the history of lineages that gave rise to present-day descendants. And, like the fossil record, phylogenies are not free from possible data or methodological biases, so they cannot necessarily be read as accurate or precise representations of history.

How do we use phylogenies to trace the pattern of diversification through time? In much the same way as we use the fossil record; we plot time against the number of lineages we infer to have existed at each time period. In the phylogenetic context, this is known as a lineages through time (LTT) plot. As an example, **Figure 10.7** shows the phylogeny of the vireos (the New World passerine bird family Vireonidae), with the corresponding LTT plot below. The plot is constructed by tracing vertical lines through time slices and counting the number of lineages that each line intersects. The tree in Figure 10.7 is not calibrated to an absolute timescale, so the time units are arbitrary (scaled from 0 to 1).

From the origin of the clade up to the present day, the growth in the number of lineages certainly looks rapid and doesn't seem to be levelling off. But to make proper sense of an LTT plot, we need to compare it with what we expect to see under different diversification processes. To generate these expected patterns we need a model of diversification (these are discussed in more detail in Chapter 9.) If we assume that a group is diversifying freely without ecological constraints, a simple model of diversification is one in which the rates of speciation and extinction (or simply the rate of branching in the phylogeny) are constant through time and equal across lineages (i.e. there is no bias in the probability of branching). A model of diversification of this kind gives rise to exponential growth in the number of lineages, which means that the number of lineages doubles at regular intervals of time. If we log-transform the number of lineages, the exponential growth curve becomes a linear relationship. So, if diversification is unconstrained and constant through time, we expect to see a straight line on the LTT plot if we log-transform the y-axis. **Figure 10.8** shows what this looks like for the vireos.

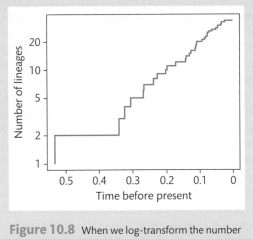

Figure 10.8 When we log-transform the number of lineages on the *y*-axis, the LTT plot for the vireos becomes more linear.

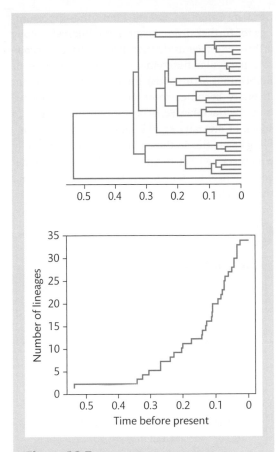

Figure 10.7 Vireos (Family Vireonidae) are a family of about 58 passerine birds found in North and South America. A sample of these are represented in the phylogeny. Below the phylogeny is the lineages through time (LTT) plot, which shows the number of reconstructed lineages present at each time point during the diversification of the clade.

Is this a straight line? From about 0.34 time units (the age of the crown node of the main radiation), the line looks fairly straight, but does seem slightly concave, perhaps curving towards an asymptote. How can we decide if this is consistent with constant-rates diversification? Several methods have been developed to compare the pattern of lineages through time in a clade with the expected pattern under a constant-rates model. The gamma statistic, for example, is based on the distribution of node heights along the time axis, and takes a negative

value if there is a slow-down, and a positive value if there is a speed-up.[11] Gamma is negative (γ = –1.06) for the vireo phylogeny, but still within the expected bounds of a constant-rates model. We could also compare the fit of the vireo phylogeny to a constant-rates model with the fit to a rate-variable model. Such a model might include an additional parameter for an upper limit to species numbers (the carrying capacity), and parameters for the way the probability of speciation decreases as this upper limit is approached. In principle, the carrying capacity itself can even be estimated for a given clade. We have to remember that methods for comparing the slow-down pattern of diversification with some theoretical expectation are only describing the pattern, and not actually measuring the ecological or evolutionary processes that might have caused the slow-down in diversification rates. We could conclude that a pattern of diversification slow-down is the result of species numbers approaching an ecologically determined upper limit, but we need to keep in mind that this is an inference of a process from indirect evidence provided by the phylogeny.

To return to our original question: how could we use phylogenetic diversification models to test whether regions, biomes, or ecosystems are full of species? In principle, it should be as easy as for the fossil diversity through time approach: we choose our taxon of interest (say, birds) and test for a global pattern of diversification slow-down. But is it really as straightforward as this? Even if speciation and extinction rates are in equilibrium and species numbers are at steady state through time, new clades will still be originating and radiating all the time. Clades that arose recently might still be in the early rapid-growth phase of their diversification curve, while older clades are perhaps more likely to have reached saturation. Even if the birds *as a whole* were at equilibrium diversity, we would undoubtedly find that within this large group some subclades show diversification slow-down, while others do not. So this presents us with another tricky thing to consider: at what phylogenetic or taxonomic level, and for which particular taxa, do we test for equilibrium versus non-equilibrium diversity dynamics? If we are testing whether the diversity of vireos is at equilibrium, the answer would probably reveal different ecological and evolutionary processes than a test of whether diversity of the order Passeriformes (which includes vireos) is at equilibrium, or a test of whether the diversity of the class Aves (all birds) is at equilibrium.

In the end, the group of organisms we choose to define the scope of our question about diversity equilibrium will depend on both biological and practical considerations. Does it actually make much ecological sense to ask whether or not the world's 'biota' is or is not at equilibrium species diversity? If we decide that the answer is yes (the world is full of species), then this could lead to ludicrous predictions; for example, it might imply that saturation of the niche space available to coniferous

trees in Canadian forests limits the opportunities for further diversification of rainforest canopy arthropods in central Africa. If we restrict the scope to a particular functional guild (such as trees or herbivorous arthropods), the notion of equilibrium diversity becomes more ecologically and evolutionarily tractable. Perhaps it is only meaningful to speak of diversity equilibria, upper limits, or carrying capacities in the context of a geographically circumscribed set of species of broad functional equivalence, for which we can truly imagine that there might exist a finite number of possible niches.

Key points

- The most common way of testing whether diversity is at equilibrium (or steady state) is to plot the inferred numbers of species at different periods of time, and examine whether the diversity through time curve has levelled off, or is still increasing. This can be done using data from the fossil record or phylogenies.

- Fossil data have the advantage of including extinct species, but may suffer from systematic biases, and are generally at coarse taxonomic, temporal, and spatial resolution.

- Lineages through time plots inferred from phylogenies allow for finer resolution and more carefully focused questions, but usually carry little information about extinct species and may also suffer from certain biases.

Testing hypotheses for tropical diversity

As we have just seen, there is (unfortunately) no simple unambiguous answer to the question of whether the world is full of species. What does seem clear is that this is probably not a question we can really even

answer on the grand global scale. On a more positive note, it may be more feasible to test for diversity equilibrium or non-equilibrium by restricting the scope of our investigations to a more ecologically

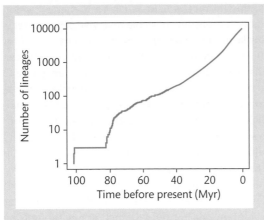

Figure 10.9 The LTT plot (with lineages log-transformed) for all the known bird species.

coherent group of organisms, such as birds. In fact, birds happen to be among the very few taxonomic groups of this rank and size for which we have species-level estimates of phylogeny. This means that we can examine the LTT plot (**Figure 10.9**) to give us an idea of the trajectory of bird diversification.

Even with the y-axis (number of lineages) log-transformed, it looks as if bird diversity has been increasing through time at a rate that has been increasing for the past 80 million years. This is supported by a positive value of gamma ($\gamma = 12.12$) that is so high that the chance that it could have been generated by a constant-rates diversification process is basically zero. Based on this phylogeny, there seems little evidence that the rate of diversification is slowing or that global bird diversity is levelling off towards an equilibrium. At this point in the history of the spectacular radiation of birds, there seems to be no constraint on the ability of evolutionary innovation to keep adding to the number of species already in existence.

It is time to return to the central issue of this chapter: the latitudinal diversity gradient. If we accept that bird diversity is still increasing, how does this help us decide what to do next? Under the scheme presented in Figure 10.5, evidence that diversity is not (yet) at equilibrium would suggest that the LDG is more likely to have a historical, rather than

ecological, explanation. But remember that this is a somewhat simplistic view. There is no logical reason that the LDG should not be influenced by history and ecology together. Maybe particular ecosystem types are saturated with diversity, but others aren't, or maybe tropical ecosystems are saturated but those at high latitudes are not—or the reverse. The challenge is to come up with tests that allow us to detect the signals of historical and ecological mechanisms: what kinds of observations are consistent with particular mechanisms? Even more powerful inference will be possible if we can also identify the kinds of observations that are inconsistent with particular mechanisms. In this way, we might be able to circle around the large set of candidate models for the LDG, supporting some and ruling out others, and gradually close in on a smaller, more plausible set of explanations.

In the remainder of this chapter, we will explore in detail one of the major hypotheses for the LDG that has been debated in recent years: that rates of diversification are higher in the tropics. We will look at the kinds of data and tests that researchers have used to support or refute this hypothesis, and we will look at some of the assumptions we need to make to ensure that the tests we apply are reasonable and appropriate.

Is diversification faster in the tropics?

As we have seen, faster evolutionary diversification is one of the two basic mechanisms by which the LDG may have arisen under the non-equilibrium scenario—the other is the 'more time' mechanism. It is worth reminding ourselves that diversification rate is the net rate at which the diversity of a clade grows: the difference between the rates of speciation and extinction. Faster diversification can be achieved by an elevated rate of speciation, a depressed rate of extinction, or both. This means that it is entirely possible for the net diversification rate to be higher in the tropics even if both speciation and extinction rates are higher there—or indeed, if they are both lower, as long as the difference between them is greater.

A number of ecological and evolutionary processes have been put forward that attempt to explain the LDG via faster diversification. Here are a few examples.

- **Faster molecular evolution** Ultraviolet solar radiation has a mutagenic effect on DNA. Under this hypothesis, in regions with greater intensity of solar radiation—typically at low latitudes where the angle of incidence is high—mutation rates are higher. It has been suggested that this generates more variation (at the molecular level) for selection to act upon, leading to more rapid adaptation and genomic divergence among isolated populations.[12]

- **Faster growth rates** Many kinds of organisms grow and develop more rapidly in higher ambient temperatures. This means that they reach reproductive maturity faster, and the turnover of generations per unit time is more rapid. Under this hypothesis, this results in a higher rate of adaptation by natural selection, elevating the speed of population divergence.[7]

- **Habitat area** The area hypothesis proposes that the geographical area of the 'tropical' eco-climatic region is larger than that of temperate or high latitude eco-climatic regions. This is largely a result of the way the tropical zones of the northern and southern hemispheres form a single contiguous region, whereas the northern temperate high latitude zones are separated from the southern ones. A rather convoluted mechanism has been suggested whereby the larger habitat area provided by the tropics allows species to achieve larger geographical ranges. This in turn elevates the probability of a species range being intersected by a geographical isolating mechanism; effectively, a species with a larger range is a bigger target for any geographical barrier that arises at random on a landscape.[8] The result of this is to elevate the rate at which new species are formed by allopatric subdivision.

We can see that faster diversification is a kind of common proximate mechanism, forming the link between higher species diversity and several 'ultimate' mechanisms such as higher mutation rates. It is entirely possible to devise separate tests for latitudinal bias in diversification rates and for the ultimate mechanisms of the kind we have just discussed. In fact, it probably makes more sense to test for faster tropical diversification before we test any of the ultimate mechanisms. If we first establish that diversification is indeed faster in the tropics, we can then move on to tests for the underlying ultimate mechanisms that may explain this. On the other hand, a test that demonstrates, for example, that the mutagenic effect of solar radiation at low latitudes leads to elevated rates of molecular evolution doesn't tell us much about the LDG by itself, until we also show that diversification rates are higher in the tropics.

As with tests for the patterns of diversity through time—and most other topics in macroevolution—the two basic approaches to investigating geographical patterns in the rate of diversification involve fossils and phylogenies. In the following sections, we will explore some examples of how each approach has been applied to the question of tropical diversity.

Faster tropical diversification: tests based on fossil data

For decades, much of the palaeontological research on the LDG has been framed around the 'cradle versus museum' metaphor, first articulated by George Ledyard Stebbins in 1974.[13] The idea is that faster diversification in tropical environments can arise through one or both of two mechanisms: elevated speciation rates (the tropics as a *cradle* of diversity, with many young lineages) or reduced extinction rates (the tropics as a *museum* of diversity, with an accumulation of ancient taxa).

Palaeontologists tend to speak of rates of 'origination' rather than 'speciation', because the low taxonomic resolution of the fossil record usually only permits analyses of diversity patterns at the level of higher taxa, not species. In some ways this doesn't matter; origination of higher taxa and species amount to the same thing if we adopt the orthodox view that the processes generating higher taxa are the same as those generating species, extended over longer periods of time. But on the other hand, structural and ecological differences among higher taxa are often a matter of kind, not just degree. This has led some researchers to the view that the origination of higher taxa is associated with evolutionary innovations that permit lineages to cross 'adaptive thresholds' and occupy novel areas of ecological space. This idea of the rate of evolution of novelty or innovation is an important theme in the palaeontological literature on the LDG.

One simple way of estimating origination rates of higher taxonomic groups from palaeontological data is to count the number of taxa that make their first appearance in the fossil record at each different period. If we consider the number of first appearances within a given period to be an indicator of the origination rate, this provides an easy measure for comparison between tropical and temperate regions. But we have to be careful—there are multiple biases that can affect the results of these kinds of comparisons. An excellent example of a study that considered and systematically accounted for these biases, revealing the geographical pattern of originations underneath, was published by David Jablonksi in 1993.[14] Jablonski used data from the fossil record of marine benthic invertebrate orders since the beginning of the Mesozoic Era (about 252 million years ago). Marine organisms typically have a more complete fossil record than terrestrial organisms because the conditions for fossilization are more favourable in the marine environment. The first issue that Jablonski addressed is that rock strata move with the continents they are attached to, so the geographical location in which a fossil is uncovered is not necessarily the same as the location inhabited by the organism that left the fossil. Fortunately, the field of palaeomagnetism provides the tools for calculating the latitude of igneous rocks at the time that they solidified. This permits the reconstruction of the palaeolatitude of land masses at different periods in the geological past, and thus the restoration of fossils to their original latitude. Online calculators that provide an easy

way to estimate the palaeolatitude of any given location at a specified time in the geological past are now available.

The next issue was preservation potential. Some invertebrate orders, such as echinoids, contain calcareous spines and shells that fossilize well, while others contain fewer hard parts and are less likely to leave abundant fossils. Jablonski identified 26 invertebrate orders with good preservation potential, and calculated their palaeolatitudes and times of first appearance in the fossil record. The majority of first appearances are in latitudes between 20° and 40° from the equator, i.e. in waters extending from the edge of the tropics into the temperate latitudes (**Figure 10.10**(a)).

For comparison, Jablonski performed the same calculations for 16 poorly preserved orders. The idea was that the number of first appearances of these groups is more likely to reflect sampling and preservation biases than the true number of originations, so they provide a kind of control or null model against which to test the pattern for the well-preserved orders. If the two patterns are very similar, this might be grounds for suspicion that the pattern for well-preserved orders is also driven by preservation and sampling bias. But the first appearances of poorly preserved orders are more temperate in distribution, occurring primarily from 30° to 50° (Figure 10.10b). The difference between the two distributions is statistically significant.

So the maximum numbers of first appearances of well-preserved orders are more tropical in distribution than expected from preservation biases. But we see in Figure 10.10(a) that there are still very few first appearances at the lowest latitudes, from the equator to 20°, which does not seems to support the hypothesis of higher origination rates in the tropics. However, we haven't yet adjusted the numbers for sampling effort. If we can assume that the number of first appearances of poorly preserved orders (Figure 10.10(b)) is a reflection of sampling intensity, then it looks as if sampling intensity is highest at mid-latitudes. This would not be surprising, since this is where the majority of palaeontologists live and work. We can then see

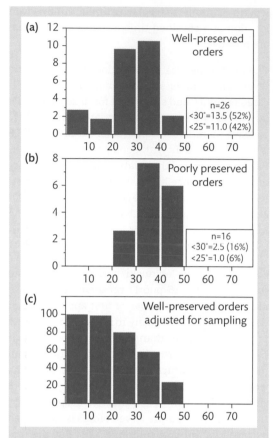

Figure 10.10 (a) Number of first appearances in the fossil record of marine invertebrate orders considered to have good preservation potential. The first appearances are grouped into latitudinal bands, from the equator (0°) up to high latitudes (70°). Numbers of first appearances are represented by the heights of the bars. (b) When the numbers of first appearances for poorly preserved orders of marine invertebrates were plotted, they were strongly clustered within mid-latitudes (30–50°). (c) Adjusting for sampling biases in numbers of first appearances of marine invertebrate orders reveals a strong skew towards tropical latitudes.[14]

From David Jablonski (1993) The tropics as a source of evolutionary novelty through geological time, *Nature* 364: 142–4. Reprinted by permission from Springer Nature.

immediately that the number of first appearances of well-preserved orders from 0° to 20° in Figure 10.10(a) is actually much higher than we would

expect from the very low sampling intensity at these latitudes. If we adjust the numbers for sampling intensity, the pattern changes dramatically (Figure 10.10(c)). Now, the number of first appearances shows a distinct increase from high latitudes towards the equator.

Stripping away the different kinds of bias in the fossil data reveals an underlying pattern of more first appearances of marine invertebrate orders in tropical waters, and a steady decline towards high latitudes. Jablonski's interpretation of this result was that evolutionary novelty—of the kind that leads to the formation of higher taxa—has arisen more frequently in the tropics.

Key points

- The number of first appearances in different time periods of groups of organisms with a good fossil record (typically marine organisms with hard body parts) can provide an indicator of rates of origination.

- However, the fossil record is biased by the preservation potential of organisms, movement of continents, and sampling intensity. In some cases, it is possible to account for biases such as these and reveal the geographical and temporal pattern of diversification beneath.

Faster tropical diversification: tests based on phylogenies

As we have seen, phylogeny-based analyses have advantages and disadvantages compared to fossil-based analyses when it comes to exploring patterns of diversification. One advantage is that phylogenies allow us to explore patterns for groups that don't have a good fossil record—such as birds. In fact, because birds have been the focus of much phylogenetic research over the past few decades, they appear frequently as case studies in phylogeny-based tests for faster tropical diversification.

It is tempting to look at the massive diversity of the tropics, and of primarily tropical evolutionary radiations such as the tyrant flycatchers (Tyrannidae) or hummingbirds (Trochilidae), and conclude that diversification must have been more rapid in tropical regions. But we know that it is unwise to accept a general hypothesis on the basis of only one or a few supportive cases. We also know that high species richness doesn't have to be the result of rapid diversification; it could also be a product of great age. So what we need are statistical methods to test for links between diversification rates and latitude that account for differences in age among clades, and provide adequate replication for sound statistical inference.

In Chapter 9, we worked through some methods for using phylogenies to estimate diversification rates and to test whether diversification rates are causally linked to the evolution of particular biological traits, such as the mode of feeding in insects. We can apply many of these same methods to test for a link between diversification rates and geographical variables such as latitude. When we do this, we are treating latitude as a species-specific 'trait'— a property of a species that we can summarize in some way. For example, we might quantify latitudinal position as a continuous numeric variable using the latitudinal midpoint of the geographical range of each species. Alternatively, we might simply treat latitude as a discrete binary variable (tropical/temperate), or we may have a more sophisticated measure to capture the 'tropicalness' of each species (e.g. the proportion of the range area that lies within the tropics).

It is important to keep in mind that phylogeny-based statistical methods make assumptions about the mode of evolution of the traits or variables we are analysing. If the geographical ranges of species don't evolve in a way consistent with these assumptions, we should think about the kind of influence this could have on the results. We will look first at what is often considered the simplest statistical method for analysing correlates of diversification rates: contrasts between sister clades. We will then explore some more complex and (at least in principle) more powerful approaches.

Contrasts between sister clades

Sister clades arose from a common ancestor, and are each other's closest relatives. Therefore the two clades are the same age, so it seems intuitively logical that if one clade has more extant species than the other, the number of species accumulated per unit time (the net rate of diversification) must have been higher in that clade. Therefore the difference or ratio of species richness between two sister clades is a measure of the increase in the average diversification rate in one clade (or the decrease in average diversification rate in the other clade) that occurred after the two sister clades diverged from one another. In **Figure 10.11**, sister clades A and B have 9 and 36 species respectively, so the species richness ratio (9/36) is 0.25. We could also calculate the contrast in other ways—for example, the difference in species richness, or the difference in log-transformed species richness. We are controlling for clade age because A and B are of equal age, having diverged simultaneously from node AB. If clade A consists of temperate-dwelling species and clade B of tropical-dwelling species, then this sister-clade contrast is one observation in support of the hypothesis that diversification is faster in the tropics.

Of course, there could be any number of other reasons why clade A is less diverse than clade

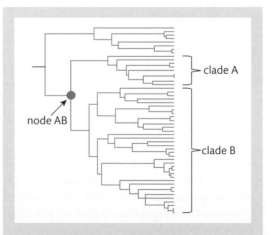

Figure 10.11 A hypothetical phylogeny showing the difference in diversity of two sister clades which have a common ancestor at node AB.

B—for example, all species in clade A might be confined to a small island. So this single contrast between A and B is not enough. We need replication to test the faster tropical diversification hypothesis statistically, and for this we need to identify multiple tropical–temperate sister-clade contrasts. Let's assume that we are able to identify 15 such contrasts. If the tropical clades are more diverse than their temperate sister clades in 12 of these contrasts, and temperate clades more diverse in the remaining three, a simple statistical test (such as a sign test) would reject the null hypothesis that there is no association between latitude and diversity, with $p < 0.05$. For a non-parametric test like the sign test, we don't even need to calculate diversity contrasts—all we need is a tally of the numbers of sister pairs in which the tropical clade has more species than the temperate clade. But we could also extract a little bit more information from the data by using a test (such as the Wilcoxon signed-ranks test) that considers the magnitudes, as well as the directions, of the contrasts.

This all seems very straightforward, but the appropriate application of the sister-clade approach needs careful thought. In practice, it is remarkably difficult to find a phylogeny of a large group of organisms from which we can identify a reasonable number of tropical–temperate sister clades. This is partly because the geographical distributions of many taxa are phylogenetically conserved; species tend to inhabit the same biomes or regions as their closest relatives. The existence of numerous tropical–temperate sister clades would suggest that there were numerous biogeographical shifts across latitudinal zones. This does occur in some large taxa, but it is not common, and it usually means choosing sister-clade contrasts deeper in the phylogeny (say, at the genus or family levels, or higher). The deeper we go in the phylogeny, the fewer contrasts there are available. Furthermore, because we are restricted to pairs of sister clades, selecting informative contrasts often means using only part of the information available in the phylogeny, particularly when our phylogeny is rather unbalanced or ladder-like in shape. In **Figure 10.12**, assume that the species

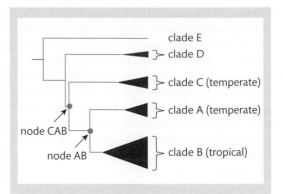

Figure 10.12 Another hypothetical phylogeny, this time showing a set of clades from different regions (tropical or temperate). The width of the triangles represents the species richness of each clade.

in clade B are all tropically distributed, and those in clades A and C are temperate. We could identify one tropical–temperate contrast at node AB, discarding the information represented by clades C, D, and E. Alternatively, we could relax the definition of a tropical clade to some extent, and identify one mostly tropical versus temperate contrast at node CAB. In this case, we still discard clades D and E.

Much of the early phylogeny-based research on diversification rates across latitudes used this simple sister-clade approach, but these studies were seriously limited by small sample sizes. The pioneers of this approach were Brian Farrell and Charles Mitter. In a 1993 study, they used sister-clade contrasts to test the hypothesis that diversification has been faster in tropical phytophagous insects, but were only able to identify five tropical–temperate contrasts, and their statistical test was equivocal.[15] A few years later, the same approach was applied to passerine birds and swallowtail butterflies.[16] Even with the availability of a large phylogeny of all bird families (and many subfamilies and tribes) and a rather relaxed definition of 'tropical' and 'temperate', only 11 suitable contrasts could be found. Nonetheless, the more tropical clade was the more diverse in 10 of these contrasts,

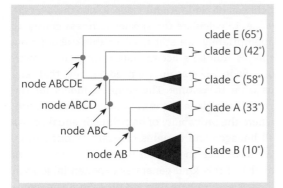

Figure 10.13 Here we quantify each clade's latitude as a continuous variable, rather than a discrete 'tropical versus temperate'. This gives us a little more flexibility in choosing sister-clade contrasts, because we can now calculate a contrast at every node in the phylogeny.

rejecting the null hypothesis under a sign test with $p = 0.006$. The null hypothesis was also rejected for swallowtails (10/13 contrasts, $p = 0.04$). This was one of the first pieces of phylogeny-based evidence in support of faster diversification in the tropics.

So far, we have been considering tests that treat latitude as a discrete variable (tropical versus non-tropical) at the clade level (even if this is based on an underlying continuously varying value such as latitudinal midpoints of species distributions). If we can treat latitude as a continuous variable, we can extract more information from a phylogeny—but we also need to make some more assumptions. In particular, we are now forced to give some thought to models of evolution—the processes by which the geographical distributions of species evolve along a phylogeny, and by which species richness accumulates over time. **Figure 10.13** shows the same phylogeny as Figure 10.11, but the latitudinal midpoint of each clade's distribution is now labelled.

Now we can calculate sister-clade contrasts in both species richness and latitude at all nodes in the tree, freeing us from the need to choose contrasts a priori, and allowing us to use information from the whole tree, rather than just selected pairs of sister clades. The contrasts in the predictor variable (latitude) can be calculated using the method of phylogenetically independent contrasts (PIC)—differences in the values of a trait between the two lineages descending from each node.[17] PICs have the important property that they are all phylogenetically independent of one another. For example, an independent contrast calculated for node AB is a measure of the evolutionary change that occurred along the two branches AB → A and AB → B. The contrast calculated for node ABC is a measure of the change along the branches ABC → C and ABC → AB. So, the two contrasts (A vs B and AB vs C) are measuring evolutionary change in non-overlapping parts of the phylogeny, which makes them evolutionarily (phylogenetically) independent.

The catch is that we now have to make assumptions about how our trait (in this case, latitude of species distributions) has evolved. This is because when we calculate an independent contrast for a trait at a given node we are, in effect, estimating an ancestral value for the trait at that node. For many continuously varying, normally distributed biological traits (e.g. body mass or wingspan), a common baseline assumption is that evolution proceeds under a 'Brownian motion' (BM) process—unbounded random drift in the value of the trait along the branches of the phylogeny, in which the value of the trait is just as likely to increase as decrease. Under this model, the values of a trait in two sister lineages are likely to drift further and further apart following their divergence. If the trait does evolve under this kind of process, the values of contrasts calculated at nodes deeper in a phylogeny will tend to be larger than those of shallower ones. We usually correct this unevenness of variance by scaling the contrasts (dividing them by the square root of the node depth).

Provided that the BM model accurately describes the evolution of the trait, this should make deep contrasts comparable in magnitude (on average) to shallow ones. But we need to be careful—if the trait hasn't evolved in a BM-like manner, we run the risk of making things worse when we scale the contrasts by increasing the unevenness of variance in contrasts and possibly biasing the results of statistical tests. This problem can be handled by first testing whether the species trait data are consistent with evolution under a BM process, using a measure such as Pagel's lambda (λ). In fact, we can transform the branch lengths of the tree according to the value of λ we obtain for our data, which then makes it safer to scale the contrasts to even up the variance. An alternative method to calculating independent contrasts is phylogenetic generalized least squares (PGLS), in which the estimation of λ and the transformation of branch lengths are done automatically before fitting a model that accounts for phylogenetic non-independence.

Does it make sense to assume that a geographical variable like the latitudes of species distributions have 'evolved' under a model normally used to describe the evolution of a genetically heritable biological trait? This is a matter of debate, but we should keep in mind that the assumption of BM (or some similar model) is made for statistical tractability, not necessarily because we believe it describes the true evolutionary process. If we calculate λ for our latitude data and find that the data are consistent with a BM model of evolution, this means that we can happily scale our latitude contrasts using the untransformed phylogeny, and our test will not violate the statistical assumption of evenness of variance. We don't need to read this as a statement that geographical distributions evolve under the same genetic processes as many life history traits.

We also need to calculate a contrast in species richness at each node. Unlike geographical distribution (or biological traits), however, it doesn't make sense to infer an ancestral value of species richness by calculating the average of the present-day values of the two sister clades. Instead, we can use the total number of species in each of the two clades descending from the node.[18] For example, in Figure 10.13, assume we are calculating a species-richness contrast at node ABC; this will be a contrast in species richness between the value for clade C and the value for node AB. The species-richness value for AB will be the summed species

richness of clade A and clade B, not the average value. To calculate the species-richness contrasts appropriately, we still need to think about how species numbers accumulate during the growth of a clade; once again, we need a model that tells us what to assume. The simplest model for the growth of a clade is one of random diversification, where the probability of branching is equal across all lineages and remains constant through time. We have seen in Chapter 9 that a constant-rates model of this kind generates exponential growth in species numbers through time. So, again, comparisons of species richness at deeper nodes will tend to be greater in magnitude than those at shallow nodes, even though diversification rates have remained (stochastically) constant. We can correct this by log-transforming species-richness values before calculating differences or ratios.

In Figure 10.13, let's say that clade A has three species and clade B has seven species. One way of calculating the species-richness contrast would be

log(species richness of clade A/species richness of clade B) = log(3/7) = –0.85.

The decision about which sister clade is the numerator and which is the denominator is arbitrary, but it is important that it remains consistent with respect to the predictor variable (latitude). In the above example, we used the clade with the higher latitude (A) as the numerator, so we must be sure to do the same for all species-richness comparisons across the phylogeny.

Limitations of sister-clade contrasts

We have focused in some detail on the way we can use comparisons between sister clades to test for a latitude–diversification rate association. This is because the sister-clade method is a conceptually simple way of understanding the kinds of information we can extract from phylogenies, and because this method has formed the basis for much of the research on correlates of diversification rate, not only with latitude but also with a large range of biological and geographical variables. But we have also seen that the sister-clade method has limitations in terms of the efficiency with which it uses the available data, especially when the predictor variable is discrete.

Another limitation of sister-clade comparisons is the central assumption that variation in diversity of clades within a phylogeny must be the result of only two things: differences in clade ages, and differences in rates of diversification. Now we need to recall our exploration earlier in this chapter of the equilibrium versus non-equilibrium diversity question. Suppose the diversity within a particular clade increases until it reaches ecological 'saturation' (or equilibrium diversity), and the clade stops growing (or, at least, its rate of growth slows down). From that point on, the association between clade age and diversity breaks down. It then becomes less clear that the difference in species richness between two sister clades can be interpreted purely as the difference in diversification rates. Instead, we now have a third possible influence on clade diversity to consider—the ecological upper limit on the clade's diversity. Unfortunately, this is something that is likely to be extraordinarily difficult to estimate. Upper limits to diversity, if they exist at all, are likely to be clade-specific and probably determined by complex interactions between the biology of species in the clade and the environments of the places they inhabit.

What do you think?

Estimating the theoretical upper limit to the diversity of a clade is a challenge. One approach has been to extrapolate from the trajectory of diversity growth through time since the clade's origin, to estimate the asymptotic species richness. Another approach has been to use some proxy for maximum possible diversity, such as the geographical area currently occupied by the clade. What are the weaknesses of these approaches? In an ideal world, what kinds of data would we need for reliable inference of the upper limit to a clade's diversity?

Estimating gradients in speciation and extinction rates separately

When we compare the extant diversity of sister clades, we are (assuming non-equilibrium diversity dynamics) estimating the relative rate of net diversification (i.e. the difference between speciation and extinction rates) averaged over the lifetime of each clade. But it would be useful if we could use phylogenetic data to produce estimates of speciation rates and extinction rates separately. If this is possible, it might allow us to address the long-standing question of whether the tropics are a cradle of diversity (high speciation rates) or a museum of diversity (low extinction rates) for taxa without a good fossil record.

Methods for separately estimating speciation and extinction rates across latitudes are based primarily on analysing the distributions of node depths in a phylogeny (also known as node heights). This introduces into the mix another variable that is a potential source of additional information, but also a source of additional error and uncertainty. Phylogenetic branch lengths (and hence node depths) are derived from the amount of evolutionary change—usually the number of DNA substitutions—inferred to have occurred between branching events. As we have seen in Chapter 6, if branch lengths and node depths are calibrated to an absolute timescale, they are also strongly influenced by the calibrations chosen (usually fossil dates) and other assumptions of the analysis. It is important to be aware of the uncertainty and the possible biases in node depths before using them to infer macroevolutionary processes.

One of the first studies that attempted to estimate separate latitudinal gradients in speciation and extinction rates came up with a somewhat surprising result. In 2007, Jason Weir and Dolph Schluter compiled phylogenetic data on divergence times of over 300 pairs of sister species of birds and mammals.[19] They found that, near the equator, the mean divergence time of sister species was 3.4Mya, but at high latitudes (above ~50°) divergences between sister species were much more recent, with most species younger than 2Mya (**Figure 10.4**).

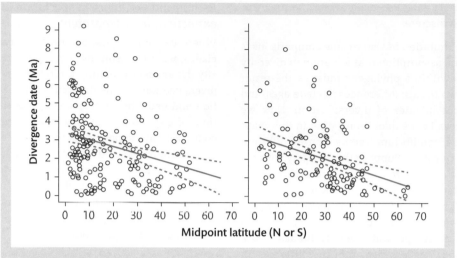

Figure 10.14 Weir and Schluter found that temperate species tend to be younger than tropical ones, as indicated by the mean divergence times between pairs of sister species.[19]

From Jason T. Weir and Dolph Schluter (2007) The latitudinal gradient in recent speciation and extinction rates of birds and mammals, *Science* 315, Issue (5818):, 1574–6. Reprinted with permission from AAAS.

This is the opposite of what we might expect to see if speciation rates were higher in the tropics, because it seems logical that a higher speciation rate should give rise to more recent divergences, on average, between sister species. Weir and Schluter confirmed this by simulating phylogenies under a constant-rates birth–death process, with a number of alternative values for the speciation and extinction rate. They used the probability distributions of the ages of sister species generated by the simulations to find the best fit of speciation and extinction rates to their bird and mammal data. They then fitted linear models that predict speciation rates and extinction rates from latitude, using maximum likelihood to obtain the estimates of the model parameters (slope and intercept). The outcome was a significant positive slope for both speciation rate and extinction rate across birds and mammals together. In other words, both speciation and extinction rates are lowest in the tropics (**Figure 10.15**).

The implication of Weir and Schluter's result is that (for birds and mammals) the tropics are not a cradle of diversity, but rather a museum, with low

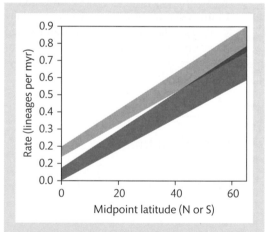

Figure 10.15 Weir and Schluter's estimated speciation rate (pink) and extinction rate (orange) for birds and mammals both show a significant positive association with latitude. In other words, their result suggests that birds and mammals occupying high latitudes speciate faster, but also go extinct faster.[19]

From Jason T. Weir and Dolph Schluter (2007) The latitudinal gradient in recent speciation and extinction rates of birds and mammals, *Science* 315, (5818): 1574–6. Reprinted with permission from AAAS.

extinction rates leading to the accumulation of old species. This might seem to conflict with the earlier studies showing that net diversification rates are highest in the tropics, but it doesn't have to. As long as there is a greater discrepancy between speciation and extinction rates at the equator (as the maximum likelihood slope estimates in Figure 10.15 seem to show), net diversification rates will be higher. At high latitudes, species originate more rapidly but are less likely to survive more than a few million years, so there is a greater turnover of species from one time period to the next. In some ways, this scenario is consistent with the picture of tropical diversity promoted by Alfred Russel Wallace and Alfred Fischer—a stable environment permitting the steady accumulation of diversity over long periods of time.

Species ages and rate of molecular evolution across latitudes

Weir and Schluter[19] determined divergence times between sister species using sequences of the mitochondrial gene cytochrome b (cytb). By assuming a uniform rate of molecular evolution of cytb across all bird taxa (a molecular clock), they were able to calibrate the amount of sequence divergence between sister species to an absolute timescale in millions of years. But what if, for some reason, rates of cytb evolution vary systematically across latitudes? As we saw earlier in this chapter, Klaus Rohde argued for exactly this kind of pattern, speculating that molecular evolutionary rates are elevated in regions of high solar radiation and environmental energy. There are also reasons to believe that molecular rates might be elevated at lower latitudes via higher metabolic rates or lower population sizes, or even a higher rate of speciation itself. In fact, there is a growing body of empirical evidence to suggest that molecular rates do tend to be higher at lower latitudes in a range of animal and plant taxa, although the evidence in birds is still equivocal.[12] If rates of cytb evolution were substantially higher in the tropics, and a universal rate of molecular evolution is assumed, divergence times of tropical sister species could be over-estimated. In this case, we would expect to see exactly the pattern of older sister-species

divergences in the tropics as found by Weir and Schluter.

How might Weir and Schluter's results change if sister-species divergence times were re-estimated using latitude-specific rates of cytb evolution? To answer this, it would first be necessary to derive a function that links the substitution rate to the latitudes inhabited by species today and, ideally, the latitudes inhabited by the ancestor of each pair of sister species. If this is not possible, another approach might be to turn the question on its head. We could ask instead how much higher the rate of cytb evolution at low latitudes would need to be before the latitudinal gradient in sister-species ages disappears. We could then compare this with what we know about cytb variation among taxa or regions to make a judgement on how realistic this amount of variation would be.

Using a complete species-level phylogeny to infer diversification rates

Only a few years ago, the idea of a phylogeny of a large taxon, thoroughly sampled and resolved to species level, was little more than a happy dream. Now this is a reality, with species-level phylogenies available for mammals, birds, and a number of other large vertebrate and plant taxa. So now we have large datasets carrying within-clade detail on phylogenetic relationships and latitudes of individual species. The analysis by Weir and Schluter is one example of how species-level data can be used to explore diversification rates across latitudes, but their approach still only made use of a subset of data represented by a sample of sister-species pairs. How might we make full use of a species-level phylogeny to address the same question?

One way is to measure diversification rate as a property of lineages, rather than clades, and estimate the rate for every lineage extending from the root of the phylogeny to each tip. This is based on the same principle as the analysis by Weir and Schluter—that the rate of speciation is inversely proportional to the times between speciation events, and hence positively associated with the number of branching events (nodes) that are

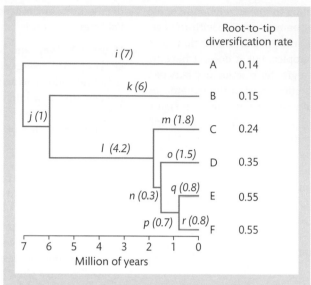

Figure 10.16 Jetz et al.[20] used a method similar to that of Weir and Schluter[19] to estimate a diversification rate for the lineage connecting each species in the bird phylogeny to the root of the phylogeny. In this hypothetical phylogeny, the lengths of every branch are shown, and these are used to calculate a root-to-tip diversification rate.

encountered as we traverse a lineage's path from the tips to the root of the phylogeny. The more nodes we encounter along a path, the higher the rate of diversification in the lineage leading to that particular species. In a way, this approach is simply an extension of Weir and Schluter's sister-species approach that calculates divergence times between sister species only.

One issue that we need to be aware of when doing this is that once we start to move deeper in the tree than the terminal nodes (those separating sister species), branches of the phylogeny are shared among several species. In **Figure 10.16**, the branches labelled *i*, *k*, *m*, *o*, *q*, and *r* are unique to species A, B, C, D, E, and F, respectively. But branch *p* is shared between species E and F, branch *n* between D, E, and F, branch *l* between C, D, E, and F, and branch *j* between all species except A. The further the distance (in number of nodes) from the tips, the more species a branch is shared between. So for the non-terminal (internal) branches, we

need to apportion their contribution to the over-all distribution of divergence times equally among their descendant species.

One method of doing this was used by Walter Jetz and colleagues in their 2012 analysis of the first species-level phylogeny of all 9993 bird species.[20] Jetz et al. calculated a measure of diversification rate for each species by summing the lengths of the branches between nodes along each tip-to-root path, downweighting each branch length according to its distance from the tip. As an example, the set of tip-to-root branch lengths for species F in Figure 10.15 is [0.8,0.7,0.3,4.2,1]. Moving from the tip towards the root, we multiply each successive branch length by one, one-half, one-quarter, one-eighth, and one-sixteenth before summing the values to obtain a measure of evolutionary distinctness known as equal splits (ES). For species F,

$$ES = (0.8 \times 1) + (0.7 \times 0.5) + (0.3 \times 0.25) +$$
$$(4.2 \times 0.125) + (1 \times 0.0625) = 1.8125.$$

The inverse of this quantity (1/ES = 0.55) is proportional to the branching rate along the path from the root to species F. By comparison, along the path leading to species B there is only one branching event, so the value of 1/ES for species B is lower (0.15).

Jetz et al. then examined whether the lineage-specific diversification rate, as represented by the quantity 1/ES, is higher for bird species living closer to the equator. They first presented a visualization of the global geographical pattern, by calculating the mean value of 1/ES for the species occupying each 110×110 km grid cell. The result is surprising: the regions of particularly high and particularly low mean diversification rates do not correspond at all to latitudinal patterns. Indeed, they seem to be geographically idiosyncratic, with high mean diversification rates throughout North America and a few smaller 'hotspots', many in temperate latitudes. Jetz et al. then tested the statistical association between their measure of diversification rate and latitude (both as a continuous measure and a discrete tropical/temperate measure). As we saw earlier in this chapter, this needs to be done in a way that accounts for phylogenetic non-independence and the assumed model of evolution, so they used PGLS for these tests. This confirmed the visual pattern: in the tests repeated for a sample of 100 phylogenies from the Bayesian posterior set (see Chapter 6), there were no significant associations between latitudinal centroid and diversification rate (**Figure 10.17**). When latitude was considered a discrete variable, there was a significant association for 31 of the 100 phylogenies: perhaps some suggestion of higher rates in the tropics, but certainly not a strong enough pattern to draw that conclusion confidently.

Faced with seemingly conflicting results from different studies that have used different approaches,

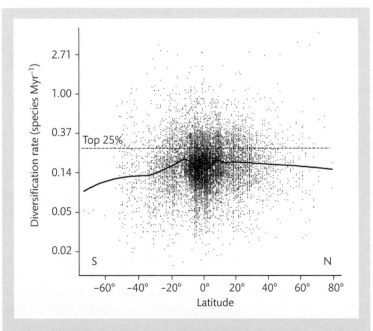

Figure 10.17 Jetz et al.'s measure of diversification rate for each lineage did not show a significant pattern of variation across latitudes.[20]

W. Jetz, G.H. Thomas, J B. Joy, K. Hartmann, and A.O. Mooers (2012) The global diversity of birds in space and time, *Nature* 491: 444–8. Reprinted by permission from Springer Nature.

what kind of conclusion should we draw about whether birds diversify more rapidly in the tropics? Unfortunately, there is no clear answer to this question. We saw earlier how the notion that variation in diversification rates underlies geographical patterns of species richness may depend on the assumption of non-equilibrium diversification dynamics. For birds in general, this assumption seems reasonable—at least, the LTT plot (Figure 10.9) provides no hint of diversification slow-down. But we also discussed the likelihood that different clades of birds across the world vary in their position on the diversification trajectory; some will be in the early stages of rapid radiation, while others may be at a more mature stage, with diversity at an equilibrium level. We don't know how prevalent diversity saturation is among subclades of birds, or even whether this can be reliably determined with the tools we currently have. How might this influence global-scale analyses of diversification rates? If it was somehow possible to filter out 'saturated' bird clades, would the geographical pattern of diversification rates look different to that in Figure 10.17?

Conclusion

The enormous variety of life in the tropics is well known to most people, and unlike the days of Humboldt or Wallace, it is now common for people from temperate countries to travel to the tropics and experience the richness of tropical biodiversity at first hand. But it comes as a surprise to many people to learn that the underlying causes for this high diversity still represent a largely unsolved mystery for scientists. For many scientists, this is indeed part of the appeal of the latitudinal diversity gradient: asking questions about a large-scale fundamental pattern of life on earth for which we still don't have complete answers.

We have seen in this chapter how the LDG is a good example of a biological problem that can only be properly addressed by taking a big-picture, macroevolutionary, and macroecological approach. There is little role for manipulative experiments in tackling the LDG. Small-scale observational field studies can provide valuable data on the way communities are structured or the way ecosystems function in different parts of the world. But the real insights come from the palaeontological record which provides a window on the changes in global diversity through hundreds of millions of years of earth's history, and from phylogenetics that allows us to reconstruct the ancestry and relationships of large groups of organisms. The challenge is to understand the limits of what we can infer from big-picture data such as these, and work within these limits to devise creative and powerful ways of testing hypotheses and comparing alternative explanations.

〇 Points for discussion

1. How important have one-off events in the history of the earth been to the development of the latitudinal diversity gradient? If we could 'replay' the geological, climatic and biological history of the earth over and over again, would we always get a build-up of biodiversity at low latitudes?

2. The concept of ecological 'saturation' plays an important part in developing explanations for high tropical diversity. But we have seen in this chapter that saturation is a difficult concept to measure, or even define. Given that every new species that arises provides new opportunities for specialization for other species, could the number of species on earth, in principle, keep increasing without limit? Or are there some constraints that will ultimately impose a hard upper limit on the number of species?

3. There are some taxonomic groups that show a 'reversed' latitudinal diversity gradient, with more species at higher than lower latitudes (examples include European bryophytes and Ichneumonid wasps). What can we learn from these exceptions to the general pattern? Do they contradict the key hypotheses for high tropical diversity, or do they provide valuable supporting information?

✳ References

1. Bass MS, Finer M, Jenkins CN, Kreft H, Cisneros-Heredia DF, McCracken SF, Pitman NC, English PH, Swing K, Villa G, Di Fiore A, Voigt CC, Kunz TH (2010) Global conservation significance of Ecuador's Yasuní National Park. *PLoS One* 5: e8767.

2. Kaufman DM, Willig MR (1998) Latitudinal patterns of mammalian species richness in the New World: the effects of sampling method and faunal group. *Journal of Biogeography* 25: 795–805.

3. Wallace AR (1878) *Tropical Nature and Other Essays.* Macmillan, London.

4. Dobzhansky T. (1950) Evolution in the tropics. *American Scientist* 38: 209–21.

5. Fischer AG (1960) Latitudinal variation in organic diversity. *Evolution* 14: 64–81.

6. MacArthur RH (1972) *Geographical Ecology: Patterns in the Distribution of Species.* Princeton University Press, Princeton, NJ.

7. Rohde K (1992) Latitudinal gradients in species diversity: the search for the primary cause. *Oikos* 65: 514–27.

8. Rosenzweig ML (1992) Species-diversity gradients: we know more and less than we thought. *Journal of Mammalogy* 73: 715–30.

9. Stevens GC (1989) The latitudinal gradient in geographic range: how so many species coexist in the tropics. *American Naturalist* 133: 240–56.

10. Sepkoski JJ (1978) A kinetic model of Phanerozoic taxonomic diversity. I: Analysis of marine orders. *Paleobiology* 4: 223–51.

11. Pybus OG, Harvey PH (2000) Testing macro-evolutionary models using incomplete molecular phylogenies. *Proceedings of the Royal Society of London Series B. Biological Sciences* 267: 2267–72.

12. Dowle EJ, Morgan-Richards M, Trewick SA (2013) Molecular evolution and the latitudinal biodiversity gradient. *Heredity* 110: 501–10.

13. Stebbins GL (1974) *Flowering Plants: Evolution Above the Species Level.* Belknap Press, Cambridge, MA.

14. Jablonski D (1993) The tropics as a source of evolutionary novelty through geological time. *Nature* 364: 142–4.

15. Farrell BD, Mitter C (1993) Phylogenetic determinants of insect/plant community diversity. In Ricklefs RE, Schluter D (eds), *Species Diversity in Ecological Communities.* University of Chicago Press, Chicago, IL.

16. Cardillo M (1999) Latitude and rates of diversification in birds and butterflies. *Proceedings of the Royal Society of London Series B. Biological Sciences* 266: 1221–5.

17. Felsenstein J. (1985) Phylogenies and the comparative method. *American Naturalist* 125: 1-15.

18. Isaac N, Agapow P-M, Harvey PH, Purvis A (2003) Phylogenetically nested comparisons for testing correlates of species richness: a simulation study of continuous variables. *Evolution* 57: 18–26.

19. Weir JT, Schluter D (2007) The latitudinal gradient in recent speciation and extinction rates of birds and mammals. *Science* 315: 1574–6.

20. Jetz W, Thomas GH, Joy JB, Hartmann K, Mooers AO (2012) The global diversity of birds in space and time. *Nature* 491: 444–8.

21. Benton MJ (1997) Models for the diversification of life. *Trends in Ecology and Evolution* 12(12): 490–5.

Case Study 10
Global dynamics of speciation, extinction, and geographical expansion

In this chapter, we have explored a number of macroevolutionary processes that might be responsible for the great disparity in species numbers between the tropics and extra-tropical regions—the latitudinal diversity gradient (LDG). These processes can be broadly categorized into equilibrium processes which assume environmental regulation of species diversity, and non-equilibrium processes which assume that the most powerful explanations for present-day diversity are to be found in the way diversity has accumulated through time. Common to all explanations for diversity patterns, however, are three fundamental processes that ultimately determine the number of species found in a region: speciation, extinction, and immigration to and from other regions. We have seen how some long-standing explanations for the LDG are based on faster diversification in tropical regions; that is, the difference between speciation rates and extinction rates is presumed to be greater. But in the past decade, two prominent hypotheses for the LDG have also assigned a key role to immigration—the expansion of the geographical distributions of species or lineages from one zone into another over evolutionary timescales. These two hypotheses include both ecological and historical dimensions, and both describe a highly dynamic scenario for the origin and continued maintenance of the LDG.

The 'out of the tropics' (OTT) model was presented in 2006 by palaeontologists David Jablonski, Kaustav Roy, and James Valentine.[1] They noted that hypotheses for the LDG based on differential diversification rates require rate estimates for different regions that are independent of one another, so they must assume that species assemblages in different regions (tropical or extra-tropical, or different latitudinal zones) consist of unique non-overlapping sets of species. Yet this assumption seems at odds with the palaeontological record, which provides plenty of evidence for shifts across latitudes in the geographical distributions of higher taxa (genus and above), and for taxa that span tropical and extra-tropical zones. Hence it is difficult to distinguish the roles of differential diversification rates on the one hand, and immigration across latitudes on the other, in producing latitudinal differences in species numbers. This suggests that we need to consider explicitly the role of shifts or expansions of latitudinal distributions of taxa, in addition to differential speciation and extinction rates, in shaping the LDG.

A long-standing debate about the LDG is whether the tropics represent a 'cradle' of diversity, with elevated speciation rates, or a 'museum' of diversity, with reduced extinction rates; either scenario is expected to result in a disproportionate accumulation of species in the tropics. The OTT model dispenses with this dichotomy because it accommodates both cradle and museum dynamics (**Figure 10.18**). Under the OTT, origination (speciation) rates are higher and extinction rates are lower in the tropics. But, additionally, there is an asymmetry in immigration across zones, with a greater number of tropical taxa expanding their distributions into extra-tropical zones without losing their tropical distribution.

Jablonski et al.[1] used data from the fossil record of marine bivalves to support all three elements of the OTT model: higher tropical origination rates, lower tropical extinction rates, and asymmetric immigration out of the tropics. As a model system for examining large-scale patterns of diversity, marine bivalves have many advantages: they are diverse and widespread, their fossil record is rich and informative, and there has been much work to standardize the taxonomy across time periods. A common approach to inferring origination and extinction rates from the palaeontological record is to tally the numbers of first and last occurrences, respectively, of taxa in different geographical zones and within particular time intervals. As we have seen earlier in this chapter, one of the pitfalls with this approach is the likelihood of geographically heterogeneous sampling and preservation biases in the numbers of known fossil taxa. In particular, tropical regions are under-sampled compared with temperate latitudes, where the majority of palaeontologists work. Jablonski et al. accounted for these biases by first identifying bivalve families with a good fossil record. They did this by selecting those living families (for which the number of genera are known) with at least 75% of their genera represented in the recent fossil record (up to 11Mya). For these well-sampled families, the number of first occurrences is significantly higher in tropical compared with extra-tropical regions, supporting a higher rate of origination in the tropics. In addition, the number of last occurrences of taxa in the fossil record is significantly greater in extra-tropical regions, supporting lower extinction rates in the tropics. Finally, asymmetric immigration across latitudes is supported by a strong pattern of expansion out of the tropics: 77% of genera with tropical first occurrences now occur outside the tropics.[2]

The OTT model is rich in other predictions that are testable with both fossil data and present-day biogeographical data. Some of these are reminiscent of John Willis's classic 'age and area' model of the 1920s (see Chapter 1). For example, OTT predicts that genera inhabiting high latitudes should have more cosmopolitan distributions than those inhabiting the tropics and their average age should be greater. Both predictions are borne out by the bivalve data: frequency distributions of genus ages show a high proportion of young genera with a long tail of old genera in the tropics, but only a small proportion of young genera in high latitude Arctic and Antarctic regions (**Figure 10.19**).[2]

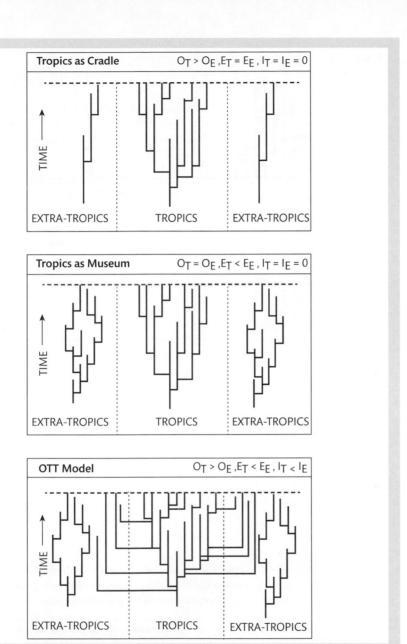

Figure 10.18 Illustration of the phylogenetic diversification patterns we would expect to find under three models for the latitudinal diversity gradient. In the 'cradle' model, the origination rate is higher in the tropics than in extra-tropical regions ($O_T > O_E$), extinction rates do not vary between zones ($E_T = E_E$), and immigration from one zone to the other is symmetrical ($I_T = I_E$). In the 'museum' model, extinction rates are lower in the tropics, so that $O_T = O_E$, $E_T < E_E$, and $I_T = I_E$. In the 'out of the tropics' (OTT) model, immigration rates are asymmetrical and origination rates are higher in the tropics ($O_T > O_E$, $E_T = E_E$, $I_T < I_E$).[1]

From David Jablonski, Kaustuv Roy, James W. Valentine (2006) Out of the tropics: evolutionary dynamics of the latitudinal diversity gradient, *Science* 314, (5796): 102–6. Reprinted with permission from AAAS.

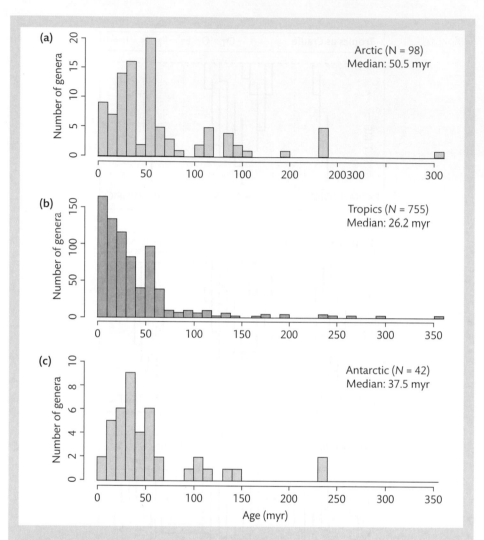

Figure 10.19 Distribution of ages of genera of marine bivalves from the fossil record. For fossils from tropical regions (middle panel), the age distribution is biased towards younger genera, with a greater proportion of genera <20Mya, compared with Arctic and Antarctic faunas.[2]

From David Jablonski, Christina L. Belanger, Sarah K. Berke, et al., (2013) Out of the tropics, but how? Fossils, bridge species, and thermal ranges in the dynamics of the marine latitudinal diversity gradient, *Proceedings of the National Academy of Sciences of the USA* 110 (26): 10,487–94.

About the same time that the OTT model was presented, John Wiens and Michael Donoghue presented another model[3] which has also become regarded as a major historical and biogeographical hypothesis for the LDG. Their model, generally known as tropical niche conservatism (TNC), shares with the OTT model an emphasis on the historical development of the LDG via the dynamics of diversification and range expansion. However, OTT and TNC differ in some important ways, which perhaps reflect the different outlooks of their authors: palaeontological in the case of OTT, and phylogenetic and ecological in the case of TNC.

Wiens and Donoghue framed the TNC in the context of integrating historical biogeography and ecology, arguing that biogeographers and ecologists had often ignored the insights of each other's disciplines in their attempts to explain patterns of diversity. A key process that links biogeography and ecology is dispersal, which determines which species occupy particular localities or habitats (ecology), and thereby the outcome of processes such as range expansion or speciation by vicariance (biogeography). But the central concept of TNC, and the main contrast between TNC and OTT, is not dispersal across latitudes but, rather, lack of dispersal. According to Wiens and Donoghue, within large clades of organisms niche conservatism tends to prevail over niche evolution; that is, the great majority of species remain adapted to the same kinds of climates and habitats that their ancestors did. This means that a lineage that arose in a tropical climate gives rise to descendants that are primarily tropical. Shifts from the tropics to non-tropical climates do happen, but they are comparatively rare. Wiens and Donoghue then showed how niche conservatism can account for the present-day LDG by considering the climatic history of earth over the past 100 million years or so. During the late Cretaceous and much of the Palaeogene Periods (around 100–34 Mya), global average temperatures were around 6–10°C higher than they are today, enough for tropical climates and tropical forest habitats to extend into high latitudes, well beyond today's tropical zone. This is also a period thought to coincide with the early diversification of many of today's large groups of organisms, such as birds, mammals, and angiosperms (but see Chapter 6 for a discussion of the uncertainty in estimating divergence times).

What this means is that many of the large clades of today are likely to have had their origins in tropical environments. Wiens and Donoghue argue that as tropical environments contracted and the temperate zone expanded during the second half of the Palaeogene, the distributions of these major clades also contracted as they remained adapted to tropical environments. The occasional infrequent adaptation of a species to extra-tropical environments allowed a few lineages to expand out of the tropics. These lineages tend to be comparatively young, and the greater tropical residence times of most clades means that diversity has had more time to accumulate in the tropics (**Figure 10.20**). In this way, the great disparity in species numbers between tropical and extra-tropical zones is the outcome of more time (see Figure 10.5 and the discussion earlier in this chapter), reinforced by niche conservatism. Higher origination or speciation rates in the tropics are not needed to produce the LDG under the TNC model, although the model does not deny that this may be the case—it is agnostic on the variation of rates across latitudes.

How can we determine whether the global biodiversity patterns we observe are more consistent with the OTT model or the TNC model?

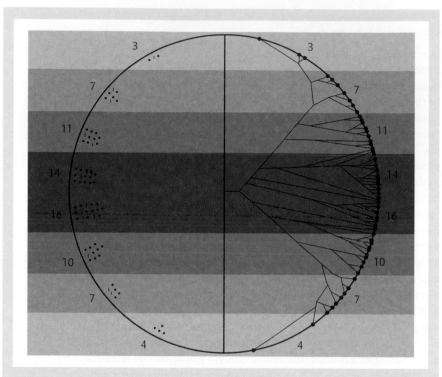

Figure 10.20 Illustration of the 'tropical niche conservatism' model for the latitudinal diversity gradient. Shaded zones across the globe indicate latitudinal bands (the darkest shade is the tropics). The phylogeny on the right has its origin in the tropics, and the majority of present-day species (the tips of the phylogeny) remain tropical in distribution. Shifts away from the tropics into higher latitudes are comparatively recent and infrequent. The phylogeny represents the historical development of the present-day pattern of species richness, represented by the numbers and sets of dots on the left.[3]

From John J. Wiens and Michael J. Donoghue, (2004) Historical biogeography, ecology and species richness, *Trends in Ecology & Evolution*, 19: 639–44. © 2004, with permission from Elsevier.

The most obvious difference between the two models is in the biogeographical shifts of taxa across latitudes. TNC predicts that shifts across latitudinal zones are comparatively rare, but not necessarily asymmetric. Under OTT, in contrast, a disproportionate number of shifts from tropical to extra-tropical zones is a key part of the model. Because TNC is linked to particular events in earth's history, it also makes predictions about the absolute timing of diversification patterns. Global temperatures declined steadily during the Eocene epoch (around 55–34Mya), and then dropped abruptly at the boundary of the Eocene and Oligocene epochs (around 34Mya). This dramatic cooling event marked a shift from a 'greenhouse world' to an 'icehouse world'. The tropical climate zone contracted to low latitudes, and a vast new temperate zone developed. Under the TNC, therefore, we would expect most of the temperate offshoots of large, predominantly tropical clades to be younger than 34Mya. OTT, on the other hand, predicts that extra-tropical clades are usually nested within larger tropical clades (so they also tend to be younger), but are not necessarily younger than 34Mya.

For taxa without a good palaeontological record, these kinds of predictions can be tested using molecular phylogenies with estimated divergence times, together with data on present-day species distributions. Two recent studies, one using angiosperms[4] and the other using birds,[5] have described these kinds of tests. Both studies used present-day latitudinal positions of species to reconstruct the latitudinal positions of ancestral nodes through the phylogeny. Once an ancestral latitude is inferred for each node, the number of transitions across latitudes between ancestral and descendant nodes can be tallied. In both angiosperms and birds, the great majority of nodes reconstructed as tropical also had reconstructed tropical ancestors, and the great majority of temperate nodes also had temperate ancestors. Although the interpretation of these patterns is rather subjective, this would seem to support the idea of a prevailing niche conservatism, as expected under the TNC model. Furthermore, the frequencies of reconstructed ancestor–descendant shifts across latitudes do not seem consistent with an 'out of the tropics' scenario; there were more shifts into the tropics than away from the tropics in both taxa. Finally, in both angiosperms and birds, nearly all clades with a reconstructed temperate ancestor tended to be younger than 34Mya, which is also consistent with the expectations of TNC.

Ancestral state reconstructions based on interpolation from contemporary species data must be treated with extreme caution. Ancestral states can only be inferred by fitting an assumed model, such as Brownian motion (see the section 'Comparisons of sister clades' in this chapter), describing the way that the character of interest (latitudinal position of ranges in this case) is thought to have evolved along the branches of the phylogeny. Both the angiosperm and bird studies compared several alternative models of range evolution, and chose the model that provided the best fit to the data. But without fossil data to 'anchor' the inferred ancestral latitudes, there is no way to verify that the best-fitting model does a reasonable job of describing the way species ranges evolved. This could lead to inferences of ancestral states that are highly misleading.

Debate over the OTT versus TNC models continues. Given that the bivalve fossil data seem consistent with OTT, and the angiosperm and bird phylogenetic data with TNC, it is worth thinking about why we see apparently contradictory patterns in these two kinds of data. Perhaps the answer lies partly in the different taxonomic resolutions; the bivalves were analysed at genus level, while the angiosperms and birds were analysed at species level. Genera are probably less likely to exhibit niche conservatism than species, especially if a genus can consist of species that are endemic to different latitudes. Or perhaps there are genuine differences in the niche conservatism of marine bivalves compared with angiosperms or birds, such that bivalve lineages are more readily adaptable to new environments. As always with the latitudinal diversity gradient, we cannot rule out the possibility of a pluralistic explanation, with different processes generating similar patterns of biodiversity in different groups of organisms.

Questions to ponder

1. The out of the tropics model assumes higher tropical origination rates combined with net immigration from the tropics to higher latitudes. If origination is faster in the tropics, is the immigration part of the model necessary to produce a latitudinal diversity gradient? How would you expect the patterns of biodiversity produced under OTT to differ from patterns produced under a model that only assumes faster tropical diversification?

2. Much of the support for OTT comes from fossil data, while much of the support for TNC comes from phylogenetic and contemporary geographical data. How can these be reconciled? How could you investigate whether the different patterns result from the nature of the data being used to test the two models?

3. What predictions would you make about future fossil discoveries at different latitudes under the OTT and TNC models?

Further investigation

The latitudinal diversity gradient is the largest geographical pattern in the spatial distribution of biodiversity, but there are other major patterns such as elevational gradients, depth gradients, and biodiversity hotspots. Can the OTT and TNC models also be applied to non-latitudinal biodiversity patterns such as these?

References

1. Jablonski D, Roy K, Valentine JW (2006) Out of the tropics: evolutionary dynamics of the latitudinal diversity gradient. *Science* 314(5796): 102–6.

2. Jablonski D, Belanger CL, Berke SK, Huang S, Krug AZ, Roy K, Tomasovych A, Valentin JW (2013) Out of the tropics, but how? Fossils, bridge species, and thermal ranges in the dynamics of the marine latitudinal diversity gradient. *Proceedings of the National Academy of Sciences of the USA* 110(26): 10,487–94.

3. Wiens JJ, Donoghue MJ (2004) Historical biogeography, ecology and species richness. *Trends in Ecology & Evolution* 19(12): 639–44.

4. Kerkhoff AJ, Moriarty PE, Weiser MD (2014) The latitudinal species richness gradient in New World woody angiosperms is consistent with the tropical conservatism hypothesis. *Proceedings of the National Academy of Sciences of the USA* 111(22): 8125–30.

5. Duchêne DA, Cardillo M (2015) Phylogenetic patterns in the geographic distributions of birds support the tropical conservatism hypothesis. *Global Ecology and Biogeography* 24(11): 1261–8.

What is the future of biodiversity?

11

Roadmap

Why study extinction and biodiversity loss?

Throughout this book we have seen that extinction is one of the key processes that have shaped the history of life and the development of biodiversity. Currently, species declines and extinctions appear to be happening much more rapidly than at any time in geological history as a result of the way human activity is modifying the world's natural systems. There is widespread international consensus that action is needed to prevent further extinctions and mitigate ongoing loss of biodiversity. In order to develop effective strategies for conservation, we need to understand how biodiversity is distributed (phylogenetically and spatially), we need objective methods to quantify the speed and extent of biodiversity loss, and we need an understanding of the processes that cause the loss of biodiversity. In developing this kind of big-picture view of conservation, we can make use of some of the tools and knowledge gained from the study of macroevolution and macroecology.

What are the main points?

- Using the fossil record, we can estimate a 'background' extinction rate for periods between mass extinctions. Current human-driven extinction rates appear to be orders of magnitude higher than background rates and, if continued, could lead to a global mass extinction event comparable to the 'big five' from the fossil record.

- The IUCN Red List is the primary scheme for classifying species according to their inferred risk of extinction. So far, a few large groups of vertebrates have been fully evaluated under the Red List scheme, but much of the world's biodiversity remains unevaluated due to lack of sufficient data.

Photograph: Panamanian golden frog. © Justin Black/Shutterstock.com.

- The central concept of biodiversity conservation planning is prioritization because of limited resources for conservation programmes and conflicting demands on the use of land. Prioritization can be species-based (which species should we prevent from going extinct?) or area-based (which areas should we set aside as protected areas?).
- Systematic conservation planning is a structured objective approach to making conservation choices, such as choosing reserve networks, which aims to meet conservation goals efficiently by minimizing the costs associated with conservation activities.
- Phylogenetic relationships among species have been incorporated into conservation prioritization by identifying species that are evolutionarily distinct (with few close living relatives), or groups of species with high phylogenetic diversity. However, it is important to be aware of the limitations of using phylogenetic branch lengths as a proxy for conservation value.

What techniques are covered?

- Using the fossil record to estimate background extinction rates.
- Using the Red List and Red List Index to estimate the extent and speed of current species declines.
- Reserve selection algorithms using the principle of complementarity.
- Calculating phylogenetic diversity (PD) and evolutionary distinctness (ED).

What case studies will be included?

- Predicting conservation status of poorly known species using comparative models.

"It's easy to think that since the extinction of the dodo we are now sadder and wiser, but there's a lot of evidence to suggest that we are merely sadder and better informed."

Douglas Adams and Mark Cawardine (1990) *Last Chance to See*. Pan Books, London.

Macroevolution and macroecology are fields of study that focus on large-scale patterns of biodiversity, and the way those patterns have been shaped by ecological, evolutionary, and biogeographical processes that play out at large spatial scales, and over long periods of time. One of those processes is extinction, and in many of the chapters in this book we have explored the key role of extinction in shaping biodiversity. For example, mass extinction events have occurred at various times in the earth's history, and are often associated with periods of massive turnover in the composition of the global biota. Together with speciation, extinction determines the speed of diversification and the growth of clades.

As well as extinctions on macroevolutionary scales, it is also clear that many extinctions are occurring in our own time as a result of human activities, and many biologists consider this to be the start of a new global mass extinction event. How do we judge if the current 'extinction crisis' is an event of significance on a geological timescale? How do we discover the underlying processes that cause species to decline towards extinction? What consequences will ongoing extinctions have for biodiversity? How can we best protect and manage ecosystems in order to limit biodiversity loss to levels that we find acceptable? The broad-scale approach we take to asking questions about macroevolution and macroecology has a role to play in answering these conservation-related questions. In this chapter we explore how we can apply to conservation some of the analytical methods and ways of thinking that we use in investigating the origins and evolution of biodiversity.

Extinctions, biodiversity loss, and the 'Anthropocene'

So troublesome a creature: the tragic demise of the thylacine

The death throes of a species are seldom well documented; most species that have gone extinct within the period of human history have passed away quietly and comparatively unnoticed. An exception is the thylacine (*Thylacinus cynocephalus*), also known as the Tasmanian tiger (**Figure 11.1**). The thylacine—a Tasmanian marsupial of remarkably dog-like appearance—was last seen in the 1930s, and its extinction is still widely lamented. This is a case in which there exists a reasonably well-recorded account of a species in the process of decline. Ironically, the policy that left a good record of the thylacine's dwindling numbers was probably also the primary agent of its extinction. For a 30-year period in the late nineteenth and early twentieth centuries, over 2000 thylacines were killed under a Tasmanian government bounty scheme, set up with aim of eradicating them from sheep-farming districts. It is very likely that this reduced the thylacine population to a level from which it could never recover.

Fossil and subfossil remains, together with indigenous rock art, tell us that thylacines were once widely distributed on the Australian mainland, but seem to have become extinct 3000–3500 years ago. The extinction of thylacines from mainland Australia followed the arrival of the dingo (*Canis lupus*) in Australia around 4000 years ago, but there is debate about the role of the dingo in mainland extinctions of thylacines. Dingoes have often been considered more efficient and competitively superior hunters compared with thylacines, driving thylacines to extinction through competitive exclusion. Whereas thylacines were supposedly solitary hunters, dingoes often hunt in packs, perhaps giving them the edge in hunting large prey such as kangaroos, or even in direct encounters with thylacines. Dingoes are also quite flexible in their dietary preferences, whereas thylacines appear to have been more specialized large-prey

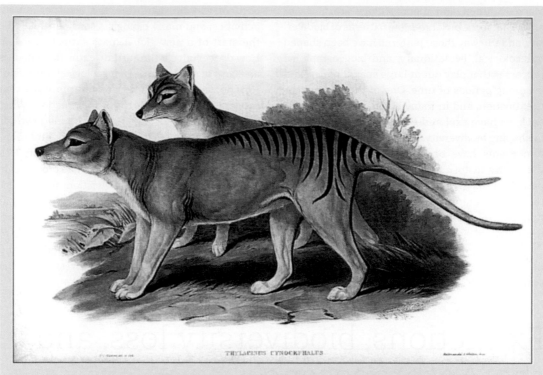

Figure 11.1 The lithographic prints by John and Elizabeth Gould (based on drawings by H.C. Richter of animals in London Zoo) are some of the most widely reproduced images of thylacines. The Goulds' prints portray a hauntingly beautiful, mysterious creature, and we might easily attribute to the artists a sensitivity for these animals and their welfare. Unfortunately, though, the sentiments expressed in their work *The Mammals of Australia* seem more typical of nineteenth-century views of thylacines: 'neither the shepherd nor the farmer can be blamed for wishing to rid the island of so troublesome a creature'.

hunters—another advantage for dingoes, particularly in times of scarcity.

But the dingo–thylacine competition hypothesis is not universally accepted, for several reasons. First, contemporary accounts suggest that thylacines were not strictly solitary hunters, but in fact often hunted cooperatively in family groups. Secondly, thylacines were heavier and more powerfully built than dingoes, so direct conflict, if there was any, may not have been so one-sided. Thirdly, the timing of thylacine extinction may not correspond to the arrival of dingoes at all. Rock art and fossil records suggest that thylacines may have disappeared from northern Australia some time before the arrival of dingoes, even if they remained in southern Australia.[1] This implies that some other

agent of decline may have caused the extinction of thylacines from mainland Australia. One candidate is hunting by expanding human populations, assisted by increasingly effective hunting technology such as the woomera (spear-thrower).

Whatever the cause of the thylacine's extinction from the mainland, they maintained a population on the island of Tasmania—isolated from mainland Australia for at least 8000 years—until the invasion and settlement by the British in the nineteenth century. The expansion of sheep grazing throughout eastern and northern Tasmania brought settlers into conflict both with the indigenous peoples of Tasmania and with some of its wildlife. Thylacines were blamed for killing sheep, although the common view today is that this was largely unjustified—certainly, wild dogs (which became common soon after British occupation of Tasmania) were likely to have been a greater threat to sheep than thylacines.[2] Nonetheless, lobbying by influential graziers prompted the government of Tasmania to introduce a bounty in 1888, under which £1 was paid for each adult thylacine presented to police stations. If we assume that the numbers of bounty payments made each year reflects the overall thylacine population, then the collective bounty payment records kept by the police stations provide a proxy record of population trends for the period 1888–1909, when the bounty scheme was in place (**Figure 11.2**).

The bounty records tell an interesting story about the thylacine population in the final decades of its existence. For the first 10 years, numbers were fairly steady. This is consistent with longer-term records of thylacines killed or captured at Woolnorth, a grazing property in the northwest of Tasmania, where numbers were also quite steady from 1874 to 1899. Then, from 1899 to 1901, there was a sudden sharp increase in the numbers of thylacines killed, both under the Tasmania-wide bounty scheme and

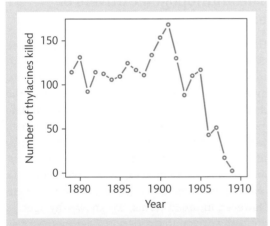

Figure 11.2 Number of thylacines presented to Tasmanian police stations for bounty payment each year between 1889 and 1909. Note the sudden spike in numbers from 1899 to 1901, followed by a precipitous fall after 1905. More than 90% of thylacines presented for bounty payment were adults, which were easier to trap and attracted a bounty of £1, compared with 10 shillings for a juvenile.

at Woolnorth. The increase was short-lived. In the space of three years from 1906 to 1909, the numbers of thylacines killed, and presumably the overall population, collapsed. The bounty scheme had done its job, and it was ended in 1909.

Was the population crash a direct result of the number of thylacines killed under the bounty scheme? On the one hand, a consistent offtake of over 100 individuals per year was maintained for many years, suggesting that this level of population depletion was sustainable. The number of juveniles presented for payment remained at a fairly consistent level, suggesting that recruitment was being maintained. On the other hand, the total population of thylacines in Tasmania was probably never very large, and perhaps the numbers of thylacines in the sheep-grazing districts were being maintained by continual immigration from outlying areas until the total population was diminished to below the level of recovery.

But there is also evidence that other agents of extinction were at play. Beginning around 1896, there are numerous reports of diseased and distressed thylacines, with symptoms including sore feet, hair loss, skin lesions, diarrhoea, and emaciation. It is clear that a major epidemic was sweeping through the thylacine population, although what this disease was or where it came from remain unknown—it may have been introduced by domestic dogs. It seems likely that the epidemic was responsible for the spike in the numbers of thylacines presented for bounty around the turn of the century, because diseased animals were easier to trap or shoot and it became common to find dead ones in the bush. Even after the population crash of 1905 and the end of the bounty scheme, small numbers of thylacines continued to be caught or killed for the zoo and museum trade. Presumably, near-extinction increased their value as live or mounted specimens. The final nail in the coffin of the species may have been a particularly virulent new disease strain that appeared in the late 1920s, killing most of the thylacines in captivity (Hobart Zoo lost nine between 1928 and 1930) and, perversely, creating another small spike in the numbers of animals captured in the bush. The last known wild thylacine was captured in 1933 and sent to Hobart Zoo. It died in 1936, 59 days after the Tasmanian government afforded the species official protection. No confirmed sightings of a thylacine have been made since.

We have begun this chapter with an account of the extinction of the thylacine partly to remind us that underlying the broad-scale quantitative patterns of extinction that will be the main focus of this chapter are numerous stories of the demise of individual species, many as tragic and preventable as that of the thylacine. But the case of the thylacine also demonstrates that even where the decline of a species is comparatively well documented and driven by a seemingly obvious cause, the ecological processes that lead to final extinction can be complex, unpredictable, and difficult to pin down. The fluctuations in the thylacine population implied by the bounty records, together with numerous contemporary accounts of thylacines in newspapers and official records, present a picture of a species suffering under the onslaught of multiple threatening processes. Unsustainable hunting and trapping for nearly a century may have been compounded by competition or conflict with wild dogs and loss of habitat as native bush was cleared for farmland. The disease epidemics of 1896–1901 and the late 1920s, perhaps resulting from the introduction of domestic dogs, arrived at times when the thylacine population was already particularly vulnerable as a result of over-hunting, and the outcome was devastating. This complexity of interacting factors is something we need to keep in mind when we try to characterize the key threatening processes for large groups of species.

Something else we can learn from the thylacine case is that extinctions caused by humans have been happening for thousands of years, and there might be informative parallels between the processes that led to extinction in prehistoric times and those occurring today. If thylacine extinction from Tasmania was caused by consistent steady hunting pressure by humans using effective hunting technology and assisted by dogs, perhaps it is plausible that similar pressures caused the prehistoric extinction of thylacines from mainland Australia, even if the two extinction processes ran their course at very different speeds.

Key points

- Species decline and extinction occurs when the ecological needs of a species conflict with human use of the natural environment and resources within its distribution.

- Most species extinctions are poorly documented, but in some cases (such as the thylacine) good records exist that allow us to trace the population patterns and the appearance of threatening processes that precede extinction.

- In many cases, final extinction results from multiple processes, which may interact, making it difficult to pinpoint an exact cause of extinction.

The anthropogenic drivers of extinction

Unfortunately, the thylacine is not an isolated case. The threatening processes that led it to extinction—hunting, disease, habitat loss, and introduced species—have also caused the decline of many other species, some to complete extinction or near-extinction, and many others to small fractions of the geographical areas they previously occupied. For the thylacine, the mechanisms that led to final extinction were complex and require some detective work to tease apart. However, in broad terms, we know why species are being driven to extinction; over the past few centuries, humans have developed the knowledge and technology to allow their own populations to expand rapidly, appropriate a large share of the world's resources, and convert large areas of natural habitat to anthropogenic ecosystems devoted to sustaining human populations. In fact, human activity has now had such a substantial impact on the world's ecosystems, energy budget, atmosphere, and climate, that some scientists propose that a new geological epoch, the 'Anthropocene,' should be recognized.[3]

Understanding the nature of the threats that species face is an important part of developing plans for conservation. The International Union for Conservation of Nature (IUCN) uses a hierarchical classification of threat types[4] that includes 12 primary threat categories, each divided into sub-categories, aiming to capture all the significant threats faced by species. A simpler, but arguably more influential, classification of threat types was Jared Diamond's 'evil quartet': habitat loss, hunting, introduced species, and extinction chains (when the extinction of a species causes the co-extinction of dependent species).[5] These four broad categories effectively capture the great majority of the processes that cause species to decline, although it has been suggested that the effects of climate change do not fit into this scheme, and should be added as a fifth category.

What do you think?

There are moves by groups of geologists and biologists to have our current age recognized as a new geological epoch, the 'Anthropocene', although this has yet to be formally accepted by the International Commission on Stratigraphy, the International Union of Geological Sciences, or the US Geological Survey. Do you think that recognition of the Anthropocene is warranted, given the extent and magnitude of human modification of the world's biological, atmospheric, and geochemical systems? To answer this question, it is worth examining what it is that defines and distinguishes other geological epochs, such as the Holocene, Pleistocene, or Pliocene. Will the case for the Anthropocene change if currently projected climate-change scenarios eventuate over the coming decades?

Taking stock: quantifying biodiversity

There may always have been a minority of people concerned about the loss of natural areas and the species they support, but only in the second half of the twentieth century did nature conservation emerge as a globally important social and political movement. The establishment of international conventions to limit environmental damage and biodiversity loss, such as the Convention on International Trade in Endangered Species or the Convention on Biological Diversity, attests to a general global consensus that action needs to be taken to mitigate ongoing biodiversity loss.

Given that there is general agreement that action is needed to limit the loss of biodiversity, all the relevant quantities and qualities involved must be defined. We need operational definitions for 'biodiversity', 'species', and 'extinction'. We need ways of determining how much biodiversity there is on earth, how biodiversity is distributed across the earth's surface, and how many species have gone extinct. We need a way of judging how likely any given species is to go extinct, and of understanding what the threatening processes are. We can only assess the scale and extent of the problem of biodiversity loss if we know what we are counting. In this chapter, we won't go into detail on the different definitions of concepts such as biodiversity or species, but it is important to be aware that there is debate about these definitions—in particular, about the concept of species and the way species are defined.

The term 'biodiversity' itself is comparatively new—it was first coined in the 1980s. Today, the term tends to be used (in both everyday speech and official contexts) as a catch-all term to describe the variety of life, without a particularly precise definition. This variety can encompass taxonomic, genetic, ecological, or other forms of diversity, but one quantity in particular plays an important role in the measurement of biodiversity—the number of species, also known as species richness. Species are widely regarded as a fundamental natural unit of biodiversity, and a tally of their numbers is usually considered an acceptable representation of biodiversity—although, again, it is important to be aware that neither of these assertions is universally accepted. But given that there are classification schemes and formal species definitions for a large proportion of the world's macroscopic biota, species richness provides a useful currency with which to quantify the amount and distribution of biodiversity and to assess the extent of its erosion. Later in this chapter we will examine alternative ways of representing biodiversity, and how they may change the conservation decisions that we make.

The Red List: what does it mean to be 'threatened'?

The use of species as the general currency of conservation means that species extinctions are one of the primary means of quantifying the erosion or loss of biodiversity. If we quantify biodiversity loss using species extinctions, assessing the risk of each species going extinct in the near future provides a way of estimating the current and potential impact of human activities on biodiversity. What this requires is a fairly precise definition of what it means for a species to be 'threatened with extinction'.

In the 1950s, the IUCN began compiling biological and ecological information on rare species, and publishing the first lists of threatened mammals, birds, and plants as 'Red Data Books'. The classification of species as threatened in the Red Data Books was quite subjective, based largely on expert opinion. This changed in the 1990s, when Georgina Mace and Russell Lande developed a set of rigorous objective criteria under which species could be assigned to different categories of extinction risk.[6] Mace and Lande set out to establish an extinction-risk classification system with a number of clear goals:

1. the system should be simple, with extinction-risk categories based on a probabilistic assessment of extinction risk;

2. it should be flexible in the data required to assign species to categories;

3. it should be applicable not only to species but to any population unit;

4. the terminology used to describe extinction risk should be appropriate and unambiguous;

5. the system should incorporate uncertainty;

6. the categories should express extinction risk on timescales relevant to conservation decision-making.

The categories and criteria for assessing extinction risk developed by Mace and Lande were adopted by the IUCN for a new classification scheme known as the Red List. Under the Red List, the scheme for classifying species into categories of extinction risk is hierarchical (**Figure 11.3**). In the first case, many species still haven't been evaluated under the Red List scheme, and for those that have, many still lack sufficient data on distribution and on population size and trends for a classification to be made—these are Data-Deficient species. Species with adequate data are placed in one of seven categories, from lowest to highest risk of extinction. Three of the categories (Vulnerable, Endangered, and Critically Endangered) constitute threatened status, so species in these three categories are officially regarded as 'threatened'.

There are different ways in which a species can be assigned to an extinction risk category. In fact, there are five criteria for listing, and a species need only satisfy one criterion to be listed in a particular category. The descriptions of the criteria are quite long and detailed,[7] but broadly the five criteria are as follows.

A. Rate of decline in population size

B. Geographical range size and decline, fragmentedness, or size fluctuation

C. Population size and rate of decline, division into subpopulations, or fluctuation

D. Population size alone

E. Quantitative modelling of extinction probability

If the Red List had existed in 1909, the Thylacine, following its population crash, would probably have qualified for listing as Critically Endangered (CR) under criterion A.2:

an observed, estimated, inferred or suspected population size reduction of ≥80% over the last 10 years or three generations, where the reduction or its causes may not have ceased.

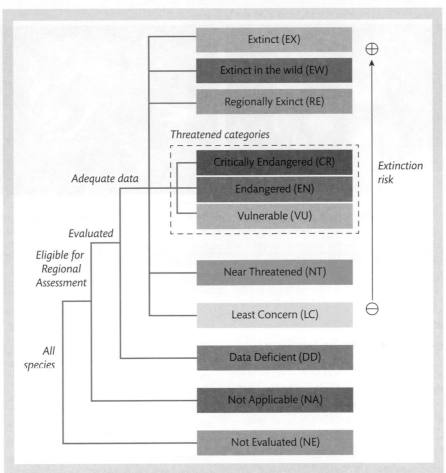

Figure 11.3 Categories of extinction risk under the IUCN Red List scheme. Species that have been evaluated are assigned to one of eight categories; those in the categories of Vulnerable, Endangered, or Critically Endangered are considered to be 'threatened'.

Source: IUCN 2018. The IUCN Red List of Threatened Species. Version 2018–2. http://www.iucnredlist.org

The thylacine might also have qualified as CR under criterion C.1:

population size estimated to number fewer than 250 mature individuals and an estimated continuing decline of at least 25% within three years or one generation.

The population reduction of the thylacine would not have been estimated from direct population counts or surveys, but inferred from the number of animals presented for bounty payments, or perhaps by interviewing graziers and trappers about their observations of thylacine numbers in different districts. This would be permissible under the Red List criteria: if no detailed population surveys exist for a species, it is preferable to use whatever proxy information is available to infer population trends, than to ignore a possible serious decline.

Figure 11.4 A surprising number of mammal species have been 'rediscovered' following long periods without being sighted. The Tonkin snub-nosed monkey (*Rhinopithecus avunculus*) (a) was presumed extinct until its rediscovery in a mountainous region of northern Vietnam in 1989. The bridled nailtail wallaby (*Onychogalea fraenata*) (b) was also presumed extinct until its rediscovery on a pastoral property in central Queensland, Australia, in the early 1970s.

(a) Photo by Quyet Le via Wikimedia Commons (CC2.0 license). (b) Photo by Bernard Dupont via Wikimedia Commons (CC2.0 license).

What does it mean to be 'extinct?'

The IUCN considers a taxon extinct when there is 'no reasonable doubt that the last individual has died'. This definition is very simple, but the concept of extinction is not completely unambiguous. The death of the last individual of a species may arrive at the end of a long period in which the species was absent from most of its original range, or absent from the wild entirely, so that it had already ceased to play any role in natural ecosystems—this is termed 'functional' or 'ecological' extinction. This was the case for the thylacine at the time the last known individual died in captivity in 1936.

In terms of the evidence required to list a species in a particular category of extinction risk, the category of Extinct is treated slightly differently from the others. It is actually very difficult to demonstrate, beyond reasonable doubt, that the last individual of a species has died. By definition, a species on the brink of extinction is extremely rare, and the last few individuals may well be confined to the more remote and inaccessible parts of the original distribution, making them hard to find.

In fact, the number of species presumed extinct, but then rediscovered, is surprisingly high. Even among terrestrial mammals, over 60 species have been rediscovered after presumed extinction, or at least after remaining undetected for many decades (**Figure 11.4**). A recent example is the hispid hare (*Caprolagus hispidus*), recorded from Chitwan National Park in Nepal in 2016, having remained undetected since 1984.[8]

The possibility of rediscovery means that a stringent standard of evidence is needed before complete extinction can be assumed. Listing as EX in the Red List requires exhaustive searches and surveys of the species habitat to have been undertaken throughout its original range, and no positive evidence of the continued existence of the species to have been found for a substantial period of time. The thylacine was last seen in 1936, but not declared extinct by the IUCN until 1982. There are potentially important implications of officially declaring a species extinct: funding for conservation programmes may be withdrawn, and search efforts may wind down (although some people still search for thylacines, in both Tasmania and mainland Australia).

What do you think?

Large sums of money are spent on conservation programmes to rescue species on the edge of extinction, such as the Iberian lynx (*Lynx pardinus*) or the gharial (*Gavialis gangeticus*). It has been argued that the chance of success of such programmes is limited. Are expensive programmes focusing on Critically Endangered species a sensible use of limited conservation funds, or would conservation goals be met more effectively by diverting more funding to less threatened species, for which the chances of recovery are higher?

Patterns and trends in the Red List

The taxonomic coverage of the Red List is continually increasing, with new species assessed and added to the list every year. To date, a number of large taxonomic groups have been fully assessed, including mammals, birds, amphibians, sharks and rays, conifers, and cycads. But that doesn't mean all species in these groups have been assigned to an extinction risk category; in mammals, amphibians, and sharks and rays, large proportions of species remain Data-Deficient. Nonetheless, an overview of the relative numbers of species listed as Threatened and non-threatened in these taxa presents a sobering picture (**Figure 11.5**) which gives us a first glance at the magnitude of the current extinction crisis. Of the 307 species of cycads, for example, around 63% are assigned to the categories VU, EN, or CR; that is, they are considered threatened with extinction. This figure is 41% for amphibians, 34% for conifers, and 25% for mammals.

In those groups that have not been comprehensively evaluated, or for which there are still many Data-Deficient species, the Red List gives us a basis for estimating the overall number of threatened species. To do this, we need to make some

assumptions about the representativeness of the species already assigned to Red List categories (this is explored in more detail in Case Study 11). In groups like insects, in which basic biodiversity knowledge is very incomplete, it is likely that the species with sufficient data for listing are relatively common, and either widely distributed or found in countries in which many entomologists are based. Narrowly distributed, uncommon insect species from tropical countries are less likely to be studied and listed, but it is precisely these kinds of species that may be under the greatest threat of extinction. This means that the proportion of evaluated insect species listed as threatened (18%) may be an under-estimate of the proportion of insects globally that face extinction. Given that thousands of previously unknown insect species are discovered each year in regions where large-scale deforestation continues, it is quite likely that many insect species are going extinct before they have even been discovered and described by biologists.

The Red List Index: how rapidly is biodiversity being lost?

In addition to estimating the current proportions of species in different taxa that are threatened with extinction, the Red List can also be used to understand how rapidly the situation is changing. The Red List Index (RLI) is a metric to quantify the aggregate extinction risk of a set of species, in either a taxonomic group or a region.[9] The RLI for a group has a value of 1 if all species in the group are Least Concern, and a value of zero if all species are Extinct. Every species that moves to a higher extinction risk category from one period to the next decreases the value of the RLI. Changes in the RLI from one assessment to the next can provide an indicator of the rate at which species in a taxon, or a region, are increasing or decreasing in extinction risk. Therefore, for taxonomic groups that have been comprehensively assessed at least twice, we can estimate their rates of movement towards extinction, and we can compare these rates between taxa, countries,

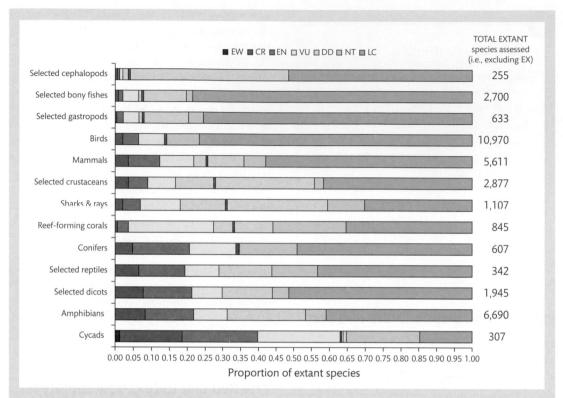

Figure 11.5 A selection of animal taxa and their proportional numbers of species assigned to the different Red List categories of extinction risk. Birds are the most comprehensively assessed group, with only 58 species listed as Data-Deficient out of 11,122 taxa (species and subspecies) evaluated. When looking at this chart, it is important to keep in mind that these are the proportions of evaluated species only, and for some groups only a small proportion of known species have been evaluated. For example, 7639 insects have been evaluated under the Red List scheme, but there are around a million described insect species.

Source: IUCN 2018. The IUCN Red List of Threatened Species. Version 2018–2. http://www.iucnredlist.org

or regions of the world. In some cases, it is even possible to 'back-cast', or to estimate the likely extinction risk category of a species for Red List assessments in the past, in which the species wasn't included (because of lack of information at the time). This can be done, for example, by extrapolating backwards from recent population trends, or using information on historical habitat loss. **Figure 11.6** shows the RLI trends for four comprehensively assessed taxa, back-cast to different dates, depending on the reliability of information available.

Key points

- Anthropogenic drivers of extinction have been summarized into four key types, known as the 'evil quartet': habitat loss, hunting, introduced species, and extinction chains.

- 'Biodiversity' is typically used in an imprecise way to refer to all aspects of variety in the living world. Species are the most widely used currency of conservation, as both the unit of extinction and the unit of biodiversity (species richness).

- The IUCN Red List is an objective, evidence-based, hierarchical scheme for classifying species according to their risk of extinction. Species in three of the categories (Vulnerable, Endangered, Critically Endangered) are considered 'threatened'.

- The final extinction of a species is difficult to demonstrate conclusively, because species nearing extinction become rare and hard to detect.

- The Red List Index tracks species changes in Red List threat status through time, providing a way of quantifying the speed at which different groups are declining.

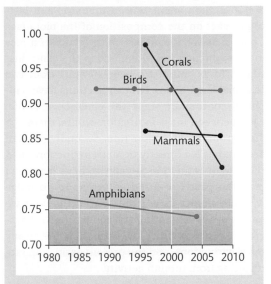

Figure 11.6 Change in the Red List Index (RLI) over time for groups with at least two comprehensive Red List assessments. A group would have an RLI value of 1 if all its species were classified as Least Concern, and a value of zero if all species were Extinct. For each group, the line shows a downward slope, indicating a worsening in the overall aggregate level of extinction risk of its species. The steepness of the slope gives an indication of how rapidly the species in a group are moving towards extinction.

Source: IUCN 2018. The IUCN Red List of Threatened Species. Version 2018-2. http://www.iucnredlist.org

The sixth mass extinction?

Extinction is a natural phenomenon. How can we be sure that the recent and current extinctions happening as a result of human activities are anything unusual in the history of life on earth? One way to answer this question is to compare current extinction rates with rates of extinction that we infer from the pre-human fossil record. Palaeontologists are uniquely well placed to understand patterns of extinction in the history of the earth's biota. As we have seen in Chapters 3 and 10, compilations of the numbers of fossil taxa known from different geological ages display periods of increase, periods of stasis, and periods of decrease. The most sudden and dramatic decreases in known fossil diversity were labelled 'mass extinctions' by David Raup and Jack Sepkoski in 1982.[10] They highlighted the five largest mass extinction events, which became known as the 'Big Five'. These events are characterized by the disappearance from the fossil record of a large proportion of the known taxa: over 60% of species, over 50% of genera, and over 17% of families. Each mass extinction occurs on the boundary between two geological ages, marked by a significant turnover in the composition of the world's biota. There is debate over whether the Big Five were qualitatively different from other extinction events that have occurred throughout earth's history, or if they were simply the largest on a continuous scale of extinction severity.

The term 'sixth extinction' has been used by biologists who believe we are currently entering a global extinction event—driven by human modification and impact on natural ecosystems—comparable in magnitude to the five largest mass extinctions of the fossil record.[11] This is quite a strong claim, given that these events were extremely rare, occurring at an average interval of 100 million years, with a dramatic and lasting impact on the composition of the biota (see Chapter 5). What evidence is the claim of a new mass extinction based on, and is the claim justified? To examine this, we can start from two reasonable premises. First, extinctions have happened continually since the emergence of life; even between mass extinction events, there is a 'background' extinction rate. Secondly, human activity has led directly to many extinctions, and the rate of extinction appears to have increased over the past few hundred years. So, to judge whether or not a mass extinction is imminent, we need ways of estimating and comparing the background extinction rate, the extinction rate during mass extinctions, and the present extinction rate resulting from human activity.

Estimates of the background rate of extinction are often expressed as the number of extinctions (E) per million species-years (MSY) (the sum of the number of species known from the fossil record in a million-year time period). Expressing extinctions as a proportion of known biodiversity is more conservative than attempting to estimate absolute numbers of extinctions, given that the global number of species is highly uncertain, and the fossil record represents only a selective fraction of this number at any point in earth's history. Expressing extinction rates as E/MSY also provides a common currency that allows background and recent rates to be compared, if we extrapolate recent rates to a million-year period. Such estimates are subject to multiple uncertainties and depend on a number of assumptions, so they will always be fairly rough approximations. We know, for example, that the fossil record under-estimates the true number of taxa present at a particular time period (because not all species will have left fossils that have been found). The degree of under-estimation increases as taxonomic rank decreases; the fossil record will under-estimate numbers of species to a greater degree than numbers of genera or families, and it may not underestimate the numbers of phyla at all. Estimates of background extinction rate that have been made for various groups of (primarily marine)

invertebrates and vertebrates point to rates in the range 0.001–0.2 E/MSY.[12]

How does this compare with recent and current extinction rates? Making these comparisons presents additional challenges. For example, species in the fossil record are necessarily distinguished on the basis of morphological differences from other species, whereas extant species are often defined using other criteria, such as the distinctness of their DNA. These differences may lead to under-estimates of species numbers and extinction rates from fossil data. On the other hand, the fossil record is likely to under-estimate the longevity of taxa, because for most taxa extinction will actually have occurred some time after their disappearance from the fossil record. This may lead to over-estimates of extinction rates.

In any case, estimates of recent and current extinction rates, expressed in units equivalent to those in the fossil record, tend to be many times higher than background rates inferred from the fossil record. For example, 102 species of birds have been recorded as having gone extinct or possibly extinct in historical times, out of a total of 10,152 known modern species. To estimate the 'duration period' of each modern species, we can use the time that has elapsed since the species was first formally described. If we do this, modern bird species represent a total of 1,911,231 species-years. The rate of extinction of bird species is then $(102/1,911,231) \times 10^6 = 53.4$ E/MSY.[12] But this figure hides an important pattern; when we calculate extinction rates separately for species that were first formally described before and after the year 1900, we get figures of 49 E/MSY before 1900 and 132 E/MSY after 1900. For other vertebrate groups, the figures are quite similar to those for birds. For amphibians, estimates are 66 and 107 E/MSY before and after 1900, and for mammals the estimates are 72 and 243 E/MSY.[12] The extinction rates are much higher for more recently described vertebrate species, probably because these tend to be rarer species with restricted distributions, often found in parts of the world in which detailed biological study has only taken place in recent times. Some estimates of extinction rates for regional

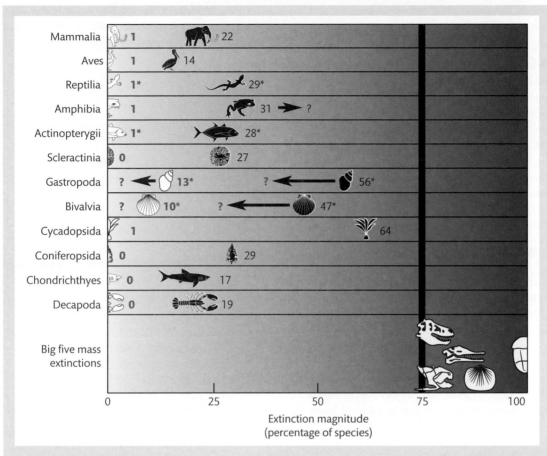

Figure 11.7 This figure illustrates the magnitude of the potential extinction event for a range of contemporary taxa, compared with the 'Big Five' mass extinctions we know from the geological record. The red numbers in each row are the percentage of species in each taxon that have gone extinct in recent times. The black numbers are the percentage of species currently listed as threatened with extinction. For taxa such as Gastropods and Cycads, extinction of all currently threatened species would represent an extinction event, for those taxa, approaching the Big Five in magnitude.[28]

From Anthony D. Barnosky, Nicholas Matzke, Susumu Tomiya, et al., (2011) Has the Earth's sixth mass extinction already arrived?, *Nature* 471: 51–7. Reprinted by permission from Springer Nature.

faunas are considerably higher than these; one of the highest is the estimate of extinction rates among freshwater gastropods in North America at 954 E/MSY.[13] Even if we acknowledge the high degree of uncertainty in these extinction rate estimates and assume only order-of-magnitude accuracy, there seems little doubt that recent extinction rates exceed background rates many times over. In the North American gastropods, it has been estimated that current rates are 9,539

times greater than background rates.[13] Based on these kinds of figures, it certainly appears that the human-driven extinctions we are currently witnessing represent an event of unusual severity in the history of life, at least during the Phanerozoic Aeon (**Figure 11.7**).

But is the current extinction event potentially severe enough to rank alongside the Big Five mass extinctions, as some biologists claim? The view of one of the teams responsible for key research on this topic, led by Anthony Barnosky, is clear.

> 66 *... a critical question is whether current rates would produce Big-Five-magnitude extinctions in the same amount of geological time that we think most Big Five extinctions spanned. The answer is yes. Current extinction rates for mammals, amphibians, birds, and reptiles ..., if calculated over the last 500 years (a conservatively slow rate) are faster than ... or as fast as ... all rates that would have produced the Big Five extinctions over hundreds of thousands or millions of years.* 99

A.D.Barnosky, N. Matzke, S. Tomiya, et al. (2011) Has the Earth's sixth mass extinction already arrived? *Nature* 471: 51–7.

Barnosky and colleagues then turn this around to ask how long it would take, given estimates of current extinction rates, for species losses in vertebrate taxa to reach 75%, equivalent in magnitude to the Big Five extinctions. If all species currently considered threatened with extinction were to go extinct within a century, and the estimated rates of extinction were maintained, extinctions in birds, mammals and amphibians would reach Big Five magnitudes within 240–540 years. In the light of estimates such as these, talk of a sixth mass extinction does not seem overstated.

Key points

- Mass extinctions have occurred periodically throughout the history of life. The five largest of these ('Big Five') resulted in the disappearance of the majority of species known from the fossil record.

- A 'background' rate of extinction between mass extinction events can be estimated from the fossil record as extinctions per million species-years (E/MSY). Estimates are typically in the range 0.001–0.2 E/MSY.

- Estimates of recent, human-driven extinct rates, extrapolated to units of E/MSY, suggest rates of 49–243 E/MSY in terrestrial vertebrates, and >900 E/MSY in North American gastropods. Extinction rates of this magnitude could result in a mass extinction of Big Five magnitude.

Big-picture conservation

What to prioritize? The goals of conservation

In the rest of this chapter we will focus on how we might use an understanding of the patterns and processes of extinction risk to design conservation interventions that will slow down the rate of species decline and extinction, and mitigate the erosion of biodiversity. Before we begin, there is one critical point to emphasize; any plan for conservation action is based on the key concept of *prioritization*. Conservation costs money; monitoring species populations, researching their ecology, surveying habitats, staffing and managing protected areas, and implementing captive breeding programmes are all expensive demands on the budget of a country, a government agency, or a non-governmental organization. Furthermore, conservation frequently conflicts with other demands on natural resources that are sources of revenue for a country; for example, farming, extractive industries such as mining or timber harvesting, or urban development. For these reasons, the budget for conservation programmes is almost always severely limited, and a great many programmes go unfunded and many areas go unprotected. This is where prioritization becomes critical; conservation planners must decide how to allocate limited resources to best meet conservation goals. In other words, decisions must be made about which species most urgently need intervention, which areas are most deserving of protection as reserves, and what kinds of management strategies are most efficient and effective.

In some ways, making decisions about the best way to allocate scarce conservation funds might not seem very difficult. We know that some species are closer to extinction than others, and the Red List gives us the means to rank species according to their risk of extinction. If we assume that focusing conservation efforts on species listed as Critically Endangered (CR) is a better way to minimize the number of extinctions than focusing on Least Concern (LC) species, then we would direct conservation funding towards programmes and strategies that will help stabilize or increase the populations of as many CR species as possible. In the following sections we will see why conservation planning is, in fact, usually more complex than this.

First, we need to think about what our *goals* for conservation are. Is the prevention of extinction the end goal, or is this a means for achieving some other conservation outcome? One argument for preventing extinctions as a goal in itself is that many people believe every species has intrinsic value,[14] and that we are morally obliged to act to allow all species to continue to exist. Certainly, it is easy to see intrinsic value in some species; large 'charismatic' species are widely perceived to have a value in themselves, independent of any instrumental, or practical, value they might also have. However, the argument of intrinsic value becomes more challenging to support when we apply it to uncharismatic and annoying species (**Figure 11.8**). How many people would care if the Australian

Figure 11.8 All species are not equal when it comes to perceived value. Large charismatic species such as the bald eagle (a), giant sequoia (b), and polar bear (c) are generally regarded as more worthy of conservation than small, little known, or harmful species, such as the paralysis tick (d). Charismatic species are often presented as 'flagships' for conservation campaigns, and may help to attract large amounts of funding from public donations.

(a) AWWE83 (CC BY-SA 3.0); (b) Jim Bahn (CC BY 2.0); (c) Iakov Filimonov/Shutterstock.com; (d) Paulo Oliveira/Alamy Stock Photo.

paralysis tick (*Ixodes holocylus*) was found to be on the verge of extinction?

An alternative to the focus on threatened species is an area-based conservation planning approach. It is important to remember that no species exists in isolation; all species are part of integrated systems of dependencies and interactions with other species. For example, if a particular rainforest tree species goes extinct because its habitat is destroyed to make way for oil palm plantations, the host-specific beetle species that use it for feeding and habitat may also go extinct. If a hummingbird species goes extinct, the plant species that depend on it for pollination may also decline. Although co-extinction (when one species goes extinct as the result of the extinction of another species) is not easy to demonstrate, the complete extinction of species is not necessary for severe disruption to the network of species connections in an ecosystem to occur. If the geographical range of a species contracts and it disappears from a particular ecosystem (ecological extinction), some of the key ecological processes and the functioning of that ecosystem might be impaired.

Why do ecological processes and ecosystem functioning matter? One reason is, again, a widespread belief that wilderness and natural areas have intrinsic value.[14] A second reason is that they have clear instrumental value; there is a growing recognition of 'ecosystem services' that support human well-being. For example, natural areas and their ecological processes play a role in pollination of crop plants, flood regulation, providing recreation opportunities, and supplying medicinal resources. A third reason is that protecting adequate areas of natural habitat is a prerequisite for preventing species extinctions. Not only are we providing sufficient habitat to support a population of a target species, but we are probably also providing habitat for other species on which it depends, as well as the broader ecosystem of which it is a part. For all the above reasons, an area-based approach to conservation planning—deciding which areas to protect, rather than which species—is likely to be an effective way both to minimize extinctions and to maintain other qualities that we value in natural ecosystems.

What do you think?

The existence and nature of intrinsic value in species is a subject of debate among environmental philosophers. For most people, it is easy to see why some species have value in themselves and should be protected from extinction, but this is less easy to appreciate for other species. Where should we draw the line between species worth saving and species not worth saving? Most people would probably like to inhabit a world in which orang-utans (*Pongo spp.*) continue to exist, but what about the Critically Endangered central rock rat (*Zyzomys pedunculatus*) from central Australia? Should all species be valued equally for their intrinsic value or contribution to biodiversity? Or are some species more valuable than others? What criteria should we use to make choices about which species to prioritize?

Systematic conservation planning

Principles of systematic conservation planning

Both species-based and area-based prioritization have a role to play in meeting general goals of conservation, such as minimizing extinctions, maintaining biodiversity, and maintaining sufficient areas of properly functioning ecosystems and their key ecological processes. Protected areas (national parks, nature reserves, etc.), in which restrictions are placed on destructive activities, are a key tool for achieving these goals. Most countries have a system of protected areas, but in many cases these systems have grown in an ad hoc, unplanned manner. In the past, many areas were designated reserves on the basis of their scenic or recreational values, rather than their biodiversity values. In many countries, the majority of reserves are in remote or unproductive regions with low economic value. When a reserve is added to the system, it is often an opportunistic acquisition, for example if a suitable property comes up for sale. Because they are unplanned, many reserve systems do a poor job of achieving the overall conservation goals of the country or region.

Over the past few decades, a more structured, objective, and goal-oriented approach to designing reserve networks has developed, generally known as 'systematic conservation planning'. Systematic conservation planning aims to provide an evidence-based framework to inform the choices and priority-setting that are part of conservation planning. The value of a systematic approach is particularly clear when we consider that an overall conservation goal (e.g. adequate representation of all major habitat types in reserves) can usually be achieved in alternative ways by choosing to include or exclude different areas in a network of protected areas. How these choices are made should be based, as far as possible, on an understanding of ecology, but choices also become important when we factor in the non-ecological constraints on establishing protected areas. These include existing land tenure, the financial and social costs of declaring and managing reserves, lack of political cooperation, and opportunity costs (lost revenue from unrealized exploitation of natural resources). According to Chris Margules and Robert Pressey, two researchers who have played a major role in the development of the systematic conservation planning approach, 'Conservation planning is therefore an activity in which social, economic and political imperatives modify, sometimes drastically, scientific prescriptions'.[15]

So what is the 'systematic' part of systematic conservation planning? Broadly, the approach is systematic in the way it works through a structured process that requires planners to consider not only the important biodiversity and ecological factors, but also the economic, social, and political constraints. Margules and Pressey[15] highlighted two general goals that underpin the systematic conservation planning approach: *representativeness* and *persistence*. Representativeness means that a reserve system should aim to protect a representative sample of the biodiversity present in the planning region; this could be defined as the species richness, the variety of higher taxa, or the variety of habitats, ecosystems, or landforms. Persistence means that the reserve system must be effective at doing the job it is meant to do: excluding threats, allowing ecosystems to function properly, and allowing species populations to remain viable. Persistence has also been referred to as 'adequacy'.

An important stage of a systematic conservation planning scheme is choosing areas to add to the conservation reserve system, also referred to as 'spatial conservation prioritization'. This is perhaps the most systematic part of the process, because it often involves the use of mathematical algorithms to select areas that achieve a pre-specified overall goal for the planning area in the most efficient way. In the following section we will examine in more detail the use of algorithms and software for reserve system design.

Designing reserve systems: algorithmic approaches

Selecting areas for inclusion in a network of reserves is essentially a mathematical problem of constrained optimization: how to maximize the representation of the desired biodiversity features efficiently (within the smallest possible area), under constraints on the choices available (imposed by factors such as economics). Most modern reserve selection methods are *dynamic*; they go through an iterative process of adding or removing areas from the system, each time recalculating and updating the score for the biodiversity quantity being optimized until some predefined target is achieved. The 'score' might be something very simple, such as the number of species weighted by the inverse of their range size, so that narrowly distributed species are prioritized over widespread species. But it can also be more complex. The widely used conservation planning software Marxan[16] calculates a function for a set of areas that combines biodiversity features with costs (economic or otherwise), and penalizes the boundary length of the reserved area so that the system does not become too fragmented.

To achieve the overall target efficiently, one principle in particular is of key importance: the principle of *complementarity*. Complementarity refers to the degree to which each new area added to a set contributes to the achievement of the overall target. For example, if the target is representation of all species (maximizing species richness), once the first area has been selected, the next area should be selected from those that include species not found in the first area (the taxonomic complement), and so on until the biodiversity target is reached. The complementarity principle addresses the problem of representativeness: choosing a reserve network that maximizes representation of species or other measures of biodiversity.

The greater efficiency of reserve selection algorithms that apply the complementarity principle has been demonstrated.[17] Under a simple rule for selecting reserves, as the combined area of the network increases, the target level of biodiversity representation is achieved far more rapidly when complementarity is used than when it is not (**Figure 11.9**). The difference in area between the two strategies at the target level (*a* in Figure 11.9) is proportional to the increased cost of purchasing land and ongoing management of reserves under the non-complementarity strategy. On the other hand, complementarity-based reserve selection requires more complete and detailed spatial biodiversity data than ad hoc selection, so that the cost of biodiversity surveys could be prohibitive. However, as long as the cost of biodiversity surveys required to implement a complementarity-based approach is lower than the value *a*, the surveys are good economic value.

As well as representativeness, a reserve network should also aim to maximize persistence (long-term viability of populations), and this requires different solutions. Early attempts to address the persistence problem were inspired by the ecological theory of island biogeography to specify universal rules about the ideal size, shape, or connectivity of reserves. These rules had limited success, largely because the best solution tends to be contextual, depending strongly on the species present and their particular ecological requirements. Persistence is now often addressed by allowing expert opinion to modify the algorithmic solutions.

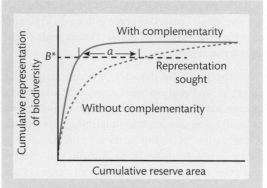

Figure 11.9 As areas are selected for inclusion in a reserve network, the total area of the network increases (*x* axis). How rapidly the representation of biodiversity by the network increases to the representation target (*B**) indicates the efficiency of the reserve selection procedure. When a complementarity-based algorithm is used to select areas, efficiency is greater and the target is reached more rapidly. The difference in total area between the two strategies at the point the target is reached (*a*) is proportional to the cost savings of adopting a complementarity method.

From Andrew Balmford and Kevin J. Gaston (1999) Why biodiversity surveys are good value, *Nature* 398: 204–5. Reprinted by permission from Springer Nature.

Key points

- The central concept of any conservation planning system is prioritization. Because resources are limited, choices must be made about which species and areas are most deserving of protection.

- Conservation planning is done in the context of explicit conservation goals for a country, region, or globally. Goals can emphasize the intrinsic value of species, natural places, or biodiversity, or the value of properly functioning ecosystems for preventing extinctions and providing ecosystem services.

- Conservation systems in many countries have developed in an unplanned ad hoc manner, so they do not meet conservation goals efficiently. Systematic conservation planning has emerged as a structured, objective approach to meeting goals in an efficient way that minimizes the costs of conservation. Systematic conservation planning aims to produce conservation systems that maximize representativeness and persistence of biodiversity.

- Reserve selection algorithms are a key part of the systematic conservation planning approach. These use principles such as complementarity to design protected area networks that meet representation targets efficiently.

Prioritization on the global scale

Biodiversity hotspots

Reserve selection algorithms have been developed and applied, by and large, at the level of countries or sub-national jurisdictions such as states or provinces. Such areas are typically under a single political system, with similar economic costs, and similar social and cultural considerations. The majority of conservation decisions are taken at these levels, but some international agencies are faced with making conservation choices across larger areas, or even globally. This introduces heterogeneity in the political, social, and economic context of the planning scheme, which could make decisions more tricky. For example, an impoverished African country and a wealthy European country are likely to have very different priorities for managing natural resources, and the costs and benefits associated with establishing reserves are likely to be quite different. The need for consensus across countries and cultures might mean that when planning on large scales, setting simple biodiversity goals is most realistic. For reasons such as these, some non-dynamic, non-algorithmic approaches to conservation prioritization have remained important. Such approaches are based on scoring and ranking areas based on a biodiversity quantity, such as number of endemic species, and then choosing a proportion of areas with the highest scores as priority areas. Although scoring approaches have overall goals, the choice of priority areas is not made in the context of meeting explicit targets.

The best known application of the 'scoring' approach is the global biodiversity hotspot scheme, administered by the non-governmental agency Conservation International. The biodiversity hotspot concept originated in the 1980s with British environmentalist Norman Myers.[18] Myers' initial focus was on tropical forests, because they contain much of the world's biodiversity and are being destroyed at a rapid rate. He identified 10 tropical forest regions that cover only 3.5% of the global tropical forest biome, but contain perhaps half of its vascular plant species. These regions had already undergone severe deforestation, which was continuing at a rapid pace. The key rationale for identifying such regions is that they provide a 'silver bullet' strategy for conservation prioritization; by focusing conservation activity on a limited number of small but highly biodiverse regions, we are more likely to protect a large proportion of the world's species (especially endemic species) in a minimal area.

In subsequent years the definition of hotspots, and their geographical boundaries, were made more explicit. Conservation International applies two criteria for a region to qualify as a hotspot: a biodiversity criterion and a threat criterion. A region must (1) contain at least 1500 vascular plant species endemic to the region, and (2) have already lost at least 70% of its original natural vegetation.[19] The focus is on vascular plant diversity as a surrogate for biodiversity more generally, for two reasons. First, vascular plant diversity tends to be more well characterized across more of the world, compared with the diversity of animals (especially invertebrates). Secondly, plants provide the habitat structure on which much animal diversity depends. This means that by using spatial data on plant diversity and habitat loss, it is possible to identify new regions that

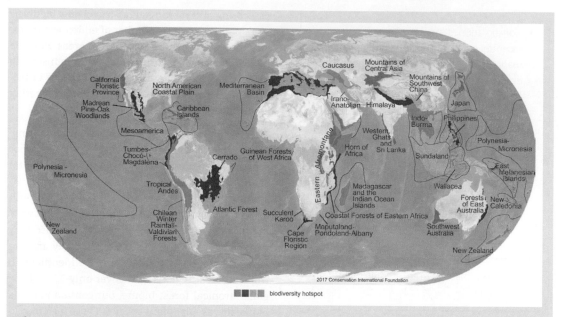

Figure 11.10 The 36 biodiversity hotspots currently recognized by Conservation International. Each area must contain at least 1500 vascular plant species and have lost at least 70% of its original natural vegetation. The remaining habitat within hotspots covers only 2.3% of the earth's surface, but contains over half of its plant species. The reason for the disproportionate representation of biodiversity within these small areas is that hotspots are centres of endemism: they contain large numbers of narrowly distributed plant species found nowhere else. A list and description of the 36 hotspots can be found at https://www.cepf.net/our-work/biodiversity-hotspots.

Source: Biodiversity Hotspots Map. © Conservation International 2017. Used with the kind permission of Conservation International Foundation (CC BY-SA 4.0).

meet the criteria for hotspots, particularly as our knowledge of plant diversity changes and the extent of habitat loss increases. The most recent additions to the hotspot network are the Forests of East Australia (added in 2011) and the North American Coastal Plain (added in 2016). The currently recognized set of 36 hotspots includes not only tropical forest regions such as Indonesia and West Africa, but drier Mediterranean-climate biomes such as Southwest Australia and California, as well as islands and mountain biomes (Figure 11.10). The hotspot scheme has been influential in the way conservation funding is distributed at an international level. For example, both Conservation International and the Critical Ecosystem Partnership Fund provide funding explicitly for conservation projects within hotspot regions.

How representative are hotspots?

The global hotspot scheme is based on a simple biodiversity goal: maximizing the number of species that are protected from extinction. This goal is addressed by focusing on the protection of narrowly distributed species endemic to single regions, because such species are considered to be at greater risk of extinction. If one of these species disappears from one region, it becomes globally extinct. Under this approach, restriction to a single region (endemism) is the only criterion under which some species are given higher priority for protection than others.

There are a few questions we can ask about whether the hotspot scheme adequately addresses the goal of minimizing extinctions and maximizing the protection of biodiversity.

1. How well does species richness of endemic vascular plants reflect the amount and distribution of biodiversity of other taxa? More generally, how congruent are the hotspots for different taxonomic groups?

2. Are there features of species, other than endemism and presence in a heavily impacted region, that distinguish species at greater and lesser risk of extinction?

3. Do all endemic species contribute equally to the overall biodiversity qualities that we wish to maintain, or are there features of species which mean that some kinds of species should be valued over others?

These were some of the questions asked by Richard Grenyer and colleagues in 2006.[20] They used large databases of species geographical distributions to map global patterns of species richness in birds, mammals, and amphibians. One of the aims of their study was simply to examine the congruence in the global species richness hotspots of these three groups: do they overlap, or are the hotspots in different places? Grenyer et al. found that spatial congruence in total species numbers among the three taxa is quite high. They defined 'hotspots' as the 1° × 1° grid cells with the top 5% of richness values. Around 18% of hotspot grid cells were hotspots for all three taxa. In other words, where there are lots of bird species, there also tend to be lots of mammal and amphibian species, with the tropical Andes and Amazon Basin in South America appearing as the major hotspot region for all three taxa. This is encouraging for the 'silver bullet' strategy, because it suggests that small areas that capture a high proportion of the species of one taxon also do so for other taxa.

Grenyer et al. then addressed the second of the questions above. They reasoned that the species most likely to be threatened with extinction are (a) 'rare' species, which they defined as those with geographical range sizes in the lower quartile of their taxon, and (b) species currently listed as threatened in the Red List. When they mapped the spatial patterns of species richness of rare species and threatened species, they found that cross-taxon congruence was much lower than it was for total species richness. For rare species, only 2.3% of hotspot cells were common across all three taxa, while for threatened species it was only 0.6%. The highest concentrations of rare and threatened species of birds, mammals, and amphibians are found in different places, so much larger areas would need to be selected as part of a priority set. Therefore if protection of rare or threatened species richness is considered an appropriate biodiversity goal, these findings lead to a less encouraging conclusion for the hotspot approach:

 We anticipate, however, that 'silver-bullet' conservation strategies based on particular taxonomic groups will not be effective because locations rich in one aspect of diversity will not necessarily be rich in others. **R. Grenyer, C.D.L. Orme, S.F. Jackson, et al.** (2006) Global distribution and conservation of rare and threatened vertebrates. *Nature* 444: 93–6.

Key points

- Conservation planning at global scales presents challenges to the algorithmic approach to designing protected area networks, such as heterogeneous political, cultural, and economic systems. Approaches based on scoring and ranking areas, chiefly the global biodiversity hotspots scheme, are important on the global scale.

- Hotspots are appealing because they capture a large proportion of the world's vascular plant diversity in a small combined area, allowing for a 'silver bullet' strategy for focusing conservation efforts on areas with high potential for saving endangered species.

- Shortcomings of the global hotspot approach include (1) lack of explicit biodiversity targets means that it is difficult to quantify the scheme's success, (2) areas are added to a priority set in a way that may not be efficient at meeting representativeness criteria, and (3) lack of congruence between areas of maximum diversity of threatened and rare species, or between hotspots identified for different major taxonomic groups.

The role of phylogeny in conservation prioritization

Evolutionary distinctness

The third question we asked above was 'Do all species contribute equally to biodiversity, or should we value some species over others?' This is not the same as asking which species we should prioritize because they are at greater risk of extinction (second question above); we are now concerned with the relative value of species independently of how rare or threatened they are. To explore this question, we will look at the role of phylogeny in conservation planning and prioritization.

That some species are 'evolutionarily distinct', with no close living relatives, has been recognized for a long time. One of the classic examples is the tuatara (*Sphenodon punctatus*), a reptile species endemic to New Zealand, the only living member of its order (Rhynchocephalia), and separated from its nearest relatives (snakes and lizards) by more than 200 million years. Other well-known examples are the coelacanths (*Latimeria*), the only survivors of the fish order Coelacanthiformes, which diverged from their nearest living relatives perhaps 390 million years ago, and the ginkgo (*Ginkgo biloba*), separated from its nearest living relative by around 300 million years (**Figure 11.11**). Such species have long been considered worthy of special conservation attention because of their uniqueness. The extinction of a species with no close relatives could mean the loss of an entire higher taxon, such as a family, or the loss of the only representative of a distinct morphological type. Evolutionarily distinct species are valued for their disproportionate contribution to representing the variety of types in the living world.

Phylogenetic diversity

In the 1990s, the concept of valuing evolutionarily distinct species gave rise to a more quantitative spin-off: *phylogenetic diversity*. In 1991, Richard Vane-Wright and colleagues proposed a simple index to quantify biodiversity, that they called taxonomic distinctness.[21] This index dispensed with the notion of species equality by considering the phylogenetic relationships among species. Essentially, a score is calculated for each species in a phylogeny based on the number of nodes that separate it from the root of the phyogeny, such that species with fewer close relatives receive a higher score. These scores can then form part of the biodiversity information used in complementarity-based algorithms to choose priority areas for conservation. Soon after, Dan Faith proposed phylogenetic diversity (PD), a biodiversity measure based on phylogenetic branch lengths rather than hierarchical groupings.[22] PD is calculated for a set of species as the sum of the branch lengths connecting all species in the set (the 'minimum spanning tree'). In this way, a set of species representing distantly related lineages will have a higher biodiversity score than the same number of species that are each other's close relatives (**Figure 11.12**).

(a)   (b)   (c)

Figure 11.11 Three classic examples of evolutionarily distinct species: tuatara (*Sphenodon punctatus*), coelacanth (*Latimeria chalumnae*), and gingko (*Ginkgo biloba*). All three are the only one or two remaining species of a higher taxonomic grouping, and are separated from their nearest living relatives by periods of hundreds of millions of years. Their extinction would mean the loss of a disproportionate amount of the world's morphological diversity.

(a) Photo by Bernard Spragg from Christchurch, New Zealand, via Wikimedia Commons; (b) Photo by Daniel Jolivet from Muséum d'Histoire Naturelle de Nantes, via Flickr (CC2.0 license); (c) Photo by Liné1 via Wikimedia Commons (CC2.0 license).

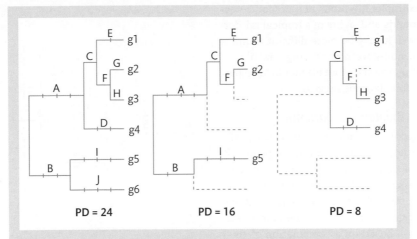

Figure 11.12 Calculation of phylogenetic diversity (PD) for a set of species g1–g6. PD is the sum of the branch lengths (indicated here by the division of branches into intervals) along the shortest paths connecting the set of species. The middle panel shows how a small subset of species can capture a large proportion of the evolutionary history of the whole group (left panel) if the paths connecting them traverse the root of the phylogeny. In the right panel, a different set of species has a much lower PD score because they are more closely related to one another.[23]

From Ana S.L. Rodrigues and Kevin J. Gaston, (2002) Maximising phylogenetic diversity in the selection of networks of conservation areas, *Biological Conservation* 105: 103–11. © 2004, with permission from Elsevier.

The introduction of taxonomic distinctness and PD represented a conceptual shift in the way evolutionary distinctness was treated in a conservation context. Instead of particular evolutionarily isolated species, such as the tuatara, being valued for their uniqueness, phylogenetic relatedness was now being used as a continuous-scale measure of conservation value. The key justification for this, given by Faith and others, is that phylogenetic diversity is likely to be a good proxy for 'feature diversity', the variety of ecological or morphological traits represented by a set of species. Preserving maximum feature diversity, in turn, is argued to provide greater 'option value'. We can never be sure what features of a species will be valued in the future, nor how different traits might allow species to adapt to changing conditions, so maximizing the variety of features is a kind of precautionary principle. Many authors have also cited the preservation of 'evolutionary history' as a conservation goal in its own right that is addressed by PD and

similar approaches, although this seems to be a more nebulous concept without a clear definition.

PD was simple and computationally undemanding to calculate, and arrived on the scene at the start of a period when ever larger, more complete, and well-resolved phylogenies were rapidly being published. Not surprisingly, the number of studies using PD to make recommendations for conservation priorities increased rapidly. PD has been applied in both area-based and species-based prioritization exercises. For example, maximizing PD has been used as a target in complementarity-based reserve selection algorithms.[23] However, most of the work on PD has remained within the academic literature, and its use in real-world conservation applications has been limited so far. There are probably a number of reasons for this, but one reason may be that it can be difficult to appreciate the connection between branch lengths on a phylogeny and the more tangible values of biodiversity that are familiar to most people: for example, the protection of iconic species, or the sounds and sights of a tropical rainforest or coral reef. It may be more difficult to make the case to a non-governmental organization or government funding body that PD should form part of a conservation planning process.

The EDGE of Existence programme

Partly in response to the limitations of PD, the Zoological Society of London has developed EDGE of Existence, a programme based on evolutionary distinctness that shifts the focus back to individual species.[24] EDGE (Evolutionarily Distinct and Globally Endangered) prioritizes species for conservation attention based on a combination of their phylogenetic position and their threat status. A species with few close living relatives and a high Red List threat status will receive a high EDGE score. Evolutionary Distinctness (ED) is measured as each species' contribution to the PD of a group; a species with lots of close relatives will have a low ED (**Figure 11.13**). The EDGE score is then calculated as

$$\log(1 + ED) + (GE \times \log 2)$$

where GE is a numerical index corresponding to the Red List status of the species: LC = 0, NT = 1, VU = 2, EN = 3, and CR = 4.

By combining evolutionary distinctness and threat status, and focusing on species, the EDGE programme aims to capture some of the important values that biodiversity represents for many people. The mammal species with the highest EDGE score of 6.48 is the Yangtze River dolphin (*Lipotes vexillifer*), while those with the lowest score (0.05) are a number of closely related, widespread murid rodent species.[24]

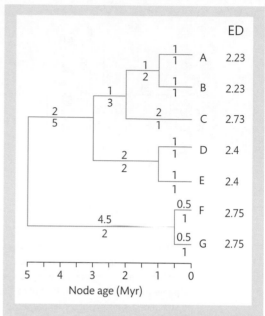

Figure 11.13 Calculation of the evolutionary distinctness (ED) of species. Numbers above each branch on this hypothetical phylogeny are the branch lengths, and numbers below each branch are the number of descendant species. The ED of a species is its share of the length of each branch connecting it to the root of the phylogeny. For species A, the deepest ancestral branch is 2Myr long and has five descendant species (A–E), so species A's share of this branch is 2/5. Working down the branches, the ED score for species A is (2/5 + 1/3 + 1/2 +1/1) = 2.23Myr. The ED score is combined with an index of threat status to produce an EDGE score. The mammal species with the highest EDGE score is the Yangtze River dolphin.[24]

From Nick J.B. Isaac, Samuel T. Turvey, Ben Collen, et al. (2007) 'Mammals on the EDGE: conservation priorities based on threat and phylogeny, *PLoS One* 2, e296 (CC BY 4.0).

What do you think?

Is evolutionary distinctness a legitimate criterion for valuing one species over another? If it is, does it follow that the combined phylogenetic diversity of the species present in different regions is a legitimate criterion for prioritizing one region over another? Suppose region A is home to eight small mammal species, all closely related rodents of the family Muridae, with a PD of 30Myr. Region B is home to only three small mammal species— one rodent, one marsupial, and one shrew—but with a PD of 350Myr. Which region contributes more to biodiversity—the one with more species, or the one with more evolutionary history?

Branch lengths: uncertainty, inaccuracy, and what do they actually mean?

PD and similar methods have been embraced as an objective and quantitative measure of conservation value that can be consistently applied across all taxa, without the need for value judgements on the worthiness of each species. By providing a quantitative measure, these metrics avoid treating all species equally and the need for subjective assessments of the value of each species. While the aim of PD may be to capture feature diversity, taxonomic distinctness, or evolutionary history, it is a proxy for these characteristics—they are not measured directly. Instead, the metrics are based on some measure of phylogenetic branch length or branching pattern in the hope that it will reflect the biodiversity that we value and wish to conserve.

Whenever we use a proxy measure—estimating one quantity under the assumption it provides a fair representation of another property—we need to be sure that our measurement really reflects the properties we care about. Does phylogenetic branch length reflect biodiversity features or values important in conservation? To answer this question we need to think about how branch lengths are usually derived. Nowadays, most phylogenies are based on DNA data. DNA sequences of the same gene in different species are compared, and the differences in base sequence between them are used to infer how long it has been since they last shared a common ancestor. The DNA sequences are observations, but the inference of history is not a direct estimate; it is a reconstruction based on a large number of assumptions about the evolutionary process.

In particular, the branch lengths of the phylogeny are based on assumptions about rates of change over time (see Chapter 6). Molecular change can accumulate at different rates in different lineages at different times. Since we never know for sure what the rates were in any particular lineage at any given point in the past, we have to make a best guess based on an assumed evolutionary model, using any calibrating information we have (such

as fossils). The branch lengths we obtain depend on the assumptions we make, and we rarely have independent evidence to tell us which assumptions are more reasonable.

A high PD score for a taxon might indicate that it is a unique taxon separated from its nearest relatives for a long period of independent evolution. But we might also get a long branch and a high PD value for a species with a very slow rate of molecular evolution (so its branch length is over-estimated), through an alignment error (so that more bases in the sequence look as if they have changed), or by using a model of molecular evolution that doesn't match the real process in the data (e.g. if the model fails to capture the degree to which rates of change vary over the genome). Furthermore both the length of the branch to the nearest neighbour and the number of nodes between a taxon and the rest of the tree will be affected by taxon sampling. If not all living species or populations have been included in the phylogeny, a species with a high PD might be in an under-sampled clade, making it seem more distinct than it really is.

More generally, while phylogenies have an appearance of objective analysis, they are in a constant state of flux. Each new molecular phylogeny published is likely to agree with previous publications in many ways, but also to differ in some points. Whenever phylogenies differ, measures of PD might also be affected. How do we know which phylogeny we should use in our conservation planning? Phylogenetic measures of conservation value seem to solve many problems in conservation prioritization. But when we adopt an indirect measure of value, such as one based on branch length, we need to evaluate how well it represents the properties we wish to target. If we intend our measures of PD to represent evolutionary time, we need to explore how robust those estimates are to changing assumptions of the phylogenetic analysis, inclusion or exclusion of particular taxa, and different calibration strategies. Otherwise we might find that the prioritization rankings change with any change in phylogenetic analysis, including changes to the data, methods, or assumptions.

Key points

- Evolutionarily distinct species (such as tuataras) have always been valued for their uniqueness. This concept has been codified and quantified through the use of phylogenetic diversity (PD), which measures the evolutionary history represented by a group of species as the sum of their phylogenetic branch lengths.

- PD has had limited influence on real-world conservation planning. The EDGE programme shifts the focus back to species by calculating a score based on species phylogenetic position and Red List threat status.

- These and similar methods assume that phylogenetic branch lengths can be considered a proxy for biodiversity features that we value. There are a number of reasons why we must be cautious about making such assumptions and using branch lengths without explicitly considering what they really represent.

Comparative analysis of extinction risk

By and large, conservation can be considered a 'responsive' or 'reactive' discipline; much of conservation activity is aimed at mitigating the immediate crisis of extinction and biodiversity loss. But in conservation, as in medicine, prevention is better than cure; action to prevent species declining in the first place is probably a cheaper and more effective means of mitigating biodiversity loss than expensive programmes to rescue species already on the brink of extinction. For this reason, some scientists are advocating a more proactive conservation planning approach, which anticipates potential future species declines by identifying species that are especially 'extinction-prone', so that preventative action can be taken. How can such species be identified?

We saw earlier in this chapter that current species declines and extinctions are (mostly) driven by one or more threatening processes that result from human activities, such as introduced species or habitat loss. So, extinction-prone species might simply be those which inhabit places that are more exposed to threatening processes. For example, these might be regions where natural habitats have been converted to heavily modified intensive agricultural landscapes, or where human population density is high. But we also know that different species respond in different ways to the same threats; some species thrive in areas heavily modified by agriculture, while others disappear from those areas and can only maintain populations in relatively undisturbed habitat. These differences in sensitivity to disturbance probably occur because of the way that different ecological or physiological traits of species mediate the effects of threatening processes. For example, an ecological generalist with broad and flexible dietary requirements might do better in an agricultural landscape than a specialist that relies on particular forest fruits.

Therefore, to identify species that are more extinction-prone than others, we need a broad general understanding of how species biology interacts with external threatening processes to affect the growth of populations. In this way, we can identify particular biological traits or characters that confer a high risk of extinction when threatening processes are encountered. The most effective way of doing this is to use the comparative method, comparing the biological traits of many different species that differ in their current extinction risk status. This is where some of the analytical tools of macroevolution and macroecology can be put to use. In Chapters 8 and 9 we looked at the issue of phylogenetic non-independence when analysing correlations between variables across clades or species. The values of biological traits for species cannot usually be considered as statistically independent observations, because closely related species are more likely to have similar trait values than more distant relatives (having inherited them from a recent common ancestor). We need to account for phylogenetic non-independence of trait values by testing the associations between variables in the context of phylogeny, using methods such as phylogenetically independent contrasts (PIC) or phylogenetic generalized least squares (PGLS). This approach is described in Chapter 8.

In many comparative models of extinction risk, the response variable is a numerically coded form of species Red List categorizations. Predictor variables often include species biological traits (such as body mass or gestation length), or geographical measures that represent threatening processes (such as human population density or forest loss) within the distribution of each species. By including in the model a phylogenetic estimate for the group of species we are analysing, the phylogenetic covariances among species can be accounted for in the calculation of the model parameters. Analyses of this kind for large groups of organisms have revealed a number of different biological traits as significant predictors of species Red List status. In mammals, for example, large body size is often associated with high extinction risk.[25] This makes sense: large mammals have a slower life history, with smaller litters, longer gestation lengths, and longer times to sexual maturity. This means that their rates of population growth are lower, and populations do not bounce back from low levels as readily as those of smaller species.

Comparative analyses have also revealed patterns of extinction risk that are less immediately obvious, but still make ecological sense. For example, in birds, the primary biological predictors of Red List status are different for species that face different kinds of threats.[26] Among species that are threatened primarily by hunting or introduced predators, the key predictors of risk are large body size and long generation time—features associated with slow population growth. On the other hand, among bird species threatened primarily by habitat loss, the key predictors are small body size and habitat specialization.

Comparative analyses of extinction risk can be valuable for several reasons. First, they provide a broad general understanding of the ecological processes involved in species declines that cannot be gained from detailed studies of single species. By comparing many species in a robust statistical framework, we can ask why species differ in their response to human impacts. Secondly, by identifying traits associated with high extinction risk, we can apply comparative models in predictive ways. For example, it may be possible to predict the extinction risk status of Data-Deficient species (for which we lack detailed information on population trends or sizes) from biological traits measured from a small sample of individuals. In Case Study 11 we look at this way of applying comparative analyses in more detail.

Another example of how comparative models have been used in a predictive way is by calculating 'latent extinction risk'.[27] This is the difference between a species' current Red List status and the status that would be predicted from its biological features, based on the results of a comparative model. A species with positive latent risk is currently less threatened than we would expect from its biology. Such species may not yet be at risk because they inhabit regions that are still little disturbed, but they could rapidly become threatened if levels of impact increase. Even though they are not currently considered at risk, species with high latent risk, such as the musk ox (*Ovibos moschatus*) of Arctic America, could be worth placing in a special category of species that deserve additional monitoring and vigilance. We can go further than this, and identify regions in which the species assemblage has a high average latent risk. When this is done for the world's mammal species,[27] the 'hotspots' of latent risk (**Figure 11.14**) are distributed quite differently from the hotspots of endemic plant diversity and existing habitat loss (Figure 11.10). How should global conservation prioritization schemes account for these different patterns?

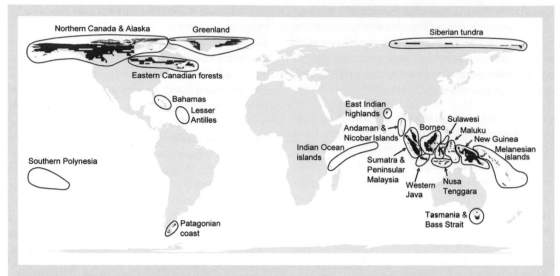

Figure 11.14 Hotspots of latent extinction risk (the difference between current and predicted extinction risk) are found primarily in Arctic North America and the island chain from southeast Asia to Melanesia. These are regions where a high proportion of the mammal fauna are not yet threatened, but have the biology that could make them sensitive to future increases in human impact.

From Marcel Cardillo, Georgina M. Mace, John L. Gittleman, and Andy Purvis (2006) 'Latent extinction risk and the future battlegrounds of mammal conservation,' *Proceedings of the National Academy of Sciences of the USA* 103 (11): 4157-4161. Copyright (2006) National Academy of Sciences, USA.

Conclusion

Conservation biology is a relatively new discipline: it formally began only in 1978, when Michael Soulé and colleagues organized a conference at the University of California, San Diego, with the aim of 'focusing the attention of the biological community, particularly population biologists, on nature conservation as an area overdue for the application of existing theory and technology'. Since then, both the research discipline and the problem it was established to address have grown massively. The original focus of conservation biology was on detailed field-based studies of single species populations; in order to make informed recommendations for interventions to ensure protection of a species, we need a detailed understanding of its population dynamics, ecology, behaviour, and the threats that it faces. Increasingly, though, conservation scientists have come to realize that there is also much to be gained by taking a 'big-picture' approach to conservation issues. As we have seen throughout this book, large-scale patterns of biodiversity can teach us much about the processes that shape biodiversity that we can't learn from more narrowly focused research.

In the same way, a large-scale perspective in the study of extinction, species decline, and biodiversity loss can provide an understanding of the extent, magnitude, and patterns of extinction that we could never achieve through studies focusing on single populations only. By taking a 'macro' approach we have been able to understand that the current extinction crisis is happening more rapidly than any previous extinction event in the history of life, as far as we are able to infer from the geological record. By mapping global patterns of biodiversity, we can develop global-scale prioritization schemes for conserving biodiversity. And by using phylogenetically informed comparative methods, we can answer the question of why species respond differently to human impacts by identifying biological traits consistently associated with high threat status. An exciting area of future research is how this can be applied to developing a more predictive approach to conservation planning, in which we have the tools to anticipate declines in species before they happen, based on biology and phylogeny, allowing us to step in to take preventative action. This must surely be a more efficient and cost-effective approach to conservation than attempting to rescue species from the very brink of extinction.

Points for discussion

1. Prioritizing species or areas for conservation involves the consideration of both ecological values and more subjective (even emotional) values. What should be the relative importance attached to these two kinds of values? Should ecological values be emphasized because they are quantifiable, objective, and more easily presented as a strong case to governments and funding bodies? Or should we acknowledge explicitly that subjective values are equally important?

2. Methods of prioritizing species or sets of species that utilize phylogenetic branch lengths are vulnerable because of the variability and uncertainty in branch-length estimates, and the lack of evidence that they represent the biodiversity values that it is assumed they do. Are there ways that more robust estimates of, for example, phylogenetic diversity could be made, which account for some of these uncertainties?

3. Imagine you are responsible for developing a conservation plan for a tropical forest province that is being opened up for timber and oil palm concessions. The provincial government has agreed to set aside 20% of the land area of the province as protected areas. What baseline information will you need to compile in order to decide on the placement of areas to include in the protected area network?

✳ References

1. Johnson CN (2006) *Australia's Mammal Extinctions: A 50,000 Year History.* Cambridge University Press, Cambridge.

2. Paddle R (2012) The thylacine's last straw: epidemic disease in a recent mammalian extinction. *Australian Zoologist* 36, 75–92.

3. Waters CN, Zalasiewicz J, Summerhayes C, and 21 other authors (2016) The Anthropocene is functionally and stratigraphically distinct from the Holocene. *Science* 351 (6269): aaad2622.

4. International Union for Conservation of Nature (IUCN) (2018) *Threats Classification Scheme (Version 3.2).* Available at: https://www.iucnredlist.org/resources/threat-classification-scheme.

5. Diamond JM (1984) 'Normal' extinction of isolated populations. In Nitecki MH (ed.) *Extinctions.* Chicago University Press, Chicago, IL: pp.191–246.

6. Mace GM, Lande R (1991) Assessing extinction threats: toward a reevaluation of IUCN threatened species categories. *Conservation Biology* 5: 148–57.

7. International Union for Conservation of Nature (IUCN) (2001) *IUCN Red List Categories and Criteria version 3.1.* Available at: https://www.iucn.org/content/iucn-red-list-categories-and-criteria-version-31.

8. Khadka BB, Yadav BP, Aryal N, Aryal A (2017) Rediscovery of the hispid hare (*Caprolagus hispidus*) in Chitwan National Park, Nepal after three decades. *Conservation Science* 5: 10–12.

9. Butchart SHM, Stattersfield AJ, Bennun LA, Shutes SM, Resit Akçakaya H, Baillie JEM, Stuart SN, Hilton-Taylor C, Mace GM (2004) Measuring global trends in the status of biodiversity: Red List indices for birds. *PLoS Biology* 2: e383.

10. Raup DM, Sepkoski JJ (1982) Mass extinctions in the marine fossil record. *Science* 215: 1501–3.

11. Ceballos G, Ehrlich PR, Barnosky AD, García A, Pringle RM, Palner TM (2015) Accelerated modern human-induced species losses: entering the sixth mass extinction. *Science Advances* 1(5): e1400253.

12. Pimm SL, Jenkins CN, Abell R, Brooks TM, Gittleman JL, Joppa LN, Raven PH, Roberts CM, Sexton JO (2014) The biodiversity of species and their rates of extinction, distribution, and protection. *Science* 344(6187): 1246752.

13. Johnson PD, Bogan AE, Brown KM, and 11 other authors (2013) Conservation status of freshwater gastropods of Canada and the United States. *Fisheries* 38: 247–82.

14. Sarkar S (2005) *Biodiversity and Environmental Philosophy: An Introduction.* Cambridge University Press, Cambridge.

15. Margules CR, Pressey RL (2000) Systematic conservation planning. *Nature* 405: 243–53.

16. Ball IR, Possingham HP, Watts M. (2009) Marxan and relatives: software for spatial conservation prioritisation. In Moilanen A, Wilson KA, Possingham HP (eds), *Spatial Conservation Prioritization: Quantitative Methods and Computational Tools*. Oxford University Press, Oxford: pp.185–95.

17. Balmford A, Gaston KJ (1999) Why biodiversity surveys are good value. *Nature* 398: 204–5.

18. Myers N (1988). Threatened biotas: hotspots in tropical forests. *Environmentalist* 8: 187–208.

19. Myers N, Mittermeier RA, Mittermeier CG, da Fonseca GAB, Kent J (2000) Biodiversity hotspots for conservation priorities. *Nature* 403: 853–8.

20. Grenyer R, Orme CDL, Jackson SF, and 13 other authors (2006) Global distribution and conservation of rare and threatened vertebrates. *Nature* 444: 93–6.

21. Vane-Wright RI, Humphries CJ, Williams PH (1991) What to protect? Systematics and the agony of choice. *Biological Conservation* 55: 235–54.

22. Faith DP (1992) Conservation, evaluation and phylogenetic diversity. *Biological Conservation* 61: 1–10.

23. Rodrigues ASL, Gaston KJ (2002) Maximising phylogenetic diversity in the selection of networks of conservation areas. *Biological Conservation* 105: 103–11.

24. Isaac NJB, Turvey ST, Collen B, Waterman C, Baillie JEM (2007) Mammals on the EDGE: conservation priorities based on threat and phylogeny. *PLoS One* 2: e296.

25. Cardillo M, Mace GM, Jones KE, Bielby J, Bininda-Emonds OR, Sechrest W, Orme CD, Purvis A (2005) Multiple causes of high extinction risk in large mammal species. *Science* 309: 1239–41.

26. Owens IPF, Bennett PM (2000) Ecological basis of extinction risk in birds: habitat loss versus human persecution and introduced predators. *Proceedings of the National Academy of Sciences of the USA* 97: 12,144–8.

27. Cardillo M, Mace GM, Gittleman JL, Purvis A (2006) Latent extinction risk and the future battlegrounds of mammal conservation. *Proceedings of the National Academy of Sciences of the USA* 103: 4157–61.

28. Barnosky AD, Matzke N, Tomiya S, Wogan GO, Swartz B, Quental TB, Marshall C, McGuire JL, Lindsey EL, Maguire KC, Mersey B (2011) Has the Earth's sixth mass extinction already arrived? *Nature* 471(7336): 51–7.

Case Study 11

Modelling extinction risk: predicting conservation status of Data-Deficient species

Because one of the primary goals of conservation is to minimize the loss of biodiversity, species threatened with extinction are an important focus of many conservation policies and programmes. The IUCN Red List is the principal global scheme for classifying species into categories of extinction risk, and plays an important role in conservation planning and priority setting. As we have seen in this chapter, the summary statistics for the Red List present a sobering picture which makes it clear that we are facing an extinction episode of great magnitude. For example, 25% of mammal species with sufficient data to be classified in the Red List are considered threatened with extinction.

But do these figures accurately indicate the proportions of species that are actually threatened with extinction? This depends strongly on the true conservation status of the species that are categorized as 'Data-Deficient'. Assigning a species to a Red List category requires us to know enough about its distribution or population size to allow it to be assessed against the Red List criteria. For many species, we simply don't have enough of this information for a listing to be made. If a species has been evaluated, but lacks sufficient data for listing, it is categorized as Data-Deficient (DD). Among the groups that have been comprehensively evaluated, with the exception of birds and cycads, the proportions of DD species are substantial, in some cases more than 50% (Figure 11.5).

If the proportion of DD species that are threatened with extinction is similar to data-sufficient species, then it doesn't change the overall proportion of a taxon that we know to be threatened. On the other hand, the DD species in a taxon might be non-representative of the conservation status of the taxon overall—they might have relatively more or relatively fewer threatened species. If this is the case, it could dramatically change the overall conservation status of the taxon. In mammals, if we assume the proportion of DD species threatened is the same as assessed (data-sufficient) species, the proportion of threatened species is 25%. However, if all DD species turned out to be non-threatened, this figure would drop to 22%; if all DD species were threatened, it would rise to 36%. There are, in fact, reasons to believe that DD species are not a random sample of a taxon with respect to conservation status. For many species, the reasons we lack data on distribution or population size is because they are narrowly distributed or rare, and therefore poorly studied. Such species are more likely to be listed as threatened if sufficient data become available.

The obvious answer to the DD problem would be to invest more funding into surveys and monitoring to provide more baseline data for poorly known species. But the cost of data collection and the sheer scale of the task makes this an unfeasible option for the foreseeable future. An alternative approach is to infer the conservation status of DD species, based on patterns that we can quantify for data-sufficient species. This is where some of the tools of macroevolution and macroecology can be put to use. Comparative analyses of the correlates and predictors of extinction-risk status tell us that, in many cases, extinction-risk status is associated significantly with species biological traits, or with geographical variables that summarize the threatening processes within species distributions.[1] If we have some information on the biology, ecology, and distribution of DD species, we may be able to use this to predict their conservation status from comparative models.

Using this kind of approach, Lucie Bland and colleagues set out to estimate the number of DD mammal species that are likely to be threatened.[2] The first step was to fit models that establish the statistical associations between extinction-risk status and a set of predictor variables. The choice of predictors is important; they must be variables that have been recorded for a large proportion of both DD and data-sufficient species, and have been shown to be associated with extinction risk. Bland et al. chose a set of predictors that included biological traits such as body size, ecological variables such as habitat breadth, environmental variables such as mean precipitation, and geographical proxies for threatening processes, including human population density.

The authors then used machine learning (ML) methods to model the associations between predictors and extinction-risk status. ML is a class of analytical methods for recognizing patterns in datasets, which can be used for prediction. The key feature of ML is that (in contrast with traditional statistical models) the computer algorithm is automatically updated and the predictive performance of the model progressively improved. This makes ML methods well suited to large datasets with a lot of variables that may interact in complex ways. Typically, an ML algorithm is developed, or 'trained', using part of the dataset, known as the training set. The predictive accuracy of the model is then evaluated by taking the predicted outcomes from the training set and applying them to another part of the dataset, known as the validation set. Bland et al. compared several alternative ML methods, and found that the best model had a high predictive accuracy; around 93% of assessed mammal species listed as threatened in the validation set were correctly predicted by the model to be threatened, on the basis of their biological, ecological, and geographical characteristics.

Bland et al. then applied this model to predicting the status of DD mammals. Of the 493 DD species in their dataset, the model classified 313 (63.5%) as threatened. This result supports the idea that DD

Figure 11.15 The golden-bellied mangabey (*Cercocebus chrysogaster*), found in the tropical forests of the Congo Basin in central Africa, is listed as Data-Deficient in the IUCN Red List. Its Red List entry states that 'This is a poorly known species, and there is a paucity of information available on its population status'. Together with over 60% of other Data-Deficient mammal species, this species is predicted by comparative models to be threatened with extinction.

Photo by Nathan Rupert (CC2.0 license).

mammal species are not a random sample of species with respect to conservation status. Species such as the golden-bellied mangabey (*Cercocebus chrysogaster*) are far more likely to be threatened than species we know well enough to place into a Red List category (Figure 11.15). If correct, these predictions substantially increase the proportion of mammal species that are threatened from 22% to 27%.

As well as being associated with biology, ecology, and geography, comparative studies tell us that extinction-risk status is often also phylogenetically and spatially autocorrelated. In other words, a species status is likely to be more similar to that of its close relatives, or other species inhabiting the same region, than to distant relatives or species that live a long way away.[3] If not accounted for in the analysis, these kinds of autocorrelation can result in misleading interpretation of the effects of predictors on extinction risk. The ML methods applied by Bland et al.[2] were not designed to handle phylogenetic autocorrelation. The authors partially accounted for phylogenetic relationships among species by including levels of the taxonomic hierarchy (genus, family, and order) in the models, but this only goes part of the way to removing phylogenetic autocorrelation. How much does this matter? We may be able to answer this question by comparing the predictions of Bland et al. with those of another study which used a different methodological approach that was able to fully account

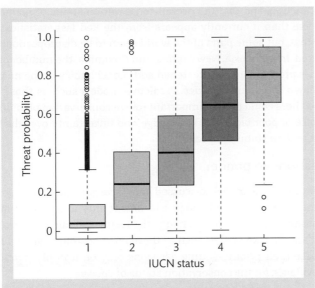

Figure 11.16 The probability that a mammal species is at risk of extinction is predicted by its biology, ecology, geography, and phylogeny. In this plot, there is a different bar for mammal species assigned to each Red List category (1, Least Concern; 2, Near Threatened; 3, Vulnerable; 4, Endangered; 5, Critically Endangered). The bars represent the distribution of threat probabilities predicted from a statistical model for species in each category (the black line is the median, the coloured box encloses 50% of the values between the lower and upper quartiles). This predictive accuracy, confirmed for well-known species, makes it possible to infer the likely conservation status of poorly known (Data-Deficient) species.[3]

From Walter Jetz, Robert P. Freckleton (2015) Towards a general framework for predicting threat status of Data-Deficient species from phylogenetic, spatial and environmental information, *Philosophical Transactions of the Royal Society B*, 370 (1662: 20140016). Republished with permission of the Royal Society.

for phylogenetic relationships. Walter Jetz and Rob Freckleton used a generalized least-squares (GLS) model that predicted extinction-risk status among mammal species from a set of predictors (again, these included biological, ecological, and geographical variables) in a way that incorporates both phylogenetic and spatial covariances.[3] Although their analytical approach was quite different from that of Bland et al., their results are surprisingly similar. For assessed species, the Jetz–Freckleton model had good power to correctly predict threat status (**Figure 11.16**). Applied to DD species, the model predicted that 331 out of 483 DD species (69%) are threatened with extinction.

Together, these two studies provide strong evidence that the mammal species we know least about are likely to be under the greatest threat of extinction, and that the overall conservation status of mammals is worse than it currently appears from the Red List statistics. Not only that, but some parts of the world seem to be disproportionately affected. In Indonesia, New Guinea, and Colombia, the number of DD species predicted to be threatened adds considerably to the number of known threatened species. Predictive models such as those discussed here could play an important role in conservation planning in the face of poor biological knowledge and uncertainty over the true conservation status of species.

Questions to ponder

1. Although the models discussed here have good predictive power, the analyses are still subject to uncertainty and rest on many assumptions that could change the results. Should conservation planners take models such as these into account, or should they base their priority-setting decisions only on firm observational evidence for the conservation status of species?

2. Criterion E of the IUCN Red List allows a listing to made on the basis of quantitative analysis of projected population trends for a species. This usually requires detailed data on a species' population dynamics and life history, so few species are listed under this criterion. Should Criterion E be extended, or an additional criterion added, to allow listing to be made on the basis of comparative models?

3. Fewer than 10,000 of the one million or so known species of insects have been evaluated under the Red List. Is there potential for inferring conservation status for the majority of insect species using comparative models?

Further investigation

Comparative models do quite a good job of assigning assessed mammal species to their correct Red List categories. But is it only mammals where extinction risk can be predicted in this way? Could we do the same for plants, which face different kinds of threats?

References

1. Cardillo M, Mace GM, Gittleman JL, Jones KE, Bielby J, Purvis A (2008) The predictability of extinction: biological and external correlates of decline in mammals. *Proceedings of the Royal Society B-Biological Sciences* **275**, 1441–8.

2. Bland LM, Collen B, Orme CDL, Bielby J (2015) Predicting the conservation status of Data-Deficient species. *Conservation Biology* 29: 250–9.

3. Jetz W, Freckleton RP (2015) Towards a general framework for predicting threat status of Data-Deficient species from phylogenetic, spatial and environmental information. *Philosophical Transactions of the Royal Society B. Biological Sciences* 370(1662): 20140016.

Appendix: Geological timescale

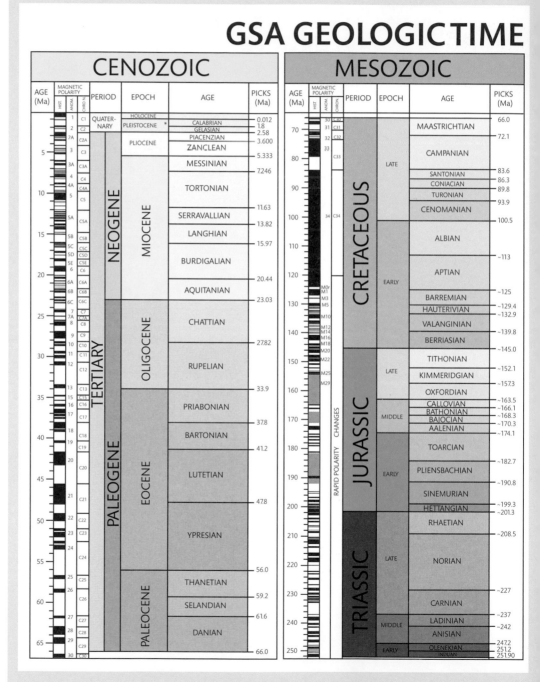

From Walker JD, Geissman JW, Bowring SA, and Babcock LE, compilers, 2018, Geologic Time Scale v. 5.0: Geological Society of America © 2018 The Geological Society of America.

SCALE   v. 5.0

PALEOZOIC

AGE (Ma)	PERIOD	EPOCH	AGE	PICKS (Ma)
	PERMIAN	Lopingian	CHANGHSINGIAN	251.90
			WUCHIAPINGIAN	254.14
260				259.1
		Guadalupian	CAPITANIAN	265.1
			WORDIAN	268.8
			ROADIAN	272.95
280		Cisuralian	KUNGURIAN	~283.5
			ARTINSKIAN	290.1
			SAKMARIAN	295.0
300			ASSELIAN	298.9
	CARBONIFEROUS	PENNSYLVANIAN — LATE	GZHELIAN	303.7
			KASIMOVIAN	307.0
		MIDDLE	MOSCOVIAN	
320		EARLY	BASHKIRIAN	315.2
				323.2
		MISSISSIPPIAN — LATE	SERPUKHOVIAN	330.9
340		MIDDLE	VISEAN	
				346.7
		EARLY	TOURNAISIAN	358.9
360	DEVONIAN	LATE	FAMENNIAN	~372.2
380			FRASNIAN	~382.7
		MIDDLE	GIVETIAN	~387.7
			EIFELIAN	~393.3
400		EARLY	EMSIAN	~407.6
			PRAGIAN	~410.8
			LOCHKOVIAN	~419.2
420	SILURIAN	PRIDOLI		~423.0
		LUDLOW	LUDFORDIAN	~425.6
			GORSTIAN	~427.4
		WENLOCK	HOMERIAN	~430.5
			SHEINWOODIAN	~433.4
440		LLANDOVERY	TELYCHIAN	~438.5
			AERONIAN	~440.8
			RHUDDANIAN	~443.8
	ORDOVICIAN	LATE	HIRNANTIAN	~445.2
			KATIAN	~453.0
460			SANDBIAN	~458.4
		MIDDLE	DARRIWILIAN	~467.3
			DAPINGIAN	~470.0
480		EARLY	FLOIAN	~477.7
			TREMADOCIAN	~485.4
	CAMBRIAN	FURONGIAN	AGE 10	~489.5
500			JIANGSHANIAN	~494
			PAIBIAN	~497
		Epoch 3	GUZHANGIAN	~500.5
			DRUMIAN	~504.5
			AGE 5	~509
		Epoch 2	AGE 4	~514
520			AGE 3	~521
		TERRENEUVIAN	AGE 2	~529
			FORTUNIAN	541.0
540				

PRECAMBRIAN

AGE (Ma)	EON	ERA	PERIOD	BDY. AGES (Ma)
	PROTEROZOIC	NEOPROTEROZOIC	EDIACARAN	541
			CRYOGENIAN	635
750				720
			TONIAN	
1000		MESOPROTEROZOIC	STENIAN	1000
1250			ECTASIAN	1200
				1400
1500			CALYMMIAN	1600
		PALEOPROTEROZOIC	STATHERIAN	1800
1750				
2000			OROSIRIAN	2050
2250			RHYACIAN	2300
			SIDERIAN	
2500	ARCHEAN	NEOARCHEAN		2500
2750				2800
3000		MESOARCHEAN		
3250				3200
		PALEOARCHEAN		
3500				3600
3750		EOARCHEAN		
4000	HADEAN			4000

Index

Tables and figures are indicated by an italic *t* and *f* following the page number.